Particle Size Measurement

Powder Technology Series

Edited by
B. Scarlett

Department of Chemical Engineering
University of Technology
Loughborough

Particle Size Measurement

TERENCE ALLEN Ph.D.

Senior Lecturer in Powder Technology
University of Bradford

THIRD EDITION

LONDON   NEW YORK
Chapman and Hall

First published 1968
Second edition 1975
Third edition 1981
published by Chapman and Hall Ltd.
11 New Fetter Lane, London EC4P 4EE

Published in the USA by
Chapman and Hall
in association with Methuen, Inc.
733 Third Avenue, New York, NY 10017

© 1968, 1975, 1981 T. Allen

Typeset and printed in Great Britain by
Spottiswoode Ballantyne Ltd,
Colchester and London

ISBN 0 412 15410 2

British Library Cataloguing in Publication Data

Allen, Terence
    Particle size measurement.–3rd ed.–(Powder
    technology series).
    1. Particle size determination
    I. Title   II. Series
    620'.43   TA418.8   80-49866
    ISBN 0-412-15410-2

Contents

viii   Contents

x   Contents

xvi   Contents

Editor's foreword to the first edition

The study of the properties and behaviour of systems made up of particulate solids has in the past received much less attention than the study of fluids. It is, however, becoming increasingly necessary to understand industrial processes involving the production, handling and processing of solid particles, in order to increase the efficiency of such systems and to permit their control. During the past few years this has led to an increase in the amount of study and research into the properties of solid particle systems. The results of this effort are widely dispersed in the literature and at the moment much of the information is not available in a form in which it is likely to influence the education of students, particularly in chemical engineering, who may later be employed in industrial organizations where they will be faced with the problems of solids' handling. It is also difficult for the engineer responsible for the design or selection of solids' handling equipment to make use of existing knowledge, with the result that industrial practice is not always the best that is achievable. It is hoped that the publication of a series of monographs on Powder Technology, of which this is the first, will help by providing accounts of existing knowledge of various aspects of the subject in a readily available form.

It is appropriate that the first monograph in this series should deal with the measurement of the size of small particles since this is the basic technique underlying all other work in powder technology. The reliability of research results, for example, on the size reduction of solid particles, cannot be better than the reliability of the particle size measurement techniques employed. Too often the difficulties and limitations of size measurement are ignored in such work, so that any conclusions become suspect. The importance of a thorough understanding of the problems involved in measuring the size of small particles for anyone working in any aspect of powder technology is therefore difficult to overestimate. It is hoped that this monograph, written by an experienced size analyst who has studied critically most of the methods described, will be of value in encouraging an informed and critical approach to the subject and that it will help in the selection of equipment and in realistic assessment of the value of particle size measurements.

Preface to the first edition

Although man's environment, from the interstellar dust to the earth beneath his feet, is composed to a large extent of finely divided material, his knowledge of the properties of such materials is surprisingly slight. For many years the scientist has accepted that matter may exist as solids, liquids or gases although the dividing line between the states may often be rather blurred; this classification has been upset by powders, which at rest are solids, when aerated may behave as liquids, and when suspended in gases take on some of the properties of gases.

It is now widely recognized that powder technology is a field of study in its own right. The industrial applications of this new science are far reaching. The size of fine particles affects the properties of a powder in many important ways. For example, it determines the setting time of cement, the hiding power of pigments and the activity of chemical catalysts; the taste of food, the potency of drugs and the sintering shrinkage of metallurgical powders are also strongly affected by the size of the particles of which the powder is made up. Particle size measurement is to powder technology as thermometry is to the study of heat and is in the same state of flux as thermometry was in its early days.

Only in the case of a sphere can the size of a particle be completely described by one number. Unfortunately, the particles that the analyst has to measure are rarely spherical and the size range of the particles in any one system may be too wide to be measured with any one measuring device. V.T. Morgan tells us of the Martians who have the task of determining the size of human abodes. Martian homes are spherical and so the Martian who landed in the Arctic had no difficulty in classifying the igloos as hemispherical with measurable diameters. The Martian who landed in North America classified the wigwams as conical with measurable heights and base diameters. The Martian who landed in New York classified the buildings as cuboid with three dimensions mutually perpendicular. The one who landed in London gazed about him despairingly before committing suicide. One of the purposes of this book is to reduce the possibility of further similar tragedies. The above story illustrates the problems involved in attempting to define the size of particles by one dimension. The only method of measuring more than one dimension is microscopy. However, the mean ratio of significant dimensions for a particulate system may be determined by using two methods of analysis and finding the ratio of the two mean sizes. The proliferation of measuring techniques is due to the wide range of sizes and size dependent properties

that have to be measured; a twelve-inch ruler is not a satisfactory tool for measuring mileage or thousandths of an inch and is of limited use for measuring particle volume or surface area. In making a decision on which technique to use, the analyst must first consider the purpose of the analysis. What is generally required is not the size of the particles, but the value of some property of the particles that is size dependent. In such circumstances it is important whenever possible to measure the desired property, rather than to measure the 'size' by some other method and then deduce the required property. For example, in determining the 'size' of boiler ash with a view to predicting atmospheric pollution, the terminal velocity of the particle should be measured; in measuring the 'size' of catalyst particles, the surface area should be determined, since this is the property that determines its reactivity. The cost of the apparatus as well as the ease and the speed with which the analysis can be carried out have then to be considered. The final criteria are that the method shall measure the appropriate property of the particles, with an accuracy sufficient for the particular application at an acceptable cost, in a time that will allow the result to be used.

It is hoped that this book will help the reader to make the best choice of methods. The author aims to present an account of the present state of the methods of measuring particle size; it must be emphasized that there is a considerable amount of research and development in progress and the subject needs to be kept in constant review. The interest in this field in this country is evidenced by the growth of committees set up to examine particle size measurement techniques. The author is Chairman of the Particle Size Analysis Group of the Society for Analytical Chemistry. Other committees have been set up by The Pharmaceutical Society and by the British Standards Institution and particle size analysis is within the terms of reference of many other bodies. International Symposia were set up at London, Loughborough and Bradford Universities and it is with the last-named that the author is connected. The book grew from the need for a standard text-book for the Postgraduate School of Powder Technology and is published in the belief that it will be of interest to a far wider audience.

Terence Allen

Postgraduate School of Powder Technology
University of Bradford

Preface to the third edition

The response to this book has been most encouraging and as a result a third edition has been written. The five years since the advent of the second edition have been full of technological changes and in order to be comprehensive it has become necessary to enlarge the edition.

With regard to the first three chapters on sampling, it is with some regret that I note that the 'golden rules of sampling' are observed more in the breach than in the commission. On the positive side it is good to see that man is becoming more aware of the need to monitor and control his environment. Chapter 4 has been greatly enlarged and a statement on the German approach to mathematical handling of size data has been presented. The chapter on centrifugal methods has been expanded to present a full statement on the disc centrifuges, and Chapter 13 has also been expanded to give a fuller statement of the Coulter principle. The chapters on surface area determination have also been enlarged due, in no small part, to the enormous interest in this parameter which has produced a considerable number of important advances. Pore size determination has been expanded to two chapters and, finally, a new chapter has been added on 'On-line particle size analysis'. My thanks go to my colleague Dr N. G. Stanley-Wood for permission to borrow extensively from his lecture notes in writing this new chapter.

It is my sincere hope that you find in this book a full statement on this particular field of analysis. Omissions and errors do have a tendency to creep in despite one's best endeavours and I would like to thank those who have drawn these to my attention in the past and hope they will continue to do so.

Terence Allen

Postgraduate School of Powder Technology
University of Bradford

Acknowledgments

I would like to express my grateful thanks to Dr Brian H. Kaye for introducing me to the fascinating study of particle size analysis. My thanks are also due to numerous workers in this field for the helpful discussions we have had. Bradford University has provided me with a well-equipped laboratory in which, in teaching others, I have learnt some of the secrets of this science. One of my students was Mr T.S. Krishnamoorthy and the chapter on gas adsorption is taken from his M.Sc. thesis. At Bradford, Mr. John C. Williams has always had the time to offer helpful advice and criticism. I make no apology for taking up so much of his time since his advice was invariably good and whatever virtue this book possesses is due, in part, to him.

My thanks are also due to holders of copyright for permission to publish and to many manufacturers who have given me full details of their products.

Finally, I would like to thank my wife for her forbearance while the writing of this book has been in progress.

Terence Allen

1 Sampling of powders

1.1 Introduction

There are many instances where estimates of population characteristics have to be made from an examination of a small fraction of that population and these instances are by no means confined to the field of powder technology. Regrettably, there are many powder technologists who still assume that sample selection procedure is unimportant. This results in the analyst being frequently presented with hastily taken, biased samples on which he devotes a great deal of attention to derive precise results which do not reflect the characteristics of the bulk powder. It is essential that the samples selected for measurement should be representative of the bulk in grain size distribution and the relative fractions of their various constituents, irrespective of whether a physical or chemical assay is to be carried out, since these characteristics are frequently inter-dependent. The magnitude of the problem may be realized when one considers that the characteristics of many tons of material are assumed on the basis of analyses carried out on grams or even milligrams.

 The probability of obtaining a sample which perfectly represents the parent distri-bution is remote. If several samples are taken their characteristics will vary and if these samples are representative, the expected variation may be estimated from statistical analysis. However, the sampling equipment will introduce a further variation which may be taken as a measure of sampler efficiency. Imposed on this there may also be an operator bias. The reduction from bulk to measurement sample may be convenient-ly divided into the four stages illustrated below.

Process or delivery of materials	Gross sample	Laboratory sample	Measurement sample
(10^nkg)	(kg)	(g)	(mg)

Bias at any of the reduction stages will adversely affect the final analysis.

1.2 Theory

The ultimate that may be obtained by representative sampling may be called the perfect sample; the difference in population between this sample and the bulk may be

ascribed wholly to the expected difference on a statistical basis.

A powder to be sampled may be considered as made up of components A and B. The probability that the number fraction (p) of the bulk in terms of A shall be represented by the corresponding composition (p_i) of a perfect sample can be computed from the number of particles of A and B in the sample (n) and in the bulk (N) :

$$\text{Var}(p_i) = \frac{p(1-p)}{n}\left(1 - \frac{n}{N}\right) \tag{1.1}$$

The theoretical standard deviation (σ_i) is equal to the square root of the variance. Assuming a normal distribution of variance, if representative samples are taken from the bulk, 68.3% of the samples should not vary from the true number fraction of A by more than one standard deviation and 95.4% should lie within two standard deviations of the mean. These percentage values are taken from a table of the area under a normal curve. These tables may be found in any statistics book [29, 30].

These areas give the probability of an event occurring and are derived from the integration of the normal probability equation :

$$d\phi = \frac{dz}{\sqrt{2\pi}.\sigma_i} \exp\left(-\frac{z^2}{2\sigma_i^2}\right) \tag{1.2}$$

where $z = (p - \bar{p})$ and $\bar{p}$ = the 50% probability level (median).

The area between $p = \bar{p}$ and $p = \bar{p} + \sigma_i$ equals 0.3413, hence 68.26% of the samples should deviate from the mean by less than one standard deviation.

Instead of the number fractions it is more convenient to assess sample and bulk compositions in terms of weight fractions P and P_i giving [2] :

$$\text{Var}(P_i) = \frac{P(1-P)}{w}[Pw_B + (1-P)w_A]\left(1 - \frac{w}{W}\right) \tag{1.3}$$

where W and w are the bulk and sample weight respectively and w_A and w_B are the weights of individual grains of components A and B.

Example

Consider a binary powder made up of equal weights of particles of weight 0.05 and 0.10 g. Determine the expected variation, assuming perfect sampling, for a 50 g sample removed from a bulk of 800 g.

From equation (1.3):

$$\text{Var}(P_i) = \frac{0.50 \times 0.50}{50}[0.50(0.10) + 0.50(0.05)]\left(1 - \frac{1}{16}\right)$$

$$= 3.52 \times 10^{-4}$$

$$\sigma_i = 1.88\%$$

From equation (1.1):

$$\left(\text{derived data; } p = \frac{1}{3}, n = 750\right)$$

$$\text{Var } (p_i) = \frac{1}{750} \left(\frac{1}{3} \times \frac{2}{3}\right) \left(1 - \frac{1}{16}\right)$$

$$= 2.78 \times 10^{-4}$$

$$\sigma_i = 1.67\%$$

Assuming a normal distribution of variance, the probability of representative samples lying within one standard deviation of the mean is 68.3%, and 95.4% should lie within two standard deviations of the mean. Three results in a thousand will fall outside three standard deviations.

On a weight basis, of the 50 g withdrawn, the weights of each component at the three probability levels above, are (in grams): 25 ± 0.94, 25 ± 1.88, 25 ± 2.82.

On a number basis, of the 750 particles withdrawn, the numbers of small particles at the above three probability levels are: 500 ± 12.5, 500 ± 25, 500 ± 37.5.

On a percentage basis by number, $p \pm \sigma_i = (66.7 \pm 1.67)\%$ of 750.

These equations may be used as a basis from which to assess the efficiency of a real, non-ideal sampling device. In this case the variance of the sample assay, Var (P_n), will be greater than Var (P_i) due to the non-ideality unless the bulk powder is homogeneous.

$$C = \frac{\text{Var } (P_i)}{\text{Var } (P_n)} \tag{1.4}$$

This should approximate to unity when sampling errors are low, hence the sampling efficiency may be defined as $100C$. Sample variance Var (P_n) can be calculated from experimental results:

$$\sigma_n = \sqrt{\frac{(x_i - \bar{x})^2}{n}} \tag{1.5}$$

where σ_n, the sample standard deviation, equals the square root of the sample variance. $\bar{x}$ is the true percentage of A or B particles by weight in the mixture, x_i is the percentage in the ith sample and n is the number of samples examined. If $\bar{x}$ is unknown, it may be approximated to, using the equation:

$$\bar{x} = \frac{\sum\limits_{i=1}^{n} x_i}{n} \tag{1.6}$$

In this case the denominator in equation (1.5) should be replaced by $(n - 1)$.

Theoretical and experimental errors may be separated using the equation:

$$\sigma^2 = \sigma_n^2 - \sigma_i^2 \tag{1.7}$$

The maximum sample error (E) can be expressed as a percentage of the bulk assay.

$$E = \pm \frac{2\sigma_n}{P} \times 100 \tag{1.8}$$

It should be noted that, since sample variations will probably be related to the particle size distribution of the powder to be sampled, it is not possible to measure the absolute efficiency of a sampling technique. Sampling errors may be of two kinds: random fluctuations and a steady bias due to faulty design of the sampling technique.

Numerical example

Any powder can be considered as being made up of two components, the fraction above and below a certain size. For a reduction of 16 to 1 during sampling with $W_A = 0.10$ g, $W_B = 0.05$ g and $w = 50$ g, equation (1.3) becomes:

$$\text{Var}(P_i) = \frac{3}{160} W_B (P^3 - 3P^2 + 2P).$$

The maximum value of the variance, derived by differentiation and equating the differential to zero, occurs for $P = (1 - 1/\sqrt{3})$ giving $\text{Var}(P_i) = 0.000361$ and $\sigma = 1.90\%$.

This means that if 50 g samples are repeatedly removed from a bulk powder of 800 g and replaced without loss (with $w_A = 0.10$ g, $w_B = 0.05$ g), the maximum standard deviation at any percentage level would be equal to or less than 1.9%, provided no sampling bias existed.

This deviation is similar to what one would expect if one took samples of balls from a well-mixed bag containing one-third red balls and two-thirds white, but otherwise identical. If 18 balls were removed $((n/N)$ very small), from equation (1.1), $\sigma_i = 11.1\%$ or, $S = \sqrt{(np(1 - p))} = \sqrt{(18 \times \frac{1}{3} \times \frac{2}{3})} = 2$ balls. That is, the minimum number of white balls one would expect would be six, i.e. the mean minus three standard deviations.

The probability of having r white balls is given by:

$$^nC_r\, p^r\, (1 - p)^{n-r}$$

If $r = 7$, $P(7,18,\frac{2}{3}) = {}^{18}C_7 \left(\frac{2}{3}\right)^7 \left(\frac{1}{3}\right)^{11} = 1.05\%$

Further details on statistical analysis can be found in standard texts [29, 30].

If 1800 balls were removed the standard deviation would fall to 1.11% (i.e. 20 balls). If the experimental deviation exceeded this, one would suspect either a non-random mix or selective sampling.

It is obvious that the larger the sample, the smaller will be the deviations. The size of a practicable laboratory sample is usually minute in relation to the whole of the material being examined. Thus the laboratory sample itself is subject to a degree of variation which is unrealistically large. There are two ways of reducing this variation;

one is to make up a bulk sample from many increments and divide them down to produce laboratory samples, and the second is to take a number of replicate samples and mix them.

1.3 Golden rules of sampling

There are many possible situations in which a sample has to be obtained and conditions often necessitate the use of inferior techniques. Some principles can however be laid down, and they should be adhered to whenever possible:

Rule 1. A powder should be sampled when in motion.

Rule 2. The whole of the stream of powder should be taken for many short increments of time in preference to part of the stream being taken for the whole of the time.

Observance of these rules coupled with an understanding of the manner in which segregation may have occurred during the previous treatment of the powder will lead to the best sampling procedure. Any sampling method which does not follow these rules should be regarded as a second-best method liable to lead to errors.

1.4 Bulk sampling

There are a very large number of possible systems from which the gross sample has to be abstracted, so it is impossible to lay down instructions which will meet all situations. The problem may be to obtain a sample from continuous streams, batches, packets, heaps or trucks. The difficulty is that when a particulate material is handled, segregation may take place. The most important segregation-causing property is particle size and the problem is enhanced with free-flowing material. When poured into a heap, the fines tend to collect at the centre of the heap. In vibrating containers a layer of coarse material tends to collect near the surface; even if a large particle is more dense than the smaller particles in which it is immersed, it can be made to rise towards the surface. This can be demonstrated by placing a one-inch diameter steel ball in a beaker which is then filled with sand to a depth of about two inches. By stroking the base of the beaker gently with one hand, the steel ball can be made to rise to the surface of the sand. Since the surface region is always rich in coarse particles, samples should never be removed from the surface region. An understanding of these tendencies to segregation prevents careless practice in obtaining samples.

When sampling is undertaken from a continuous stream, the sampling may be continuous or intermittent. In continuous sampling a portion of the flowing stream is split off and frequently further subdivided subsequently. In intermittent sampling the whole stream is taken for small increments of time at fixed time intervals. These increments are usually compounded and samples for analysis taken from this gross sample. Consignment sampling is carried out on a single consignment (e.g. a truck-load or wagon-load).

A general rule in all sampling is that whenever possible the sample should be taken when the powder is in motion. This is usually easy with continuous processes; with consignment sampling it may be possible during the filling or emptying of storage containers.

1.4.1 Sampling from a moving stream of powder

In collecting from a moving stream, care should be taken to offset the effects of segregation. For example, the powder may be sampled as it falls from the end of a conveyor; this is one of the best methods of sampling and should be adopted whenever possible. The powder on the conveyor will probably show two forms of segregation. If the powder was charged on to the conveyor belt from a centrally placed feeder or hopper outlet, the fines will tend to concentrate at the centre of the

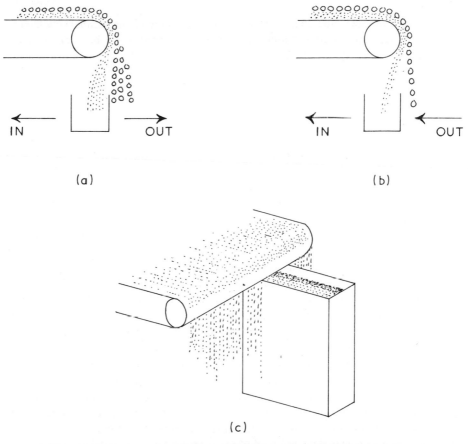

Fig. 1.1 Sampling from moving streams. (a) Bad sampling technique (b) Good sampling technique (c) Sampling procedure to be adopted for mass flow rate.

belt and the coarse particles will roll to the outer edges. If there has been any vibration of the belt, larger particles will tend to rise to the top of the bed of powder.

Each increment should be obtained by collecting the whole of the stream for a short time. Care must be taken in putting the sampler in and out of the stream. Figure 1.1 shows correct and incorrect methods of doing this.

Unless the time, during which the sample receiver is stationary in its receiving position, is long compared with the time taken to insert and withdraw the sampler, the method shown in figure 1.1(a) will lead to an excess of coarse particles as the surface region of the stream, usually rich in coarse particles, is sampled for a longer time than the rest of the stream. The method shown in figure 1.1(b) is not subject to this objection. If the method shown in the figure (b) is not possible due to some obstruction, the ratio of stationary time to moving time for the receiver should be made as large as possible.

In many cases it is not possible to collect the whole of the stream as this would give too large an amount to be handled. The best procedure is to pass through the stream a sample collector of the form shown in figure 1.1(c).

The width of the receiver, b, will be chosen to give an acceptable weight of sample but must not be made so small that the biggest particles have any difficulty

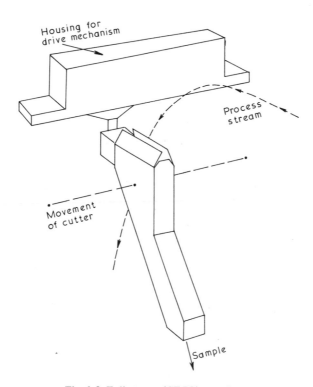

Fig. 1.2 Full-stream (GECO) sampler.

in entering the receiver. Particles that strike the edges of the receiver are likely to bounce out and not be collected so that the effective width is $(b - d)$, where d is the diameter of the particles. The effective width is therefore greater for small particles than for large particles. To reduce this error to a reasonable level, the ratio of box width to the diameter of the largest particle should be made as large as possible with a minimum value of 20 : 1. The depth a of the receiver must be great enough to ensure that it is never full of powder. If the receiver fills up before it finishes its traverse through the powder, a wedge-shaped heap will form a size-selective particle collector. As more material falls on the top of the heap the fine particles will percolate through the surface of the heap and be retained, whereas coarse particles will roll down the sloping surface and be lost. The length of the receiver c should be sufficient to ensure the full depth of the stream is collected.

A proprietary example of this type of sampler is the GECO manufactured by John Smith & Co. Ltd, Horsham, Sussex, diagrams of which are shown in figures 1.2 and 1.3.

This equipment is satisfactory for many applications but it has limitations which restrict its use. These are:

(1) Although comparatively readily designed into a new plant, it is frequently difficult and expensive to add to an existing plant.

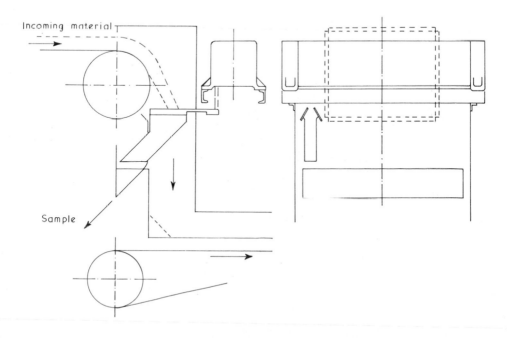

Fig. 1.3 Traversing-type sampler in a chute.

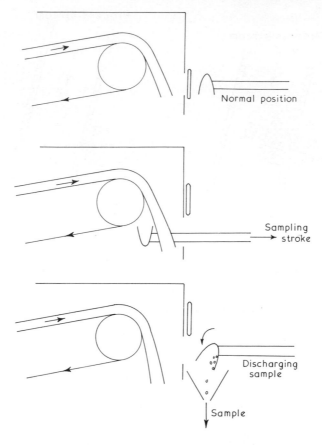

Fig. 1.4 Full-stream trough sampler.

(2) The quantity of sample obtained is given by the product of slit width and plant rate divided by the cutter speed and is independent of stream shape or area. This proportionality to the plant rate can be inconvenient when the plant rate is subject to wide variations and a fixed sample is required. On the other hand, where a plant daily average is required, this is a necessary requirement of the sampler. Also the quantity of sample may be inconveniently large.

(3) It is difficult to enclose the sampler to the extent required to prevent the escape of dust and fume when handling a dusty product.

Figure 1.4 shows a sampler designed by ICI [1] to sample a dusty material, sampling taking place only on the return stroke. This is suitable provided the trough does not overfill. For this reason the constant volume sampler, the action of which can be readily understood from figure 1.5 cannot be recommended. The slide valve sampler (figure 1.6), developed by the same company, also has defects which render it unsuitable for collecting size-representative samples.

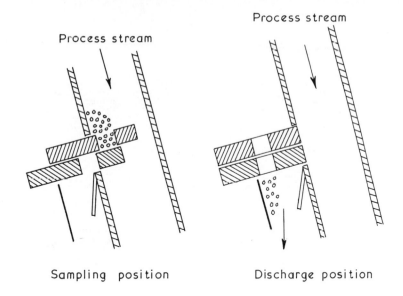

Sampling position   Discharge position

Fig. 1.5 Constant volume sampler.

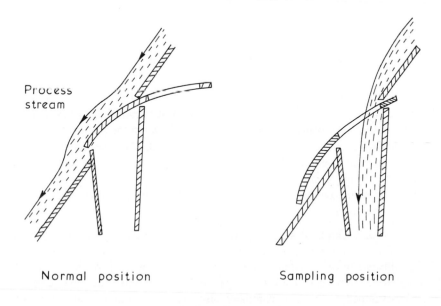

Normal position   Sampling position

Fig. 1.6 Slide valve sampler.

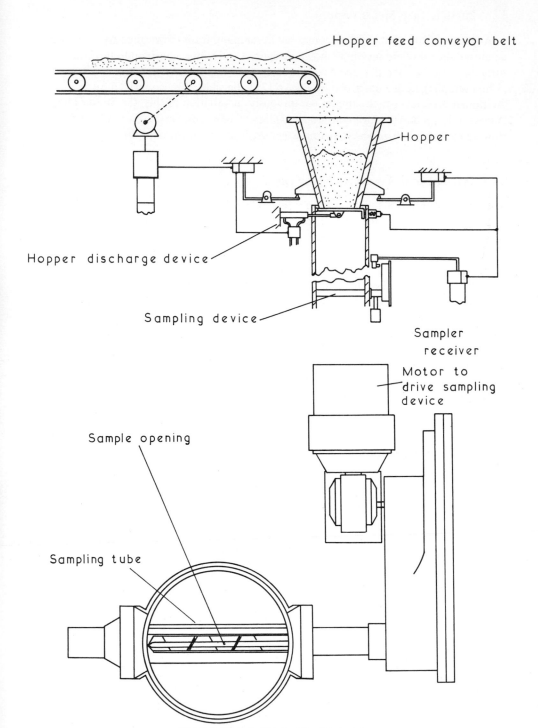

Fig. 1.7 Automatic sampling from a hopper.

A variant of this problem is encountered in sampling from preweighed batches; such an automatic sampling system is shown in figure 1.7. From a consideration of this diagram, it can be seen that the device is liable to be highly size-selective.

In a sampling device described by Clarke [21] the slide valve is replaced by an Archimedean screw which samples continuously. A variation of this, the Simon Limpet Autosampler, is designed for continuous collection of a representative sample of free-flowing material such as wheat, maize, other seeds and their flows or other products.

1.4.2 Sampling from a conveyor belt or chute

When a sample is to be collected from a conveyor belt, the best position for collecting the increments is where the material falls in a stream from the end of the belt. If access at such a point is not possible, the sample must be collected from the belt. The whole of the powder on a short length of the belt must be collected, and again it must be borne in mind that the particles at the edge of the belt may not be the same size as those at the centre, and particles at the top of the bed may not be the same as those at the bottom. If the belt can be stopped, the sample can be collected by inserting into the stream a frame consisting of two parallel plates shaped to fit the belt; the whole of the material between the belts is then swept out. With the belt in motion, the same procedure could be adopted but it is difficult to do this by hand. Some mechanical samplers are available for collecting a sample from a moving conveyor belt (figure 1.8); before using such a device its action should be carefully examined to ensure that it collects the whole of the stream and does not, for example, leave a thin layer on the belt.

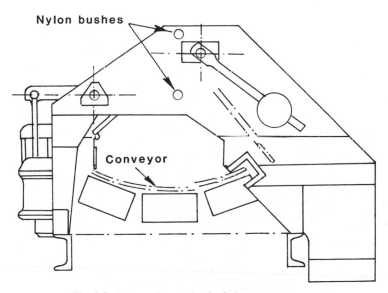

Fig. 1.8 Automatic sampler for belt conveyor.

1.4.3 Sampling from a bucket conveyor

In sampling from a bucket conveyor no attempt should be made to collect part of the material from a bucket. Each increment should consist of the whole of the contents of a bucket. If this is too large an amount for convenient handling, it should be reduced in size by one of the methods described later.

1.4.4 Bag sampling

Suppose an analysis is required from several tons of material which is available in sacks. Should a sack be selected at random with the hope that it is representative of the bulk? Having selected a sack how should a sample be withdrawn for subsequent measurement?

In an analysis of this problem Kaye [4] suggests that several sacks should be selected in order to obtain a representative sample. These may be selected systematically, i.e. the 100th, 200th, 300th and so on or, preferably, using a table of random numbers.

The sacks may be examined individually in order to determine whether the variation between sacks is of an acceptable level or combined in order to determine an average. In either case it is necessary to obtain representative samples from each sack. For this purpose Kaye recommends a thief sampler but warns that this may easily give a biased sample.

The best methods are undoubtedly those that conform with the 'golden rules of sampling' such as the LADAL Bulk Spinning Riffler and the expense of installing and using this type of equipment has to be balanced against the cost of rejecting good material or accepting poor material by using a more inexpensive technique.

1.4.5 Sampling spears

Scoop sampling is the most widely used method of sampling because of its simplicity. A presupposition is that the powder, at the point at which the scoop is inserted, is representative of the bulk. Accuracy should be increased by taking more than one sample. These should be examined separately in a preliminary investigation and combined in later investigations if the variance between the samples is at an acceptable level. This technique is improved if samples from the body of the material are included and this may be effected with the aid of a sampling spear. Three types are available (figure 1.9); in type 1 the sampling chamber runs the full length of the spear; in type 2 the sampling chamber is at the end of the spear and in type 3 there are separate sampling chambers along the length of the spear. The spear is thrust into the powder with the inside chamber closed off. When in position the inner tube is rotated to allow powder to fall into the inner chamber. When the sampling chamber is full the inner tube is turned to the closed position and the spear is withdrawn. Possible segregation throughout the bed may be investigated with type 3, an average value for the length of the spear with type 2 and a spot sample obtained with type 1.

A biased sample may be obtained, if the bulk has a wide size range, due to perco-
lation of fines through the coarse particles leading to an excess of fines in the withdrawn sample. Particles may get trapped between the two tubes making it difficult to
close them. Fragile material may suffer particle fracture and fine material may become
compacted so that it does not flow into the tube opening(s). The accuracy (or in-
accuracy) of the sampling spear is comparable to the accuracy of scoop sampling and,
on the whole, its use is to be deprecated.

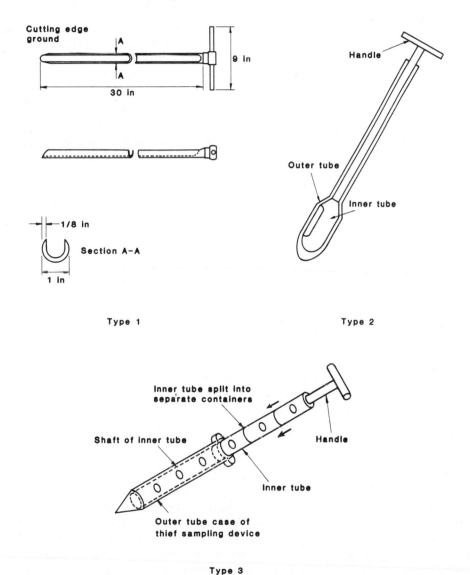

Fig. 1.9 Sampling spears.

1.4.6 Sampling from wagons and containers

It is very difficult, in fact practically impossible, to obtain a satisfactory sample from a wagon or container, because of the severe segregation that would almost certainly occur in filling the wagon and in its subsequent motion. A method for removing samples that avoids some of the worst errors is described in [5]. No increments are to be collected at less than twelve inches below the surface; this avoids the surface layer in which extreme segregation will probably have occurred due to vibration. In removing the samples there must be no surfaces down which particles can slide; this is achieved either by pushing in a sampling probe which extracts the sample (figure 1.9), or by removing particles by hand in such a way that no sliding occurs; when the sample region is exposed the sample can be extracted.

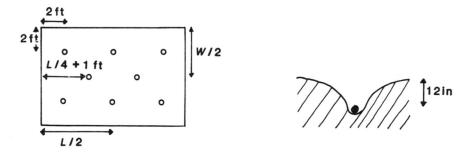

Fig. 1.10 Sampling points for a wagon or truck (sampling from bottom of granular excavation).

Increments should be extracted at eight points as shown in figure 1.10. This method of obtaining a sample is mentioned here because it is possible that there may be circumstances in which there is no alternative but to use it, but this must not be taken to imply that such methods will give satisfactory sampling. Every effort should be made to avoid this method and to use one that satisfies the two 'golden rules'. If a powdered material is in a container, the container has been filled and presumably is going to be emptied. At both these times, the powder will be in motion, and a more satisfactory sampling procedure can then be used.

1.4.7 Sampling from heaps

There is only one sound piece of advice to give regarding sampling from a heap. 'Don't! Never!' Examination of the cross-section of a heap of powder containing particles of different sizes shows that there is a very marked segregation, the fine particles being concentrated in a region near the axis of the heap, with the coarse particles in the outer part of the heap.

The photograph (figure 1.11) of the cross-section of a heap of granules of two sizes shows the segregation of powder when it is poured into a heap. Although segregation

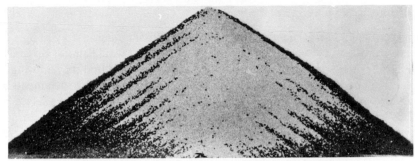

Fig. 1.11 Cross-section of a binary powder showing how powders 'unmix' when poured into a heap, the finer particles tending to remain in the centre. Sizes: coarse (black) approximately 1 mm diameter; fine (white) approximately 0.2 mm diameter.

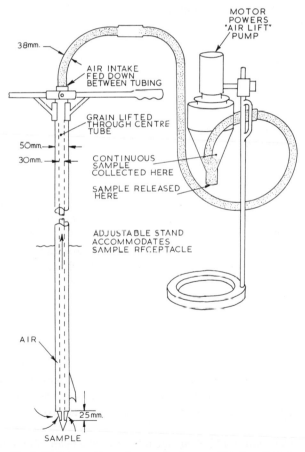

Fig. 1.12 Pneumatic probe sampler for sampling from heaps.

does not occur on such a scale with sub-sieve granules, it indicates the care that must be taken when sampling from a heap. Sometimes a sample has to be obtained from a powder at rest, an indication that the sampling is taking place at the wrong time, since it must at some time have been in motion. In this case, the sample should be compounded from incremental samples and a lower sampling efficiency is to be expected. Figure 1.12 shows a pneumatic probe for extracting such samples.

In one case a binary mixture of equal weights of particles of two sizes, having a diameter ratio of 2.8, was poured into a heap and the concentration of fine particles in different parts of the heap was found to vary between 1 and 70%. Any attempt to find the composition of the whole heap from measurements based on sampling from a heap has, therefore, little chance of giving an accurate answer.

The solution to the problem of sampling from a heap lies in the fact that the powder must have been poured to form the heap, and this is the time when the sample should be collected; increments can be withdrawn from the falling stream by one of the recommended methods.

1.5 Slurry sampling

Clarke [21] discusses the problems of sampling from liquid streams and describes a parallel-sided scoop for extracting a sample size which is proportional to flow rate in an open channel.

Hinde and Lloyd [25] are more interested in extracting samples for continuous on-line analysis. They state that the process streams from industrial wet classifiers can vary in volume flow rate, solids concentration and particle size distribution. Any sampling technique should be capable of coping with these variations without affecting the representativeness of the extracted sample.

For batch sampling, automatic devices are available where a sampling slot is traversed intermittently across a free-falling slurry. Unfortunately it is difficult to improvise with this technique for continuous sampling since such samplers introduce pulsating flow conditions into the system.

Osborne [26] described a sampler which consists of a narrow slot continuously rotated on an axis parallel to the slurry flow (figure 1.13(a)). Cross [27] used a slotted pipe mounted vertically in the overflow compartment next to the vortex finder of a hydrocyclone (figure 1.13(b)).

Since most continuous size analysers require a small constant volume flow rate further subdivision is often necessary. Osborne's solution (figure 1.13(c)) was to feed the sample stream to a well-agitated sampling tank and withdraw a representative sample at a controlled flow rate.

Autometrics [28] analyse the whole of the extracted sample (figure 1.14(a)). Gaps are left below the weirs to prevent sanding. Sub-division of the sample stream (for calibration purposes) can be accomplished very successfully by syphoning from a vertical fast-flowing stream (figure 1.14(b)).

Preferably the whole process stream should be used for on-line analysis as discussed in Chapter 19.

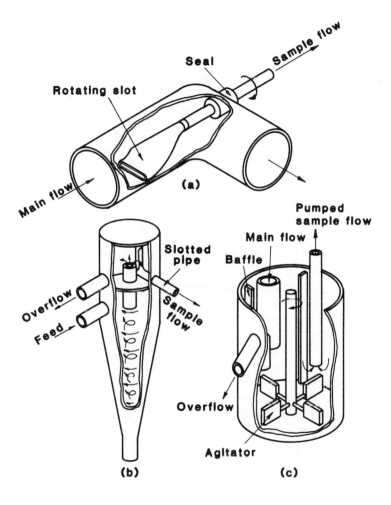

Fig. 1.13 Sampling devices: (a) rotating slot of Osborne [26]; (b) slotted pipe of Cross [27]; (c) sampling tank of Osborne [26].(From [25] with permission)

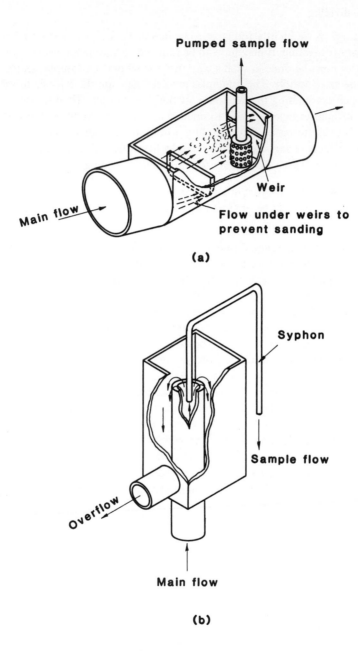

Fig. 1.14 More sampling devices: (a) Autometrics design for direct sampling of process stream; (b) Autometrics design for extracting small calibration samples [28]. (From [25] with permission)

1.6 Sample dividing

The gross sample is frequently too large to be handled easily, and before sending it to the laboratory for testing, it may have to be reduced to a more convenient weight. Obviously this must be done in such a way that the laboratory sample has the same size grading as the gross sample. The principles involved here are the same as those already discussed when considering the collection of the gross sample. The first step is to realize the difficulties involved, and to understand the ways in which the segregation of the powder is likely to occur. The sampling procedure must then be designed so as to minimize the effects of segregation. The two 'golden rules' of sampling mentioned earlier apply equally here. The best method of sample dividing is:

(1) Get the powder into motion in a stream.
(2) Each increment should be obtained by collecting the whole of the stream for a short time.

Increments are collected at equal time intervals and put together to form the laboratory sample. To obtain the best results, the stream should be made as homogeneous as possible. If complete homogeneity of the stream is achieved, then clearly every increment will have the same size grading as the whole of the material and, in such an imaginary case, it would be enough to collect one increment. If there are considerable changes in the size grading of the material from one part of the stream to another, the probability of getting increments with the same size grading as the whole is reduced. Everything must be done to make the process of sample dividing easier and, by giving thought to the handling of the powder fed to a sample divider, more accurate results can be obtained. For example, if the sample divider is fed from a hopper, this hopper should have steep sides (at least 70°) so that mass flow is likely to occur when it is emptied, and the hopper should be filled in such a way that size segregation does not occur; this can best be done by moving the pour point about so that the surface of the powder in the sample divider is always more or less horizontal and no conical heap is formed. Several sample-dividing devices have been recommended and some of these are in wide use. Each method will be described briefly and its action discussed.

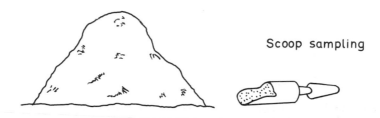

Scoop sampling

Fig. 1.15(a) Scoop sampling.

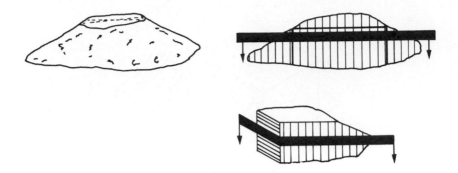

Fig. 1.15(b) Cone and quartering.

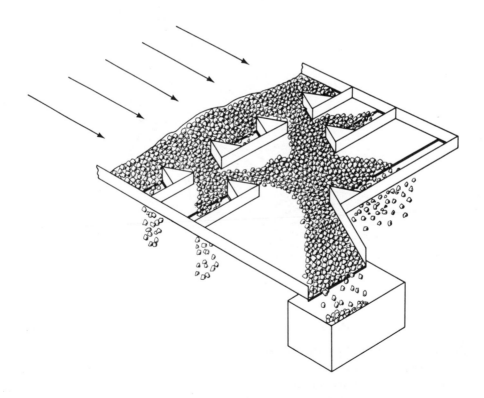

Fig. 1.15(c) Table sampler.

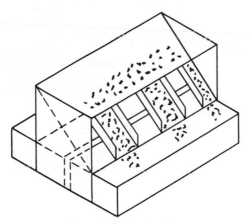

Fig. 1.15(d) Chute splitter.

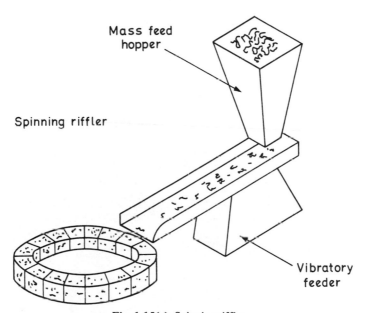

Mass feed
hopper

Spinning riffler

Vibratory
feeder

Fig. 1.15(e) Spinning riffler.

1.6.1 Scoop sampling (figure 1.15(a))

The method consists of plunging a scoop into the batch and withdrawing a sample. This is particularly prone to error since the whole of the sample does not pass through the sampling device, and since the sample is taken from the surface, where it may not be typical of the mass. In an attempt to eliminate segregation produced in previous handling, it is usual to shake the sample vigorously in a container before sampling.

Figure 1.16 shows five modes of shaking which were investigated by Kaye [8, 4], using six operators in order that both operator and technique bias could be evaluated. The samples consisted of coarse and fine particles mixed in known proportions.

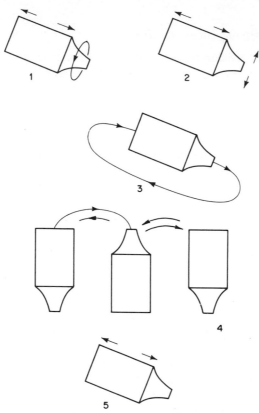

Fig. 1.16 Five modes of shaking a container prior to removing a sample for analysis.

The various models were as follows:

(1) In the vertical plane the bottle is moved rapidly around a circle. Superimposed on this motion is a brisk horizontal shaking in the direction of the arrows.
(2) Brisk shaking in the vertical and horizontal planes; no swirling action.
(3) The bottle is placed in the palm of the hand and then shaken in the manner shown with a brisk jerky action.
(4) Starting in a vertical position the bottle is rotated to the right until it is inverted and then the rotation is reversed to the left until the bottle is again inverted. The cycle was carried out in two seconds.
(5) The bottle is gripped firmly and then shaken in the direction of the arrows.

The first and third methods introduced no bias and the others a definite bias thus demonstrating that the efficiency is dependent on the previous history of the powder.

Kaye also tested a standard laboratory shaker 'the microid pipette flask shaker' (Griffin and George Ltd), and found that its efficiency depended on its speed but even at peak efficiency it was no better than manual shaking.

1.6.2 Coning and quartering (figure 1.15(b))

This method of sample dividing consists of pouring the material into a conical heap and relying on its radial symmetry to give four identical samples when the heap is flattened and divided by a cross-shaped metal cutter (figure 1.15(b)). This method would give reliable results if the heap were symmetrical about a vertical axis and if the line common to the two cutting planes coincided with this axis. In practice the heap is unlikely to be symmetrical, and the second condition, symmetry of cutting, would be very difficult to achieve without precision equipment. Since severe size segregation will certainly occur in forming the heap, departure from symmetry in the cutting will lead to differences in the size of the four portions into which the heap is cut. The method is very dependent on the skill of the operator, and should not be used. If coning and quartering is possible, this implies that the quantity of material to be divided is such that it can easily be moved by hand; it is just as easy to feed it into the hopper of a device such as a rotary sample divider in which increments are collected from a stream in an acceptable manner.

1.6.3 Table sampling (figure 1.15(c))

In a sampling table the material is fed to the top of an inclined plane in which there are series of holes. Prisms placed in the path of the stream break it into fractions. Some powder falls through the holes and is discarded, while the powder remaining on the plane passes on to the next row of prisms and holes, and more is removed, and so on. The powder reaching the bottom of the plane is the sample. The objection to this device is that it depends on the initial feed being uniformly distributed, and complete mixing after each separation, a condition not in general achieved. As it relies on the removal of part of the stream sequentially, errors are compounded at each separation, hence its accuracy is low [4].

1.6.4 Chute splitting (figure 1.15(d))

The chute splitter consists of a V-shaped trough along the bottom of which is a series of chutes alternately feeding two trays placed on either side of the trough. The laboratory sample is poured into the chute and repeatedly halved until a sample of the desired size is obtained.

When carried out with great care on a laboratory scale this method can give satisfactory sample division, but if the trough is filled in such a way that segregation occurs the results are liable to be misleading. The method is particularly prone to operator

bias, which is frequently detectable by unequal splitting of the sample.

In a review of factors affecting the efficiency of chute rifflers Batel [6] stated that there is little to be gained by increasing the number of chutes. Kaye [4] points out that this is only true if the chute widths are kept constant; increasing the number of chutes by reducing the chute widths will increase the efficiency. He tested this hypothesis, using four rifflers of the same overall width having 4, 8, 16 and 32 chutes, and found that the efficiency increased as the number of chutes increased.

It follows that two narrow, oppositely directed chutes intercepting the flow of powder would provide a very efficient sampling device. An equivalent selection procedure is to have one narrow powder stream oscillating between two reception bins. This in fact forms the basis of the British Standards oscillating hopper sample divider [7] (figure 1.17) and the ICI oscillating paddle divider (figure 1.18).

In the oscillating hopper sample divider, the feed hopper is pivoted about a horizontal axis so that it can oscillate during emptying.

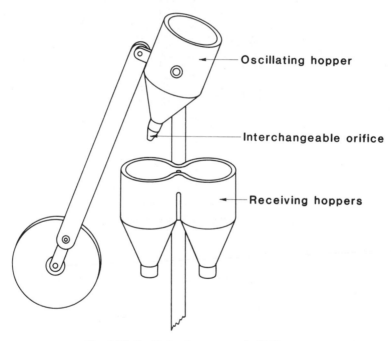

Fig. 1.17 Oscillating hopper sample divider.

Two collectors are placed under the hopper outlet so that the powder falls into them alternately. The contents of one box are retained; at each step the weight of sample is thus halved.

Kaye [4] suggests that samples of a given size may be obtained directly by controlling the amount of time, during each oscillation, that the feed is directed to each hopper or container. For efficient sampling the number of increments should be high.

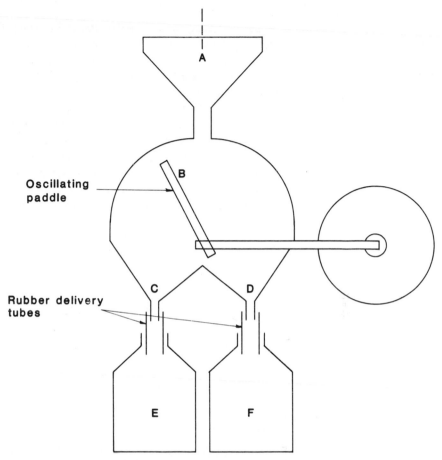

Fig. 1.18 Oscillating paddle sample divider.

1.6.5   The spinning riffler (figure 1.15(e))

The rotary sample divider or spinning riffler was first described in 1934 [9] and conforms to the golden rules of sample dividing. The preferred method of using this device is to fill a mass flow hopper, in such a way that little segregation occurs, by avoiding the formation of a heap (figure 1.19). The table is then set in motion and the hopper outlet opened so that powder falls into the collecting boxes. The use of a vibratory feeder is recommended to provide a constant flow rate. In 1959 Pownall [10] described the construction and testing of a large laboratory spinning riffler and a year later Hawes and Muller described the construction of a small instrument [11]. In 1964 Allen [12] described how six operators carried out identical experiments using three techniques. Little variation between operators is found with the spinning riffler but two-fold differences occur with chute splitting and cone and quartering. Several other investigations have also been carried out (see section 1.10).

Fig. 1.19 The LADAL Spinning Riffler.

The turntable type of sampler recommended in BS 3406 is very similar to the spinning riffler and a commercial version is available from Pascall (figure 1.20). In this instrument the powder falls through a hopper outlet on to a cone whose position may be varied to vary the outlet aperture. The powder slides down the cone into containers on a revolving table.

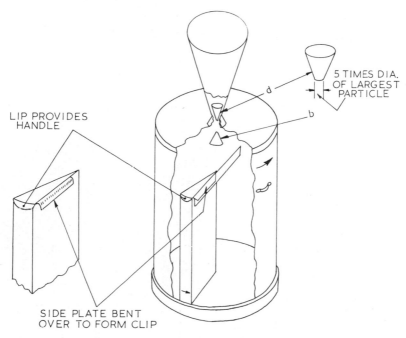

LIP PROVIDES
HANDLE

5 TIMES DIA.
OF LARGEST
PARTICLE

SIDE PLATE BENT
OVER TO FORM CLIP

Fig. 1.20 Rotary sample divider.

1.7 Miscellaneous devices

The moving-flap sample divider (figure 1.21) consists essentially of a flap which is
pivoted about a horizontal axis so that it can rest in either of two positions. The stream
of powder to be divided either falls down the process line to store or is diverted to the
sampling section. The time during which the powder is sampled can be varied by varying the time during which the flap is in each of the positions and this may be done
automatically. Since the diverted stream is still too large it is further divided on a table
sampler and most of it is returned to store. Although the moving flap forms an efficient sampling device the table sampler does not, hence the combination will give a
biased sample.

A splitter described by Fooks [13] (figure 1.22), consists of a feeder funnel through
which the sample is fed. It passes on to the apex of two resting cones, the lower fixed,
the upper adjustable by means of a spindle. Segments are cut from both cones and by
rotation of the upper cone the effective area of these slits may be varied to vary the
sampling proportion. Material which falls through the segmental slots is passed to a
sample pot. The residue passes over the cone and out of the base of the unit.

A similar unit may be installed in a feedpipe down which particulate material is flowing. In this unit the splitter cones are mounted within the feedpipe and the sample falling through the segmental slots passes out of a side pipe, while the remainder flows
over the cone and continues down the feedpipe.

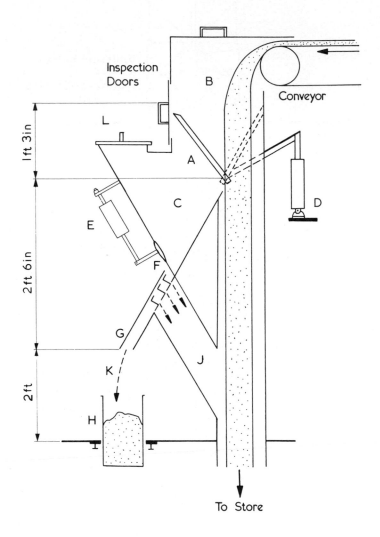

Fig. 1.21 Moving-flap sample divider.

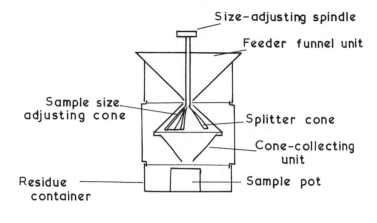

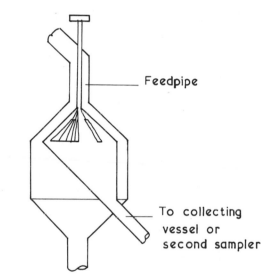

Fig. 1.22 Fooks's sample splitter.

Since the whole of the powder stream is not taken for many short increments of time this particular design is not recommended.

1.8 Reduction from laboratory sample to analysis sample

The methods described so far are suitable for the reduction of a gross sample collected from a process to a weight that can conveniently be sent to the size analysis laboratory, normally of the order of one kg. To obtain the amount required for the actual measurement a further reduction is necessary, and again careful consideration must be given to the problem of obtaining an analysis sample with the same size grading as the laboratory sample. The amount of material required as an analysis sample depends on the size

measurement technique being used.

For sieving it would normally be 25 to 50 g; for sedimentation it would depend on whether the technique was cumulative (0.5 g) or incremental (8 – 12 g); for elutriation 0.5 g.

It is sometimes suggested that a satisfactory sample division can be made by shaking the bottle containing the laboratory sample in order to mix it, then extracting the analytical sample with a scoop [4, 8]. For accurate work this method should not be used for the following reasons:

(1) Shaking a bottle containing a powder is not an effective method of mixing particles of different sizes. Anyone who doubts this should try the experiment of putting into a glass bottle equal volumes of two powders of different sizes and colours and examining the mixture after shaking. Pockets of segregated material form and cannot be broken up by further shaking. In particular the surface will be made up predominantly of the larger particles. A sample removed by a scoop must include part of the surface, where the composition is very likely to be different from that of the whole of the sample.

(2) A further objection to the use of a scoop is that it is liable to be size-selective, favouring the collection of fine particles. The reason for this is that, when the scoop is removed from the material, some particles will flow down the sloping surface of the powder retained in the scoop; the finer particles tend to be captured in the surface craters and retained, whereas coarse particles are more likely to travel to the bottom of the slope and be lost. This effect is particularly important if a flat blade (such as a spatula) is used for the removal of the sample. Such methods should never be used.

The spinning riffler is the equipment recommended for subdivision down to a gram, but the powder needs to be free-flowing. Coning and quartering a paste [14] has been used for preparing samples down to about 20 mg in weight but the efficiency of the method is operator-dependent. Sampling a stirred suspension with a syringe is quite a good method for fine powder suspensions, but with suspensions of coarser powders, concentration gradients and segregation of particle sizes due to gravitational settling and the centrifugal motion of the liquid caused by the stirring action are more likely to arise. For these reasons Burt *et al.* [15] have developed a suspension sampler which is shown in figure 1.23. The sampler consists of a glass cylinder closed at either end by stainless-steel plates. Around the periphery of the base plate are ten equidistant holes leading to ten centrifuge tubes via stainless-steel capillary tubes. The cover plate has a central hole through which passes a stirrer, a sealable inlet for introduction of the powder suspension, and a gas orifice which enables the gas pressure to be increased or decreased. In operation 100 cm^3 of suspension are introduced, stirred, then blown into the test tubes under a small applied pressure. Burt analysed a mixture of barium sulphate (0.25 g) and calcium carbonate (0.25 g) in 100 ml of water. The percentage of barium sulphate in each of the ten tubes was then determined and found to give significantly better results than syringe withdrawal, i.e. approximately 1.0% standard deviation as opposed to 3.0%.

Lines [31] compared three methods of sampling prior to Coulter analysis and found a spread of results of 6% with cone and quartering dry powder, a standard deviation of ± 3.25% with random selection and ± 1.25% with coning and quartering a paste.

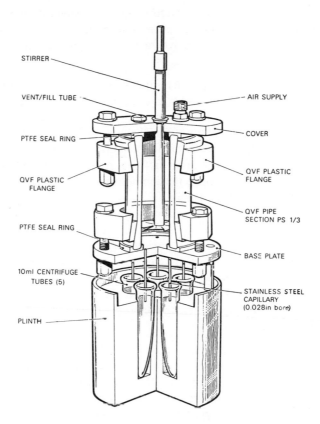

Fig. 1.23 Burt's suspension sampler.
(a) Suspension sampler, 10 ml to 2 ml

1.9 Reduction from analysis sample to measurement sample

In microscopy the required sample consists of a few milligrams of material. This may be extracted from the analysis sample by incorporating it in a viscous liquid in which it is known to disperse completely. The measurement sample may then be extracted with a measuring rod [7]. An alternative technique is to disperse the analysis sample in a liquid of low viscosity with the addition of a dispersing agent if necessary. The measurement sample may be withdrawn using a pipette or, preferably, a sample divider as illustrated in [16].

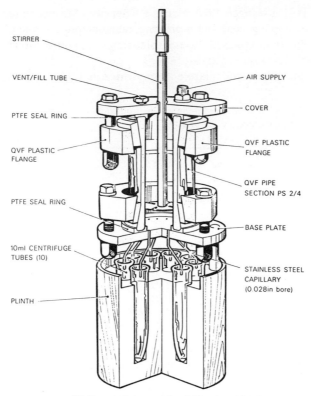

STIRRER

VENT/FILL TUBE

PTFE SEAL RING

QVF PLASTIC
FLANGE

PTFE SEAL RING

10ml CENTRIFUGE
TUBES (10)

PLINTH

AIR SUPPLY

COVER

QVF PLASTIC
FLANGE

QVF PIPE
SECTION PS 2/4

BASE PLATE

STAINLESS STEEL
CAPILLARY
(0.028in bore)

(b) Suspension sampler, 100 ml to 10 ml

1.10 Experimental tests of sample-splitting techniques

In 1960 Hawes and Muller critically [11] examined the spinning riffler to find how various factors influence its efficiency. They used quartz and copper sulphate crystals of the same size and designed experiments to investigate the relative effect of the different factors. Their conclusions were that the efficiency:

(a) Is dependent on the relative proportions of the mixture (it increased when the proportion of copper sulphate was raised from 1% to 5%).
(b) Increases with increasing particle size.
(c) Is reproducible under similar experimental conditions.
(d) Is not affected by combination of variables.
(e) Is not affected by the number of volume units (a volume unit consists of the single presentation of a sample container to the feed. Since the minimum number of such units was 100, this statement applies only to larger values than this).

The powders mixed in these experiments contained particles all of the same size, hence the main cause of segregation was not present, and the conclusions may not apply in the more usual case where size segregation occurs.

In 1968 Khan [17] investigated the relative efficiencies of five common sampling procedures: scoop sampling, cone and quartering, table sampling, chute riffling and spinning riffling. Binary mixtures of different size ranges and different densities were sampled and the results examined statistically.

The results for the binary sand mixture are presented in Table 1.1 [18].

Table 1.1 Reliability of selected sampling methods (60 : 40, coarse: fine, sand mixture; $P = 0.60$)

Method	Standard deviation of samples (%) σ	Var (P_n) (%)^2	Estimated maximum sample error (%) E	Efficiency (%) C
Cone and quartering	6.81	46.4	22.7	0.013
Scoop sampling	5.14	26.4	17.1	0.022
Table sampling	2.09	4.37	7.0	0.13
Chute riffling	1.01	1.02	3.4	0.56
Spinning riffling	0.125	0.016	0.42	36.3
Random variation	0.076	0.0058	0.25	

It was deduced that very little confidence can be placed in the first three techniques and the spinning riffler was so superior to all other methods that it should be used whenever possible.

Since the spinning riffler was far superior to the other techniques this was examined further using factorial design experiments in order to determine its optimum operating conditions. It was found that a minimum of 35 presentations is required to give optimum results. If the speed of rotation is made too great the efficiency will fall again since powder will be lost due to the air currents set up by the spinning disc.

In 1968 a comparison between the spinning riffler and the Jones chute splitter indicated that the latter introduced bias into the samples [19]. This was later confirmed by Kaye [4].

Several other articles [19 – 24] conclusively prove the superiority of the spinning riffler over all other methods investigated, and this method is recommended in ASTM Standard F577B – 78.

In 1978 Hatton [23] deduced theoretically that Var $(P_n) \propto \left(1 + \dfrac{L}{V}\right)^{\frac{1}{2}}$ where L is the linear flow rate of the feed and V the peripheral velocity of the spinning disc. He found that the equation fitted Allen's data, which suggests that in order to reduce variations it is necessary to reduce L/V which may be effected by reduction of L.

References

1 Hulley, B.J. (1970), *Chem. Engr.*, CE410 – CE413.
2 Stange, K. (1954), *Chem. Engr. Tech.*, **26**, 331.
3 Cordell, R.E. (1969), Automatic Sampling System, US Patent 3 472 079.
4 Kaye, B.H. and Naylor, A.G. (1972), *Particle Technol.*, 47 – 66.
  Proc. Seminar, Madras (1971), Indian Institute of Technology.
5 ASTM D451 – 63 (1963): *Sieve Analysis of Granular Mineral Surfacing for Asphalt Roofing and Shingles.*
6 Batel, W. (1960), *Particle Size Measuring Technique*, Springer Verlag, Germany.
7 BS 3406 (1961): *Methods for the Determination of Particle Size of Powders*, Part 1, Sub-Division of Gross Sample Down to 0.2 ml.
8 Kaye, B.H. (1961), Ph.D. Thesis, London University.
9 Wentworth, C.K., Wilgers, W.L. and Koch, H.L. (1934), A rotary type of sample divider. *J. Sed. Pet.*, **4**, 127.
10 Pownall, J.H. (1959), *The Construction and Testing of a Large Laboratory Rotary Sampling Machine*, AERE – R – 2861, Harwell, Oxfordshire, England.
11 Hawes, R.W.M. and Muller, L.D. (1960), *A Small Rotary Sampler and Preliminary Study of its Use*, AERE – R – 3051, Harwell, Oxfordshire, England.
12 Allen, T. (1964), *Silic. Indust.* **29**, 12, 409 – 15.
13 Fooks, J.C. (1970), Sample splitting devices. *Br. Chem. Engr*, **15**, 6, 799.
14 Burt, M.W.G. (1967), *Powder Technol.*, **1**, 103.
15 Burt, M.W.G., Fewtrell, C.A. and Wharton, R.A. (1973), *Powder Technol.*, **7**, 6, 327 – 30.
16 BS 616 (1963): *Methods for Sampling Coal Tar and its Products.*
17 Khan, A.A. (1968), M.Sc. Thesis, Bradford University.
18 Allen, T. and Khan, A.A. (1970), *Chem. Engr.*, 238, CE108 – CE112.
19 Montgomery, J.R. (1968), *Analyt. Chem.*, **40**, 8, 1399 – 1400.
20 Kaye, B.H. (1962), *Powder Metall.*, **9**, 213.
21 Clarke, J.R.P. (1970), Sampling for on-line analysis. *Measurement and Control*, **3**, 241 –4.
22 Scott, K.J. (1972), *The CSIR Rotary Powder Sample Divider*, CSIR Special Report Chem. Pretoria, S.A., p. 206.
23 Hatton, T.A. (1978), *Powder Technol.*, **19**, 227 – 33.
24 Charlier, R. and Goossens, W. (1970/71), *Powder Technol.*, **4**, 351 – 9.
25 Hinde, A.L. and Lloyd, P.J.D. (1975), *Powder Technol.*, **12**, 37 – 50.
26 Osborne, B.F. (1972), *CIM Bull.* **65**, 97 – 107.
27 Cross, H.E. (1967), Automatic Mill Control System, Parts I and II. *Mining Congress J.*, S.A., 62 – 7.
28 Hinde, A.L. (1973), *J.S. Afr. Inst. Min. Metall.*, **73**, 8, 258 – 68.
29 Hayslett, H.T. and Murphy, P. (1968) *Statistics Made Simple*, W.H. Allen.
30 Herdan, G. (1960), *Small Particle Statistics*, Butterworths.
31 Lines, R. (1973), *Powder Technol.* 7,3, 129 – 36.

2 Sampling of dusty gases in gas streams

2.1 Introduction

Legislation requires that manufacturers carefully monitor and control their particulate gas discharge effluent. Pollution today is a very 'dirty' word and visible discharge from chimneys with deposition of soot and sulphurous compounds creates a barrier between the industrialist and an environment-conscious public. The industrialist who accepts his duty to society and the environment equips his factories with dust-arresting plant such as filters, cyclones, electrostatic precipitators and scrubbers, balancing the cost against his profit margins. His less public-spirited counterpart keeps this non-productive plant to the minimum required by law. In general, financial and technical resources and an incentive to act are necessary for results.

> The dust-generating industries have shown little incentive to devote resources, or spontaneously show any resolve to control, except where such control could be directly reflected for economic reasons [1].

The main economic incentives to control and measure are to reduce production losses and production costs. Government health inspectors must also be satisfied and the risk of successful legal actions minimized.

In the UK, the Public Health Act (1936) can be invoked. Health is defined as a state of well-being, both physiological and psychological, and the concept of a healthy environment includes the protection of inanimate objects from damage.

It is therefore necessary that dusty gas streams be regularly sampled to ensure both that the plant is working efficiently and that legal requirements are being met. Procedures for sampling particulate matter suspended in a gas stream have been described [2 – 8, 73]. These all emphasize the need to obtain representative samples. The samples must be taken across the cross-section of the duct to represent the total dust burden. The size distribution is usually required so that the deposition pattern of the discharge may be estimated, the larger particles falling more rapidly than the smaller ones. Since the particles are very fragile, the sampling device often incorporates size-grading sections. Representative samples are extremely difficult to obtain, a single series of tests may well take a week or more and a complete survey on a large plant may take several months.

It is generally accepted that any obstruction to flow will create turbulence and an asymmetrical dust burden (figure 2.1), but even in the absence of obstacles the dust

concentration is non-uniform [9]. It is therefore imperative to sample at many points over the cross-section.

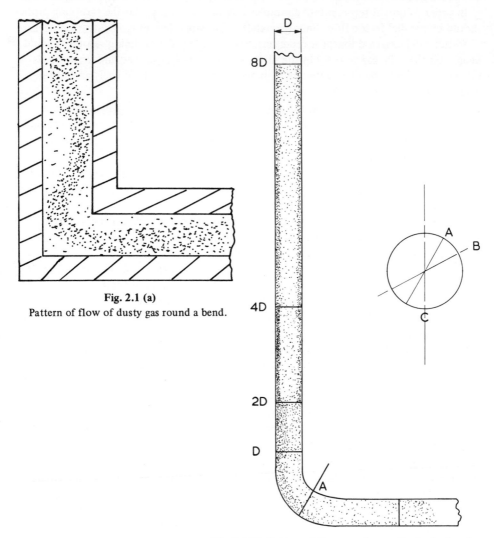

Fig. 2.1 (a)
Pattern of flow of dusty gas round a bend.

Fig. 2.1(b) Pattern of flow round a gently sloping bend.

In vertical pipes the concentration can be assumed axisymmetrical and it is permissible to take samples along one radius of cross-section only; in other cases it is necessary to cover the whole cross-section [10]. Problems are also encountered in gaining access to the gas stream [11, 12].

There are then difficulties in removing a sample which is representative of the gas stream at this point, since a sample will be representative only if the velocity of aspiration into the nozzle is equal to the gas velocity at that point. This is known as *isokinetic*

sampling. Under isokinetic conditions there is no disturbance of the gas streamlines and all the particles approaching the nozzle, and only those, will enter it [13 - 16].

In order to obtain representative samples it is also necessary that the sampling probe should be parallel to the flow lines (isoaxial), otherwise a loss of concentration occurs.

Walter [17] asserted that, for representative sampling, the sampling velocity should be greater than the gas velocity in the duct. This was later disproved and Walter's results were shown to be due to the use of sampling probes with thick walls and obtuse angles [10].

Rouillard and Hicks [70], on the basis of experimental work, suggest that for a sampling system to cause negligible disturbance of the velocity profile, sampling should be isokinetic using a knife-edge nozzle with little external bevel and the stem of the probe should be at least eleven diameters downstream of the nozzle.

Dust that settles in the probe should be collected and a gravity correction should be applied if the duct is vertical, but this is usually not done due to the complexity of the correction.

As it is not possible to collect increments of the whole emission it is necessary to select suitable sampling positions across the cross-sectional area perpendicular to the direction of gas flow. Since the quantity of emission varies with time, results can only give an average for the time of tests. Tests should be duplicated and averages taken. If sampling is carried out in accordance with BS 3405 a minimum of four points are sampled, each for a minimum duration of two minutes. It is necessary to ensure that the actual sampling time is adequate to give a quantity of dust which can be accurately weighed. Agreement should be better than 50% and the average taken [2]. Increasing the number of points to eight should improve the likely accuracy to 15% and an increase to 24 with a sampling time of ten minutes should give an accuracy of 5% [3].

Greater discrepancies may indicate a faulty test or that plant conditions have varied, hence further tests are required. The plant cycle should also be observed to establish loading conditions. Averages do not normally include short periods of high dust emissions as during soot blowing. Attempts to include these could well render it impossible to get any measure of agreement.

When sampling hot aerosols containing condensible vapours it is necessary to have a heated box for an external filter otherwise the vapour condenses and clogs the filter. Removal of the vapour can lead to dust losses.

When the gas to be sampled is at a temperature in excess of 400°C problems arise in the selection of sampling equipment which is often not designed for such extreme conditions. For example, pitot-static tubes, though manufactured from stainless steel, have silver-soldered joints [71]. One solution is to use water-cooled tubes and probes.

These problems are circumvented with the use of internal sampling and replaced by the problems of corrosion and the difficulty of installing bulky equipment in the duct without interfering with flow. On the whole, external sampling is preferable [18, 19].

A wide range of duct velocities can be encountered; hence, in order to maintain isokinetic conditions, a series of nozzles of different diameters is necessary. These are particularly useful when the sampling volume flow rate can only be adjusted over a limited range or when the collecting device exhibits a constant pressure drop.

2.2 Basic procedures

(1) Select a suitable access position.
(2) Conduct temperature and velocity surveys.
(3) Assemble and test the sampling apparatus.
(4) Carry out isokinetic sampling at preselected point for preselected times.
(5) Remove the apparatus containing deposited solids.
(6) Repeat the velocity and temperature surveys.
(7) Repeat steps (4), (5) and (6).
(8) Determine the mass and size distribution of collected samples; make the necessary calculations and fill in report.

2.2.1 Sampling positions

An important factor in the choice of the sampling position is the ease of access. It may be necessary to assemble equipment at the test site with only a limited space in which to work and in inclement weather conditions. Changing nozzles, extending the probe and handling filters may be impossible under such conditions. Where a long-term sampling programme is envisaged it may be worthwhile to install permanent sampling ports with clamping devices and easy access. For both inflammable and toxic gases, entry into the duct should be made through a double entry valve. The exhaust from the sampling system should be returned to the gas main or a point of safety [71]. The ideal sampling position is in the flue immediately prior to the discharge point in order to determine the actual emission. Since this is often impracticable, the position should be as near to the outlet as possible. Further, the dust burden is seldom distributed uniformly over the cross-section of the flue and this non-uniformity is aggravated by gas turbulence. Disturbances are caused by inlets, outlets, bends, constrictions and dampers and, in order to minimize the effects of these, it is preferable to select a site eight to ten duct diameters downstream, three to five diameters upstream from a disturbance. Often this is physically impossible and compromises must be made. The sampling position should be downstream of any settling chamber, dust-collecting plant or long horizontal duct where grit may collect, otherwise the amounts collected in the dust traps must be determined separately and subtracted from the amounts of collected dust.

Table 2.1 Mean values of relative particle concentrations (radius of bend, $2D$; gas velocity, 40 ft s^{-1})

Number of diameters from bend	A1	A2	A3	A4	B1	B2	B3	B4	C1	C2	C3	C4
H	0.66	0.72	0.51	0.84	0.64	0.56	0.53	0.90	0.65	0.83	1.19	1.12
1	0.33	0.24	0.20	0.78	0.35	0.15	1.35	2.13	0.27	0.22	0.13	0.68
2	0.49	0.23	0.34	1.01	0.34	0.15	0.98	1.42	0.25	0.37	0.76	1.46
4	0.67	0.47	0.47	1.13	0.28	0.26	0.57	1.15	0.33	0.38	0.76	1.28
8	0.78	0.59	0.54	0.73	0.73	0.72	0.56	0.70	0.62	0.57	0.71	0.79
16	0.78	0.59	0.54	0.73	0.73	0.72	0.56	0.70	0.62	0.57	0.71	0.79

H indicates the horizontal cross-section; A is a diameter through the elbow of the bend; B and C are at 30° to diameter A.

Sansone [5] investigated the patterns of flow of solids downstream from a 90° bend. Particles were fed into an airstream in a horizontal duct and concentrations were determined in a vertical section using a twelve-point sampling pattern at the centre of annuli of equal areas. Some of his results are given in Table 2.1.

The data in the above table reveal that concentrations are generally higher immediately downstream of a bend along the duct wall farthest from the centre of curvature of the elbow. Similarly the concentrations are low along the duct wall nearest the centre of turning. In negotiating the bend, particles resist change of direction and are thrown to the outer wall. There is also a loss in concentration due to assuming that particle concentrations at centres of equal areas are representative of these areas. These effects are enhanced with a gas velocity of 80 ft s^{-1}.

2.2.2 Temperature and velocity surveys

These should be carried out at ten equally spaced points along a sampling line, excluding the 50 mm nearest the flue walls. If temperature variations exceed 10%, a leak

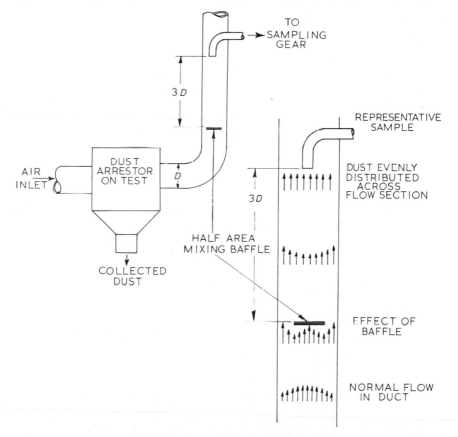

Fig. 2.2 Dust arrestor on test; laboratory rig.

should be suspected and this should be plugged otherwise the position is unsuitable.

If the ratio of maximum to minimum velocity exceeds 2, the four-point method of sampling is unsuitable. Gas flow is rarely uniform across the cross-section of a duct. For streamline flow (*Re* less than 2000), the velocity on the axis is twice the mean velocity, and for fully turbulent flow (*Re* greater than 2500), the axial velocity equals the mean velocity at $D/6$ from the walls, and it is claimed that a sample extracted at this point will be representative, but this is discounted by Stairmand [4], who states that the dust distribution rarely follows the gas distribution. Hawksley [20] also discusses variation of dust and gas flow across a duct.

2.2.3 Sampling points

In BS 893 the use of 24 sampling points, with at least three access holes, is recommended. The sampling lines should form the diameters of a hexagon with sampling access at alternate ends and sampling points at $0.07r$, $0.21r$, $0.39r$, $0.65r$, $1.35r$, $1.61r$, $1.79r$, $1.94r$ from the access point (r being the radius of the flue). These points are at the centres of annuli of equal area. Rectangular ducts should be divided up into 24 rectangles of equal area with sampling points at the centres of each rectangular element.

In BS 3405 four or eight points are recommended, again at the **centres of annuli of** equal area. The four points lie on mutually perpendicular diameters at a distance $0.7r$ from the centre of the duct. With eight sampling points, four lie on mutually perpendicular diameters a distance $0.5r$ from the centre of the duct and four on the same diameters $0.87r$ from the centre of the duct. A reduction to a single sampling point has been made by Stairmand (figure 2.2) [4, 34, 38, p. 583] who uses a half-area centrally located obstruction (a mixing baffle), three duct diameters upstream of the sampling probe, leaving at least two duct diameters of straight pipe downstream of the sampling probe. He claims that the sampling baffle makes the concentration of dust at the sampling position fairly uniform.

A similar device, 'the ring and doughnut flow straightener', has also been described [11]. Fuchs [10] says of Stairmand's half-area baffle: 'a considerable part of the disperse phase will be deposited on the screen and afterwards will pour down or be blown off the screen. The method is useless for particle size determination but may, under favourable conditions, measure **concentration**.'

A device for routine sampling, for comparative purposes only, has also been developed [22] (figure 2.3). This consists of a suitable probe inserted into the flue duct. The open end faces the gas stream where it flows either horizontally or downwards so that the momentum of the grit carries it into the sampler. The device was used to detect failure of electrostatic precipitators and multicyclones installed on boilers earlier than would otherwise have been possible without expensive testing.

2.3 Sampling equipment

The accepted method of sampling flue gases in the UK is given in BS 893 [3]. The method is lengthy and complicated, but in spite of development and experience, more recently issued standards of other countries [23] differ only in detail. A more recently

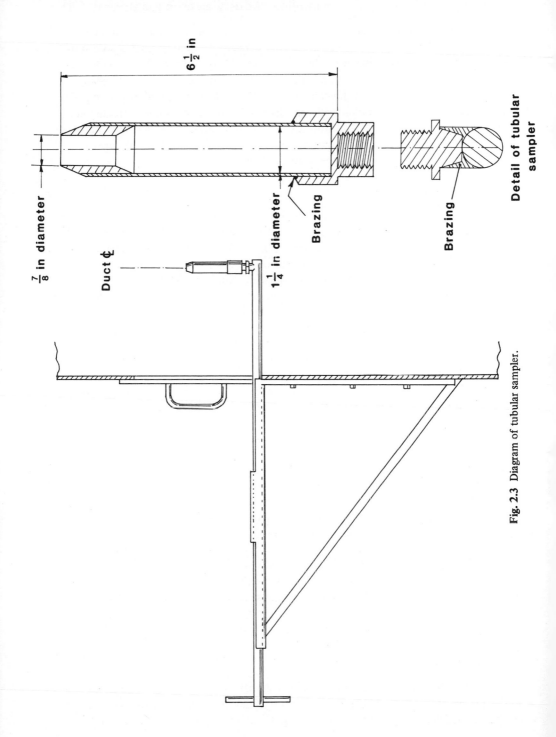

$\frac{7}{8}$ in diameter

$6\frac{1}{2}$ in

Brazing

$1\frac{1}{4}$ in diameter

Duct ₵

Detail of tubular sampler

Brazing

Fig. 2.3 Diagram of tubular sampler.

issued British Standard (BS 3405) [2] gives simplified procedures for quick investigations.

The basic equipment consists of a set of nozzles attached to a sampling tube; a miniature collector to retain the particles; means for determining the amount of gas sampled and the gas velocities in the ducts; means for withdrawing samples of gas through the system; thermometers, manometers, cooling coils, catchpots, connecting tubing and stopwatches.

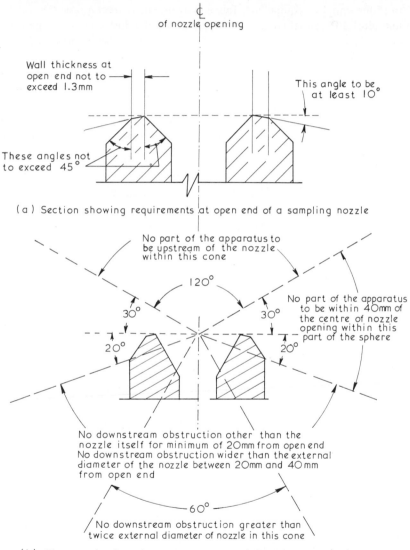

(a) Section showing requirements at open end of a sampling nozzle

(b) Diagram showing clearances necessary between open end of sampling nozzle and other parts of the apparatus

Fig. 2.4 Requirements for sampling nozzles (BS 3405).

2.3.1 Nozzles [3, 4, 12]

The nozzles may vary in shape and construction but they should be of circular cross-section, the open end having a thickness less than 1.3 mm and the internal and external surfaces having an inclination not greater than 45° to the axis of the nozzle (figures 2.4 and 2.5). With blunt-edged probes, a damming effect occurs upstream from the probe which deflects particles away from it [24 - 28]. The nozzle size must be chosen so that, with isokinetic sampling, the maximum flow rate will be within the capacity of the sampling equipment. Tables showing the correct nozzle size for the maximum pitot differential are usually supplied with the dust-sampling apparatus.

When using these simple sampling nozzles the flow field must be previously measured with a pitot tube.

An alternative solution is to use a pump-free system. The Cegrit (figure 2.6) is an example of this type of device [19]. The device consists of a heated cyclone through which a sample is continuously drawn by an ejector device operated by the suction of the flue itself, the dust collecting in a detachable glass container. The great ingenuity of the device lies in its continuous isokinetic operation irrespective of variations in gas flow.

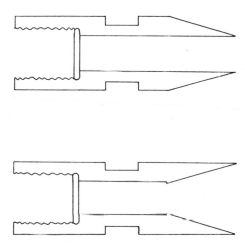

Fig. 2.5 Alternative designs for sampling nozzles.

Null-type nozzles have double walls in which there are two ring-like channels connected by holes with the inside and outside of the probe. For isokinetic sampling the pressure difference in these channels should be zero (see figure 2.7) [29 - 32].

Velocity-sampling nozzles are a combination of sampling probe and pitot tube. The flow velocity is determined by the difference between the dynamic and static pressure. A drawback is the clogging of holes at high dust concentration. These probes were extensively studied by Noss [33].

Rammler and Breitling [18] examined the various available systems and preferred the velocity-sampling probes.

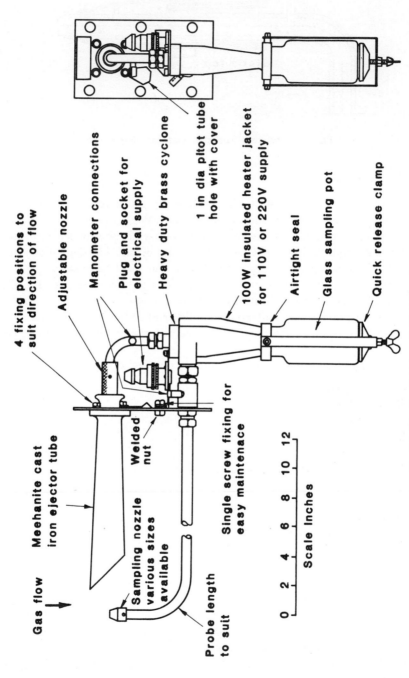

4 fixing positions to suit direction of flow

Adjustable nozzle

Manometer connections

Plug and socket for electrical supply

Heavy duty brass cyclone

1 in dia pitot tube hole with cover

100W insulated heater jacket for 110V or 220V supply

Airtight seal

Glass sampling pot

Quick release clamp

Meehanite cast iron ejector tube

Gas flow

Sampling nozzle various sizes available

Welded nut

Single screw fixing for easy maintenace

Probe length to suit

0   2   4   6   8   10   12

Scale Inches

Fig. 2.6 The Cegrit Automatic Dust Sampler.

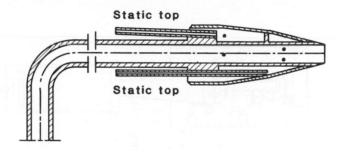

Static top

Static top

Fig. 2.7 A typical balanced **pressure-sampling** nozzle.

2.3.2 Dust-sampling collector

The purpose of the sampling collector is to remove the particulate matter from the gas sample extracted from the main duct. Where it is only desired to determine concentration, the choice is fairly easy; where composition or size distribution has to be determined, more precautions have to be taken. Stairmand describes a range of collectors in [4] and discusses their performance; other descriptions are to be found in [2].

The perfect collector would offer little resistance to gas flow even when it had collected a considerable amount of dust. Further, the dust should be easily separable from the collector without residual contamination. The only collector without these requirements is the electrostatic precipitator, but difficulties arise in the use of this equipment [34].

The range of collectors described by Stairmand consists of:

(a) Cyclone separators.
(b) Glass-wool filters.
(c) Ferrule glass-wool filters.
(d) Slag-wool filters.
(e) Composite filters.
(f) Terylene and superfine glass-wool filters.
(g) Soluble filters.
(h) Volatile filters.
(i) Soxhlet filters.
(j) Ceramic thimble filters.
(k) Electrostatic filters.
(l) Impingement filters.

A high-efficiency cyclone (figures 2.8 and 2.9) forms a useful sample collector, since it can collect large quantities of dust without increase of resistance. In many cases it will trap more than 95% of the suspended matter and may be used without a backing filter. When the dust is fine or complete removal is required, the cyclone may be backed by any of the range of filters annotated above.

A typical 'absolute' filter is shown in figure 2.10, an electrostatic filter in figure 2.11 and an impingement filter in figure 2.12.

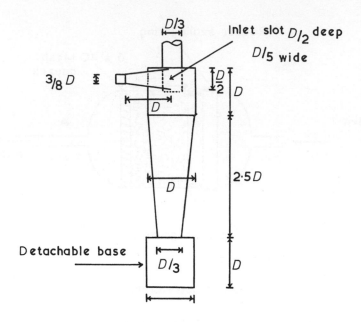

Fig. 2.8 Proportions of sampling cyclone.

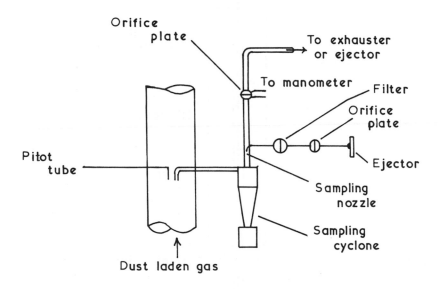

Fig. 2.9 Arrangement of sampling cyclones.

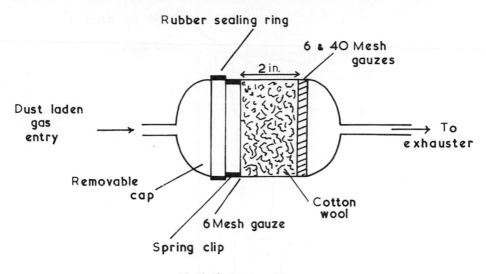

Fig. 2.10 Glass-wool filter.

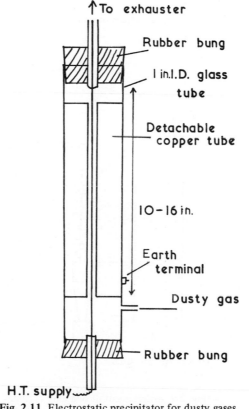

Fig. 2.11 Electrostatic precipitator for dusty gases.

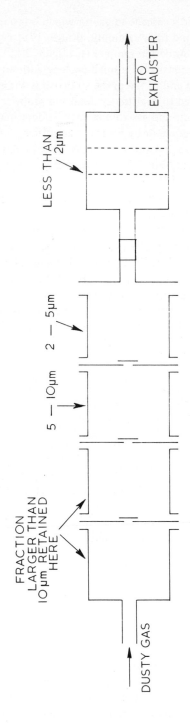

Fig. 2.12 Train of impingement samplers.

Pilat *et al.* [35] described a cascade impactor which is inserted into the stack in series with an inlet nozzle and filter assembly. Bridger [71] describes the instrument in use and prefers it to the Andersen sampler (see figure 2.15).

The simplest equipment consists of a small cyclone without a series filter. Locating the cyclone near the nozzle end of the probe so that it is within the flue, is a further simplification, since the need to heat the cyclone and probe to prevent condensation is eliminated. The sampling rate and gas volume may be determined by measuring the pressure drop across the cyclone or by means of an orifice plate.

A refinement of the above apparatus is to pass the gas from the cyclone through a filter, usually of packed glass fibres, thus permitting the collection of particles that have escaped through the cyclone [36, 37].

Proprietary equipment based on the above design are used by BCURA [42], CEGB, NCB and others [38, p. 604].

Other equipment has heated filters on the outward end of the probe or unheated filters on the inward end of the probe [25]. Equipment developed by BISRA for rapid sampling using filters is described by Granville and Jaffrey [36].

The ICI sampler forms one of the most useful of the composite equipment [4] (figure 2.13). It consists of a nozzle attached to a heated probe, glass-wool filter, cooling coil, catchpot, orifice plate flowmeter and an ejector. Appropriate manometers and thermometers are provided so that the necessary calculations for correct isokinetic sampling can be made.

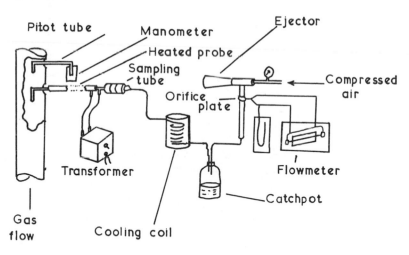

Fig. 2.13 Equipment for sampling dust from flowing gases.

With the CEGB Mark II dust sampler the dust is collected in a glass fabric or paper thimble. Details of design and operation are given in [38, p. 605].

With the impingement-type filters, the dust is classified according to its aerodynamic properties. This is preferable to microscopic methods or sieving since these properties may more easily be related to particle behaviour on leaving the stack, the approximate

area of deposition and the probable point of respiratory deposition. Further, the classification is carried out *in situ* which reduces the probable breakdown of floccu-lates.

The Day isokinetic dust sampler is used to measure dust particles discharged from dust control equipment as well as dust in dust control ducts. It is basically a thimble and a nozzle with a rotameter and pump.

With the Staksampler, sampling is isokinetic with the sample passing through a glass sampling train where the particulates and gases are collected.

The sample from the Dust Difficulty Determinator is drawn, by means of an exhaust fan, through two wet scrubbers. The first scrubber has a pressure drop of 5 in water and the pressure drop in the second can be varied over the range 5 to 100 in water.

Farthing [21] recently reported on a conference at which seventeen papers were presented on improved instruments and techniques for measuring the size, number and composition of particles in process streams.

Recent studies were presented on cascade inpactors, cyclones and diffusion batter-ies and several methods of making *in situ* size or mass determination by light scatter-ing were reported.

2.3.3 Ancillary apparatus

(a) Pitot tube (BS 1042). This should face directly into the gas stream with a maximum deviation (angle of yaw) of 10°.
(b) Pressure-measuring instrument(s). An inclined gauge reading 0.05 to 0.30 mbar within ± 0.01 mbar.
(c) A probe tube to which nozzle is attached maintained at a temperature sufficient to prevent condensation. If the dust collector is fitted immediately after the nozzle and before the probe, this requirement does not hold. In the former case the probe should be constructed of stainless steel.
(d) Flow and temperature-measuring device to measure v to 5% and T to 5 K.
(e) A control valve to regulate flow through the equipment.
(f) A method for withdrawing sample at required rate for isokinetic sampling.

2.3.4 On-line dust extraction

An on-line isokinetic particle sampler for continuously sampling finely divided material being transported in a flow system is the subject of a patent [39] (figure 2.14). The particulate material flows through conduit (1) and is sampled at nozzle (2), which is positioned high enough into the flow stream so that it is not affected by any disturb-ances in the flow due to housing (3). An artificial atmosphere of dry nitrogen gas is maintained within sampling tube (4) to control the pressure within the tube as well as to transport particles out of it. The nitrogen exits at (5) creating a partial vacuum at the intersection with (2), thus drawing the particulates into the system isokinetically. The pressure is maintained at the pressure in conduit (1) by outlet (6), which is con-

nected to the automatic differential pressure control (4). When a difference in pressure is noted, a signal is fed to the automatic valve which varies the valve position to maintain equal pressure.

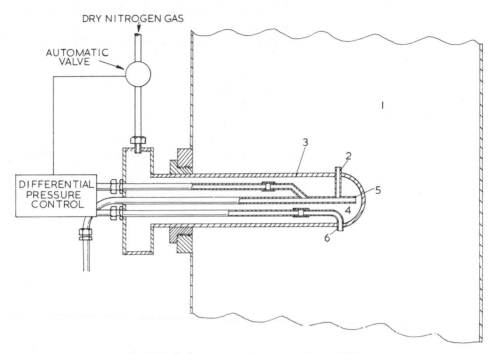

Fig. 2.14 Isokinetic particle sampler (Lynn [39]).

2.3.5 *The Andersen stack sampler*

The Andersen stack sampler [40] has a cascade impactor-type sampling head which, in operation, is mounted inside the flue (figure 2.15).

   The sampler contains nine jet plates, each having a pattern of precision-drilled orifices. The nine plates, separated by 2.5 mm stainless-steel spacers, divide the sample into eight fractions of particle size ranges. The jets on each plate are arranged in concentric circles which are offset on each succeeding plate. The size of the orifices is the same on the given plate, but is smaller for each succeeding downstream plate. Therefore, as the sample is drawn through the sampler at a constant flow rate, the jets of air flowing through any particular plate direct the particulates toward the collection area on the downstream plate directly below the circles of jets on the plate above. Since the jet diameters decrease from plate to plate, the velocities increase such that whenever the velocity imparted to a particle is sufficiently great, its inertia will overcome the aerodynamic drag of the turning airstream and the particle will be impacted

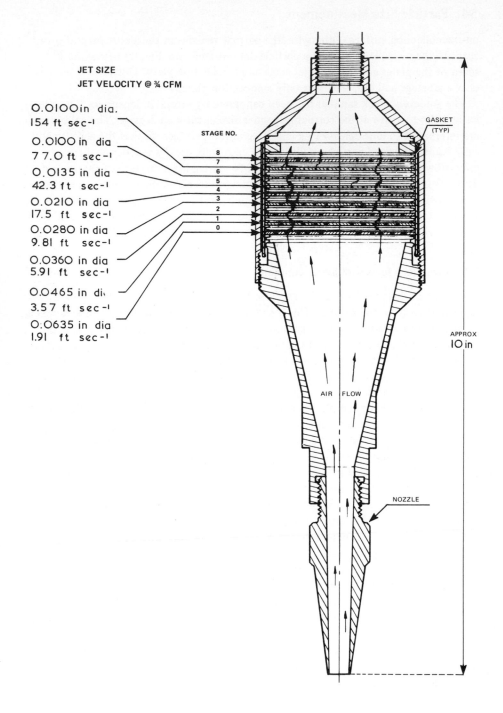

JET SIZE
JET VELOCITY @ ¾ CFM

0.0100 in dia.
154 ft sec⁻¹

0.0100 in dia
77.0 ft sec⁻¹

0.0135 in dia
42.3 ft sec⁻¹

0.0210 in dia
17.5 ft sec⁻¹

0.0280 in dia
9.81 ft sec⁻¹

0.0360 in dia
5.91 ft sec⁻¹

0.0465 in dia
3.57 ft sec⁻¹

0.0635 in dia
1.91 ft sec⁻¹

STAGE NO.

8
7
6
5
4
3
2
1
0

GASKET (TYP)

APPROX 10 in

AIR FLOW

NOZZLE

Fig. 2.15 The Andersen stack sampler.

on the collection surface. Otherwise, the particle remains in the airstream and proceeds to the next plate. Since the particle deposit areas are directly below the jets, seven of the plates act as both a jet stage and a collection plate. Thus, No. 0 plate is only a jet stage and No. 8 plate is only a collection plate.

The Andersen stack sampler has been calibrated by several independent laboratories in order to arrive at the correct respective size cuts for each stage. The calibrators are referenced to unit density (1 g cm^{-3}), spherical particles, so that the aerodynamically equivalent-sized particles collected on each stage are always identical for any given flow rate. For this reason, a stack sample containing a mixture of shapes and densities is fractionated and collected according to its aerodynamic characteristics and is aerodynamically equivalent in size to the unit density spheres collected on each specific stage during calibration. A seven-stage Andersen impactor has been used for soot blowing and for non-isokinetic as well as isokinetic sampling [72].

2.4 Corrections for anisokinetic sampling

A measure of dust content can be expressed as:
(a) The mass flow rate of dust. This is the mass of dust passing per unit time across an element of area of the gas stream. In isokinetic sampling, the quantity directly determined is the amount of dust passing, during the time of sampling, across an area in the gas stream equal to the area of the nozzle opening.
(b) The concentration of dust. This is the mass of dust per unit volume of gas, and is usually calculated from the ratio of the mass of dust collected to the volume of gas sampled. When sampling is isokinetic, the concentration of dust in the sample is equal to the concentration in the gas stream at the point of sampling.

When the sampling velocity differs from that of the gas stream (anisokinetic sampling), the gas streamlines are disturbed, causing some particles to be deflected from their original direction of motion, so that the quantity of particles entering the probe per unit time will differ from that entering under isokinetic conditions (see figure 2.16). It will be less when the sampling velocity is less than that of the gas stream, because some of the gas that should enter the probe flows past it carrying some particles with it. Conversely, if the sampling velocity is higher the amount of particles entering the probe will be too high. It is easy to envisage what happens in the case of very fine or very coarse particles. When the particles are very fine, they follow closely the deflected gas streams and the amount entering is proportional to the sampling velocity. That is, the *concentration* of fine particles in the sample is equal to that of the gas stream, irrespective of the sampling velocity. When the particles are coarse, their inertia is so great that they persist in their original direction of motion; the amount entering the probe is independent of the sampling velocity. That is, the *mass flow rate* of coarse particles is measured correctly irrespective of the sampling velocity, a fact first reported by Hemeon and Haines [63].

If the particles cover a range of sizes, anisokinetic sampling affects not only the mass of particles in the sample, but also their size distribution. During prolonged sampling the flow rate of the gas stream may vary throughout the test, and it may not be

possible to vary the sampling rate to compensate for this. A method for correcting this is described later.

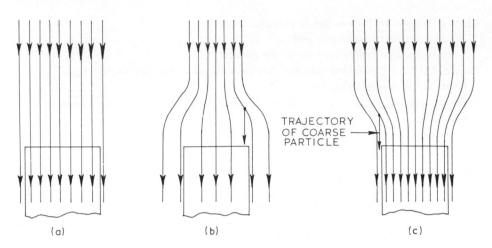

TRAJECTORY
OF COARSE
PARTICLE

(a)   (b)   (c)

Fig. 2.16 (a) Isokinetic sampling; representative concentration and grading.
　　　　　　(b) Sampling velocity too low; excess of coarse particles.
　　　　　　(c) Sampling velocity too high; deficiency of coarse particles.

Seldon [41] suggests that deviations from isokinetic sampling conditions give rise to errors in concentration which may be summarized in the general formula:

$$A = 1 + \left(\frac{1}{w} - 1\right) F \tag{2.1}$$

where
$$A = \frac{C_s}{C} \tag{2.2}$$

C_s is the solids concentration in the sampled gas
C is the solids concentration in the gas stream.

$$w = \frac{v_s}{v} \tag{2.3}$$

v_s is the sampling velocity
v is the gas stream velocity
F is a function of the dimensionless groups.

The Reynolds number $Re = \dfrac{\rho_f v d}{\eta}$ $\tag{2.4}$

where ρ_f is the fluid density,
　　　η the fluid viscosity and
　　　d the particle diameter.

The Stokes number $Stk = \dfrac{v_f v}{gD}$ (2.5)

where g is the acceleration due to gravity,

$\quad$ D is the nozzle diameter and

$\quad$ v_f is the free-falling velocity of the particle.

For $Re < 0.5$ the free-falling velocity may be replaced by the Stokes velocity.

$\quad$ Other terms in general use are:

λ, the particle range, the distance the particle will travel in still air before coming to rest;

τ, the particle relaxation time in the gas.

$$Stk = \frac{\lambda}{D} = \frac{\tau v}{D}$$ (2.6)

From equations (2.5) and (2.6) For $Re < 0.5$

$$\tau = \frac{\rho_s d^2}{18\eta}$$ (2.7)

ρ_s is the particle density.

$\quad$ Badzioch [26] stated that the mass of particles M_s entering the nozzle at a sampling velocity v_s is related to the mass M entering the nozzle when the sampling velocity v equals the gas velocity by the equation:

$$\frac{M_s}{M} = (1 - \alpha)w + \alpha$$ (2.8)

where:

$$\alpha = 1 - \frac{\lambda}{L} \exp\left(-\frac{L}{\lambda}\right)$$ (2.9)

L is a length representative of the distance upstream of the nozzle over which there is a disturbance in the gas stream [27].

Internal diameter of nozzle (in)	$\frac{3}{8}$	$\frac{1}{2}$	$\frac{5}{8}$	$\frac{3}{4}$	$\frac{7}{8}$	1
L (cm)	4.5	4.0	3.5	3.0	2.5	2.0

For fine particles $\alpha = 0$ $\quad$ and $\quad \dfrac{M_s}{M} = \dfrac{v_s}{v}$ (2.10)

as stated above.

For coarse particles $\alpha = 1$ $\quad$ and $\quad \dfrac{M_s}{M} = 1$ (2.11)

as stated earlier, i.e. the correct flow rate is obtained whatever the value of sampling velocity to gas velocity, w [28].

Badzioch's work is questioned by Fuchs [10] who states that the scatter of $A \left(= \dfrac{C_s}{C} \right)$ values obtained by him is very large and differs considerably from later data.

A versus w values have been calculated by Vitols [42] at different Stokes numbers.

Isokinetic sampling is only critical in a narrow size range, the limits of which are conveniently expressed in terms of Parker's [43] inertia parameter ψ

$$\psi = CStk \tag{2.12}$$

where C is the Cunningham correction factor.

Parker stated that isokinetic sampling was necessary for:

$$0.05 < \psi < 50 \tag{2.13}$$

Svarovsky [44] shows the effect of sampling, nozzle diameter and flow velocity on the upper and lower limits of particle size, in graphical form, using Parker's inequality relationship.

Ruping [45] measured the flow field by means of a probe and calculated the resulting particle trajectories. His results deviate widely from the latest experimental data.

Voloshchuk and Levin [46] based their work on earlier work by Vandrey [47] and, although they made some sweeping assumptions, they obtained reasonable agreement with experimental work.

The most relevant measurements were made by Zeuker [48] who found that:

$$\frac{M_s}{M} = 1 + K(w - 1) \tag{2.14}$$

for $0.4 < w < 2.5$.

K is a coefficient depending only on the Stokes number and equal to 1 at $Stk = 0$ and to 0 at $Stk = \infty$.

As M_s is proportional to $C_s v_s$ and M to Cv.

$$A = K + (1 - K)/w \tag{2.15}$$

Other experiments were performed by Belyaeu and Levin [49, 50] and their results can be approximated to with the formula:

$$A = 1 + \left(\frac{1}{w} - 1 \right) \frac{(2 + 0.62w)Stk}{1 + (2 + 0.62w)Stk} \tag{2.16}$$

for $0.18 < w < 5; 0.18 < Stk < 2$.

Their results are in fair agreement with Zeuker's [10].

Comparing Badzioch's equation (2.8) with Seldon's general equation (2.1) gives F as a function only of Stk. Equation (2.16) gives F as a function of w and Stk. Seldon suggests that F is a function of three-dimensionless groups and proposes an equation similar to equation (2.16) which, rearranged, gives:

$$N = \frac{Aw - 1}{w - Aw} \tag{2.17}$$

$$= F(Stk, Re, w)$$

$$= (Stk^a\ Re^d)/(b + cw) \tag{2.18}$$

For equation (2.16) $a = 1$, $b = 2$, $c = 0.62$, $d = 0$.

Seldon proposes $a = -0.918$, $b = 1.53$, $c = 0.617$, $d = 0.355$.

The error due to anisokinetic sampling is determinable as follows: The true dust concentration in the gas stream is given by:

$$C = \frac{M}{av} \tag{2.19}$$

where a is the area of the sampling nozzle. It follows that, when sampling fine dusts, the observed concentration will be nearly correct whatever the sampling velocity, provided it is calculated from the rate of dust collection M_s divided by the volume flow rate of sampling av_s.

$$C_s = M_s/av_s \tag{2.20}$$

When sampling coarse dusts anisokinetically, a correct value is obtained if M_s is divided by av, i.e.:

$$C'_s = \frac{M_s}{av} \tag{2.21}$$

When the fine dust method is used, the percentage error E is given by:

$$E = 100\,\frac{C_s - C}{C}$$

$$= 100(A - 1)$$

$$= \left(\frac{100}{N + 1}\right)\left(\frac{1 - w}{w}\right) \tag{2.22}$$

Similarly when the coarse dust method is used, the apparent dust concentration is given by:

$$E' = 100\,\frac{C'_s - C}{C}$$

$$= 100\,\frac{(w - 1)N}{N + 1} \tag{2.23}$$

The latter method will give more accurate results if E' is less than E, if:

$$N > \frac{v}{v_s}$$

The magnitude of the errors involved in using these formulae can be obtained from Table 2.2. For example, with $D = 1.25$ cm, $v = 800$ cm s^{-1}, $v_s = 1600$ cm s^{-1}, $v_{Stk} = 1.53$ cm s^{-1} making $\lambda = 1.25$ cm and $Stk = 1$, the error is $E = -38.2\%$ or $E' = -76.4\%$.

Table 2.2 Percentage errors in concentration (E), lower triangle and (E'), upper triangle, as a function of Stokes number and relative velocity for anisokinetic sampling.

			Stk	
w	0.2	0.3	1	2
0.2	4.8 / 119	2.1 / 206	42.2 / 212	109 / 272
0.5	3.2 / 31.6	13.4 / 53.6	34.9 / 69.8	82.2 / 82.2
1	0 / 0	0 / 0	0 / 0	0 / 0
2	−7.8 / −19.6	−30.9 / −30.9	−76.4 / −38.2	−173 / −43.3
5	−1.3 / −13.1	−144 / −57.5	−335 / −66.9	−729 / −72.9

2.5 Probe orientation

An essential detail of isokinetic sampling is that the entry nozzle axis should be aligned with the wind direction (isoaxial sampling). A sampling probe making an **angle θ with** the wind direction (angle of yaw) suffers a diminution of effective area related to the cosine of the yaw angle.

May [51] considers that the efficiency of entry of very small particles, which follow the air flow, is unaffected by yaw whereas the efficiency of entry of large particles obeys the cosine relationship. He defines large as referring to particles that have sufficient inertia to ignore bends in the streamlines and considers that they would be about 100 μm in diameter at low airspeeds; he defines very small as less than 1 μm in diameter.

As regards entry efficiency, as distinct from collection efficiency, the effect of yaw is not very serious; even with large particles and an angle of yaw of 25° the loss is only about 10% and for 5 μm particles May showed that the error would be about 7% with a yaw angle of 45°.

In most instruments the particles must pass down a tube immediately after entry and an important effect of yaw is that they may then be impacted into the side of the duct facing the wind direction. Losses can be very serious; they are related to the yaw angle, the duct width and the stopping distance of the particle. If the product of the particle's stopping distance and the sine of the yaw angle is greater than the duct width then impaction is likely to occur [52].

Watson's data [53] for non-isoaxial sampling are considered by Fuchs [10] to be unreliable. According to Fuchs, the correction for small θ is:-

$$A = 1 - \frac{4}{\pi} Stk \sin \theta \qquad (2.24)$$

2.6 Radiation methods [62]

For *in situ* measurements the absorption of visible light, gamma rays and acoustical energy has also been investigated. The most widely attempted techniques have used absorption of visible light [54 – 56]. The two problems associated with this type of device are that turbidity is a function of the projected surface area of particles within the light beam, whereas most requirements are for a mass-dependent technique, and that the light absorbed by a particle is not linearly related to size, i.e.:

$$\text{turbidity} = kcLS_wK$$

where k is a shape factor,

   c is the concentration of particles in the beam,

   L is the length of the light path in the suspension,

   S_w is the weight-specific surface of the particles in the beam,

   K the extinction coefficient, is the ratio of the light cut off by the particles to the
      light which would be cut off if the laws of geometric optics held.

Despite these objections the technique is useful for continuous monitoring.

Several instruments for measuring particulate concentration and size in smoke stacks have been produced commercially but, due to the breakdown in the laws of geometric optics, they are only suitable for comparison purposes.

The Nebetco continuous smoke monitor is designed to measure the optical turbidity across a 24 in or 36 in circular stack.

In the Leeds and Northrop smoke sampler, the smoke is extracted from the stack and passed through a light beam.

The Bailey smoke/dust density transmitter measures the turbidity across a stack.

The Edison visibility monitor is a forward light-scattering photometer having a receptor core from 12° to 70° using an ultra-violet source.

Lucas *et al.* [65, 66] discuss some of the limitations of smoke recorders (opacimeters) and state that the inaccuracies could be reduced by the use of a steel tube to maintain alignment, by the use of Everclean windows [67] and the use of purging air to record the effective instrument zero. They also discuss an instrument for coarse dust measurement (impacimeter) so called because it measures the opacity of dust after it has settled by impaction. The opacimeter is referred to as the CERL Self-Checking Smoke Recorder (commercially available as the SEROP) and the impacimeter as the CERL Flue Dust Monitor Mark II.

The LTV monitor [69] combines laser illumination, TV imaging and signal-processing electronics to determine particle emissions in the range 0.2 to 10 μm. The machine is portable and shows good agreement with EPA filter trains for measuring mass loads.

Harris *et al.* [68] in a paper entitled 'Turbulence Noise and Cross-Correlation Techniques Applied to Measurement in Stacks' applied the expertise they had gained in developing an on-line sizing instrument to the problems of stack sampling. The instrument they describe monitors forward, 30° and 90° scattering and processes the signals representing random fluctuations and discards information in the form of mean values of source radiation variations, drift or optical surface degradation. The signals are stored

on magnetic tape and analysed off-line using cross-correlation functions. The original instrument has a collimated white light source and an alternative using a laser source is also described.

The Gelman Stack Monitor is a system designed for industrial stack monitoring. It can be used to sample, measure and record automatically the mass of emitted particles from a stack or duct system. The gas from the sampling probe is drawn through a paper tape filter and the weight deposited determined using β-ray attenuation. The RAC Automatic Stack Monitor is similar in principle to the Gelman instrument.

Nuclear Measurements Corporation manufacture a series of filtration samplers which incorporate a variety of radiation detectors including α-scintillation, Geiger–Müller and α_2 – β-proportional counters.

The MAP-1B is a moving filter monitor for the continuous monitoring of the radio-activity of airborne particulate matter using a filter paper and detecting system.

An alternative approach to light attenuation utilized gamma radiation from a nuclear source and a Geiger-tube detector [57]. Absorption of the energy is again a function of path length, concentration and absorption factor. In this case the absorption factor is proportional to the atomic number, hence the mass of the particles, provided they are reasonably uniform in chemical composition.

A further technique is the absorption of acoustical energy [58]. The response is influenced by individual particle characteristics and gas density and gives very poor sensitivity to particle concentration.

Continuous analyses of extracted particulates have also been investigated. Several of these are a direct measurement of the mass collected and the primary problem is related to representative sampling. A system based on an electrobalance has been used as a direct measure of mass. The extracted particulate is deposited on a collection medium, which is then automatically cut off, transferred to the balance and weighed. The approach is really semi-continuous and would have application primarily where high accuracy on variable particulate is required. A second technique for direct mass measurement is based on frequency of oscillation as a function of the mass of material deposited on the surface. The most widely investigated extraction approach is based on collection of the particulate on a filter medium and detection of the mass through absorption of low-energy β-particles [59]. Absorption of the energy is a function of the total mass collected and chemical composition of the sample. Indirect measurements based on electrical devices, light absorbance and light reflectance have been employed. The electrical devices measure current transferred by the particles from a charging zone to discharge collection plates [60, 61]. Light absorbance and reflectance have also been used.

The application of continuous monitoring instrumentation presents fundamental difficulties. In particular the techniques do not satisfy usual requirements. They may be used as process monitors after complete characterization of the system using manual methods. However, the engineer should be fully aware of the complexities introduced by the interaction between particulate properties and potential measurement approaches.

References

1 Train, D. (1976), *Chemy Ind.*, **15**, 621 - 4.
2 BS 3405 (1961): *Simplified Methods for Measurement of Grit and Dust Emission from Chimneys.*
3 BS 893 (1940): *Methods of Testing Dust-Extraction Plant and the Emission of Solids from Chimneys.*
4 Stairmand, C.J. (1951), Sampling of dust-laden gases. *Trans. Inst. Chem. Engrs*, **29**, 31.
5 Sansone, E.B. (1969), *Am. Ind. Hyg. Assoc. J.*, Sept. – Oct., 487 - 93; also (1967), Sampling Airborne Solids in Ducts Following a 90° Bend, Ph.D. Thesis, University of Michigan.
6 American Society of Mechanical Engineers (1957), Determining Dust Concentration in a Gas Stream, Power Test Code No. 27.
7 Wolfe, E.A. (1961), Gas Flow Rate and Particulate Matter Determination of Gaseous Effluents, Bay Area Air Pollution Control District 1480, Mission Street, San Francisco, California, USA.
8 Los Angeles Air Pollution Control District (1963), Source Testing Manual, 434, San Pedro Street, Los Angeles 13, California, USA.
9 Sehmel, G. (1970), *Am. Ind. Hyg. Assoc. J.*, **31**, 758 - 71.
10 Fuchs, N.A. (1975), *Atmos. Envir.* **9**, 697 - 707.
11 Chatterton, M.H. and Lund, I.E. (1976), *Chem Ind.*, **15**, 637 - 40.
12 Davis, I.H. (1972), *Air Sampling Instruments*, 4th edn, Am. Conf. Governmental Industrial Hygienists.
13 Badzioch, S. (1957), Members Information Circular, No. 174, BCURA Leatherhead, UK.
14 Davies, C.N. (1954) *Dust is Dangerous,* Faber and Faber, London.
15 Watson, H.H. (1954), *Am. Ind. Hyg. Assoc. Quart.* **15**, 1.
16 Haines, G.F. and Hemeon, W.C.L., Information Circular, No. 5, American Iron and Steel Institute.
17 Walter, E. (1957), *Staub,* **53**, 880.
18 Rammler, E. and Breitling, K. (1957), *Chem. Technik,* **9**, 636 - 45.
19 Hawksley, P.G.W., Badzioch, S. and Blackett, J.H. (1958), *J. Inst. Fuel,* **31**, 4, 147 - 56.
20 Hawksley, P.G.W., Badzioch, S. and Blackett, J.H. (1961), *Measurement of Solids in Flue Gases,* BCURA, Leatherhead, UK.
21 Farthing, W.E. (1978), *J. Air Poll. Control Assoc.,* **28**, 10, 999 - 1001.
22 Lees, B. and Morley, M.C. (1960), A routine sampler for detecting variations in the emission of dust and grit, *J. Inst. Fuel,* **33**, 90 - 5.
23 ASTM D2928 (1970), *Method for Sampling Stacks for Particulate Matter,* American Society for Testing and Materials.
24 Walter, E. (1957), *Staub,* **53**, 880.
25 Szabolcs, G. (1959), *Energia es atom technika,* **9**, 12.
26 Badzioch, S. (1960), Correction for anisokinetic sampling of gas-borne dust particles, *J. Inst. Fuel,* **33**, 106 - 10.
27 Howells, T.J., Beer, J.M. and Fells, I. (1960), Sampling of gas-borne particles, *J. Inst. Fuel,* **33**, 512.
28 Whiteley, A.B. and Reed, L.E. (1959), The effect of probe shape on the accuracy of sampling flue gases for dust content, *J. Inst. Fuel,* **32**, 316.
29 Dennis, R. *et al.* (1952), *Chem. Engng,* **59**, 196.
30 Narjes, L. (1965), *Staub,* **25**, 148 - 53.
31 Toynbee, P.A. and Parkes, W. (1962), *Int. J. Air Water Poll.,* **6**, 113 - 20.
32 Wasser, R.W. (1958), *Am. Ind. Hyg. Assoc. J.,* **19**, 6.
33 Noss, P. (1955), *VDI-Berichte,* **7**, 5 - 10.
34 Anon. (1941), Sampling of gas-borne particles, *Engineering,* **152**, 141, 181.
35 Pilat, M.J., Ensor, D.S. and Bosch, J.C. (1971), *Am. Ind. Hyg. Assoc. J.,* **32** (8).
36 Granville, R.A. and Jaffrey, W.G. (1959), Dust and grit in flue gases engineering, *Engineering,* **187**, 4851 (Feb), 285 - 8.

37 Overbeck, E.M. and Thayer, K.B. (1970), Review on particulate sampling, ASME Paper 70, PET - 1.
38 Nonhebel, G. (ed.) (1964), *Gas Purification Processes,* Newnes.
39 Lynn, L.G. (1969), Isokinetic Particle Sampler, US Patent 3 473 388 21.
40 Andersen, A.A. (1958), *J. Bacteriol.*, 76, 471 - 84.
41 Seldon, M.G. Jr. (1977), *J. Air Poll. Control Assoc.,* 27, 3, 235 - 6.
42 Vitols, V. (1966), *J. Air Poll. Control Assoc.,* 16, 79 - 84.
43 Parker, G.J. (1968), *Atmos. Envir.,* 2, 477 - 90.
44 Svarovsky, L. (1976), *Chemy Ind.*, 15, 626 - 30.
45 Ruping, G. (1968), *Staub Reinhalt Luft,* 28, 137 - 44.
46 Voloshchuk, V.M. and Levin, L.M. (1969), *Trans. Inst. Exp. Met. (Trudy IEM),* No. 1, 84 - 105.
47 Vandrey, E. (1940), *Ing. Arch.,* II, 432 - 7.
48 Zeuker, P. (1971), *Staub Reinhalt Luft,* 31, 252 - 6.
49 Belyaeu, S.P. and Levin, L.M. (1972), *J. Aerosol Sci.,* 3, 127 - 40.
50 Belyaeu, S.P. and Levin, L.M. (1974), *ibid.,* 5, 325 - 8.
51 May, K.R. (1967), 17th Symposium of the Society for General Microbiology, Imperial College, London, University Press, Cambridge.
52 Maguire, B.A. and Harris, G.W. (1975), Airborne dust sampling for industry, Safety in Mines Research Establishment, Sheffield, England, *Protection,* 12, (6), 3 - 7.
53 Watson, H.H. (1954), *Am. Ind. Hyg. Assoc. Quart.,* 15, 21.
54 Dorizin, V.G. and La Mer, V.K. (1959), *J. Colloid Sci.,* 14, 74 .
55 Kerker, M. and Matijevic, E. (1958), *J. Air Poll. Control Assoc.,* 18, 665.
56 Wolber, W.G. (1968), *Res. Devel.,* 12, 18.
57 Holzhe, J. and Demmrich, H. (1969), *Neue Huė He,* 14, 198.
58 Mitchell, R.L. and Engdahl, R.B. (1968), *J. Air Poll. Control Assoc.,* 18, 216.
59 McShane, W.P. and Bulba, E. *ibid.*
60 Grindell, D.H. (1960), *Proc. Instn Elec. Engrs,* 34.
61 Konig, W. and Rock, H. (1960), *Staub,* 20, 212.
62 Dorsey, J.A. and Burckle, J.O. (1971), Particulate emissions and process monitors, *Chem. Eng. Prog.,* 67, 8, 92 - 6.
63 Hemeon, W.C.L. and Haines, G.F. (1954), The magnitude of errors in stack dust sampling, *Air Repair,* 4, 159.
64 Badzioch, S. (1959), Collection of gas-borne particles by means of an aspirated sampling nozzle, *J. Appl. Phys.,* 10, 26.
65 Lucas, D.H., Snowsill, W.L. and Crosse, P.A.E. (1972), *Measurement and Control,* 5, 9 - 16.
66 Lucas, D.H. (1976), *Chemy Ind.*, 15, 630 - 3.
67 Lucas, D.H., Crosse, P.A.E. and Snowsill, W.L. (1961), *J. Inst. Fuel,* 34, 250, 403 - 50.
68 Harris, D.B., Llewellyn, G.J. and Beck, M.S. (1976), *Chemy Ind.*, 15, 634 - 7.
69 Tipton, D.F. (1976), *Powder Technol.,* 14, 245 - 52.
70 Rouillard, E.E.A. and Hicks, R.E. (1978), *J. Air Poll. Control Assoc.,* 28, 6, 599 - 603.
71 Bridger, P.J. (1978) Practical applications of sampling dust in gas streams. Lecture presented at the Particle Size Analysis Course, University of Bradford, England.
72 Goldfarb, A.S. and Gentry, J.W. (1978), *Stud. Envir. Sci. (Atmos. Poll.),* 161 - 5.
73 Paulus, H.J. and Thron, R.W. (1976), Stack sampling in *Air Pollution,* (ed. A.C. Stern), 3rd **edn,** 3, Academic Press.

③ Sampling and sizing from the atmosphere

3.1 Introduction

Sampling from the atmosphere is carried out mainly to monitor health and nuisance hazards. The air is full of particulates, even in a non-industrial environment, but these are not usually noxious. The unenviable task of the legislator is to balance the needs of industry to pollute the atmosphere against the public demand for clean air. In Britain, the Clean Air Act of 1956 requires the discharge of effluent into the atmosphere to be reduced to a minimum using 'the best practical means' and health inspectors have been appointed to enforce this nebulous legislation. In the United States the Environmental Protection Agency (EPA) standards are more precise which makes the inspectors' job easier but raises other problems. The EPA standards for ambient air are that the annual geometric mean for particulate matter should not exceed 75 μg m^{-3} and that the maximum concentration should not exceed 260 μg m^{-3} more than once a year. Now if one compares Mobile, Alabama (124 μg m^{-3}, mean) with New York (105 μg m^{-3}, mean) it would seem that New York should have less difficulty in meeting this standard than Mobile. However, the reverse is true, since Mobile has no particulate matter controls whereas all controls possible have been installed in New York [1].

The inapplicability of this type of blanket regulation is highlighted by the Los Angeles requirement that emission from any one source must not exceed 20 lb hr^{-1}. This cannot be met by States which have, for example, cement plants; indeed one would not expect the same standards to apply to heavy industrial areas as apply to areas having only light industry [2].

It is not only the concentration of particulate contamination that is important; it is their nature as well. Since from where I sit I can see Hebden Valley, the tragedy of Hebden Bridge comes readily to mind. In Acre Mill they wove asbestos with inadequate safeguards and much of the population, even those who had never worked at the mill, contracted asbestosis. There are all too many similar cases.

The problems facing inspectors include identifying the source of particulate matter, and determining whether it is liable to be injurious to health and whether it constitutes a nuisance. Physical and chemical assay may be required to resolve these problems.

Airborne dust in quarries, factories and mines has been causing increasing concern in recent years [257]. This has led to the design of a number of dust-sampling instruments

which are now commercially available. Airborne dust is usually measured in terms of number or mass concentration. Historically, for a number count, the particles were collected in such a way that a microscope count of them could be made. This has been made easier with the introduction of automatic microscopy; the introduction of sensing zone (e.g. light-scattering) instruments has, in many cases, rendered the method obsolete. In 1959, the Johannesburg International Conference on Pneumoconiosis [3] recommended that the mass concentration of respirable dust was the best single descriptive parameter to measure in order to assess the hazard of pneumoconiosis from coal dust; for quartz dust the surface area of the respirable dust was thought to be the best parameter to measure.

As a result of these recommendations, a whole range of gravimetric dust-sampling instruments has been developed for use in coal mines, quarries and other industries with an airborne dust problem. There has been little work on surface area dust-sampling equipment although some work has been carried out on light-scattering techniques [4, 5].

Anyone living and working in a dusty environment inhales dust with the air he breathes [209]. Most of the dust is exhaled again but some is retained and can cause lung disease. The walls of the respiratory passages are lined with mucous membrane which is covered with hair-like cilia that keep up a constant ordered rhythmical movement with the force directed towards the mouth and nose. Since the factor controlling deposition is sedimentation, it is unlikely that any unit density particles larger than about a hundred μm will penetrate beyond this region. The total capacity of the lung is about 5500 cm³ but, even during activity, only about 2500 cm³ of air are breathed in and out. This is called 'tidal air' and the exchange of oxygen, carbon dioxide and particulates with the residual air is by diffusion. Dust particles smaller than about one μm diffuse very slowly and only a few per cent of these particles are deposited since they penetrate only a short distance into the residual air region [6 - 8]. Davies [9] shows that only particles within the size range 0.25 to 10 μm (called respirable dust) enter the lung and only a percentage, which varies with the size of the particle and the person's retention characteristics, are retained.

In gravimetric sampling, the weight of the dust less than 0.25 μm in size is usually negligible compared with the total weight of respirable dust, hence there is no need to prevent this very fine dust from being collected. However the presence of a very few particles larger than 10 μm in size would be significant, hence a size selector, which prevents the collection of such dust, is necessary in all equipment which assess dust as a health hazard [10 - 12, 215].

Irregular particles, coarser than 10 μm, are often found in the lung and this is due to their low sedimentation velocity. Asbestos can readily break into bundles of fibres having a diameter of only 0.02 to 0.12 μm but a length of 1 to 150 μm. Amosite and crocidolite have straight fibres and are more likely to penetrate the respiratory tract than chrysolite which has curved fibres. Elmes [13] considers that this is why the last-named is less likely to cause lung damage than the others.

The recommended maximum threshold values in the UK are [256]:

chrysolite, amosite	2 fibres cm^{-3} (0.10 mg m^{-3})
crocidolite	0.2 fibre cm^{-3} (0.01 mg m^{-3})
gypsum, limestone	10 mg m^{-3} total dust
quartz	10 mg m^{-3}/(per cent respirable + 2)
coal dust	8 mg m^{-3} respirable

Although the incidence of asbestosis can be regulated at these levels, inhalation of a single fibre of crocidolite can cause cancer.

The only statutory regulations are for coal dust (1975) and at this level a 1% level of pneumoconiosis is incurred. No compensation is available to coal miners who contract bronchial disease since it is not accepted that this is caused by working in a dusty environment. The Federal Coal Mine Health and Safety Act of 1969 defines an upper level of 2 mg m^{-3} for the States as compared with the British level of 8 mg m^{-3}.

Urban industrialization and increased use of motor vehicles have led to an increasing amount of toxic material in the atmosphere. The consequent hazards and need for effective monitoring are well documented [14-16].

There is also some evidence that non-respirable dust may be a factor in the development of bronchitis [17] and in a recent survey at Hamilton, Ontario, a strong relationship was found between the atmospheric pollution index and hospital admissions for respiratory diseases [18].

Fine tobacco smoke cannot be deposited to cause bronchial carcinoma but if the particles are inhaled and held in the lung for a few seconds they can, by absorbing moisture, increase in size to about 3 μm and be readily deposited [9].

Care should be taken that the sample accepted by any instrument is representative of the environment and that, where samples are taken from a moving airstream, the sampling should be isokinetic (Chapter 2). For non-isokinetic sampling, at a wind velocity of 4 m s^{-1}, the collecting efficiency has been found to be less than 15% for particles coarser than 10 μm. External shields and covers were found to have a large effect on efficiency [19]. Efficiency may also be impaired due to re-entrainment [206].

Davies [26] gave the criteria for accurate sampling in still air:

$$10\left(\frac{Q\tau}{4\pi}\right)^{\frac{1}{3}} \leq D \leq \frac{2}{5}\left(\frac{Q}{\pi g\tau}\right)^{\frac{1}{2}} \tag{3.1}$$

where Q is the suction flow, τ the relaxation time and D the probe diameter. The first term expresses error due to particle inertia and the second sedimentation error ($<4\%$) which is not relevant when the nozzle is horizontal. The first term reduces to $Stk < 0.016\frac{v_s}{v}$ which is the criterion for an inertial sampling error of less than 1%. The corresponding criteria for 5% and 10% error give the numerical constant as 0.085 and 0.179 respectively.

Stk the Stokes number, is defined as

$$Stk = \frac{\rho_s d^2 v}{18\eta D}$$

where ρ_s is the particle density, d the particle size, η the gas viscosity and v the wind velocity.

Gibson and Ogden [276] state that particles smaller than 40 μm are collected efficiently in calm air by sharp-edged orifices of diameters 3 to 10 mm aspirated at 5 to 20 1 min^{-1} pointing upwards or horizontally. Errors arise due to sedimentation for flow rates less than 5 1 min^{-1}, in agreement with Davies. They state that the extrapolation by Davies of an empirical equation by Belyav and Levin is unjustified.

The effect of probe bluntness on sampling particulates from air was investigated by Sansone and Christensen [278] who found that increasing bluntness produced a coarser distribution at a sampling velocity of 40 cm s^{-1}. They found no effect for sampling velocities of 20 and 4 cm s^{-1}. Bluntness was defined as $(D/d)^2 - 1$ where D and d are the outside and inside probe diameter respectively.

In order to obtain meaningful data two types of instrument have been designed. Fixed-point samplers are designed to sample dust in the body of the air; the results can only be significant if the site selected is representative.

In British coal mines such instruments are sited 70 m down the return airway from the coal face [20] but they are supplemented by personal dust-sampling instruments. These latter instruments are usually carried for the whole of the working day and must therefore be small and light. Further, they should have an inlet near to the worker's nose and, as far as possible, simulate the person's breathing.

Fixed-point samplers are frequently used in controlled environments. These environments are essential in many areas of modern precision manufacturing technology where the presence of particulates can affect the reliability and performance of the product. This is especially true in aerospace equipment, electrical components and integrated circuits. Dust can also affect quality in the pharmaceutical and food-processing industries. Extraneous matter can cause defects in photographic emulsions. Air contamination monitoring is also essential in hospital operating theatres.

In America the EPA specify a high-volume sampling method [293] while the Occupational Safety and Health Administration (OSHA) personnel use personal monitoring to determine particulate levels in a plant.

Duvall and Bourke [21] compared a gravimetric personal sampler having about a thousandth of the flow rate of the high-volume samplers and found good correlation with EPA regulations for a 'clean' laboratory area, outside ambient air and an industrial location.

Particle size determination of collected dust is also necessary for determining the grade efficiency curves for dust-arresting plant. The techniques involved here are usually quite different to those used for sizing discharge from chimneys and flue gases, since the quantities of collected material are usually much greater and the individual particles are more robust.

Methods of sampling and measuring the size distribution of aerosols have been studied

extensively [22-27, 188] and detailed surveys of equipment available have been presented [28, 29]. The broad objective is to obtain a representative sample, either in such a way that the health or nuisance hazard may be estimated immediately or in such a way that it may be examined further without altering the characteristics of interest. For these reasons some sampling devices have been developed which automatically size-grade the collected particles and others which collect the particulate matter so that further examination may be carried out with the minimum of effort, e.g. gravimetric assay for concentration determination and microscopic analysis for size grading.

Many aerosol particles consist of flocs or aggregates and in collecting and redispersing them the identity of the original distribution is often lost. This may be of little important if the toxic effect is being examined, but of fundamental importance when settling behaviour is being studied. The statistical accuracy of the collected samples is usually very low since small samples are collected from enormous inhomogeneous volumes of gases. For this reason the expense of sophisticated collecting devices cannot always be justified, and collecting devices which are expensive in terms of highly qualified operators may be similarly discarded. It is also found that particular industries have developed their own equipment. For these and other reasons a wide range of sampling devices is available. They tend, however, to fall into one of the following categories:

(1) Inertial techniques, impingement, impaction and sedimentation.
(2) Filtration.
(3) Electrostatic precipitation.
(4) Thermal precipitation.
(5) Light scattering.

3.2 Inertial techniques

Simple sedimentation techniques are suitable for high particle concentrations. With low concentrations a deep cell is required to give a sufficiently dense deposit, but the time of sampling becomes impractically long. Sedimentation cells have the useful property that the particles are not altered physically in the course of collection; usually there is no size selection [30, 31]. Particles much smaller than 1 μm in diameter are too small to be collected efficiently by sedimentation techniques. Aitken developed a method of increasing the size of particles by condensing water vapour on them thus enhancing the rate of sedimentation. The Aitken Nucleus Counter is still occasionally used [238].

Size distribution measurements may be made using the Casella settlement dust counter, in which a volume of air is enclosed in a cylinder at the base of which is an arrangement for exposing a number of microscope slides in sequence. By timing the exposures and counting the number of particles collected on each slide, a particle size distribution may be obtained.

Hallworth and Hamilton [207] describe the use of a settling drum followed by microscope analysis of slides with a Quantimet 720 for sizing of pressurized suspension.

The earliest methods of sampling gas-borne particles have incorporated impingement devices [32 - 34]. The principle is shown in figure 3.1. The large jet with low velocity

will deposit the larger particles on the slide, while the small jet will deposit the smaller particles on a subsequent slide.

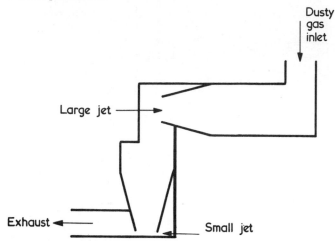

Fig. 3.1 Impingement device for collecting airborne particles.

All impactors are designed to separate particles from gas streams by their inertia which causes them to impact against a collecting plate when the gas streams are made to abruptly change direction. Owens [35] and Ferry [36] used single jets and a dry glass slide to collect the samples which were then analysed by microscopy. A comprehensive review of early techniques used for air-pollution studies is to be found in [35].

Modern impactors both sample and classify in order to reduce the subsequent labour of the microscope examination. The most versatile and popular aerosol sampler and grader is the 'cascade impactor' developed by May [37], and many variations of the instrument exist (figure 3.2). The May cascade impactor samples liquid or solid particles from 0.5 to 50 μm diameter and deposits them in four size fractions on separate glass discs. The intake velocity is kept deliberately low in order that delicate particles may be deposited without damage; for this reason it is particularly useful for establishing the efficacy of sprays for spreading insecticides, fine mists, atomized liquids, as well as pollen, spores, smokes and other delicate structures. This use of low velocities has been criticized on the grounds that high Reynolds numbers are required to give sharp cut-off characteristics [38].

The glass collecting discs are coated with a medium suitable for the material to be collected. For solid particles, a non-drying, sticky film is used. For liquid particles, dyes etc., other media have been devised. The coated discs are loaded into each of the four stages; a filter paper, if required, is put into the fifth stage for the collection of fines and a source of suction is connected to the outlet pipe. After running for a given time, the discs are unloaded and mounted on glass slides. The slides are evaluated by strip counting or, for more accurate analysis, a Porton graticule (see Chapter 6) is used in the microscope eyepiece. A modified impactor extending into the sub-micron range was described by Sonkin [244].

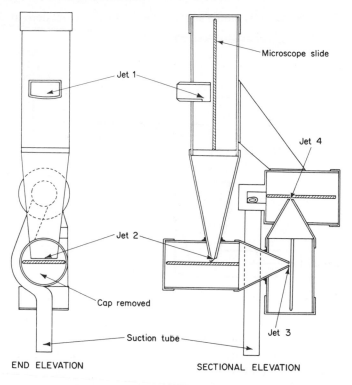

Fig. 3.2 The cascade impactor (after May [18]).

The impaction process was examined theoretically by May and others [39, 40], all
of whom showed that the impaction efficiency is related to a dimensionless parameter
ψ given by:

$$\psi = \frac{\rho v d^2}{\eta W} \times \text{constant} \tag{3.2}$$

(In a later paper May [41] quotes the constant as $C/18$ where C is Cunningham's slip
correction factor.) ρ is the material density, v is the gas jet velocity at the orifice, d is
the particle's aerodynamic diameter, η is the air viscosity and W is the diameter for a
circular jet or the width for a long slit.

May's impactor has been used by Laskin [42] on heavy aerosol particles. Other
writers have carried out extensive theoretical and experimental work with cascade
impactors [43, 44]. Characteristics and designs of impactors have also been improved
[45 - 47]. Special duty impactors have also been developed, e.g. Brink's five-stage cascade impactor for sampling and grading acid mists in the size range 0.3 to 3.0 μm [48].
Davies and Aylward [39] in a theoretical study of rectangular jets, showed that the
sharpness of the cut should improve if the jet-to-slide clearance is reduced to approximately three-eights of the jet width.

The Unico cascade impactor is designed to operate between 2 and 40 l min⁻¹ and uses standard microscope slides for its five impaction surfaces. A special feature is the manual slide-moving mechanism which allows the slides to be moved in $\frac{1}{16}$ in increments up to $\frac{1}{2}$ in total, permitting the collection of nine times more sample than is possible on a non-moving slide.

The Casella cascade impactor samples at the rate of 17.5 l min⁻¹, on four 1-in diameter glass discs. Soole [56] tested it, with particular reference to its calibration. Impactor calibration was discussed by Berner [50] and data handling and computer evaluation were described, for the Casella counter, by Picknet [51].

Stevens and Stephenson [258] describe a single-stage size-selective dust sampler in which the particle size deposition curve is nearly identical to that for the pulmonary compartment of the lung as proposed by the ICRP Task Group on Lung Dynamics (1966). Casella market this instrument as a personal sampler operating at 2 l min⁻¹.

Stevens and Churchill [259] later describe a high-flow-rate version (35 l min⁻¹) which can be operated as an environmental sampler. This instrument is manufactured by Bird and Tole Ltd.

Bucholz [52, 53] describes a low-pressure cascade impactor and Bürkholz [54] uses one for droplet size determination in the 0.5 to 10 μm range. Jacobson *et al.* [55] describe a new midget impactor.

The Greenburg–Smith impinger [32] employs a jet and plate immersed in a liquid of low volatility so that the arrested particles are retained in suspension in a liquid [57 - 64]. ASTM Standard D 3365-74T describes the use of this instrument in conjunction with a Coulter Counter (see Chapter 13).

Kotrappa [65] discusses the determination of mass median diameters using particle shape factors. Pelassy [66] calculated the theoretical efficiency and deposition pattern for a Casella Mark 2 impactor. Marple [67] considered the fluid mechanics of the laminar flow impactor and compared theory and practice for round and rectangular jets. Hänel [68] considered jet impactor characteristics versus relative humidity and found that the cut-off size increased with increasing humidity. In 1975, May [41] examined the possibility of preconcentrating particles down the axis of an air jet in order to increase efficiency. Other parameters examined were the origin, magnitude and elimination of halo deposits; the effect of nozzle shape on impaction performance and the validity of the characteristic impactor parameter. He also introduced an 'ultimate' cascade impactor [70]. From the three alternatives available to him he preferred slots to single or multiple circular jets.

Thomas [71] preferred the horizontal elutriator to the cascade impactors using several, of different capacities, in parallel. They have the advantage, over cascade impactors, in that the cut-off size can be determined theoretically from the efficiency equation by putting E equal to unity:

$$E = \frac{bLnv_p}{q_t} \tag{3.3}$$

where E is the efficiency, b the slit width, L the slit length, v_p the particle velocity and q_t is the volume flow rate through n channels.

The relative merits of dry and coated impaction surfaces were investigated by Dzubay *et al.* [72], who found that dry surfaces of aluminium foil, glass-fibre material and wax-coated paper gave distorted particle size distributions when compared with grease-coated surfaces, because particles bounced off the dry surfaces and were collected at a later stage.

Rao and Whitby [271] found that the retention efficiency of a single-stage impactor to polystyrene latex in the size range 0.79 to 1.1 μm rises from 0 to 100% for oil-coated plates; for uncoated plates the rise is from 0 to 50% followed by a fall. They also found that glass-fibre filter collectors minimize bounce but modify collection characteristics [272] and that some Whatman filters are unsuitable as collection surfaces.

Conventional impactors utilize a solid treated or untreated surface from which particles rebound, re-entrain and agglomerate leading to erroneous results. In 'virtual impactors' the solid surface is replaced with a slowly pumped stagnant air void [273 – 275].

Marple and Rubow [73] designed an impactor specifically for calibrating optical counters. A single calibration point is obtained by connecting the impactor on to the inlet of the optical counter and comparing the size distribution with that obtained without the impactor. Various calibration points were obtained by using different diameter nozzles. The possibility of using two impactors in series was also investigated but found superfluous for calibration purposes.

The Brink Model B Cascade Impactor can be used under ambient conditions or at elevated temperatures and pressures encountered in process gas streams. A glass cyclone is used upstream of the impactor to remove all particles greater than 7 μm. The impactor then collects the rest in five steps down to 0.3 μm.

In the Konimeter [74] a measured volume of air is sucked through a narrow aperture at high velocity by means of a spring-loaded piston. The particles impinge on a slide coated with adhesive, so that the particles may be counted microscopically. The collecting efficiency of this instrument is reported to be low, aggregation on the slide makes counting difficult, and the instrument cannot be recommended for size analysis [75]. Sizing may be speeded up with this instrument using a newly developed scanner [75].

The Sartorius Konimeter is recommended for routine and control measurements in factories where dust containing quartz, cristobalite, tridymite, asbestos etc., are present. Quantitative separation of particles in the size range 0.7 to 5 μm is effected and a built-in microscope permits fast evaluation of the dust hazard.

In the Owens jet counter [35] and the Bausch and Lomb dust counter [76] the air passes through a humidifying chamber and then expands through a narrow slit. Moisture condenses on the particles, thus aiding the deposition. Such devices have a very variable efficiency depending on the type of aerosol being examined [75].

Cunningham *et al.* [77] report on a programme for classifying airborne particulate material with respect to time and size using inertial impaction. The major chemical constituents were measured using infrared spectroscopy and showed variations of the chemistry with size and time. This programme was initiated as a result of the Pasadena Smog Aerosol Study [78] and followed a similar study by Whitby *et al.* [79 – 81], who

concluded that in a photochemical smog sub-micrometre particles arise from condensation and coagulation of photochemical reaction products, whereas larger particles arise from reflotation or other mechanical sources. Cunningham used a four-stage Lundgren impactor [82, 83] which uses rotating drums as the impaction surface in each stage. The impactor operates at a flow rate of 6.8 m³ hr⁻¹ and collects samples with mass-median diameters of 12.3, 1.2 and 0.3 μm. Using a continuous drive mechanism, it was operated for 21 continuous hours and later, using a stepwise mode of operation, for 3 days with 3 hours at each step. The particles were deposited on a Mylar film and the size distribution determined by weighing. The Lundgren impactor has also been used in comparison studies with a light-scattering counter [84] and to determine variation of particle size concentration and composition with time [85].

Leary [86] has used an autoradiographic technique in which radioactive particles are sized through their emission. A filter carrying a deposit of radioactive particles is placed in contact with a photographic plate for selected exposure times. The number of tracks is counted and an equation applied to calculate the size of each particle. The technique can only be applied to small particles (0.1 to 20 μm) and has the advantage that aggregates may be distinguished from single particles. Cowan [87] has reported on a simpler method which depends on the formation of a visible spot on the film.

In the Cascade Centripeter [88 – 90] air is sampled at 30 l min⁻¹. When the airstream passes through an orifice, the central flow lines pass right through, and the peripheral flow lines are deflected sharply and particles separated from them by their inertia. A series of three orifices of diminishing diameter are used to give cuts at 14, 4 and 1.2 μm.

This has been used, together with a Casella impactor and a multi-stage liquid impinger to analyse inhalation aerosols [208].

In order to determine the concentration of respirable dust in a cloud it is necessary to remove large particles which do not reach the lung and efficiently collect the fines. In a study of collection through vertical and horizontal ducts (elutriator), Walton [91] showed that, with the former, the sampling efficiency is $1 - (f/v)$, where the sampling efficiency is defined as the fractional number of particles of falling speed f collected with an intake velocity v; with the latter, the sampling efficiency for a volume flow rate Q is $1 - fA/Q$, where A is the horizontally projected flow area of the duct. As a result of these studies, a horizontal elutriator was developed with a cut at a falling speed of 0.15 cm s⁻¹ which is equivalent to a 7 μm sphere of density 1 g cm⁻³ falling in air; later instruments were constructed to give a cut at 5 μm [137]. The sample is collected on filter paper and may be examined gravimetrically or by using a microscope. In the discussion following Walton's paper Dawes drew attention to field work carried out by the Safety in Mines Research Establishment (SMRE) with a horizontal elutriator [92] in which it was found that the size selection characteristic was not stable with time, but that over an eight-hour period the cut size rose from 7 to 10 μm due, it was suggested, to redispersion of dust collected on the elutriator plates caused by the entrance of large particles. This effect was not noted by Wright of the Pneumoconiosis Research Unit of the Medical Research Council in Cardiff, who was working on an elutriator, now available as the Hexlet, which was similar to that of Walton. In this instrument, unit density spheres larger than 7 μm diameter are collected in a Soxhlet thimble and the weight is

determined gravimetrically. The aspiration rate is 50 l min^{-1} compared to the 2.5 l min^{-1} of the gravimetric dust sampler [247].

Ashford and Jones [248] compared the Hexlet and Conicycle airborne dust samplers and standard and long-running thermal precipitators at three collieries and concluded that the Hexlet was unsuitable for prolonged underground work.

The SMRE work resulted in the development of several instruments for sampling respirable dust in coal mines and these are marketed by Casella. The Simgard [93 – 95] is based on a parallel plate being collected on a membrane filter. Suction is effected by ejecting a stream of carbon dioxide gas from a small nozzle into the diffuser throat.

In the Simped [96, 97], the parallel plate elutriator is replaced by cyclones. The instrument weighs only 805 g and samples at the rate of 1.85 l min^{-1}, taking its power from the miner's cap lamp battery.

The BCIRA [246] gravimetric size-selecting personal dust sampler and the Unico respirable dust sampler which consist of a cyclone and a filter, are used by the Factory Inspectorate [29] in the manner described by Crosby and Hamer [292]. Lippmann [249] discusses the relative merits of elutriators and cyclones for removing the non-respirable dust fraction and prefers the latter.

A multi-stage impactor was described by Andersen [98] for use primarily with biological material in which the gas stream passes through small circular holes in a plate and impinges on a prepared surface in a Petri dish. A calibration of this impactor has been reported [99], and a simplified application to the study of airborne fungus [100]. A later model was designed for respiratory health hazard assessment [237].

Andersen Inc. make a range of these multi-orifice samplers including one small enough for in-stack sampling (see Chapter 2). The Hi-Vol impactor designs permit fractionation in five, three or two size ranges from 7 down to 0.01 μm. This instrument has been tested at an EPA site and subsequently adopted for use in their Community Health Effects Surveillance System (CHESS) sampling stations [101]. A procedure of changing filters to minimize losses has also been devised [102], a description of calibration procedures published [103] and its use in characterizing suspended airborne particulates presented [104, 269]. Andersen also manufacture an 'Ambient Sampler' for in-plant use together with a 'Mini Sampler', all based on the multi-orifice principle. Rao and Whitby [272] describe the calibration of a four-stage Lundgren and a six-stage Andersen with oil particles. For the former, calibration is in good agreement with theory; for the latter, the calibration curves are less steep and are shifted to a higher aerodynamic diameter. The peak efficiency is very sensitive to the jet used and to some extent particle size.

May [261] modified the Andersen sampler to have a single jet and inlet tube at the topmost stage. This was done to improve the large-particle performance over the known poor efficiency of the standard multiple-hole arrangement. May [260] also suggested that the efficiency would be improved by replacing the conventional 400 holes by 200 holes of 1.85 mm diameter arranged in a radial pattern.

A cascade impactor was designed by Newton et al. [277] according to basic principles. By defining flow characteristics at each stage, within defined ranges, the collection efficiency of each stage remained sharp throughout the impactor and effective cut-off

diameters were in accord with those predicted. Another advantage is a high-flow-rate impactor. Suggested criteria were: $1 < s/w < 2 : 1 < T/w < 5 : 500 < N_{Re} < 3000$: for round hole jets, where s = jet-to-plate separation; w = jet diameter; T = jet-throat length.

P.L. Anderson [69] described a small centrifugal aerosol collector with a design that included multi-orifice cascade impactor plates. Spectral calibration for the size range 3.0 to 0.16 μm is also included. The centrifuge is thermally protected and small enough for in-stack applications. The particles are collected individually for electron microscopy and microanalytical techniques.

Other commercially available equipment include the Heming Aerosol Spectrometer which samples airborne particles, classifies them according to their Stokes velocity and deposits them on glass slides for later examination. The instrument is intended for long-term sampling and a variant is available for use in coal mines.

The Rotorod [105] consists of a 20-cm length of 0.159-cm square section brass rod bent to form a U-shape with two vertical arms 6 cm long, 8 cm apart. The centre of the straight and level part is fitted to the shaft of a 6 – 12 V, 2500-rpm electric motor. The two vertical arms sweep out 120 l min^{-1} air and collect particles by impact on the leading edge. The Rotorod is simple and portable with a tiny power requirement, high sampling rate, indifference to wind speed and direction and turbulence. It has the disadvantages that the collecting surface is easily overloaded and awkward to handle for later microscope examination. It is of little use at sizes smaller than 7 μm at which size the collection efficiency is 50% [106].

The Hirst spore trap [107] is for long-period sampling of airborne spores and pollens. The Edwards sampler [108] consists of a holder for a 1.8-cm diameter filter mounted on a servo-operated wind vane to keep the filter facing upwind. The Tilting impinger [109], designed for high recovery of large airborne particles, consists of a cylindrical flask, 2 cm in diameter and 5 cm long, with a 6.5-mm hole near the bottom through which air is aspirated.

May [110] carried out an investigation of sixteen sampling techniques for windborne particles: (1) The Rotorod, (2) Hirst spore trap, (3) Casella Impactor Mark I, (4) Casella Impactor Mark II with a guard tube, (5) Cyclone (to dimensions of Decker *et al.* [111]), (6) Cyclone (as above with smaller (1 cm) diameter nozzle), (7) Edwards sampler, (8) Tilting impactor, (9) Ideal sampler [109], (10) wind-operated impactor [112], (11) glass ribbon on a wind vane, (12) multi-stage all-glass liquid impactor [113], (13) MCS sampler (modified Andersen cascade impactor), (14) modified Andersen lying horizontal, (15) modified Andersen lying vertical and (16) Pagoda [260].

May concluded that the Casella Impactor Mark II with a guard tube was the best instrument, having a high efficiency of down to 1 μm. In the absence of electric power May recommended the use of the Rotorod. The Hirst spore trap had a low and inconsistent efficiency and was, in general, disappointing. The glass ribbon on a wind vane was the simplest and the Tilting impactor was the only reasonably satisfactory low-sampling-rate instrument. The Andersen cannot be recommended for large-particle sampling.

The Conifuge, which consists of a cone and shell rotating at 3000 rev min^{-1}, draws in 25 ml min^{-1} dusty air (see figure 3.3). The indrawn cloud is winnowed by an intern-

ally circulating stream of clean air in such a way that the particles are classified according to their settling velocities and deposited on a glass slide in a continuously graded sample [114]. Unit density spheres in the size range 0.8 to 30 μm can be handled, the

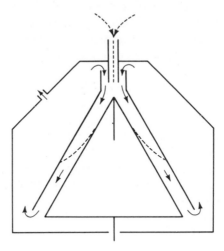

Fig. 3.3 The Conifuge (after Sawyer and Walton [114]). The solid arrows represent air flow, the broken arrows represent particle trajectories.

size grading being carried out by microscope count. Mathematical and graphical solutions relating deposition point and particle size are provided. A description of the application of this instrument for measuring the distribution and concentration of cigarette smoke is given by Keith [115] and for measuring the particle size distribution of nebulizers by Hauk [287].

The Conifuge was the first centrifugal device capable of separating aerosols into a continuous spectrum of sizes. Later another principle of centrifugal precipitation was introduced with the Goetz aerosol spectrometer [116, 117] which consisted of a centrifugal helix to collect the airborne particles and a console to monitor and control the instrument. Particles are graded and deposited on a removable strip from which they may be removed for analysis by titration, or a microscopic examination may be carried out. The desired size separation was, to some extent, sacrificed to permit a high sampling rate. Due to the complexity of the deposition pattern great difficulties were encountered in devising a mathematical expression for accurate size data [119].

A number of publications describe instruments based on the Conifuge system [119 – 123]. In all these designs the aerosol is fed into a spinning annular duct with a coaxial flow of clean air which is traversed by the aerosol particles according to their aerodynamic diameter. Stöber and Flaschbart [125] critically reviewed these efforts and concluded that although the sampling rate could be improved over Sawyer and Walton's original design, the resolution was confined to one order of sizes and in order to extend this range the Conifuge principle would have to be abandoned.

As a result of their deliberations Stöber developed a centrifugal version of the gravitational horizontal duct separator [125]. In the spiral centrifuge aerosol flow rates of 3 l min⁻¹ can be fed into the rotor at the centre of rotation where it is entrained in a laminar flow of clean air. Best results are obtained at a speed of 3000 rpm to give a deposit, in the size range 0.08 to 5 μm, on a 180-cm long strip foil which forms the outer wall of a spiral duct cut into a plane disc mounted on the centrifuge rotor.

The calibration procedure was outlined in a later paper [126] in which is stated that a number or a weight distribution was determinable. Porstendörfer [127] applied the instrument using radioactive markers, and later compared the results [128] with a laser air spectrometer LASS [129] finding reasonable agreement, with the centrifuge results a little low. Kops [130] found that a correction factor had to be applied in order to obtain reliable results. Oeseburg [131] determined the particle size distribution of dioctyl phthalate aerosol and found good agreement with High Order Tyndall Spectra (HOTS). An examination of the spiral centrifuge revealed that the deposition of particles in the inlet was higher than previously reported and that the experimental results were not compatible with theory [124].

Berner and Reichelt [132, 133] introduced a new concept in their cylindrical aerosol spectrometer which greatly simplified the construction and operation. They allowed the complete inlet system to rotate with the spectrometer thus eliminating the need for rotating seals. When the spectrometer is rotated, a pressure drop across the inlet to outlet nozzles draws in air, the flow rate being controlled by nozzle size and speed of rotation. For their 'clean' air, however, they used room air which restricts the use of the instrument [134]. To overcome this limitation Hochrainer and Brown [135] designed two centrifugal spectrometers, one cylindrical and one conical in design, having an inlet system similar to Berner and Reichelt's as well as a method for furnishing clean air.

This design was further modified by Matteson and Boscoe [136] who designed a cylindrical-type spectrometer into which aerosol particles entered a horizontal centrifugal field through two ports, a pin-hole orifice opposite a larger port. The aerosol through the larger port is removed in a cumulative size deposition on half a foil and serves as clean air for the aerosol through the other port which is deposited in a thin streak.

In 1975 Oeseburg and Roos [270] improved the bearing system of the Stöber aerosol spectrometer by the introduction of bearing rings to reduce leakage.

3.3 Filtration

Most filtration methods are unsuitable for collecting solid aerosol particles for microscopic analysis, since aggregation tends to occur on the filter if concentrated aerosols are sampled, while smaller particles tend to penetrate deeply between the fibres of the usual fibrous material used as filtering media so that they cannot be microscopically counted.

Filtration is the simplest method of removing particles from the atmosphere for subsequent analysis; however, removal of particles from a filter and dispersion prior to size

analysis will usually completely alter the size characteristics of the sample due to the breakdown of flocs. The most efficient filter media for this purpose are the membrane filters which can exhibit almost 100% collection efficiency for particle size above 0.01 μm with the bulk of the material deposited on or near the surface. These filters consist of cellulose esters which are soluble in acetone, so that it is possible to dissolve away the filter and transfer the deposited material to prepared surfaces for further examination (Kalmus [138]). The deposited particles may be examined *in situ*, using reflection microscopy, or the membrane can be made transparent by adding a few drops of cedar oil and examination carried out by transmission microscopy.

Although filtration, followed by sizing, is tedious and inaccurate, the method is extremely useful for determining mass concentration. This is usually carried out gravimetrically although other methods are also employed. Krost *et al.* [139] for example used β-absorption to obtain a sensitivity of 8 μg cm^{-2} on their teflon filter which they recommend as superior to the Millipore-type filter. Luke *et al.* [210] collect particles on a Millipore filter and ash it; the amount of metallic particulates is then determined by X-ray spectrometry.

Handling of filters can present problems and to overcome these many manufacturers use a cassette filter system. Instead of changing only the filter, the whole cassette is removed for subsequent analysis under laboratory conditions.

Rotheroe and Mitchell market a wide range of dust samplers which incorporate Millipore membrane filters. The L30, L60, L90 are suitable for continuous operation at flow rates of 30, 60 and 90 l min^{-1} respectively. The L2C is a pocket-sized personal air sampler (weight 1.14 kg, flow rate 0.5 to 3.0 l min^{-1}) to sample air close to the worker's face.

C.F. Casella also use Millipore membrane filters in their range of instruments whereas Sartorius use their own membrane filters. The Sartorius Dust Sampler EM100 is a high-performance dust sampler with continuously adjustable air throughputs of up to 100 m^3 hr^{-1} and a filtration area of up to 415 cm^2. Isokinetic sampling in closed aerosol systems and flues is possible with instant read-out of momentum flow rate, total flow and gas temperature, so that volumes can be converted to standard m^3. Applications include efficiency tests on cyclones and electrostatic filters, checking of dust concentration in exhaust flues, immission monitoring and supervision of spray-drying plants.

The Sartorius Dust Sampler Gravicon is a dust-collecting instrument for taking dust samples for gravimetric, chemical and mineralogical analysis. The Porticon is a small, portable battery-operated instrument. The air intake velocity corresponds to that of inhalation through the nose. Dust samples, taken at work stations, therefore show a particle size distribution relative to that in the respiratory tract.

The Sartorius Dust Sampler Gravicon VC25 classifies dust into coarse (>5 μm) and fine (<5 μm) and collects the two samples separately on a filter which can be evaluated to monitor dust conditions at individual working sites.

The Heming air pollution monitor monitors atmospheric pollution by sucking the air through a moving filter paper at a rate of 42 cm^3 s^{-1} using a diaphragm vacuum pump. The pressure drop across the filter increases with deposition and is a measure of pollution level.

The Frieseke and Hoepfner type FH62A dust monitor samples air at fixed times and the particulate matter is deposited on a filter tape and the mass deposited determined by β-ray adsorption from a krypton-85 source of activity 50 mCi.

3.4 Electrostatic precipitation

When a charged particle passes between two electrodes carrying a high electric potential, it will move normally to the direction of flow under the force of the electric field.

The electrostatic precipitator, based on this principle, consists of an ionizing cathode at a high potential surrounded by a collecting anode; typically, these consist of concentric cylinders, the inner one often being a single wire. The dust passes between the cylinders, picks up a charge and travels to the anode where the charge is deposited. For extremely small particles, the charge consists of an excess or deficiency of one electron, and the transfer of electrons from one electrode to the other constitutes an electric current which may be measured with a micro-ammeter. The magnitude of the current indicates the number of particles and their size may be found by varying the flow rate or the applied potential [140 – 142]. Details of the design of an ion chamber are given by Hurd and Mullins [143]. This particular equipment has two flat circular plates parallel to each other and maintained at a fixed potential difference.

The trajectories of particles in an electrostatic field depend on their size and their electron charge. Since particles larger than about 0.01 μm may have an excess or deficiency of more than one electron, more than one size of particle may be deposited at the same point. An instrument that uses this technique with some success has been developed by Yoshikawa [144]. In order that a microscope count may be carried out the particles are collected on membrane filters.

Podol'skü [145] describes an instrument based on classification in an electrostatic field by differences in charges connected explicitly to particle size. Seven fractions are produced in the size range 0.4 to 40 μm with a reproducibility better than 3% and an analytical time of 3 min.

Instruments based on this principle have been used for aerosols of bacteria [146, 147], and an improved instrument has been described by Morris et al. [148]. This instrument consists of a cylindrical glass tube 12 in long and of 2 in diameter with a central electrode. The inner surface of the cylinder is coated with a suitable material to act as the other electrode and to collect the samples. With the central electrode at 10 kV and a flow rate of 10 to 20 1 min^{-1}, the collection efficiency was found to be 100%. In one version the cylinder is rotated and contains up to 10 ml of liquid, so that collection in liquid could be effected directly. The principal advantages of this type of instrument are high collection efficiency over a wide size range, low resistance and high flow-rate capacity [149]. It is not very suitable for number concentration measurements. Microscope analysis of the deposited particles may be simplified by placing electron microscope grids [150] or transparent plastic over the anode. Several other studies of this technique have been carried out (see [151, 152]). Byers et al. [211] use a Bendix Model 959 electrostatic precipitator to collect samples which they subsequently analyse by

scanning electron microscope using a computerized technique to determine projected area.

In 1938 Barnes and Penney [240] described a dc precipitator capable of efficient sample collection at a flow rate of 3 ft³ min⁻¹. Commercial versions are available from Bendix and Mines Safety Appliances.

Two commercial versions of high-volume electrostatic precipitators described by Decker *et al.* [111] are available as the LEAP and the Litton high-volume (10 000 ft³ min⁻¹) samplers.

Other instruments are available from Gardner Associates (the electrostatic bacterial air sampler) and Del Electronics (the high-volume electrostatic precipitator). Collection efficiency determinations of the latter have been reported [242].

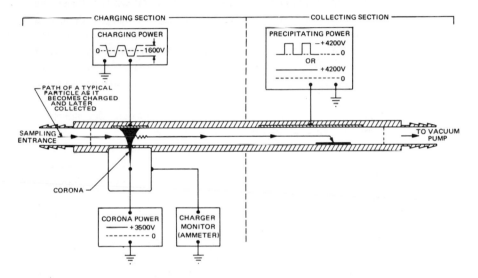

Fig. 3.4 The Thermo-Systems Model 3100 Electrostatic Aerosol Sampler.

The Thermo-Systems 3100 Electrostatic Aerosol Sampler (figure 3.4) consists of a charging section and a collecting section. A vacuum pump draws the aerosol through the system at a constant rate. As the aerosol passes through the charging section, the particles are subjected to alternating pulses of positive ions generated by a corona discharge from a fine wire. The positively charged particles then flow through the collecting section. A positive voltage periodically applied to the upper plate drives the particles to the lower surface. After sufficient time to deposit all charged particles, the voltage on the upper plate is shut off. The continuous aerosol flow through the chamber again fills it with charged particles. The unique separation of the charging section and the collecting section, together with the pulsed precipitating voltage produces a uniform, representative sample, which may be evaluated without bias due to particle characteristics. The instrument allows precise determination of the sampled aerosol quantity because this volume is completely independent of the flow rate through the instrument.

Volume depends only on the number of precipitating pulses or cycles and the volume directly above the sample surface. These characteristics are in marked contrast to normal electrostatic samplers. A second mode of operation facilitates the collection of a sample when classification due to particle characteristics is tolerable. In this mode, a constant dc precipitation voltage collects all charged particles soon after they enter the collection section. The size range is 0.02 to 10 μm at a flow rate of 4 to 10 l min^{-1} (6.7 to 167 cm^3 s^{-1}).

The point-to-plant electrostatic precipitator [243] was designed to collect aerosol samples for particle size analysis by electron microscopy. Its function is therefore different from the other instruments described above, which were designed to collect large samples for mass concentration analysis.

3.5 Electrical charging and mobility

The Electrical Aerosol Analyser was described by Whitby in 1966 [262]. Calibration was effected by capturing the particles after they had passed through the instrument and measuring them by microscopy [263], hence problems were caused due to the difference between the projected area diameter and the aerodynamic diameter. In a second method, a polydisperse aerosol was generated and passed through a differential mobility analyser, where monodisperse particles are separated from the airstream and used for calibration. A version of this instrument is manufactured by Thermo-Systems (see figure 3.5); its performance has been examined [212, 264, 265, 279] and its use, together with various other aerosol monitors, has been described in an airborne particle monitoring facility [266]. The mode of operation is as follows:

As a vacuum pump draws the aerosol through the analyser, a corona generated by a high-voltage wire within the charging section gives the sample a positive electrical charge. The charged aerosol flows concentrically from the charger to the analysing tube section which is an annular cylinder of aerosol surrounding a core of clean air. A metal rod, to which a variable, negative voltage can be applied, passes axially through the centre of the analyser tube. Particles smaller than a certain size (with high electrical mobility) are drawn to the collecting rod when the voltage corresponding to that size is on the rod. Larger particles pass through the analyser tube and are collected by the current-collecting filter. The electrical charges on these particles drain off through an electrometer, giving a measure of current. A step increase in rod voltage will cause particles of a discrete larger size to be collected by the rod with a resulting decrease in electrometer current. This decrease is related directly to the number of particles in the aerosol between the two discrete particle sizes. A total of 11 voltage steps divide the 0.0032 to 1.0 μm size range of the instrument into 10 equal geometrical size intervals. The 11 internal boundaries are 0.0032, 0.0056, 0.0100, 0.0178, 0.0316, 0.0562, 0.1000, 0.1780, 0.3160, 0.5620 and 1.0000 μm. Different size intervals can be programmed via an optional plug-in memory card.

The instrument can be operated either automatically or manually. In the automatic mode, the analyser steps through the entire size distribution range. For size and concentration monitoring over an extended period of time, the analyser may be intermit-

tently triggered by an external pulse from a timer. The size distribution can therefore be automatically monitored at preset time intervals. The standard read-out consists of a digital display within the control circuit module. This module programmes the analyser voltages and also provides command logic for the automatic counting mode.

Reviews of electrical measuring devices have recently been published [213, 286] together with descriptions of their applications [216, 217].

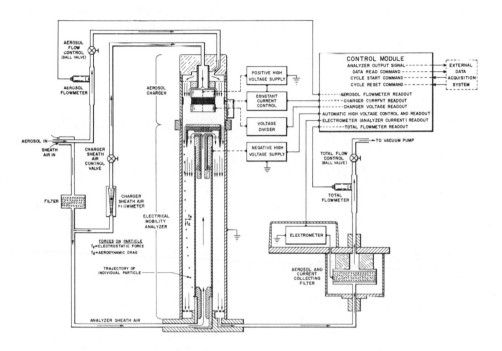

Fig. 3.5 Thermo-Systems Model 3030 Electrical Aerosol Size Analyser.

3.6 Thermal precipitation

Dust particles in suspension move away from hotter to colder regions [153, 154]. This effect was noted first by Tyndall and later by Lord Rayleigh, and Aitkin showed that the force that caused this movement was thermal in origin.

For particles of the same order of size or greater than the mean free path of the gas molecules, an equation may be developed using 'thermal creep' theory. This is based on the force set up at the gas – solid interface between a particle and the surrounding gas. When the gas temperature increases along the surface, the molecules leaving the surface will have a greater component of velocity in the direction of the temperature increase than when they arrived at the surface. The net result is a creeping flow of gas from the colder to the warmer regions along the surface of the particle giving a net force in the cold direction. This motion is called thermophoresis [239].

Epstein [155] developed the following equation for the thermal force:

$$F_t = \frac{9\pi\eta^2 d}{2\rho_f T \left(2 + \dfrac{x_g}{x_p}\right)} \frac{\mathrm{d}T}{\mathrm{d}x} \tag{3.4}$$

where d is the particle diameter; η, the gas viscosity; ρ_f, the gas density; T, the gas temperature; $\mathrm{d}T/\mathrm{d}x$, the thermal gradient; and x_g and x_p are the thermal conductivities of the gas and particle respectively.

Equating with Stokes resistance, including Cunningham's correction, gives the velocity in the thermal gradient as:

$$U_t = -\frac{3C\eta}{2\rho T \left(2 + \dfrac{x_g}{x_p}\right)} \frac{\mathrm{d}T}{\mathrm{d}x} \tag{3.5}$$

where $C = (1 + 2A\lambda/d)$, see equation (7.18). This equation is found to give good agreement with experimental results.

For very small particles, equations based on the kinetic theory of gases have been developed [156]. Waldmann's equation may be written:

$$U_t = -\frac{1}{\left(5 + \dfrac{\pi}{2}\right)} \frac{x_g}{P} \frac{\mathrm{d}T}{\mathrm{d}x} \tag{3.6}$$

where P is the gas pressure. For further details readers are referred to [157, p.417].

This principle is applied in thermal precipitation where the hot body is an electrically heated wire placed between two collecting plates. The thermal precipitator developed by Watson [158] (figure 3.6) consists of a channel 0.51 mm wide between two microscope cover glasses with an axially situated wire heated electrically to about 100°C above ambient.

A dust sample, about 9 mm long and 1 mm wide, is collected on each plate. The normal flow is 7 cm^3 min^{-1} and collection efficiency is high for sub-5 μm particles. Larger particles are collected elsewhere in the instrument. The collecting device may be modified so that the sample is collected directly on an electron microscope grid, although transfer of particles from the collecting plate to a suitable film and thence to a grid is a more usual procedure. This instrument is manufactured by C.F. Casella & Co., and also distributed by Mines Safety Appliances [177].

Direct measurement of collection efficiency indicates that this type of instrument collects virtually all particles from 5 to 0.01 μm in size [159 – 162]. Sampling efficiency has been found to be unaffected by airspeeds up to 6 m s^{-1} [163].

Modifications to the traditional design [164] include a means of centring the wire in position [165], the substitution of a ribbon for the wire to give a more uniform deposit [166], the provision of an inlet elutriator to exclude coarse particles [167], and the attachment of a rotating magazine containing six pairs of cover glasses to avoid the con-

tamination arising when glasses have to be transferred from the sampling head to a slide box [168].

The instrument of Beadle and Kitto [169] deposits 12 dust strips on a slide and also employs an inlet elutriator and a bellows driven by clockwork. A reversible water aspirator, much smaller and more convenient than the standard form, has been developed by Wright [170] and is manufactured by Adams. Long-running thermal precipitators may be operated for periods of hours rather than minutes. In that of Camber *et al.* [171] the sample slide rotates continuously; in Walton's [172] the slide oscillates and in that of Orr and Martin [173] the slide is replaced by a transparent tape.

Hamilton's long-period (8 hours) dust sampler is the standard instrument for determining dust concentrations in British coal mines [174, 182] (Casella & Co.). The air enters through an inlet elutriator, then passes over a horizontal slide where particles down to 1 or 2 μm are deposited by gravity settlement; the fine particles are deposited at the far end of the slide by a hot ribbon. Aspiration at 2 cm^3 min^{-1} is by an electric pump energized by the battery that heats the wire.

Balashov *et al.* [175] have used a composite instrument to obtain samples at very low concentrations. Thermal precipitators sampling large volumes of air and providing dust deposits large enough for weighing have been developed [176]. Casella market one instrument for short-period sampling in the size range up to 20 μm for evaluation of health hazards, such as pneumoconiosis and another for long-period sampling of respirable dust [177].

The Thermopositor (figure 3.7) collects samples of dust, smoke, fog, bacteria, pollen and spores. Air at up to 1 l min^{-1} is drawn into the space between two discs, the upper

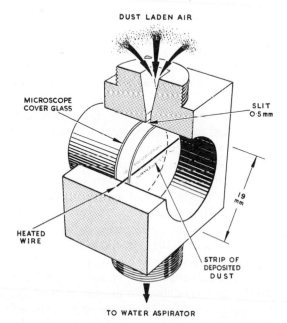

DUST LADEN AIR

MICROSCOPE
COVER GLASS

SLIT
0·5 mm

19
mm

HEATED
WIRE

STRIP OF
DEPOSITED
DUST

TO WATER ASPIRATOR

Fig. 3.6 Sampling head of standard thermal precipitator.

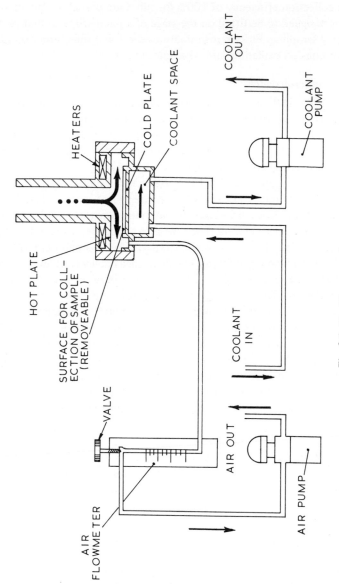

Fig. 3.7 The Thermopositor.

one being heated and the lower one cooled. The particulate matter is deposited on a substrate of glass, paper, metal foil or membrane.

The Konisampler is a compact thermal precipitator which operates at up to 50 cm^3 min^{-1} with a collection efficiency of 100% for sub-5 μm particles. The microthermal precipitator is designed to be fitted on the stage of a microscope so that particles may be observed during sampling. Ficklen manufactures these and three other thermal precipitators: 'Continuous', 'Oscillating' and 'Gravimetric'.

3.7 The quartz microbalance

In 1971 Olin [178 – 180] described a system in which aerosols are drawn into an electrostatic precipitator. A piezoelectric quartz oscillating crystal forms one electrode, the resonance frequency of which decreases linearly with increasing total mass of the particles thereby weighing them. Several commercial versions are now available and papers have been published on their applications [214].

With the Thermo-Systems 3200 Series Particle Mass Monitor, aerosol is drawn in at a flow rate of up to 1.0 l min^{-1} (17 cm^3 s^{-1}) and deposited on to a crystal oscillating at its resonant frequency. A second sensor, the reference quartz crystal, compensates for possible changes in ambient effects. The time rate of change of the output frequency signal is a measure of the total mass concentration. The sensitivity is 1 Hz, equivalent to 0.006 μg with a total loading of 50 μg. The instrument collects particles in the size range 0.01 to 10 μm approximately at concentrations between 2 to 20 000 μg m^{-3}.

The Celesco Portable Particulate Mass Monitors are very similar to the above. Model PM-39 also has a small rotating disc which collects a particulate sample over a predetermined period to provide a time-correlated record. Size range is 0.1 to 100 μm.

The Berkeley/Celesco Model C100A QCM cascade uses ten stages of quartz crystal microbalances and inertial impaction nozzles to measure the mass distribution directly in the size range 0.05 to 25 μm at concentration levels from 0.1 to 65 mg m^{-3}. The instrument may be coupled with a Texas programmable computer T159 and PC-100A printer to quickly monitor and compute stage concentrations and plot particle size distributions.

3.8 Light scattering

The scattering of light by particles suspended in a fluid has been widely used in the development of instruments for sizing aerosols. The basic requirement is that particles pass through the sensing volume in single file, hence the aerosols have to be very dilute. The radial distribution of scattered intensity is a function of particle size and shape, the wavelength of the incident radiation, the refractive index of the particle and the relative refractive index between the particle and its surroundings.

The modern theory of light scattering as developed by Mie is dealt with in detail by van de Hulst [181]. Two types of instruments have evolved, the right-angle light-scattering system and the near-forward light-scattering system. The former has the advantage of being very sensitive to small changes in particle size but has the disadvantage that the scattering is strongly dependent on the particle refractive index making it unsuitable for use with mixed aerosols. Particle refractive index has little effect on the

intensity of the forward-scattered light but this tends to be rather weak for particles smaller than 1.5 μm.

In an early instrument developed by Gucker *et al.* [182, 183] a stream of aerosol particles in a sheath of filtered air is passed through a light beam and the light scattered in a forward direction, between 1° and 20°, by each particle in transit is collected by an optical system which incorporates a photocell. The resulting electrical pulses are amplified, sized and sorted using a multi-channel pulse height selector. The sheath of filtered air plays a dual part in that it reduces the concentration of the aerosol and maintains it in a central position in the sensor and it reduces contamination to the sensor windows. In later instruments [184], right-angle scattering was used.

Right-angle scattering is used in the Aerosoloscope [185] which draws in air at the rate of 1.8 l min^{-1} and dilutes it to give a maximum count rate of 2000 particles per minute. The resulting electrical pulses are graded into 12 size channels corresponding to sizes in the 1 to 60 μm range at initial concentrations of up to several thousand particles per cubic centimetre.

Oeseburg [186] calculated that particle size response characteristics for a sphere of refractive index 1.491 and size parameter $0.1 < (\pi d/\lambda) < 30$ are best for a forward receiver angle of 20° to 120°.

There is a boom in this type of instrument at the present time due to pending and enacted legislation. Many controlled environment operations are conducted according to USA Federal Standard 209 (1973). This classifies rooms according to the maximum allowable number of particles per unit volume with equivalent diameters equal to or greater than 0.5 and 50 μm. Another application is the inspection of 'clean benches', i.e. work surfaces continuously bathed in ultrapure air to ensure operations are carried out in a class 100 environment (no more than 100 particles greater than 0.5 μm in a cubic foot of air). Instruments are also used in air pollution studies, medical research and industrial hygiene.

Gucker [251 – 252] also designed a photometer for rapid measurement of angular patterns from individual particles. A nearly complete 360° scattering pattern is recorded in 20 ms from particles in an aerosol stream flowing through a 632.8-nm laser beam. The instrument has been used to characterize latex spheres in the size range 0.3 to 11 μm [253]. Data are stored on magnetic tapes and the individual curves are fitted to Mie theory diagrams to obtain a 'best fit'.

Optical particle counters based on light scattering by single particles are widely used in aerosol research. In the LASS a fine filament of aerosol is drawn through a laser beam and the scattered light is picked up at a mean angle of 40°. Pulse height is related to particle size by means of a calibration curve size range 0.05 to 0.70 μm at particle concentrations of up to 10^6 cm^{-3} [241]. A description of the LASS and a low-angle scattering instrument (LASI) is given by Gebhart *et al.* [218].

Lasers have replaced the traditional white light source in some recently introduced commercial instruments. The particles are illuminated with a source of radiation nearly 10 times more intense than is possible with the hottest incandescent source known. This results in systems fully capable of sizing particles of 0.1 μm diameter using solid state silicon detectors, 0.08 μm using a photomultiplier and 0.05 μm using a photomul-

tiplier and a curved mirror [187, 205]. Kratel [289] describes the application of an intracavity laser to HEPA filter testing in the size range 0.06 to 3 μm. Allen [267] describes an optical system for asymmetric light-scattering determination by He – Ne laser to determine the size and shape of polystyrene latex, soot and outside aerosol particles. Some of the commercial light-scattering instruments are described below.

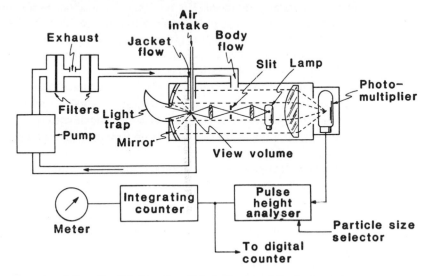

Fig. 3.8 Functional block diagram of the dust counter.

The Bausch and Lomb dust counter 40-1 uses the near-forward light-scattering principle [189] (figure 3.8). The direct light is captured by a light trap and completely absorbed, and light scattered through an angle of from 24° to 40° is reflected via a parabolic mirror to a photomultiplier. The instrument is notable for its built-in calibration system and is designed to operate in the following size ranges: 0.3, 0.5, 1.0, 2.0, 5.0 and 10 μm in seven increments. Incorporated in the instrument is a dilution system which permits counting at dust concentrations of up to 10^6 particles ft^{-3}. The counting rate is up to 35 particles s^{-1} at a flow rate of 3 cm^3 s^{-1}. Brown [191] added circuitry to the system to allow unattended, automatic measurements to be determined.

The Royco Airborne Dust Counters are available in both modes, forward and right-angle scattering. Model 220, designed for clean room requirements, employs right-angle scatter (figure 3.9). Sampling rate is 47 cm^3 s^{-1} at a maximum concentration of 10^8 particles m^{-3}; counts are recorded of particles larger than 0.5 and 5 μm; other sizes are also available.

Models 225/245 use near-forward scatter. Model 225 is a high-concentration aerosol counter designed to operate at concentrations up to 3.5 x 10^9 particles m^{-3} at flow rates of 0.01 or 0.10 ft^3 min^{-1} (4.7 or 47 cm^3 s^{-1}) and record particles larger than 0.5 μm. Model 245 is a high-flow-rate version designed to operate at concentrations up to 3.5 x 10^8 particles m^{-3} at a flow rate of 470 cm^3 s^{-1}. Model 218 is a portable version operating at concentrations up to 3.5 x 10^9 particles m^{-3} at a flow rate of 4.7 cm^3 s^{-1}.

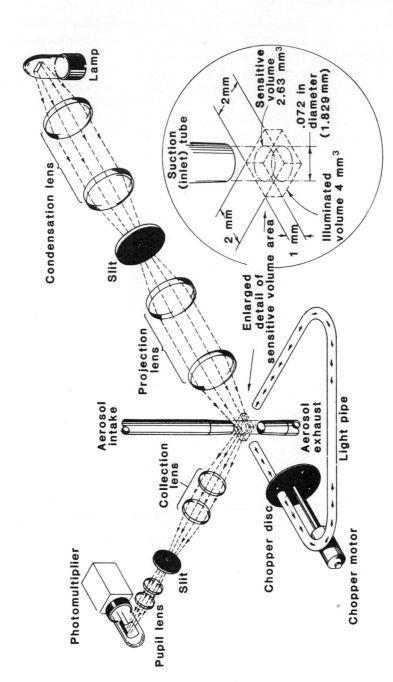

Fig. 3.9 Royco Model 220 Airborne Particle Monitor.

Lamp

Condensation lens

Slit

Projection lens

Aerosol intake

Collection lens

Photomultiplier

Pupil lens

Slit

Chopper disc

Aerosol exhaust

Light pipe

Chopper motor

Enlarged detail of sensitive volume area

Suction (inlet) tube

2 mm

Sensitive volume 2.63 mm³

.072 in diameter (1.829 mm)

Illuminated volume 4 mm³

2 mm

1 mm

The calibration of the Royco using cascade impactors was described by Marple and Rubow [73]. Jones and Khaled [192] carried out comparison tests with a thermal precipitator and Jaenicke [193] considered some of its defects. Its uses as a clean air monitor in an operating theatre [194], a filtration monitor [195], for filter testing using dioctyl phthalate [288] and a sizer of aerosols have also been described [196].

Climet manufacture a range of aerosol counters, together with two liquid-borne counters, all operating under the same principle. The sampling volume is at the focus of an elliptical mirror (figure 3.10) which accepts the scattered light and focuses it on to a

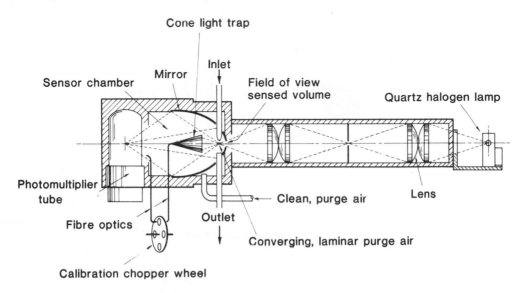

Fig. 3.10 Climet 208 optical system.

photomultiplier. The light is picked up over the scattering angles $15°$ to $105°$ making up a solid angle of 5.1 sr. This maximizes the amount of scattered light picked up and minimizes errors due to particle shape, colour and orientation. Standard particle size ranges are 0.5, 1, 3, 5, 10 μm (0.3 μm available). Particle concentrations of up to 10^9 ft^{-3} may be studied. Sample flow rate is from 1 to 20 ft^3 h^{-1}. Standard options and accessories are available. The Climet 208 has been successfully used to study the growth of a monodisperse NaCl aerosol as a function of relative humidity [197]. A microcomputer has also been attached to give direct print-out [280].

The Kratel Partoscope model A is a forward white-light-scattering instrument operating at a flow rate of 47 cm^3 s^{-1} and a maximum concentration of 300 particles cm^{-3}. Various size options are available: greater than 0.3 or 0.5 μm and greater than 5 μm; a five-channel model at 0.3, 0.5, 1.5, 3.0, 5.0 μm or 0.5, 0.7, 1.4, 3.0, 5.0 μm. Model R is a low-flow-rate (5 cm^3 s^{-1}) version.

The Kratel Aerosol Spektrometer MR is a right-angle scattering collector operating at concentrations up to 10^5 cm^{-3} with a lowest detectable size of 0.3 μm. Read-out is

digital in two or four channels or **multi-channel** analyses, minicomputer with printer, plotter or teletype.

The Kratel Partoscopes LA and LC are laser forward-scattering instruments for the sub-micron and respirable range respectively. These are claimed to have advantages over conventional light systems. In particular the use of a 50% beam splitter to produce two image planes for two detectors provides a means of precisely defining a particle's position in the sampling volume. This system also allows complete control of the sensing volume which can be varied by varying the magnification and depth of fluid of the optics.

A low-angle laser light-scattering (LALLS) instrument for particles smaller than 10 μm is also available from Beckman [198].

In the Malvern Instruments' droplet size analyser, Fraunhofer diffraction is applied to droplet size analysis [268]. The light source comprises a 1 mW He – Ne laser and the scattered light is focused on to a multi-element detector. For a given particle the position of the maximum in the diffraction pattern is determined by its size so that the light intensities at different radii from the lens axis are size-dependent. The multi-element detector consists of a series of concentric annuli positioned at the maxima of certain particle diffraction patterns. Their output is monitored to give the size distribution by weight at each of 30 diameters ranging from 5.51 to 472 μm.

Most of the above instruments have not been adopted for routine use in mines because of the stringent power and weight considerations imposed by practical considerations. For these reasons the above principles have been applied by the Centre for Air Environmental Studies in the construction of a battery-driven portable counter [202]. The CAES portable instrument can handle 1500 particles cm^{-3} at several size levels with instantaneous readout. Since no light trap is employed a compromise viewing angle of 135° to the incident radiation is used.

The Rotheroe and Mitchell Digital Dust Indicator is a battery-operated, portable monitor of airborne particles and uses the principle of right-angle scattering. The instrument gives the total count above a preset limit (0.3 or 0.5 μm) at an upper concentration level of 100 mg m^{-3} or a count rate of 167 s^{-1}.

Particle Measurement Systems manufacture a range of Knollenberg Aerosol Spectrometer Probes comprising a Classical, Active, Forward and Axially Scattering system. These counters have been examined by Pinnick *et al.* [281] and their resolution has been found to be significantly less than claimed.

In the HC 15 [290] the aerosol, at velocities in the range 0.2 to 15 m s^{-1}, passes through a measuring volume where it is illuminated by a halogen lamp. Light scattered at 90° is picked up by a detecting system and the signals from the detecting photomultiplier are passed to a channel analyser where they are measured and stored in 64 channels. The stored sizes are in a logarithmic distribution in the size range 0.3 to 60 μm.

The HIAC automatic portable counter employs light blocking, the attenuation pulses being measured as particles are drawn through the measurement zone (figure 3.11). Five standard models are available with one to five size channels with a claimed size range of from 2 to 9000 μm and flow rates from 20 to 225 000 cm^3 min^{-1} depending on the size range.

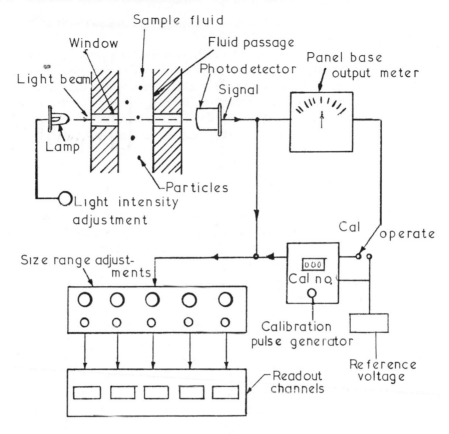

Fig. 3.11 HIAC automatic portable counter.

The Coulter Model 550 contamination counter operates on the forward light-scattering principle. The makers claim that the quartz-iodide lamp, by giving a more uniform illumination system than the conventional tungsten filament lamp, provides a more accurate system. Counts are recorded, at the 0.5 and the 5 μm level, at a flow rate of 1 ft^3 min^{-1} (472 cm^3 s^{-1}) and a maximum concentration of 3.5 particles cm^{-3}. The aerosol is sheathed in clean air to reduce contamination. Field calibration is effected by using a simulated particle count.

In the Saab photometer, black plates, on to which particles have been deposited by exposure to atmosphere, are illuminated horizontally. The reflected light gives contamination level and an approximate mean size.

The RAC reflective particle monitor is used to assess the degree of **contamination by** measuring the percentage reflected from a paper tape. In the RAC reflectance transmission monitor, the reflectance is measured after the sample is taken and the transmission while it is being taken.

Techecology make a range of optical counters which operate using **near-forward** light scattering with an elliptical mirror to collect the scattered light. In the Model 200 the

aerosol, flowing at about 8 cm^3 s^{-1}, is sheathed in clean air flowing at ten times this rate. The sheath is intended to constrain the particles and prevent contamination of the optics. The aerosol is collected on a membrane filter. Size levels are 0.5, 1.0, 2.0, 5.0, 10.0, > 10 μm at particulate levels between 3.5 x 10^5 and 3.5 x 10^8 particles m^{-3}. Calibration is by directing light pulses on to the sensor to simulate particles. Prime calibration is carried out at the factory using polystyrene spheres.

Air Technology manufacture a forward-scattering photometer, TDA-2C, for the determination of total atmospheric particulate mass.

Dynac manufacture an airborne particle monitor, M-201C, for measuring airborne particulate contamination in the selectable size ranges + 0.15, 1.0, 3.0, 5.0 μm with an optional + 0.3 μm at flow rates and count levels up to 47 cm^3 s^{-1} and 770 particles cm^{-3}.

The MRI Integrating Nephelometer has sufficient sensitivity to measure gas scattering; it has also been used to measure aerosol mass concentrations up to 3800 μg m^{-3}.

The Digital Dust Indicator is a portable instrument operating with 90° scattered light. Air is drawn through at 10 l min^{-1} and the coarse particles removed by an impaction system. Light pulses from particles passing through the beam are counted by a photomultiplier circuit and recorded.

The Gardner Small Particle Detector depends for its operation on the fact that small particles act as condensation centres for water vapour under suitable conditions. This allows for measurement of particles with radii as small as 10^{-7} cm. Its range is 2 x 10^2 cm^{-3} to 10^7 cm^{-3}.

The GEC condensation nuclei counter and the RAC condensation nuclei counter operate on similar principles.

The Sartorius Spectral Counter is an instrument for quantitative and, under certain conditions, qualitative aerosol analysis. It operates on the principle of scintillation spectral analysis [250]; the aerosol particles are brought into a heating chamber and on attaining the excitation temperature each particle emits a light impulse which is evaluated by means of an electronic-optical system. The intensity of the impulse is a measure of particle size while the spectral composition shows the chemical composition of the particle. Theoretical and practical considerations show that this instrument is capable of more accurate measurement of aerosol parameters than any other available unit. The unit displays the mass concentration in mg cm^{-3}, the particle concentration in particles cm^{-3} and the particle size distribution in ten ranges from 0.01 to 100 μm. Aerosol flow rate is 3 cm^3 s^{-1} at a maximum concentration of 20 particles cm^{-3}. A selection of monochromators is available for the measurement of single elements only.

In the Sinclair-Phoenix Forward Scattering Photometer, light is brought to a focus at the sample cell, with a diaphragm stop placed in the optical path so that a diverging cone of darkness encompasses the light-collecting lens of the photomultiplier housing. The only light reaching the detector is that scattered in the forward direction by particles in the aerosol under test. The particles are introduced into the light beam through an air inlet and pass through the beam sheathed in clean air. They can then be collected on a molecular filter for subsequent examination. Air flow can be varied up to 1 ft^3 min^{-1} (472 cm^3 s^{-1}). Solids concentration as low as 1 μg m^{-3} may be detected, the mass

concentration at any instant being displayed on a meter or plotted on a recorder [199].

The Brice-Phoenix Universal Light Scattering Photometer [200, 201] was originally designed to measure scattered light, transmittance, fluorescence and reflectance but it was refined to measure turbidity, angular dependence on Rayleigh ratio, dissymmetry and polarization ratio. It operates in the size range 1 to 5000 nm; the light source is a regulated high-pressure mercury vapour lamp in front of which one or more neutral density filters may be stationed. The beam width is adjustable and available wavelengths comprise 365, 405, 436, 546, 578 nm plus 633 nm if the mercury lamp is replaced with a helium–neon source. The instrument has been adapted to determine the particle size distribution of aerosols.

The Sinclair-Phoenix aerosol photometer is a near-forward scattering aerosol photometer employing an axisymmetric optical arrangement. Light scattered from particles within 4° to 36° of the forward direction indicates concentration. Air is drawn through the instrument at 1 ft^3 min^{-1} and is sheathed in clean air. The upper and lower concentration limits are 100 and 10^5 μg l^{-1}.

The Sartorius aerosol photometer S is a 45° scattering photometer for mass concentration measurement, designed for continuous measurement in the fields of emission control and clean room supervision. The source is a 200 W mercury arc lamp and the direct beam to the photomultiplier is interrupted by a chopper to correct for source intensity variations. Operation is over six orders of magnitude at concentrations from 10^8 to 5×10^2 particles cm^{-3}.

The Leitz Tyndalloscope is a small light-scattering photometer in which the angle of observation is 30°. A chamber is opened to admit ambient air and its intensity matched using crossed polarizing filters and a split eyepiece.

3.8.1 Discussion

The polar light-scattering diagrams for small particles are extremely complex and highly dependent upon relative refractive indices and the wavelength of the incident radiation. The wide adoption of these instruments in the field of health and safety is due to the size dependency being greater in the respirable range. It must be remembered, when using these instruments, that there is no universal light-scattering curve. These instruments have to be calibrated and if they are calibrated with PVC spheres they should be able to size PVC spheres accurately. They will not accurately size spheres having optical properties different from those of PVC. They will not accurately size non-spherical coal dust particles but this raises the thorny problem as to how one defines size for an irregularly shaped particle.

In order to obtain meaningful results one should calibrate using the particles of interest as West does with the HIAC (see section 2.1) and Marple and Rubow [73] with the Royco and Bausch & Lomb.

The errors with light-scattering particle counters have been discussed by Kratel [203] and instruments have been compared by Rimberg and Keafer [204] who find little agreement between them. Whitby and Vomela [205] compared the Royco PC200, the Bausch & Lomb, and the SR1 with microscopy and also found significant variations

and a tendency to undercount. Heintzenberg [255] compared the two-channel Royco, a condensation counter and a four-channel integrating nephelometer for sizing aerosols in the size range 0.05 to 2 μm.

3.9 Miscellaneous techniques

In the hot-wire anemometer [219], aerosol particles are drawn past a fine, short, hot filament, from which heat is extracted due to impingement on the filament and as a result of turbulence due to the presence of particles. KLD Associates manufacture a version for use in pesticide disbursement, fog and cloud analysis and aerosol monitoring in the size range 1 – 600 μm at counts up to 500 s^{-1} [cited in 254].

A diffusion battery [220, 221] has also been described in which small particles in a gas are subjected to molecular bombardment, which causes them to move in an erratic manner. Airborne particles passing through a narrow capillary tend, therefore, to collide with the capillary walls, and this property may be used for size determination of sub-micrometre particles. A discussion of data interpretation has been given [222] and a comparison with electron microscopy gave reasonable agreement. Knutson [cited in 13] coupled a diffusion battery to a data-processing system which allowed unattended runs of several days with hourly sampling. Kasper *et al.* [282] describe a six-stage miniature diffusion battery with 'collimated hole structure' plates. The penetration of salt crystals in the size range 0.2 to 0.6 μm was in good agreement with theory. Soderholm [283] gives a brief review of the use of diffusion batteries. Accurate formulae are included for the penetration function of circular tubes and parallel plate cells. A method of obtaining a computer solution is outlined and illustrated with experimental data.

The amplitude of vibration of airborne particles in an intense sound field has been used for determining particle size [223 – 225]. The relationship between the amplitude of particle vibration and particle size has been given by Brandt *et al.* [223] and suitable equipment has been designed by Schultz and Gucker [224, 225].

A β-absorption impactor aerosol mass monitor has also been described [226]. New techniques include the use of ionization [227], gas chromatography [228] and a hot-wire anemometer [229] for aerosol droplets.

The Eberline Model AIM - 3 Air Monitor uses interchangeable crystals to allow either α or β–γ monitoring.

The Radioactive air monitor monitors air with a dual phosphor probe for α and β radiation simultaneously.

A constricted glow discharge has been used for sizing particles in the + 10 μm range. The particles are drawn through a 153-μm aperture situated between two electrodes. The magnitude of the perturbation in the discharge current was found to be proportional to particle size [230].

The time-of-flight aerosol spectrometer shows promise as a method for continuous particle size determination from a few Ångströms up to a micron in diameter [231]. The aerosol is expanded through a nozzle into a vacuum and the particle velocity in the resulting flow reflects the distribution and allows size fractionation.

Mallove and Hinds describe an aerosol disc centrifuge [232]. The time variations of scattered light intensity at fixed radial locations characterize the distribution. They describe the sedimentation–diffusion theory on which the instrument is based and the applied data inversion techniques. Experimental results are presented for a mono- and a polydisperse aerosol. The technique is very rapid (< 5 min) and ideally suited for liquid droplets.

Some authors [233 - 236] have used holography to measure particle size and velocity. Also, a three-dimensional scanner, for quantitative particle analysis from a real holographic image is commercially available (Optronic's Holoscan). The Holographic particle analyser uses a switched ruby rod to produce a hologram of the viewing volume containing the aerosol particles. Particle size, using the hologram, can be studied manually or automatically using television screens. The lower size limit is 0.1 μm.

Specialized sampling devices are also available such as the plutonium dust sampler [137] and the airborne bacteria sampler for sampling bacteria for subsequent counting after incubation (Casella).

Zhulanov et al. [284] propose a method for the determination of the concentration and the degree of dispersion of sub-micron aerosols using a He–Ne resonator for uniform illumination. The method is applicable to particle radii less than 0.03 μm and concentrations less than 5×10^4 cm^{-2}.

Galli and Madelaine [285] present two size distribution analysers, one based on classical detection principles and the other a photon-counting device. The respective size ranges (μm) and concentrations are: $r > 0.2, c < 3 \times 10^4 ; 0.03 < r < 0.2,$ $c < 2 \times 10^6$.

References

1 Anon., (1971), EPA Standards for ambient air, *Environ. Sci. Technol.*, **5**, 6, 503.
2 McCabe, L.M. (1970), *Environ. Sci. Technol.*, **4**, 210–13.
3 Orenstein, A.J. (ed.) (1960), *Proc. Pneumoconiosis Conf.*, Johannesburg, 1959, Churchill, London.
4 Seaney, R.J., Halpin, R.K. and Maguire, B.A. (1973), *Staub*, **33**, 312.
5 Breuer, H. *et al., ibid.*, 187.
6 Heyder, J. *et al.* (1975), *J. Aerosol Sci.*, **6**, 5, 311–28.
7 Shah, M.A. *et al.* (1977), *Powder Technol.*, **18**, 53–64.
8 Heyder, J. *et al* (1973), *J. Aerosol Sci.*, **4**, 3, 191–208.
9 Davies, C.N. (1974), *Chemy Ind.*, 441.
10 Hamilton, R.J. and Walton, W.H. (1961), *Inhaled Particles and Vapours* (ed. C.N. Davies), Pergamon Press, p.465.
11 Hamilton, R.J. and Smith, D.S.G. (1967), *Coll. Guard.*, 471.
12 Hatch, T.F. and Gross, P. (1964), *Pulmonary Deposition and Retention of Inhaled Aerosols*, Academic Press, NY.
13 Elmes, P.C. (1977), *Royal Soc. of Health J.*, 97, 3, 102–5.
14 Bart, D. (ed.) (1973), Proc. Int. Symp. on *Environmental Health Aspects of Lead*, Luxembourg Commissions of The European Communities.
15 Natusch, D.F.S. and Wallace, J.R. (1974), *Science*, 186, 4165, 695–9.
16 Saltzmann, B.E. and Burg, W.R. (1977), *Analyt. Chem.* 49, 5, 1R–16R.
17 Rae, S., Walker, D.D. and Attfield, M.D. (1971), Inhaled particles III, In: Proc. Int. Symp. Br. Occup. Hyg. Soc. (ed. W.H. Walton), Unwin Bros., London.

18 Levy, D., Gent, M. and Newhouse, M. (1974), Commission Eur. Communities (Rep.) EUR 5360, Int. Symp. *Recent Adv. Assess. Health Eff., Envir. Poll.*, **3**, 1263–76.
19 Pattenden, N.J. and Wiffen, R.D. (1977), *Atmos. Envir.*, **11**, 8, 677–81.
20 Maguire, B.A. and Harris, G.W. (1974), *Airborne Dust Sampling for Industry*, Safety in Mines Research Establishment, Health & Safety Executive, Sheffield, England.
21 Duvall, P.M. and Bourke, R.C. (1974), *Environ. Sci.*, **8**, 8, 765–7.
22 Green, H.L. and Lane, W.R. (1957), *Particulate Clouds, Dusts, Smokes and Mists*, Spon.
23 Drinker, P. and Hatch, T.F. (1954), *Industrial Dust*, McGraw-Hill, NY.
24 Yaffe, C.D., Byers, D.H. and Hosey, A.S. (1956), *Encyclopedia of Instrumentation for Industrial Hygiene*, Univ. of Michigan.
25 White, P.A.F. and Smith, S.E. (1964), *High Efficiency Air Filtration*, Butterworths.
26 Davies, C.N. (1968), *Brit. J. Appl. Physics* (J. Phys. D.), Ser. 2, **1**, 921–32.
27 Critchlow, A. and Maguire, B.A. (1968), *Environ. Engng*, **35**, 12.
28 Clarke, M.G. and Bradburn, J.A. (1970), 'Survey of Methods of Monitoring Particulate Contaminants', Inf. paper 122 and 122A. Welwyn Hall Research Establishment, 11 White Lion House, Town Centre, Hatfield, Herts.
29 ACGIH (1972), *Air Sampling Instruments for Evaluation of Atmospheric Contaminants*, Ann. Conf. Governmental Industrial Hygienists, Cincinnati, Ohio, USA.
30 Schicketanz, W. (1969), *Staub*, **29**, 10, 417–20.
31 Wright, B.M. (1960), BP 841 698.
32 Greenburg, L. and Smith, G.W. (1922), US Bureau of Mines Report. Investigation 2392.
33 Hatch, T., Warren, H. and Drinker, P. (1932), *J. Ind. Hyg. Toxicol.* **114**, 301.
34 Katz, S.H. *et al.* (1925), *US Public Health Bulletin*, **144**, 69.
35 Owens, J.S. (1922), *Proc. R. Soc.*, **18**, A101.
36 Ferry, R.M., Farr, L.E. and Hartmann, M.G. (1949), *Chem. Rev.*, **44**, 389.
37 May, J.R. (1945), *J. Scient. Instrum.*, **22**, 187–95.
38 Marple, V.W. and Liu, B.Y.H. (1975), *J. Colloid Interfac. Sci.*, **53**, 1, 31–4.
39 Davies, C.N. and Aylward, M. (1951), *Proc. Phys. Soc.*, **B64**, 889.
40 Ranz, W.E. and Wong, J.B. (1952), *A.M.A. Arch. Ind. Health*, **5**, 464–77.
41 May, K.R. (1975), *J. Aerosol Sci.*, **6**, 6, 403–11.
42 Laskin, S. (1949), *Pharmacology and Toxicology of Uranium Compounds* (eds. C. Voegtlin and H.G. Hodge), McGraw-Hill, NY.
43 Gillespie, G.R. (1953), *Eng. Expt. Sta., University of Illinois Technical Report No. 9*, 50–1010.
44 Gillespie, G.R. and Johnstone, H.F. (1955), *Chem. Eng. Prog.*, **51**, 74F.
45 Pilcher, J.M., Mitchell, R.L. and Thomas, R.E. (1955), *Proc. Chem. Spec. Man. Ass., NY.* 1039–42.
46 Wilcox, J.D. (1953), *A.M.A. Arch. Ind. Hyg. Occup. Med.*, **7**, 376–82.
47 Wilcox, J.D. and Van Antwerp, W.R. (1955), *A.M.A. Arch. Ind. Health*, **11**, 422.
48 Brink, J.A. (1958), *Ind. Eng. Chem.*, **50**, 645–58.
49 Einbinder, H. (cited in [95]).
50 Berner, A. and Preinig, O. (1964), *Staub*, **24**, 8, 295.
51 Picknet, R.G. (1972), *J. Aerosol Sci.*, **3**, 3, 185–98.
52 Bucholz, H. (1970), *Staub Reinhalt Luft*, **30**, 4, 159–61.
53 Bucholz, H., *ibid.*, 5.
54 Bürkholz, A. (1970), *Chem. Ingr. Techn.*, **42**, 5, 299–303.
55 Jacobson, M. *et al.* (1970), *Am. Ind. Hyg. Assoc. J.*, **31**, 4, 442–5.
56 Soole, B.W. (1971), *Aerosol Sci.*, **2**, 1, 1–14.
57 Littlefield, J.B., Schrenk, H.H. and Feicht, F.L. (1937), Bureau of Mines Impinger for Dust Sampling, Report Investigation 3360.
58 Dubois, E. *et al.* (1967), *Assessment Airborne Radioactivity*, Proc. Symp., Vienna 1967, 351–77.
59 Glowiak, B. and Pilezynski, R. (1969), *Ochrona Powietrza*, **3**, 2, 10, 12–15.
60 Goetz, A. (1969), *Environ. Sci. Technol.*, **3**, 154–60.

61 Hänel, G. (1969), *Atmos. Environ.,* **3,** 69–83.
62 Noll, K.E. (1970), *ibid.,* **4,** 1, 9–19.
63 O'Donnell, H., Montgomery, T.L. and Corn, M., *ibid.,* 1–7.
64 Renshaw, F.M., Bachman, J.M. and Pierce, J.O. (1969), *Am. Ind. Hyg. Assoc. J.,* **30,** 113–6.
65 Kotrappa, P. (1973), *J.Aerosol Sci.,* **4,** 1, 47–50.
66 Pelassy, P. (1974), *ibid.,* **5,** 6, 531–49.
67 Marple, V.A., Liu, B.Y.H. and Whitby, K.T. *ibid.,* **5,** 1, 1–16.
68 Hänel, G. and Gravenhorst, G., *ibid.,* 47–54.
69 Anderson, P.L. (1976), *Environ. Sci. Technol.,* **10,** 2, 145–50.
70 May, K.R. (1975), *J. Aerosol Sci.,* **6,** 6, 413–9.
71 Thomas, J.W., *ibid.,* 583–7.
72 Dzubay, T.G., Hines, L.E. and Stevens, R.K. (1976), *Atmos. Environ.,* **10,** 3, 229–34.
73 Marple, V.A. and Rubow, K.L. (1976), *J. Aerosol Sci.,* **7,** 425–33.
74 Green, H.L. and Watson, H.H. (1935), MRC Sp. Rep. Ser. No. 199, HMSO.
75 Herdan, G. (1960), *Small Particle Statistics,* Butterworths.
76 Gurney, S.W., Williams, S.R. and Meigs, R.R. (1938), *J. Indust. Hyg.,* **20,** 24.
77 Cunningham, P.T., Johnson, S.A. and Yang, R.T. (1973), *Variations in the Chemistry of Airborne Particulate Material with Particle Size and Fine,* Chem. Eng. Div. Argonne National Laboratories, Illinois.
78 Hidy, G.M. (ed.) (1972), *Aerosols and Atmospheric Chemistry,* Academic Press, NY.
79 Whitby, K.T., Liu, B.Y.H., Husar, R.B. and Barsic, N.J. (1972), *J. Colloid Interfac. Sci.,* **39,** 1, 136–64.
80 Whitby, K.T., Husar, R.B. and Liu, B.Y.H., *ibid.,* 177–204.
81 Husar, R.B., Whitby, K.T. and Liu, B.Y.H., *ibid.,* 211–24.
82 Lundgren, D.A. (1967), *J. Air Poll. Control Ass.,* **17,** 225–8.
83 Lundgren, D.A. (1971), *Atmos. Environ.,* **5,** 8, 645–51.
84 Lundgren, D.A. and McFarland, A.R. (1970), *Am. Ind. Hyg. Assoc. J.,* **31,** 2, 36.
85 Lundgren, D.A. (1971), *Atmos. Environ.,* **5,** 8, 645–51.
86 Leary, J.A. (1951), *Indust. Engng. Chem. analyt. Edn,* **23,** 853.
87 Cowan, M., Sandia Corporation SCR–296 (cited in [4]).
88 Hounam, R.F. (1964), *AERE Rep. M132,* 8.
89 Hounam, R.F. and Sherwood, R.J. (1965), *Am. Ind. Hyg. Assoc. J.,* **26,** 122.
90 O'Connor, D.T. (1971), UKAEA, AHSB-RP-R-108, HMSO.
91 Walton, W.H. (1954), The physics of particle size analysis symposium, *Brit. J. Appl. Phys.,* Suppl. 3, S29–S39.
92 Barker, D., O'Connor, D.T. and Winder (1954), Safety in Mines Research Establishment Report No. 93, Sheffield, HMSO.
93 Critchlow, A. and Proctor, T.D. (1965), *Coll. Guard.,* **211,** 208–9.
94 Dunmore, J.H., Hamilton, R.J. and Smith, D.S.L.E. (1964), *J. Scient. Instrum.,* **41,** 669.
95 Harris, G.W. and Proctor, T.D. (1966), *Coll. Guard.,* **213,** 690–1.
96 Harris, G.W. and Maguire, B.A. (1968), *Ann. Occup. Hyg.,* **11,** 195–201.
97 Maguire, B.A. and Barker, D. (1969), *ibid.,* **12,** 197–201.
98 Andersen, A.A. (1958), *J. Bact.,* **76,** 471.
99 Heneveld, W.H. (1959), Fifth Occupational Health Conference, University of Texas.
100 Solomon, W.R. (1970), *J. Allergy,* **45,** 1.
101 Burton, R.M. *et al.* (1973), *J. Air Poll. Control Ass.,* **23,** 4, 277–81.
102 New York State, Department of Environmental Conservation, Division of Air Resources (1972), Report No. BTS-3, Dec. (Available from Andersen 200 Inc.)
103 Eaton, W.C. *et al.* (1973), EPA, Human Studies Lab., Bioenvironmental Measurements Branch, Research Triangle Park, N.C.
104 Kozel, W.M. *et al., ibid.*
105 Perkins, W.A. (1957), Second Semi-annual Report, Aerosol Lab., Dept. Chem. Chem. Engng, Stamford Univ., p. 186.

106 May, K.R. and Clifford, R. (1967), *Ann. Occup. Hyg.*, **10**, 83.
107 Hirst, J.M. (1952), *Ann. appl. Biol.*, **39**, 257.
108 Edwards, A.P. (1967), *Ann. Occup. Hyg.*, **10**, 67.
109 May, K.R. (1960), *ibid.*, **2**, 93.
110 May, K.R., Pomeroy, N.P. and Hibbs, S. (1976), *J. Aerosol Sci.*, **7**, 1, 53–62.
111 Decker, H.M., Buchana, L.M. and Dahlgren, C.M. (1969), *Contamin. Control*, August.
112 May, K.R. (1967), *Airborne Microbes* (eds P.M. Gregory and J. Monteith), CUP, London, pp. 60–80.
113 May, K.R. (1966), *Bact. Rev.*, **30**, 559–70.
114 Sawyer, K.F. and Walton, W.H. (1950), *J. Scient. Instrum.*, **27**, 272–6.
115 Keith, C.M. and Derrick, J.C. (1960), *J. Colloid Sci.*, **15**, 340–56.
116 Goetz, A.H. and Kallai, T. (1962), *J. Air Poll. Control. Ass.*, **12**, 479.
117 Ludwig, F.L. and Robinson, E. (1965), *J. Colloid Sci.*, **20**, 571–84.
118 Stöber, W. and Zenack, U. (1964), *Staub*, **24**, 8, 295.
119 Berner, A. (1968), Aerosol Research, 1st Physics Institute, University of Vienna, Status Report, Jan.
120 Hank, H. and Schedling, J.A. (1968), *Staub*, **28**, 18–21.
121 Stöber, W. (1967), *Assessment of Airborne Radioactivity*, IAEA, Vienna, 393–404.
122 Tillern, M.I. (1967), *Assessment of Airborne Radioactivity*, IAEA, Vienna, 393–404.
123 Timbrell, V. (1954), *Br. J. Appl. Phys.*, **5**, Suppl. 3, 86–90.
124 Ferron, G.A. and Bierzhuizen, H.W.J. (1976), *J. Aerosol Sci.*, **7**, 1, 5–12.
125 Stöber, W. and Flaschbart, H. (1969), *Environ. Sci. Technol.*, **3**, 1280.
126 Stöber, W., Flaschbart, H. and Boose, C. (1972), *J. Colloid Interfac. Sci.*, **39**, 1, 109–10.
127 Porstendörfer, J. (1973), *J. Aerosol Sci.*, **4**, 4, 345–54.
128 Heyder, J. and Porstendörfer, J. (1974), *ibid.*, **5**, 4, 387–400.
129 Heyder, J., Weels, A.C. and Wiffen, R.D. (1970), *Atmos. Environ.*, **4**, 149.
130 Kops, J., Hermans, L. and van de Vate, J.F. (1974), *J. Aerosol Sci.*, **5**, 4, 379–86.
131 Oeseburg, F.M., Benschop, F.M. and Roos, R. (1975), *ibid.*, **6**, 6, 421–31.
132 Berner, A. and Reichelt, H. (1968), *Staub*, **28**, 158.
133 Berner, A. and Reichelt, H. (1969), *ibid.*, **29**, 92–95.
134 Hochrainer, D. and Brown, P.M. (1969), *Environ. Sci. Technol.*, **3**, a, 831–5.
135 Hochrainer, D. (1971), *J. Colloid Interfac. Sci.*, **36**, 191–4.
136 Matteson, M.J. and Boscoe, G.F. (1974), *Aerosol Sci.*, **5**, 1, 71–79.
137 Tait, G.W.C. (1956), *Nucleonics*, **14**, 53.
138 Kalmus, E.E. (1954), *J. Appl. Phys.*, **25**, 87–89.
139 Krost, K.J., Sawicki, C.R. and Bell, J.P. (1977), *Anal. Lett.* **10**, 4, 333–5.
140 Junge, C. (1951), Nuclei of atmospheric condensation, In: *Compendium of Meteorology*, AMS. Boston, Mass., USA, pp 182–91 (cited in [3]).
141 Daniel, J. and Brackett, F. (1951), *J. Appl. Phys.*, **22**, 542–54.
142 Orr, C. and Dallavalle, J.M. (1960), *Fine Particle Measurement*, Macmillan, NY, p. 96.
143 Hurd, F.K. and Mullins, J.C. (1962), *J. Colloid Sci.*, **17**, 2, 91–100.
144 Yoshikawa, H.H., Swartz, G.A., MacWaters, J.T. and Fite, W.L. (1956), *Rev. Scient. Instrum.*, **359**.
145 Podol'skü A.A. and Kalakutskü, L.I. (1976), *Kolloid. Z.*, **37**, 6, 1198–1202.
146 Howink, E.H. and Rolwink, W. (1957), *J. Hyg. Camb.*, **55**, 544.
147 Agafonova, N.I. and Matalyavickus, V.P. (1968), *Hyg. Samit.*, **33**, 221–3.
148 Morris, E.J. *et al.* (1961), *J. Hyg. Camb.*, **59**, 487.
149 Beadle, D.G., Kitto, P.H. and Blignaut, P.J. (1954), *Archs. Ind. Hyg. Occup. Med.*, **10**, 487.
150 Lauterbach, K.E. *et al.* (1954), *Archs. Ind. Hyg. Occup. Med.*, **9**, 69.
151 Benarie, M. and Bodin, D. (1969), *Staub*, **29**, 3, 49.
152 Hanson, D.N. and Wilkie, C.R. (1969), *Ind. Engng Chem.*, **8**, 3, 357–64.
153 Gordon, M.T. and Orr, C. (1954), *J. Air Poll. Control Ass.*, **4**, 1.
154 Cartwright, J. (1956), *Br. J. Appl. Phys.*, **7**, 91.

155   Epstein, P.S. (1929), *Z. Phys.*, **54**, 537.
156   Waldmann, L. (1959), *Z. Naturf.*, **14A**, 589.
157   Strauss, W. (1966), *Industrial Gas Cleaning,* Pergamon.
158   Watson, H.H. (1936), *Trans. Inst. Min. Metall.,* **46**, 176–87.
159   Walton, W.H. and Harris, W.J. (1947), Technical Paper 1. CDRE, Porton, Hants., England.
160   Prewett, W.G. and Walton, H.H. (1948), Technical Paper 63. CDRE, Porton, Hants., England.
161   Schadt, C.F. and Cadle, R.D. (1957), *J. Colloid Sci.,* **12**, 356–62.
162   Watson, H.H. (1958), *Br. J. Appl. Phys.,* **2**, 78–79.
163   Hodkinson, J.R., Critchlow, A. and Stanley, N. (1960), *J. scient. Instrum.,* **37**, 182–3.
164   Hodkinson, J.R. (1962), *Air Sampling Instruments,* American Conf. Governmental Industrial Hygienists, 1014 Broadway, Cincinnati 2, Ohio, USA.
165   Donague, J.K. (1953), *J. Scient. Instrum.,* **30**, 59.
166   Walkenhorst, W. (1952), *Beitr. Silikoseforsch.,* **18**, 29–62.
167   Burdenkin, J.T. and Davies, J.G. (1956), *Br. J. Ind. Med.,* **13**, 196–201.
168   Gruszka, J. (1961) (*loc. cit.* [88]).
169   Beadle, D.G. and Kitto, P.M. (1952), *J. Chem. Met. Min. Soc., S. Afr.,* **52**, 284–311.
170   Wright, B.M. (1954), *J. Scient. Instrum.,* **31**, 263–4.
171   Camber, H., Hatch, T. and Watson, J.A. (1953), *Am. Ind. Hyg. Assoc. Quart.,* **14**, 191–4.
172   Walton, W.H. (1950), *J. R. Microsc. Soc.,* **70**, 51.
173   Orr, C. and Martin, R.A. (1958), *Rev. Scient. Instrum.,* **29**, 129–30.
174   Hamilton, R.J. (1956), *J. Scient. Instrum.,* **33**, 395–9.
175   Balashov, V., Bradwig, J.G. and Rendall, R.E.G. (1961), *J. Min. Vent. Soc. S. Afr.* **14**, 98–100.
176   Wright, B.M. (1953), *Science,* **118**, 195.
177   Anon. (1976), *Processing,* **22**, 2, 11.
178   Olin, J.G., Trautner, R.P. and Gilmore, J. (1971), 64th Ann. Mtg., Air Pollution Control Assoc.
179   Olin, J.G., Sem. G.J. and Christenson, D.L. (1971), *Am. Ind. Hyg. Assoc. J.,* **32**, 209–20.
180   Olin, J.G. (1971), ISA Conf. & Exhib., Chicago, Ill.
181   van de Hulst, H.C. (1957), *Light Scattering by Small Particles,* Wiley, NY.
182   Gucker, F.T., O'Konski, C.T., Pickard, H.B. and Pitts, J.H. (1949), *J. Am. Chem. Soc.,* **69**, 2422.
183   Gucker, F.T. and O'Konski, C.T. (1949), *J. Colloid. Sci.,* **4**, 541.
184   O'Konski, C.T. and Doyle, G.J. (1955), *Analyt. Chem.,* **27**, 694.
185   Fisher, M.A., Katz, S. and Lieberman, A. (1955), *Proc. 3rd Nat. Air Poll. Symp. Pasadena, California, USA.*
186   Oeseburg, F.M. (1972), *J. Aerosol Sci.,* **3**, 4, 307–11.
187   Pinnick, R.G., Rosen, J.M. and Hoffman, D.J. (1973), *Appl. Optics,* **12**, 37.
188   Farthing, W.F. (1978), *J. Air Poll. Contr. Ass.,* **28**, 10, 999–1001.
189   Martens, A.E. and Keller, J.D. (1968), *Am. Ind. Hyg. Assoc. J.,* **29**, 257.
190   Randall, K.M. and Keller, J.D. (1965), AIHA Conference, Houston, Texas, May.
191   Brown, P.M. (1969), *Environ. Sci. Technol.,* **3**, 8, 768–9.
192   Jones, I.O. and Khaled, M.J. (1971), *Solut. J.,* **2**, 6, 6–7.
193   Jaenicke, R. (1972), *J. Aerosol Sci.,* **36**, 2, 95–111.
194   Scott, C.C. (1971), *Lancet,* 1288–91.
195   Matthew, R.A. (1964), *Filtration,* July/Aug., 204–8.
196   Kratel, R. (1965), *Staub,* **25**, 11, 504–8.
197   Tang, I.N., Munkelwitz, H.R. and Davis, J.G. (1976), Report BNL-21432, NTIS.
198   Kaye, W. (1973), *J. Colloid Interfac. Sci.,* **44**, 2, 384–6.
199   Sinclair, D. (1953), *Air Repair,* **3**, 51.
200   Brice, B.A., Haliver, M. and Speiser, R. (1950), *J. Opt. Soc. Am.,* **40**, 11, 768–78.
201   Brice, B.A. and Haliver, M. (1954), *ibid.,* **44**, 4, 340.
202   Moroz, W.J., Withstandley, V.D. and Anderson, G.W. (1970), *Rev. Scient. Instrum.,* **41**, 7, 978–83.
203   Kratel, R. (1970), *Staub Reinhalt Luft.* (Engl. ed.), **30**, 5, 40.

204 Rimberg, D. and Keafer, D. (1970), *J. Colloid Interfac. Sci.,* **33,** 4, 628.
205 Whitby, K.T. and Vomela, R.A. (1967), *Environ. Sci. Technol.,* **1,** 10, 801–14.
206 Dupox, J. and Charvav, J. (1976), *Atmospheric Pollution,* Proc. Int. Colloq., **12,** 361–81.
207 Hallworth, G.N. and Hamilton, L.R. (1976), *J. Pharm. Pharmac.,* **28,** 12, 890–7.
208 Hallworth, G.N. and Andrews, V.G. (1976), *ibid.,* 898–907.
209 Fry, J.A. and Black, A. (1973), *J. Aerosol Sci.,* **4,** 2, 113–24.
210 Luke, C.L. *et al.* (1972), *Environ. Sci. Technol.,* **6,** 13, 1105–9.
211 Byers, R.L. *et al.* (1971), *ibid.,* **5,** 6, 517–21.
212 Knutson, E.O. and Whitby, K.T. (1976), *J. Aerosol Sci.,* **6,** 6, 443–51.
213 Whitby, K.T. (1975), *Fine Particles,* Proc. Symp. (1976), (ed. B.Y.H. Liu), Academic Press, NY, pp. 581–624.
214 Lundgren, A.D., Carter, L.D. and Daley, P.S., *ibid.,* pp. 485–510.
215 Lippman, M., *ibid.,* pp. 287–310.
216 John, W., *ibid.,* pp. 649–67.
217 Knutson, E.O., *ibid.,* pp. 739–62.
218 Gebhart, J. *et al., ibid.,* pp. 793–815.
219 Goldschmidt, V.W. (1965), *J. Colloid Sci.,* **20,** 617.
220 Thomas, J.W. (1955), *ibid.,* **10,** 246–55.
221 Thomas, J.W. (1956), *ibid.,* **11,** 107.
222 Mercer, T.T. and Green, T.D. (1974), *J. Aerosol Sci.,* **5,** 3, 251–5.
223 Brandt, O., Freund, M. and Heidemann, E. (1937), *Z. Phys.,* **104,** 511–33.
224 Cassel, H.M. and Schultz, M. (1952), *Air Pollution* (ed. L. McCabe), McGraw-Hill.
225 Gucker, F.T. (1949), *Proc. 1st Nat. Air Poll. Symp.* Sandford, California, USA.
226 Lilienfeld, P. (1970), *Amer. Ind. Hyg. Assoc. J.,* **31,** 35.
227 Gourdine Systems Inc. (1969), B.P. 1 161 190.
228 Kovar, V. (1970), *Ochrona Ousdusi,* 7, 104–11.
229 Goldschmidt, V.W. and Householder, M.K. (1969), *Atmos. Environ.,* **3,** 643.
230 Frederick, N.A., Bell, A.T. and Hanson, D.H. (1976), *Environ. Sci. Technol.* **6,** 13, 117–8.
231 Schwartz, M.H. and Andres, R.P. (1976), *J. Aerosol Sci.,* **7,** 4, 281–96.
232 Mallove, E.F. and Hinds, W.C. (1976), *ibid.,* **7,** 5, 409–23.
233 Thompson, B.J., Ward, J.H. and Zinky, W.R. (1967), *Appl. Optics,* **6,** 519.
234 Hickling, R. (1969), *J. opt. Soc. Am.,* **59,** 1334.
235 Trolinger, J.D., Belz, R.A. and O'Hare, J.E. (1972), *Prog. Astronaut. Aeronaut.,* **34,** 249.
236 Royer, H. (1977), *Optics Comm.,* **20,** 73.
237 Andersen, A.A. (1966), *Am. Ind. Hyg. Assoc. J.,* **27,** 160.
238 Cabrol, C. *et al.* (1972), *J. Aerosol Sci.,* **3,** 4, 281–7.
239 Matteson, M.J. and Keng, E.Y.H. (1972), *ibid.,* **3,** 1, 45–53.
240 Barnes, E.C. and Penney, G.W. (1936), *J. Indust. Hyg. Toxicol.,* **18,** 3, 167–72.
241 Roth, C., Gebhart, J. and Heigever, G. (1976), *J. Colloid Interfac. Sci.,* **54,** 2, 265–77.
242 Lippmann, M. *et al.* (1965), *Am. Ind. Hyg. Assoc. J.,* **26,** 485–9.
243 Morrow, P.E. and Mercer, T.T. (1964), *ibid.,* **25,** 1, 8–14.
244 Sonkin, L.S. (1946), *J. Indust. Hyg. Toxicol.,* **28,** 269.
245 Davies, C.N. (1976), *Chemy Ind.,* 624–6.
246 Higgins, R.I. and Dewell, P., BCIRA Report 908, British Cast Iron Research Assoc., Alvechurch, Birmingham, England.
247 Wright, B.M. (1954), *Br. J. Ind. Med.,* **11,** 284.
248 Ashford, J.R. and Jones, C.O. (1964), *Ann. Occup. Hyg.,* **7,** 85.
249 Lippmann, M., Air Sampling Instruments, In: *Ann. Conf. Industrial Hygiene,* G1–G17.
250 Binck, B. *et al.,* (1967), *Staub,* **27,** 379–428.
251 Tuma, J. and Gucker, F.T. (1971), *Proc. 2nd Int. Clean Air Congress,* Academic Press, NY, p. 463.
252 Gucker, F.T. *et al.,* (1973), *Aerosol Sci.,* **4,** 389.

253 Marshall, T.R., Parmenter, C.S. and Seaver, M. (1976), *J. Colloid Interfac. Sci.,* 55, 3, 624–36.
254 Davies, R. and Turner, H.E. (1977), *Proc. Conf. Particle Size Analysis* (ed. Groves), Chem. Soc., Analyt. Div. (1978), Heyden.
255 Heintzenberg, J. (1975), *J. Aerosol Sci.,* 6, 5, 291–304.
256 *Coal Mines (Respirable Dust) Regulations* (1975), SI No. 1433, HMSO. Hygiene Standards for: Chrysolite; Asbestos Dust (1968): Amosite; Asbestos Dust (1973); Pergamon Press.
257 Lovering, P.E. (1977), *Proc. 4th Int. Powder Technology and Bulk Solids Conf.,* Powder Techn. Publ. Series No. 7, Heyden, pp. 95–8.
258 Stevens, D.C. and Stephenson, J. (1972), *J. Aerosol Sci.,* 3, 15–23.
259 Stevens, D.C. and Churchill, N.L. (1973), *ibid.,* 4, 85–91.
260 May, K.R. (1964), *Appl. Micro.,* 12, 37.
261 May, K.R. (1973), *Airborne Transmission and Airborne Infection* (ed. J.F. Ph. Hers), Oosthoek, Utrecht, p. 29.
262 Whitby, K.T. and Clark, W.E. (1966), *Tellus,* 18, 573–86.
263 Liu, B.Y.H., Whitby, K.T. and Pui, D.Y.H. (1974), *J. Air Poll. Control Ass.,* 24, 1067.
264 Sem, G.J., Blumenthal, D.L. and Anderson, J.A. (1974), *67th Ann. Mtg Air Pollution Control Ass.,* June 9–13, Paper No. 74, 266.
265 Lui, Y.H.B. and Pui, Y.P.D. (1975), *J. Aerosol Sci.,* 6, 3/4, 249–64.
266 Hobbs, P.V., Radke, L.F. and Hindman, H.E.E. (1976), *ibid.,* 7, 3, 195–211.
267 Allen, J. and Husar, R.B. (1976), *Proc. Int. Conf. Colloid Interfacial Science* (ed M. Kerker), Academic Press, NY, 2, 141–53.
268 Swithenbank, J. *et al,* Report No. HIC 245, University of Sheffield, Dept. Chem. Engng. Fuel Technol., England.
269 Wadley, M.W. *et al.,* (1978), *J. Air Poll. Control Ass.,* 24, 4, 365–7.
270 Oeseburg, F.M. and Roos, R. (1975), *Atmos. Environ.,* 9, 9, 859–60.
271 Rao, A.K. and Whitby, K.T. (1978), *J. Aerosol Sci.,* 9, 2, 77–86.
272 Rao, A.K. and Whitby, K.T. (1978), *ibid.,* 87–100.
273 Ravenhall, D.G. Forney, L.J. and Jazaifi, M. (1978), *J. Colloid Interfac. Sci.,* 65, 1, 108–17.
274 Yoshida, H. *et al.* (1978), *Kagaku Kogaku Ronbunshu,* 4 2, 419–24.
275 Yoshida, H. *et al.* (1978), *ibid.,* 123–8.
276 Gibson, H. and Ogden, T.L. (1977), *J. Aerosol Sci.,* 8, 5, 361–6.
277 Newton, G.J., Raabe, O.G. and Mokler, B.V. (1977), *ibid.,* 339–48.
278 Sansone, E.B. and Christensen, W.D. (1978), *Am. Ind. Hyg. Assoc. J.,* 39, 1, 12–16.
279 Pui, D.Y.H. and Liu, B.Y.H. (1976), Univ. Minnesota Report C00-1248-52, NTIS.
280 Lewis, C.W. and Lamothe, P.J. (1978), *J. Aerosol Sci.,* 9, 5, 391–8.
281 Pinnick, R.G., Auvermann, H.J. and Yue, G.K. (1979), *ibid.,* 10, 1, 55–74.
282 Kasper, G., Preining, O. and Matteson, M.J. (1978), *ibid.,* 9, 4, 331–8.
283 Soderholm, S. (1979), *ibid.,* 10, 2, 163–75.
284 Zhulanov, Yu. V., Sadovskii, B.F. and Nevskii, I.A. (1977), *Kolloid. Z.,* 39, 6, 1064–9.
285 Galli, S.E.L. and Madelaine, G. (1975), *Water Air Soil Poll.,* 5, 1, 11–38.
286 Whitby, K.T. (1975), Electrical measurement of aerosols, In: *Proc. Symp. Fine Particles* (ed. B.Y.H. Liu), Academic Press (1976), pp. 581–624.
287 Hauk, H. (1972), *J. Aerosol Sci.,* 3, 1, 31–37.
288 Jesson, U., Haslop, D. and Leiberman, A. (1976), *Proc. Sem. High Eff. Aerosol Filters,* Nuclear Ind. (1977), CEC, Luxembourg, pp. 641–7.
289 Kratel, S.A. *ibid.,* pp. 649–59.
290 Weinmann, R. (1978), *Powder Metall. Int.,* 10, 2, 94–5.
291 Lovering, P E. (1977), *Powtech 77,* Heyden.
292 Crosby, M.T. and Hamer, P. (1971), *Ann. Occup. Hyg.,* 14, 65.
293 EPA (1971), *Fed. Register,* 36, (April 30) 84.

4 Particle size, shape and distribution

4.1 Particle size

The size of a spherical homogeneous particle is uniquely defined by its diameter. For a cube the length along one edge is characteristic, and for other regular shapes there are equally appropriate dimensions. With some regular particles, it may be necessary to specify more than one dimension. For example: cone, diameter and height; cuboid, length, width and height.

Derived diameters are determined by measuring a size-dependent property of the particle and relating it to a linear dimension. The most widely used of these are the equivalent spherical diameters. Thus, a unit cube has the same volume as a sphere of diameter 1.24 units, hence this is the derived volume diameter.

If an irregularly shaped particle is allowed to settle in a liquid, its terminal velocity may be compared with the terminal velocity of a sphere of the same density settling under similar conditions. The size of the particle is then equated to the diameter of the sphere. In the laminar flow region, the particle moves with random orientation, but outside this region it orientates itself to give maximum resistance to motion so that the free-falling diameter for an irregular particle is greater in the intermediate region that in the laminar flow region. The free-falling diameter, in the laminar flow region, becomes the Stokes diameter. Stokes' equation can be used for spherical particles, up to a Reynolds number of 0.2, at which value it will give a diameter under-estimation of about 2%. Above 0.2 corrections have to be applied. Corrections may also be applied for non-spherical particles, so that the derived diameter is independent of settling conditions becoming purely a function of particle size. These diameters are particularly useful for characterizing suspended particles in the atmosphere and other cases where the settling behaviour of suspended solids is being examined.

For irregular particles, the assigned size usually depends upon the method of measurement, hence the particle sizing technique should, wherever possible, duplicate the process one wishes to control. Thus, for paint pigments, the projected area is important, while for chemical reactants, the total surface area should be determined. The projected area diameter may be determined by microscopy for each individual particle, but surface area is usually determined for a known weight or volume of powder. The magnitude of this surface area will depend on the method of measurement, permeametry, for example, giving a much lower area than gas adsorption. The former gives the surface

area accessible to the gas molecules and, therefore, depends on the size of the gas molecules if the solid contains very small pores.

Table 4.1 Definitions of particle size

Symbol	Name	Definition	Formula
d_v	Volume diameter	Diameter of a sphere having the same volume as the particle	$V = \frac{\pi}{6} d_v^3$
d_s	Surface diameter	Diameter of a sphere having the same surface as the particle	$S = \pi d_s^2$
d_{sv}	Surface volume diameter	Diameter of a sphere having the same external surface to volume ratio as a sphere	$d_{sv} = \dfrac{d_v^3}{d_s^2}$
d_d	Drag diameter	Diameter of a sphere having the same resistance to motion as the particle in a fluid of the same viscosity and at the same velocity (d_d approximates to d_s when Re is small)	$F_D = C_D A \rho_f \dfrac{v^2}{2}$ where $C_D A = f(d_d)$ $F_D = 3\pi d_d \eta v$ $Re < 0.2$
d_f	Free-falling diameter	Diameter of a sphere having the same density and the same free-falling speed as the particle in a fluid of the same density and viscosity	
d_{Stk}	Stokes' diameter	The free-falling diameter of a particle in the laminar flow region ($Re < 0.2$)	$d^2 = \dfrac{d_v^3}{d_d}$
d_a	Projected area diameter	Diameter of a circle having the same area as the projected area of the particle resting in a stable position	$A = \dfrac{\pi}{4} d_a^2$
d_p	Projected area diameter	Diameter of a circle having the same area as the projected area of the particle in random orientation	Mean value for all possible orientations $d_p = d_s$ for convex particles
d_c	Perimeter diameter	Diameter of a circle having the same perimeter as the projected outline of the particle	
d_A	Sieve diameter	The width of the minimum square aperture through which the particle will pass	
d_F	Feret's diameter	The mean value of the distance between pairs of parallel tangents to the projected outline of the particle	
d_M	Martin's diameter	The mean chord length of the projected outline of the particle	
d_R	Unrolled diameter	The mean chord length through the centre of gravity of the particle	$E(d_R) = \dfrac{1}{\pi} \displaystyle\int_0^{2\pi} d_R d\theta_R$

The sieve diameter, for square-mesh sieves, is the length of the side of the minimum square aperture through which the particle will pass, though this definition needs modification for sieves which do not have square apertures. In a sieving operation, such a particle will not necessarily pass through the appropriate mesh, particularly if it will only pass through when presented in a particular orientation as with elongated particles. For all such particles to pass through, the sieving time would approach infinity. There is also a range of aperture sizes in any sieve mesh and certain particles may only pass through the largest apertures.

Microscopy is the only widely used particle-sizing technique in which individual particles are observed and measured. A single particle has an infinite number of linear dimensions, and it is only when they are averaged that a meaningful value results; this is similar for a large number of particles. When a linear dimension is measured parallel to some fixed direction (Martin, Feret, unrolled or shear diameter), the size distribution of these measurements reflects the size distribution of the projected areas of the particles. These are called statistical diameters. Comparing the projected area of the particle with series of circles, gives a diameter which describes that particle for the orientation in which it is measured. In microscopy, this is usually the projected area diameter in stable orientation but, in certain cases, the particle may rest in an unstable position to give a lower value. Some definitions of particle size are given in Table 4.1

These diameters are usually applied to an assembly of particles and distributions are quoted in terms of the measured or derived diameters. Particles having the same diameter can have vastly different shapes, therefore one parameter should not be considered in isolation.

A single particle can have an infinite number of statistical diameters, hence these diameters are only meaningful when sufficient particles have been measured to give average statistical diameters in each size range.

For a single particle, the expectation of a statistical diameter and its coefficient of variation may be calculated from the following equations [1] (see figure 4.1):

$$E(d_R) = \frac{1}{\pi} \int_0^{2\pi} R d\theta_R \tag{4.1}$$

$$\sigma_R = \sqrt{\left[E(R^2) - \frac{E^2(R)}{E(R)} \right]} \tag{4.2}$$

$$E(d_F) = \frac{1}{\pi} \int_0^{\pi} d_F d\theta_F \tag{4.3}$$

$$\sigma_F = \sqrt{\left[E(d_F{}^2) - \frac{E^2(d_F)}{E(d_F)} \right]} \tag{4.3a}$$

$$E(d_M) = \frac{1}{\pi} \int_0^\pi d_M d\theta_M \tag{4.4}$$

$$\sigma_M = \sqrt{\left[E(d_M{}^2) - \frac{E^2(d_M)}{E(d_M)} \right]} \tag{4.4a}$$

Fig. 4.1 (a) Definition of unrolled diameter d_R and radius R; (b) unrolled curve [1].

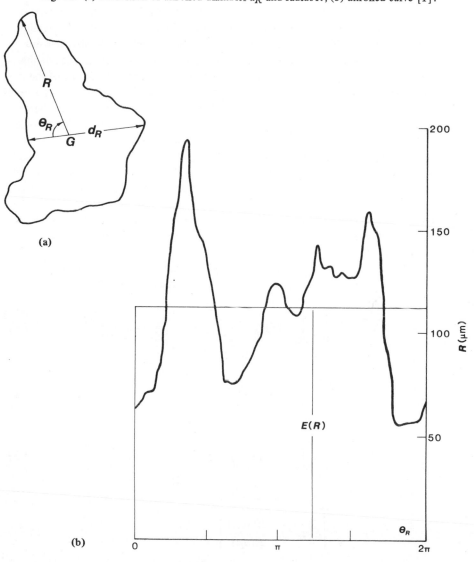

(a)

(b)

4.2 Particle shape

It is known that particle shape influences such properties as the flowability of powders, packing, interaction with fluids and the covering power of pigments, although little quantitative work has been carried out on these relationships. Qualitative terms may be used to give some indication of the nature of particle shape and some of these, extracted from the British Standard 2955: Glossary of Terms Relating to Powders, are given in Table 4.2.

Table 4.2 Definitions of particle shape

Acicular	needle-shaped
Angular	sharp-edged or having roughly polyhedral shape
Crystalline	freely developed in a fluid medium of geometric shape
Dentritic	having a branched crystalline shape
Fibrous	regularly or irregularly thread-like
Flaky	plate-like
Granular	having approximately an equidimensional irregular shape
Irregular	lacking any symmetry
Modular	having rounded, irregular shape
Spherical	global shape

Such general terms are inadequate for the determination of shape factors that can be incorporated as parameters into equations concerning particle properties where shape is involved as a factor. In order to do this, it is necessary to be able to measure and define shape quantitatively.

In recent years there has been an upsurge of interest in particle shape analysis using pattern recognition techniques [79–81] in which input data are categorized into classes. The potential use of these techniques [82] and the use of the decision function in morphological analysis have been introduced.

There are two points of view regarding the assessment of particle shape. One is that the actual shape is unimportant and all that is required is a number for comparison purposes. The other is that it should be possible to regenerate the original particle shape from the measurement data.

The numerical relations between the various 'sizes' of a particle depend on particle shape, and dimensionless combinations of the sizes are called shape factors. The relations between measured sizes and particle volume or surface are called shape coefficients.

4.2.1 Shape coefficients

There are two especially important properties of particles, surface and volume, and these are proportional to the square and cube respectively of some characteristic dimension. The constants of proportionality depend upon the dimension chosen to characterize the particle; the projected area diameter is used in the following discussion.

$$\text{Surface of particles, } S = \pi d_s^2 = \alpha_{s,a} d_a^2 = x_s^2 \qquad (4.5)$$

$$\text{Volume of particle, } V = \frac{\pi}{6}d_v^3 = \alpha_{v,\,a}d_a^3 = x_v^3 \tag{4.6}$$

where α_s and α_v are the surface and volume shape coefficients, the additional suffix denoting that the measured diameter is the projected area diameter. The symbol x denotes size, as opposed to diameter, and includes the shape coefficient. This artifact is found to be very useful for general treatment of data.

The surface area per unit volume (the volume-specific surface) is the ratio S to V:

$$S_V = \frac{\alpha_{sv,\,a}}{d_a} = \frac{6}{d_{sv}} = \frac{1}{x_{sv}} \tag{4.7}$$

Further, the volume-specific surface by microscopy is defined as:

$$S_{V,\,a} = \frac{6}{d_a} \tag{4.8}$$

One line of approach to the use of shape coefficients was given by Heywood [2] who recognized that the word 'shape' in common usage refers to two distinct characteristics of a particle. These two characteristics should be defined separately, one by the degree to which the particle approaches a definite form such as a cube, tetrahedron or sphere, and the second by the relative proportions of the particle which distinguish one cuboid, tetrahedron or spheroid from another of the same class.

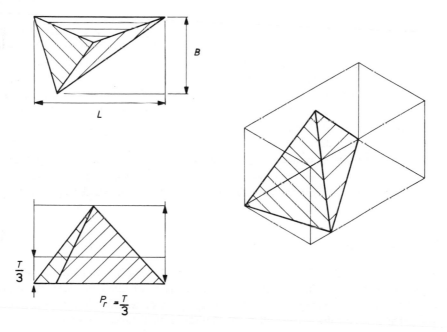

Fig. 4.2 Heywood's dimensions.

When three mutually perpendicular dimensions of a particle may be determined, Heywood's ratios [3] may be used:

$$\text{elongation ratio} \quad n = L/B \tag{4.9}$$

$$\text{flakiness ratio} \quad m = B/T \tag{4.10}$$

where

(a) the thickness T is the minimum distance between two parallel planes which are tangential to opposite surfaces of the particle, one plane being the plane of maximum stability.

(b) the breadth B is the minimum distance between two parallel planes which are perpendicular to the planes defining the thickness and are tangential to opposite sides of the particle.

(c) the length L is the distance between two parallel planes which are perpendicular to the planes defining thickness and breadth and are tangential to opposite sides of the particle.

Consider a particle circumscribed by a rectangular parallelepiped of dimensions L by B by T, then:

$$\text{projected area of the particle } A = \frac{\pi}{4} d_a^2 = \alpha_a BL \tag{4.11}$$

where α_a is the area ratio.

Volume of particle equals projected area by mean thickness:

$$\alpha_{v,a} d_a^3 = \alpha_a BL p_r T \tag{4.12}$$

where p_r is the prismoidal ratio (see figure 4.2).

Combining equations (4.11) and (4.12) gives:

$$\alpha_{v,a} = \frac{\pi\sqrt{\pi}}{8} \frac{p_r}{m\sqrt{\alpha_a}n} \tag{4.13}$$

If the particle is equidimensional, i.e. $B = L = T$ and $n = m = 1$, then the volume coefficient takes on a special value α_e where:

$$\alpha_e = \frac{\pi\sqrt{\pi}}{8} \frac{p_r}{\sqrt{\alpha_a}} \tag{4.14}$$

Thus, α_e may be used to define particle form. When the particle is not equidimensional, the appropriate value of $\alpha_{v,a}$ is $\alpha_e/m\sqrt{n}$ which substantiates the reasoning given earlier that shape is a combination of the proportions and geometrical form.

Heywood classified particles into tetrahedral, prismoidal, sub-angular and rounded. Values of α and p_r for these classes are given in Table 4.3 [4].

$\alpha_{v,a}$ can be calculated using equation (4.12) combined with direct observation to determine the shape group into which the particle fits and the values of m and n. This is practicable down to sizes as small as 5 μm by measurements on the number,

mean size, weight and density of closely graded fractions. Indeed, $\alpha_{v,a}$ may be determined directly by weighing a known number of particles of known mean size.

Table 4.3 Values of α and p_r for particles of various shapes

Shape group		α_a	p_r
Angular	{ tetrahedral	0.5–0.8	0.4–0.53
	{ prismoidal	0.5–0.9	0.53–0.9
Sub-angular		0.65–0.85	0.55–0.8
Rounded		0.72–0.82	0.62–0.75

$\alpha_{s,a}$ is more difficult to determine, but Heywood developed the following relationship on the basis of a large number of experimental measurements:

$$\alpha_{s,a} = 1.57 + C\left(\frac{\alpha_{ea}}{m}\right)^{4/3}\frac{n+1}{n} \tag{4.15}$$

in which C is constant depending upon geometrical form. Table 4.4 shows the values of α_{ea} and C for various geometrical forms and also for irregular particles.

Table 4.4 Values of α_{ea} and C for various geometrical forms and also for irregular particles

Shape group		α_{ea}	C
Geometrical forms			
tetrahedral		0.328	4.36
cubical		0.696	2.55
spherical		0.524	1.86
Approximate forms			
angular	{ tetrahedral	0.38	3.3
	{ prismoidal	0.47	3.0
sub-angular		0.51	2.6
rounded		0.54	2.1

4.2.2 Shape factors

If it is required to define the size of a particle by a single dimension, it is usual to do so by expressing the size in terms of one of the diameters defined in Table 4.1. The variation between these diameters increases as the particles diverge more from the spherical shape, and hence shape is an important factor in the correlation of sizing analyses made by various procedures.

One of the earliest defined shape factors is the sphericity (ψ_w) which was defined by Wadell [5–8] as:

$$\psi_w = \frac{\text{surface area of a sphere having the same volume as the particle}}{\text{surface area of the particle}}$$

$$\psi_w = \left(\frac{d_v}{d_s}\right)^2 \tag{4.16}$$

At low Reynolds number and with convex particles, the drag diameter equals the surface diameter and the Stokes diameter (d_{Stk}) is defined as:

$$d_{st} = \left[\frac{d_v^3}{d_s}\right]^{1/2}$$

$$= \psi_w^{1/4} \, d_v \tag{4.17}$$

Further, the surface-volume mean diameter is given by:

$$d_{sv} = \frac{d_v^3}{d_s^2}$$

$$= \psi_w \, d_v \tag{4.18}$$

For microscope analysis Laird [19] prefers the definition:

$$\psi_L = \frac{S_o}{S_p} \tag{4.19}$$

where S_o is the surface area of a sphere with a diameter equal to the equivalent diameter and S_p is the surface area of the particle computed from the measured surface area.

For rounded images, whose principal dimensions in two directions at right angles to each other are a and b, Heywood [2] quotes the semi-empirical formula for the equivalent diameter:

$$d_e = \left(\frac{4}{\pi} \times 0.77ab\right)^{1/2} \tag{4.20}$$

For rectangular images the following equation yields a result which is only 1% different from the one calculated above (4.19).

$$d_e = (ab)^{1/2} \tag{4.21}$$

Particles rest on microscope slides in the position of greatest stability. Cylindrical particles, of length kd where d is the cross-sectional diameter, would be expected to rest with the axis horizontal for $k > 1$ and vertical for $k < 1$. Laird found that this was so and that a region existed, $0.85 > k > 1.5$, where both orientations were adopted.

For discs or cylinders with $k < 0.85$:

$$\psi_L = \frac{2k}{1 + 2k} \tag{4.22}$$

For cylinders with $k > 1.5$:

$$\psi_L = \frac{2}{1 + 2k} \tag{4.23}$$

(see Table 4.5). Laird also determined sphericity from sieving and sedimentation studies.

Krumbein's [9, 10] definition of sphericity is:

$$\psi_K^3 = \left(\frac{C}{B}\right)\left(\frac{B}{L}\right)^2 \tag{4.24}$$

where L is the longest dimension of the particle, the breadth B is measured perpendicular to this, and C is the particle thickness. (Note: these definitions are different to Heywood's.)

Table 4.5 Relationship between shape coefficients and particle shape for cylindrical particles of unit diameter cross-section

k	d_v	d_s	d_{st}	$\psi_w^{-1} = \dfrac{\alpha_{sv,c}}{6}$	$\dfrac{\alpha_{sv,st}}{6}$	ψ_L^{-1}
0.125	0.572	0.791	0.485	1.912	1.620	5.00
0.25	0.721	0.866	0.658	1.443	1.316	3.00
0.50	0.909	1.000	0.867	1.210	1.156	1.50
1	1.145	1.225	1.107	1.145	1.107	—
2	1.442	1.581	1.377	1.202	1.148	2.50
4	1.817	2.121	1.682	1.363	1.262	4.50
8	2.289	2.915	2.028	1.622	1.437	8.50

Hausner [12] proposed a method of assessing particle shape by comparing the particle with an enveloping rectangle of minimum area. If the rectangle length is a and its width is b, three characteristics are defined:

The elongation ratio $x = a/b$ $\tag{4.25}$

The bulkiness factor $y = A/ab$ $\tag{4.26}$

The surface factor $z\quad = C^2/12.6A$ $\tag{4.27}$

where A is the projected area of the particle and C is its perimeter.

Medalia [13] represents the particle in three dimensions as an ellipsoid with radii of gyration equal to those of the particle and defines an anisometry in terms of the ratios of the radii.

Church [14] proposed the use of the ratio of the expected values (see equations 4.3 and 4.4) of Martin's and Feret's diameters as a shape factor for a population of elliptical particles. Cole [15] introduced an image-analysing computer (the Quantimet 720) to compare longest chord, perimeter and area for large numbers of particles. Many other methods have been proposed and reviewed by Pahl *et al.* [16], and Davies [17], Beddow [18] and Laird [19].

4.2.3 Applications of shape factors and shape coefficients

If an analysis is carried out by two different techniques, the two results can be brought into coincidence by multiplying by a shape factor provided that particle shape does not change with particle size. For example, if the median size by Coulter analysis is 32 μm and by gravitational sedimentation is 29.2 μm, multiplying the sedimentation (Stokes) diameters by [32/29.2] will yield the Coulter distribution.

This ratio is in itself a shape factor, but it can be extended by writing the alternative form of Stokes' diameter (see Table 4.1):

$$\left(\frac{d_{Stk}}{d_c}\right)^2 = \frac{d_v^3}{d_s} \frac{1}{d_v^2} = \frac{d_v}{d_s} = \left(\frac{29.2}{32}\right)^2$$

Hence, from equation (4.16):

$$\psi_w = 0.693$$

If microscopic examination reveals the form of the particles to be cylindrical, for example, of length kd where d is the cross-sectional diameter, then the relationship between k, α_{sv} and ψ can be found since:

$$S = (\tfrac{1}{2} + k)\pi d^2 = \pi d_s^2$$

$$V = \frac{\pi}{4} kd^3 = \frac{\pi}{6} d_v^3$$

Hence, the Coulter diameter, $d_c = d_v = 3d \sqrt{(\tfrac{3}{2}k)}$

the surface diameter, $d_s = d\sqrt{(k + 0.5)}$

the Stokes diameter, $d_{Stk} = \sqrt{\dfrac{d_v^3}{d_s}} = \left(\dfrac{9k^2}{2(1 + 2k)}\right)^{1/4} d$

the sphericity, $\psi_w = \dfrac{(18k^2)^{1/3}}{(1 + 2k)}$

the surface-volume shape coefficient from sedimentation data,

$$\alpha_{sv, Stk} = \left(\frac{2 + 4k}{k}\right)\left(\frac{d_{st}}{d}\right) = \left[\frac{3}{k}\right]^{1/2} \left[2(1 + 2k)\right]^{3/4}$$

the surface-volume shape coefficient from Coulter data,

$$\alpha_{sv, c} = \left(\frac{2 + 4k}{k}\right)\left(\frac{d_v}{d}\right)$$

$$= \left[\frac{3}{2k^2}\right]^{1/3} (2 + 4k)$$

It can be shown that:

$$\frac{\alpha_{sv,c}}{6} = \frac{1}{\psi_w}$$

Data, for unit diameter cylinders, are presented in Table 4.5. Data for other shapes are presented in Tables 4.6 and 4.7. From these it can be seen that $\alpha_{sv}/6$ and ψ_w^{-1} are both at a minimum when the shape is most compact ($k = 1$) and increase as the particles become rod-shaped ($k > 1$) or flaky ($k < 1$).

For the numerical illustration, when $\psi^{-1} = 1.44$, $\alpha_{sv,c}/6 = 1.32$ and $k = 0.25$ or 5.5. A visual examination will reveal whether the particles are flaky or rod-shaped.

Table 4.6 Calculated values of shape coefficients

Form	Proportions	Linear dimension used as d_r	$\alpha_{s,r}$	$\alpha_{v,r}$	$\alpha_{sv,r}$
Sphere		diameter	3.14	0.52	6.00
Spheroid	$1:1:2$	minor axis	5.37	1.05	5.13
	$1:2:2$	minor axis	8.67	2.09	4.14
	$1:1:4$	minor axis	10.13	2.09	4.83
	$1:4:4$	minor axis	28.50	8.38	3.40
Ellipsoid	$1:2:4$	shortest axis	15.86	4.19	3.79
Cylinder	height = diameter	diameter	4.71	0.79	6.00
	height = 2 diameters	diameter	7.85	1.57	5.00
	height = 4 diameters	diameter	14.14	3.14	4.50
	height = $\frac{1}{2}$ diameter	diameter	3.14	0.39	8.00
	height = $\frac{1}{4}$ diameter	diameter	2.36	0.20	12.00

Table 4.7 Measured values of surface-volume shape coefficient

Material	Approximate sizes	α_{sv}	Specific surface method	Particle size method
Alumina	15–45	16		
Coal	15–90	12–17	permeametry	microscope (d_a)
Dolomite	25–45	11		
Silica	15–70	11		
Tungsten carbide	15–45	14		
Coal	15–90	10–12	permeametry	Coulter counter (d_v)
Silica	15–70	9		
Coal	0.5–10	9–11	light extinction	weight count (d_v)
Diamond	0.5–12	8		
Quartz	0.5–10	9		

Tables 4.6 and 4.7 are from BS 4359 (1970): Part 3, reproduced by permission of the British Standards Institution, 2 Park Street, London W.1, from whom copies of the complete standard may be obtained.

Example
Consider two cuboids of similar shape but different sizes, i.e. side lengths $1:2:3$ and $2:4:6$:

maximum projected areas	$6 + 24; A = 30$
total surface areas	$22 + 88; S = 110$
total volume	$6 + 48; V = 54$

projected area diameter $\quad A = \dfrac{\pi}{4}\, n\bar{d}_a^2$ $\qquad ;\quad \bar{d}_a \;\; = 4.36$

mean surface diameter $\quad S = \pi n\bar{d}_s^2$ $\qquad ;\quad \bar{d}_s \;\; = 4.18$

surface shape coefficient $\quad S = \alpha_{s,a} n\bar{d}_a^2$ $\qquad ;\quad \alpha_{s,a} = 2.9$

mean volume diameter $\quad V = \dfrac{\pi}{6}\, n\bar{d}_v^3$ $\qquad ;\quad \bar{d}_v \;\; = 3.72$

volume shape coefficient $\quad V = \alpha_{v,a} n\bar{d}_a^3$ $\qquad ;\quad \alpha_{v,a} = 0.326$

surface-volume mean diameter $\quad d_{sv} = \dfrac{d_v^3}{d_s^2}$ $\qquad ;\quad \bar{d}_{sv} = 2.94$

volume-specific surface $\quad S_v = \dfrac{S}{V} = \dfrac{6}{d_{sv}}$ $\qquad ;\quad S_v \;\; = 2.04$

volume-specific surface by microscopy $\quad S_{v,a} = \dfrac{6}{\bar{d}_a}$ $\qquad ;\quad S_{v,a} = 1.37$

(See BS 4359 (1970): Part 3, for further examples.)

The surface-volume shape coefficient has been determined for quartz and silica from surface area measurements using nitrogen adsorption giving $14 < \alpha_{sv} < 18$ with no significant variation with particle size. Fair and Hatch [21] found that by measuring smoothed surfaces, values of $\alpha_{sv,a}$ as low as 7 were found ($\alpha_{sv,a} = 6$ for spheres). Crowl [22] found that with Prussian blue, the specific surface by nitrogen adsorption applied to the primary particles of which each single particle is made up (see figure 4.3). With red iron oxide, however, the specific surface applied to the single primary particle.

The mean volume shape may be determined from a knowledge of the number, mean size, weight and density of the particles composing a fraction graded between close limits. Further, if the surface area is determined by permeametry, a surface shape may be evaluated though this will differ from that obtained from the gas adsorption surface area. Hence, when any shape is quoted, the method of obtaining it should also be given.

Ellison [23] obtained the value 0.9 for the ratio of the sizes of silica particles determined by settling experiments and mounted in agar in random orientation. Hodkinson

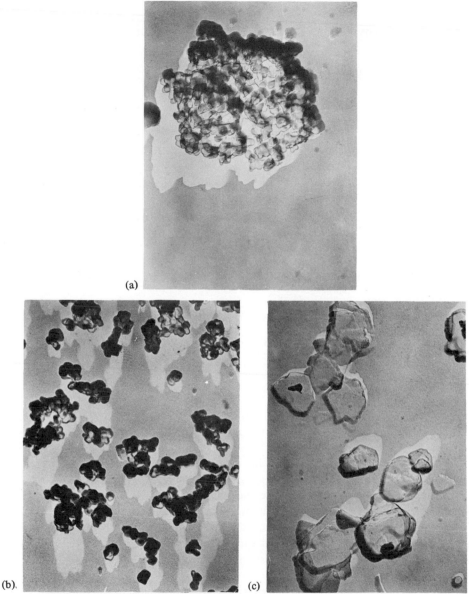

(a)

(b)

(c)

Fig. 4.3 Electron photomicrographs of two paint pigments, showing how particles can be aggregates of finer particles (Crowl [22]). (a) A single particle of Prussian blue about 1 μm in diameter. The nitrogen adsorption surface area is 61.3 m^2 g^{-1} from which the surface-volume mean diameter is 0.051 μm. This is seen to be the diameter of the individual primary particles of which the aggregate is made up. Similarly, the micronized Prussian blue (b) has approximately the same surface-volume mean diameter. (c) With the red oxide the diameter is 0.21μm which is approximately the same as the solid particle seen in the micrograph.

[24] found, from measurements of quartz particles by light scattering, a diameter ratio of 0.8 between particles in a liquid suspension and settled particles. Cartwright [57] attempted to find the magnitude of the differences in mean projected diameters between particles of quartz in random and stable orientation by microscopy. He used four different mounting techniques and found no significant differences. This, he attributed to the difficulty in mounting particles in random orientation. These factors for the mean ratio of projected diameter for random and stable orientation are indicative of the properties of the powder and are therefore of use to the analyst.

Respirable coal mine dust samples from three different US mines were classified into four fractions using a Bahco [68]. Shape factors were determined as ratios of the following diameters from microscopy:

$$\bar{d}_a = \left[\frac{\Sigma \, nd_a^2}{\Sigma \, n} \right]^{1/2}$$

A_p from photosedimentation with correction for extinction coefficient, N_w from microscopy:

$$\bar{d}_p = \left[\frac{4A_p}{\pi N_w} \right]^{1/2}$$

$$\left(A_p = \frac{\pi}{4} \, \Sigma \, nd_p^2 \right)$$

S_w from krypton gas adsorption, N_w from microscopy:

$$\bar{d}_{BET} = \left[\frac{S_w}{\pi N_w} \right]^{1/2}$$

p from density measurement, N_w from microscopy:

$$\bar{d}_{sv} = \frac{3}{2\rho A_p}$$

where N_w = number of particles/unit weight

A_p = projected area of particles, in random orientation, per unit weight. See Table 4.1 for definitions of other symbols.

It was postulated that a relationship might exist between shape and the incidence of pneumoconiosis.

A method, for shape determination, has been described in which the image analysis system consists of an Optronics scanning microdensitometer and a computer program [58]. It performs quantitative measurements of particle size distribution, shape and edge texture from photographs, where shape is defined as perimeter divided by projected area. A full analysis takes about three and a half man-hours and its most successful application has been in the analysis of silica sands in deep bed air filters for the removal of radioactive particulates. A relationship was found between collection efficiency and edge texture [59].

A device has been described to measure the degree of equi-axial deviation of powder particles as a ratio between longitudinal and transverse axes [69]. The determination is based on measuring the flow of rarified gas in two mutually perpendicular directions.

Kaye [70] describes a method using spatial filtering by which a particular particle of known shape and size can be detected among a background of other particles.

An examination of the shape distribution of sugar crystals has also been carried out using a modified Zeiss-Endter particle size analyser [71]. Shape was defined by Wadell's shape factor and by length to width ratio.

4.2.4 Shape indices

Tsubaki and Jimbo [1, 25, 26] argue that many of the proposed shape factors had little practical applicability to the analysis of real powders until the advent of electronic techniques and the computer. They define six shape indices based on ratios of the following diameters (see Table 4.1): d_a, d_c, d_F, d_R. The shape indices are: $\psi_{ac}, \psi_{aF}, \psi_{aR}, \psi_{cF}, \psi_{cR}, \psi_{FR}$, where, for example, $\psi_{ac} = d_a/d_c$. They defined the statistical diameters and coefficients of variation according to equations (4.1) to (4.4).

Furthermore, the elongation was used for study.

$$Z = d_{F\pi/2}/d_{Fmin} \tag{4.28}$$

where $d_{F\,min}$ = the minimum value of the Feret diameter and $d_{F\pi/2}$ = the diameter perpendicular to this.

They later added three more indices $\psi_{\bar{F}a}$, ψ_{Stkv} and κ. The arithmetic average of breadth and length is defined as follows:

$$d_{\bar{F}} = \tfrac{1}{2}\left(d_{Fmin} + d_{F\pi/2}\right) \tag{4.29}$$

The dynamic shape factor κ is defined as the ratio of the resistance to motion of a given particle divided by the resistance of a spherical particle of the same volume. When the particle is settling under laminar flow conditions:

$$\kappa = \left(\frac{d_v}{d_{Stk}}\right)^2 \tag{4.30}$$

Since this is a squared term which for comparison purposes they wished to reduce to unit power they introduced the shape factor $\psi_{Stkv} = d_{Stk}/d_v$.

[Note $\kappa = \psi_w^{-1/2}$ where ψ_w is Wadell's sphericity factor.]

For non-re-entrant particles, according to Cauchy's theorem, $d_a = E(d_F)$ and ψ_{Fa} has a maximum value of 1.0 for circles, rectangles and other convex shapes; it is therefore very useful for indicating the extent of concavities.

$\psi_{Ra}, \psi_{aF}, \psi_{RF}, \sigma_R, \sigma_F$ and Z were found to mainly show the slimness of the particles, the best indicators being σ_F and Z in that order.

ψ_{Hc} and ψ_{Rc} were found to correspond poorly with particle morphology.

4.2.5 Shape regeneration by Fourier analysis

Briefly, this method consists of finding the particle outline and its centre of gravity from which a polar co-ordinate system is set up. Standard Fourier transform techniques are then used to generate the Fourier coefficients A_n and their associated angles θ_n.

In 1969 Schwarcz and Shane [27] published a paper where Fourier transforms were used to analyse beach sand silhouettes. In 1969 and 1970 Meloy [28, 29] presented papers in which fast Fourier transforms were used to process particle silhouette as a signal and this work was extended in 1977 [30, 31]. One of their main conclusions is that particles have 'signatures' which depend on A_n and not on θ_n and they propose the equation:

$$A_n = A_I \left(\frac{1}{n}\right)^S \tag{4.31}$$

A plot of $\log A_n$ against $\log n$ yields a straight line which has a slope S which depends on particle shape, rounder particles having lower slopes. Beddow [18] showed how a number of particle silhouette shapes could be analysed and reproduced by Fourier transforms. Gotoh and Finney [32] proposed a mathematical method for expressing a single, three-dimensional body by sectioning as an equivalent ellipsoid having the same volume, surface area and average projected area as the original body.

4.2.6 Fractal dimension characterization of textured surfaces

Mandelbrot has introduced the term 'fractal dimension' to describe curves which have no unique perimeter. In his book [33] he describes how Richardson estimated the length of the coast line of various countries by stepping along a map of the coast line with a pair of dividers. He found that the estimated perimeter (P_λ) tends to increase without limit as the step size λ decreases. He concluded that a polygon of side n (n = number of steps) approximating a coast line would have $k\lambda^{1-D}$ sides. The perimeter estimate would be:

$$P_\lambda = n\lambda \tag{4.32}$$

$$P_\lambda = k\lambda^{1-D} \tag{4.33}$$

Hence a plot of $\log P_\lambda$ against $\log \lambda$ will have a slope $1 - D$. The parameter D was found to be characteristic of the particular coast line measured and was called by Mandelbrot the fractal.

Kaye [34] demonstrated that this technique could be applied to the determination of the ruggedness of carbon black agglomerates. Another procedure he utilized was to study the frequency of intersections made by the boundary of the profile with a grid placed at random under the profile using grids with a range of grid sizes. A third procedure he adopted was to determine the number of squares on the particle profile, accepting those that were largely inside the profile and rejecting others. The efficiencies of the three techniques were found to be in the order that they are presented above.

His main conclusion was that this technique was too time-consuming for manual operation and he recommended the use of image-analysing computers, a recommendation that was taken up by Flook [35] using a Quantimet 720. Flook considered the perimeter to be made up of a series of closely spaced points. A series of circles of radius λ is drawn up with their centres on each of the points in turn thus describing a path of width 2λ covering the curve. The area of the path divided by its width gives an estimate of its length. The process is repeated with decreasing λ.

This method was applied to a circle which, being a Euclidean curve, has a fractal equal to unity. Two mathematical models examined consisted of geometrically constructed islands taken from Mandelbrot's book. The first was a Triadic Koch Island and the second a Quadric Koch Island (figure 4.4). Measurements were also made on a typical carbon black aggregate and a simulated carbon floc, taken from a paper by Medalia [36]. Excellent agreement was found between experimental and theoretical results.

Fig. 4.4 (a) A Triadic Koch Island.

(b) A Quadric Koch Island [Reproduced from 84].

4.3 Determination of specific surface from size distribution data

4.3.1 Number distribution

Let particles of size $d_{x,r}$ constitute a fraction m_r of the total number N so that:

$$m_r = \frac{n_r}{N} \quad \text{and} \quad N = \Sigma n_r$$

$$S = \alpha_{s,x} \quad N \Sigma m_r d_{x,r}^2$$

$$V = \alpha_{v,x} \quad N \Sigma m_r d_{x,r}^3$$

$$S_v = \alpha_{sv,x} \quad \frac{\Sigma m_r d_{x,r}^2}{\Sigma m_r d_{x,r}^3} \tag{4.34}$$

4.3.2 Surface distribution

Let particles of size $d_{x,r}$ constitute a fraction t_r of the total surface S so that:

$$t_r = \frac{S_r}{S} \quad \text{and} \quad S = \Sigma S_r$$

$$St_r = \alpha_{s,x} \quad Nm_r d_{x,r}^2$$

$$V = \frac{S}{\alpha_{sv,x}} \quad \Sigma t_r d_{x,r}$$

$$S_v = \frac{\alpha_{sv,x}}{\Sigma t_r d_{x,r}} \tag{4.35}$$

4.3.3 Volume distribution

Let particles of size $d_{x,r}$ constitute a fraction q_r of the total volume V so that:

$$q_r = \frac{V_r}{V} \quad \text{and} \quad V = \Sigma V_r$$

$$Vq_r = \alpha_{v,x} \quad Nm_r d_{x,r}^3$$

$$S_v = \alpha_{sv,x} \quad \Sigma \frac{q_r}{d_{x,r}} \tag{4.36}$$

Equation (4.34) is used whenever a number count is taken and yields a specific surface if the shape coefficient is known and vice-versa. If α_{sv} is assumed equal to 6 (this is assuming spherical particles) a specific surface is obtained. For microscope counting by projected area this is written $S_{v,a}$.

Similar arguments apply to equations (4.35) and (4.36). The latter is frequently applied to sieve analyses where q_r is the fractional weight residing between two sieves of average aperture $d_{A,r}$.

4.4 Particle size distribution transformation between number, surface and mass

There are several methods of particle size measurement in which the raw data are collected in the form of a number distribution: microscopy, electrical and light-sensing zone methods are three widely used examples.

Transition is simple if one obtains a straight line on log-probability paper since the graphs for surface and weight are straight lines, parallel to the number distribution, but with the 50% value, corresponding to the surface mean x_{gs} and volume mean x_{gv} displaced (equations 4.129 and 4.135).

Considerable difficulties are presented if the log-normal law does not hold. The obvious way of calculating the proportion of surface or volume corresponding to particular particle diameters consists of weighting by the corresponding squared or cubed diameters [76]. The results are then summed to yield the cumulative surface or weight distribution. Conversion from number to surface or surface to volume is best carried out

graphically, in order to smooth out experimental errors, as described in the section on photosedimentation. Nomograms have been developed to reduce the tedium and possible calculation errors involved in number to weight conversion [72, 73]. However, it is not in general likely that the conversion from number to weight will be accurate on the following grounds:

(1) The weighted proportions are too sensitive to sample fluctuations since the method requires the raising of experimental errors to higher powers. (In a 1 to 10 μm distribution an error of one 10-μm particle has the same weighting as an error of a thousand 1-μm particles.)

(2) The grouping of the observed diameters into frequency intervals, which in practice may be broad and non-uniform, makes an accurate estimation of the average for each size range impossible.

The mean diameter for each range is usually taken as the arithmetic mean ($\bar{x} = \frac{x_1 + x_2}{2}$ or less frequently the geometric mean $\bar{x} = \sqrt{(x_1 \, x_2)}$.

Thus, the fraction by weight of particles larger than D_r is given by:

$$W_r = \sum_{D_r}^{D_{max}} D^3 dN \qquad (4.37)$$

where $\sum\limits_{D_{min}}^{D_{max}} D^3 dN = 1$, i.e. the distribution is normalized.

Integration by parts, however, yields the following equivalent relationship [74, 75]:

$$W_r = D_r^3 N_r + 3 \sum_{D_r}^{D_{max}} D^2 dNdD$$

$$= D_r^3 N_r + 3 \left[(D_r^2 N_r + D_{r+1}^2 N_{r+1}) \left(\frac{D_{r+1} - D_r}{2} \right) \right.$$

$$+ (D_{r+1}^2 N_{r+1} + D_{r+2}^2 N_{r+2}) \left(\frac{D_{r+2} - D_{r+1}}{2} \right)$$

$$\left. + \ldots \right] \qquad (4.38)$$

This is applied in Table 4.8 and the results are compared with weighting with cubed diameters.

In using this method the particle counts may be classified into fairly broad intervals thereby reducing experimental time. A computer program for carrying out this conversion has also been published [77]. The conversion is much more accurate, however, if the procedure laid down in the chapter on microscopy is followed. The errors are also reduced with electronic counting devices where the total count is large and the size intervals are small.

Conversion from weight to surface (e.g. sieving) may be carried out using the following expression:

$$\Delta S = \sum_{r-1}^{n+1} \frac{\Delta W_r}{d_{Ar}}$$
(4.39)

where ΔS is the relative surface of a weight of powder W_r residing between sieve sizes d_{Ar+1} and d_{Ar-1}.

Table 4.8 Conversion from a number to a weight distribution

Particle size (D)	Cumulative number frequency oversize	Cumulative weight frequency oversize	Cumulative weight percentage oversize	$\Delta N . D^{-2}$	$\%$ $\Sigma N D^3$
(μm)	(N)	($W \div 5^3$)			($\div 2.5$)
$50 = D_m$	$0 = N_m$	$0 = W_m$	0	.0	
$45 = D_{m-1}$	$1 = N_{m-1}$	$850.5 = W_{m-1}$	6.6	.6860	6.4
40	2	1459	11.3	.4913	11.0
35	3	1876.5	14.6	.3375	14.2
$30 = D_r$	$7 = N_r$	3021	23.6	.8788	22.4
25	16	4487	35.0	.11979	33.6
20	50	7487	58.4	.24786	56.7
15	112	10023	78.2	.21266	76.6
10	261	12165	95.0	.18625	94.0
5	380	12593	98.3	3213	97.0
1	400	12810	100.0	41	100.0

(1)　$W_m = 0$

(2)　$W_{m-1} = 45^3 \times 1 + 3 \left[(45^2 \times 1 + 0) \left(\frac{50 - 45}{2} \right) \right]$

$= 5^3 \left[9^3 \times 1 + 3(9^2 \times 1 + 0) \left(\frac{10 - 9}{2} \right) \right] = (729 + 121.5) \times 5^3 = 850.5 \times 5^3$

(3)　$W_{m-2} = 5^3 \left[8^3 \times 2 + 3(8^2 \times 2 + 9^2 \times 1) \left(\frac{9 - 8}{2} \right) \right] = (1024 + 313.5 + 121.5) \times 5^3$

$= 1459 \times 5^3$

and so on.

4.5 Average diameters

The purpose of an average is to represent a group of individual values in a simple and concise manner in order to obtain an understanding of the group. It is important, therefore, that the average should be representative of the group. All average diameters

are a measure of central tendency which is unaffected by the relatively few extreme values in the tails of the distribution. Some of these are illustrated in Tables 4.9 and 4.10 and figures 4.5 and 4.6.

Table 4.9 Cumulative percentage undersize distribution

Particle size (μm)	Cumulative percentage undersize
x_2	$\phi = \sum\limits_{0}^{x} d\phi$
5	1.4
9	9.4
11	18.0
14	32.0
17	49.5
20	64.0
23	76.0
28	88.0
33	94.0
41	98.0
50	99.4
60	99.9

ϕ, the frequency function $= \Sigma dN$ for a number distribution
$= \Sigma x dN$ for a size distribution
$= \Sigma x^2 dN$ for an area distribution
$= \Sigma x^3 dN$ for a volume or weight distribution where dN
is the percentage of the total number of particles lying in the size range x_1 to x_2.

The most commonly occurring value, the mode, passes through the peak of the relative frequency curve, i.e. it is the value at which the frequency density is a maximum. The median line divides the area under the curve into equal parts, i.e. it is the 50% size on the cumulative frequency curve. The vertical line at the mean passes through the centre of gravity of a sheet of uniform thickness and density cut to the shape of the distribution. Hence, for the mean, the moment of the sum of all the elementary areas of thickness δx about the ordinate equals the sum of all the moments:

$$\bar{x} \sum \frac{d\phi}{dx} \delta x = \sum x \frac{d\phi}{dx} \delta x$$

$$\bar{x} = \frac{\Sigma x d\phi}{\Sigma d\phi} \tag{4.40}$$

For a weight distribution $d\phi = x^3 dN$ giving:

$$\bar{x} = \frac{\Sigma x^4 dN}{\Sigma x^3 dN} \tag{4.41}$$

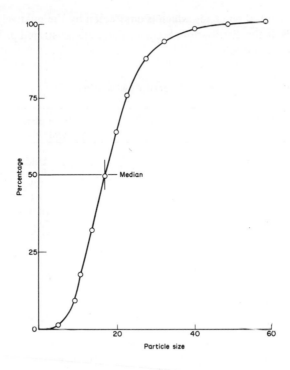

Fig. 4.5 The cumulative percentage frequency curve.

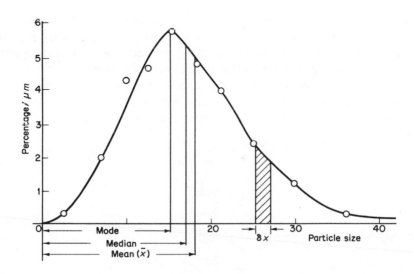

Fig. 4.6 The relative percentage frequency curve.

The mode and the median may be determined graphically but the above summation has to be carried out for the determination of the mean. However, for a slightly skewed distribution, the approximate relationship mean − mode = 3 (mean − median) holds. For a symmetrical distribution, they all coincide. In the illustration, the values are: mode = 15.0; median = 17.2; yielding mean = 18.2, as compared with the summed value of 18.47 (see Table 4.10).

Table 4.10 Relative percentage frequency distribution: tabular calculation of mean size

Particle size range x_1 to x_2	Interval dx	Average size x	Percentage in range $d\phi$	Percentage per micrometre $d\phi/dx$	$xd\phi$
0 to 5	5	2.5	1.4	0.3	4
5 to 9	4	7.0	8.0	2.0	56
9 to 11	2	10.0	8.6	4.3	86
11 to 14	3	12.5	14.0	4.7	175
14 to 17	3	15.5	17.5	5.8	271
17 to 20	3	18.5	14.5	4.8	268
20 to 23	3	21.5	12.0	4.0	258
23 to 28	5	25.5	12.0	2.4	306
28 to 33	5	30.5	6.0	1.2	183
33 to 41	8	37.0	4.0	0.5	148
41 to 50	9	45.5	1.4	0.2	64
50 to 60	10	55.0	0.5	0.1	28

$$\Sigma xd\phi \quad 1847$$

$$\text{Mean size} = \frac{\Sigma xd\phi}{\Sigma d\phi} = 18.47$$

The characteristics of a particle distribution are its total number, length, area and volume. A system of unequally sized particles may be represented by a system of uniformly sized particles having two, and only two, characteristics of the original distribution. The size of the particles in the uniform system is then the mean size of the non-uniform system with respect to these two properties.

The sizes may be expressed mathematically by dividing the system of particles into small intervals of size δx with assumed diameters of $x_1, x_2, \dots$. The symbol x is used because the method of measurement for individual particles is not specified, and all particles are assumed to have the same shape. Let the numbers of particles in these groupings be $\delta N_1, \delta N_2 \dots$ respectively. Then the aggregate length, surface and volume of the particles in each grouping are $x\delta N$, $x^2 \delta N$ and $x^3 \delta N$ and the total for the system is the summation of these expressions. Table 4.11 is a summary of the mathematical expressions for the various mean diameters [3].

The method of sizing may also be incorporated into the symbol. Hence, for particle sizing by microscopy, the arithmetic mean diameter becomes $d_{a, NL}$. The surface-volume diameter calculated from the results of a sedimentation experiment is $d_{Stk,sv}$.

The mean value of a cumulative weight percentage curve obtained by sieving would be $d_{A,\,vm}$ or $d_{A,\,wm}$.

Table 4.11 Definitions of mean diameters

Number, length mean diameter	x_{NL}	$= \dfrac{\Sigma dL}{\Sigma dN} = \dfrac{\Sigma x dN}{\Sigma dN}$
Number, surface mean diameter	x_{NS}	$= \sqrt{\left(\dfrac{\Sigma dS}{\Sigma dN}\right)} = \sqrt{\left(\dfrac{\Sigma x^2 dN}{\Sigma dN}\right)}$
Number, volume mean diameter	x_{NV}	$= \sqrt[3]{\left(\dfrac{\Sigma dV}{\Sigma dN}\right)} = \sqrt[3]{\left(\dfrac{\Sigma x^3 dN}{\Sigma dN}\right)}$
Length, surface mean diameter	x_{LS}	$= \dfrac{\Sigma dS}{\Sigma dL} = \dfrac{\Sigma x^2 dN}{\Sigma x dN}$
Length, volume mean diameter	x_{LV}	$= \sqrt{\left(\dfrac{\Sigma dV}{\Sigma dL}\right)} = \sqrt{\left(\dfrac{\Sigma x^3 dN}{\Sigma x dN}\right)}$
Surface, volume mean diameter	x_{SV}	$= \dfrac{\Sigma dV}{\Sigma dS} = \dfrac{\Sigma x^3 dN}{\Sigma x^2 dN}$
Volume, moment mean diameter	x_{VM}	$= \dfrac{\Sigma dM}{\Sigma dV} = \dfrac{\Sigma x^4 dN}{\Sigma x^3 dN}$
Weight, moment mean diameter	x_{WM}	$= \dfrac{\Sigma dM}{\Sigma dW} = \dfrac{\Sigma x dW}{\Sigma dW} = \dfrac{\Sigma x^4 dN}{\Sigma x^3 dN}$

Consider the system illustrated (figure 4.7) consisting of one particle of size 1, 2, 3, 4, 5, 6, 7, 8, 9, 10. Hence ten particles, each of length 5.50, will have the same total length as the original distribution (Table 4.12). Similarly ten particles, each of length 6.21, will have the same total surface as the original distribution.

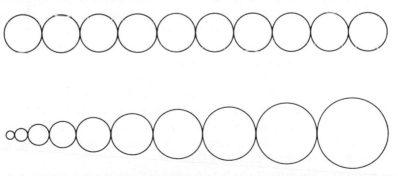

Fig. 4.7 The homogeneous distribution that represents in number and length a heterogeneous distribution of ten particles of size 1 to 10 with unit separation in size.

Each of these mean diameters characterizes the original distribution in two properties only. For example, the length-surface mean diameter is 7.00. Therefore, the uniform system contains $N = (L/x) = (S/x^2)$, which is 7.87 in each case. Hence, the uniform system consists of 7.87 particles, each of length 7.00. Thus, the total length and the total surface of the particles are the same as in the original distribution, but the total number, volume and moment are all different, e.g. $V = x^3 N = 343 \times 7.87 = 2700$.

Table 4.12

x_{NL} = 5.50	x_{LS} = 7.00	x_{VM} = 8.37	ΣdV = 3025
x_{NS} = 6.21	x_{LV} = 7.43	ΣdL = 55	ΣdM = 25335
x_{NV} = 6.71	x_{SV} = 7.87	ΣdS = 385	

The arithmetic mean is the sum of the diameters of the separate particles divided by the number of particles, it is most significant when the distribution is normal:

$$x_A = x_{NL} = \frac{\Sigma x dN}{\Sigma dN} \qquad (4.42)$$

The geometric mean is the nth root of the product of the diameters of the n particles examined; it is of particular value with log-normal distributions:

$$x_g = (\Pi x^{dN})^{1/N} \qquad (4.43)$$

$$N \log x_g = \Sigma dN \log x$$

$$\log x_g = \frac{\Sigma dN \log x}{N} \qquad (4.44)$$

The harmonic mean is the number of particles divided by the sum of the reciprocals of the diameters of the individual particles; this is related to specific surface and is of importance where surface area of the sample is concerned [37].

$$x_H = \frac{\Sigma dN}{\Sigma dN/x} \qquad (4.45)$$

4.6 Particle dispersion

The spread of the distribution data may be expressed in terms of the *range*, i.e. the difference between the minimum and maximum sizes; the interquartile range $(_{25}x_{75})$, i.e. the difference between the 75% and 25% sizes; an interpercentile range, e.g. $(_{20}x_{80})$, the difference between the 80% and 20% sizes or the standard deviation.

The least significant of these is the first, since a stray oversize or undersize particle can greatly affect its value. The most significant is the last, since every particle has a weighting which depends on the difference between its size and the mean size.

The standard deviation (σ) is defined as:

$$\sigma = \sqrt{\left(\frac{\Sigma(x-\bar{x})^2 \Delta\phi}{\Sigma \Delta\phi}\right)} \tag{4.46}$$

Hence:

$$\sigma^2 = \frac{\Sigma x^2 \Delta\phi}{\phi} - \bar{x}^2 \tag{4.47}$$

where σ^2 is called the variation.

4.7 Methods of presenting size analysis data

An example of the tabular method of presenting size distribution data is shown in Table 4.13. The significance of the distribution is more easily grasped when the data are presented pictorially, the simplest form of which is the histogram. The data in Table 4.13 give the size grading of 1000 particles in 12 class intervals which are in a geometric progression. The choice of class widths is of fundamental importance, the basic requirement being that the resolution defined as the class interval divided by the mean class size should be kept fairly constant. With narrowly classified powders, an arithmetic distribution is acceptable but it is more normal to use a geometric progression.

Table 4.13 Size data

Particle size range x_2 to x_1	Interval dx	Average size x	Number frequency in range dN	Percentage in range $d\phi$	Percentage per micron $\frac{d\phi}{dx}$	$\frac{d\phi}{d \log x}$
1.4 to 2.0	0.6	1.7	1	0.1	0.2	1
2.0 to 2.8	0.8	2.4	4	0.4	0.5	3
2.8 to 4.0	1.2	3.4	22	2.2	1.8	15
4.0 to 5.6	1.6	4.8	69	6.9	4.3	46
5.6 to 8.0	2.4	6.8	134	13.4	5.6	89
8.0 to 11.2	3.2	9.6	249	24.9	7.8	167
11.2 to 16.0	4.8	13.6	259	25.9	5.4	173
16.0 to 22.4	6.4	19.2	160	16.0	2.5	107
22.4 to 32.0	9.6	27.2	73	7.3	0.8	49
32.0 to 44.8	12.8	38.4	21	2.1	0.2	14
44.8 to 64.0	19.2	54.4	6	0.6	0.0	4
64.0 to 89.6	25.6	76.8	2	0.2	—	1

$N = 1000$

where $y = \dfrac{d\phi}{dx}$ and $d\phi = 100\dfrac{dN}{N}$

Consider an analysis of a sub-sieve powder. For an arithmetic progression of sizes, let the intervals be 2.5 to 7.5, 7.5 to 12.5, and so on, to 67.5 to 72.5 μm. The resolution

will then vary from 1 to 0.071 as the particle size increases. A geometric progression with the same number of size intervals is 0.14 to 1.18, 1.18 to 1.68, 1.68 to 2.36, and so on, to 53.7 to 75.5 with geometric means of 1, $\sqrt{2}$, $2\sqrt{2}$, 4 to 64. The resolution for each size range is constant at 0.34. If there is a constant error in defining the class intervals, say, a 1 μm undersizing, the effect of this error will be dependent on the size with an arithmetic progression being greater for small particles, whereas with a geometric size interval, the effect is independent of particle size.

Three methods of presenting the histogram are available. In the first, a rectangle is constructed over each class interval, the height of which is proportional to the number of particles in that interval (figure 4.8).

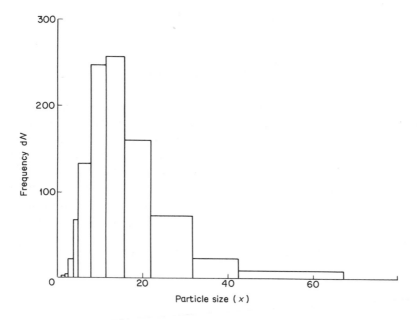

Fig. 4.8 Number–frequency histogram.

A far more useful way is to construct rectangles whose areas are proportional to the numbers of particles in the intervals. The total area under the histogram is equal to the number of particles counted, and it is useful to reduce this number to 100 by making the areas under the rectangles equal to the percentages of particles in the intervals so that histograms may be compared irrespective of the number of particles counted (figure 4.9).

If a sufficient number of particles has been counted, a smooth curve may be drawn through the histogram to give a frequency distribution. It is usual to have more than twelve intervals for this reason, with an upper limit of about twenty in order that the number of particles in each interval remains high and the work involved does not become too great.

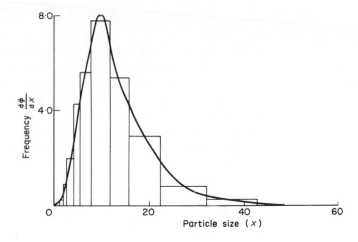

Fig. 4.9 The relative percentage frequency distribution by number.

It is often more convenient to plot the information as a cumulative distribution; the abscissa is particle size and the ordinate, the percentage smaller than or larger than the size. This method has the advantage that the median size and the percentage between any two sizes may be read off directly. The cumulative curve does often conceal detail and for comparison of similar size gradings, the relative percentage frequency should be used. If the range of particle size is very great, particularly if the intervals are in a geometric progression, it is advisable to use a logarithmic scale. In order that the distribution be plotted according to an equidistant log scale $x(d\phi/dx)$ is plotted against $\ln x$ instead of $d\phi/dx$ against x as with the distribution curve using a linear abscissa [38, p. 90]. Alternatively, $d\phi/d\log x$ may be plotted against $\log x$ (figure 4.10).

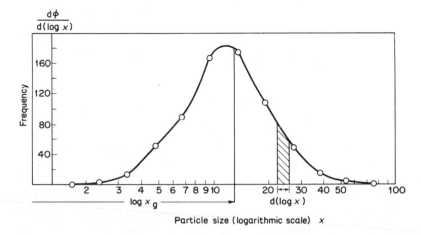

Fig. 4.10 A log-normal distribution plotted as a relative percentage frequency distribution using a logarithmic scale for particle size.

The mean size in this case is the geometric mean given by:

$$\log x_g \sum \frac{d\phi}{d \log x} \cdot d \log x = \sum \log x \, \frac{d\phi}{d \log x} \cdot d \log x$$

$$\log x_g = \frac{\Sigma \log x \, d\phi}{\Sigma \, d\phi} \tag{4.48}$$

(cf. equation (4.40)). ϕ is a general term for the variables W, S and N, i.e. weight, surface and number.

4.8 Devices for representing the cumulative distribution curve as a straight line

4.8.1 Arithmetic normal distributions

It is common practice to plot size distribution data in such a way that a straight line results, with all the advatanges that follow from such a reduction. This can be done if the distribution fits a standard law, such as the normal law. This distribution occurs when the measured value of some property of a system is determined by a large number of small effects, each of which may or may not operate. If a large number of the measurements of the value are made, and the results plotted as a frequency distribution, the well-known, bell-shaped curve results.

Although it might be expected that this type of distribution would be common, it seems to occur only for narrow size ranges of classified material. Actual distributions are skewed, usually to the right.

The equation representing the normal distribution is:

$$y = \frac{1}{\sigma\sqrt{(2\pi)}} \exp \left[-\frac{(x - \bar{x})^2}{2\sigma^2} \right] \tag{4.49}$$

where $y = \dfrac{d\phi}{dx} = f(x)$

and $\displaystyle\int_{-\infty}^{\infty} f(x) \, dx = 1$ (i.e. the distribution is normalized),

σ is the standard deviation (σ^2 is the variance), $\bar{x}$ is the mean size and ϕ is a general term for the frequency, being number, length, surface or volume.

Let $\quad t = \left(\dfrac{x - \bar{x}}{\sigma} \right)$

then $\sigma dt = dx$.

Equation (4.49) becomes:

$$\frac{d\phi}{dt} = \frac{1}{\sqrt{2\pi}} \exp \left(-\frac{t^2}{2} \right)$$

Hence:
$$\int_0^\phi d\phi = \frac{1}{\sqrt{2\pi}} \int_{-\infty}^x \exp\left(-\frac{t^2}{2}\right)^2 dt \tag{4.50}$$

A plot of $\dfrac{d\phi}{dt}$ against t results in the well-known 'dumb-bell' shape of the normal probability curve (figure 4.11).

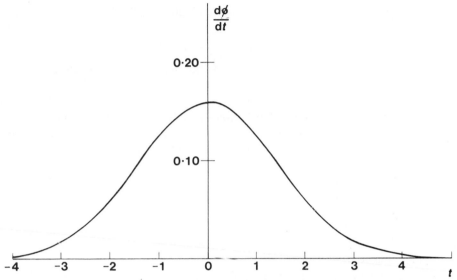

Fig. 4.11 The normal probability curve. Relative frequency against standard deviation [68.26% of the distribution lies within 1 standard deviation $-1 < t < +1$ of the mean].

Equation (4.50) is the basis for arithmetic probability graph paper and the integral is tabulated in books on statistics (Table 4.14).

Table 4.14 Integration of the normal probability equation

t	Integral
0	0.5000
0.5	0.6915
1.0	0.8413
1.5	0.9332
2	0.9772
3	0.9987
4	0.99997

The integral from Table 4.14, for $t = 1$ is 84.13%, or for $t = -1$ is 15.8%; therefore, the standard deviation:

$$\sigma = \left(\frac{x - \bar{x}}{t}\right) = x_{84.13} - x_{50} = x_{50} - x_{15.87}$$

·A powder whose size distribution fits the normal equation can therefore be represented by two numbers, the mean value and the standard deviation. The fraction of particles lying between given sizes can then be found from the tables giving the areas under the graph between any two ordinates; such tables are published in Herdan's book [38, p.77]. One advantage of this method of plotting is that experimental and operator errors may be smoothed out and the median and standard deviation can be read off the graph. In the illustrated example (figure 4.12), the median size, i.e. the 50% size, is 35 μm, and the standard deviation is half the difference between the 84% and the 16% sizes (15 μm):

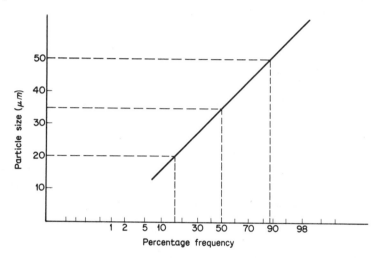

Fig. 4.12 A normal distribution plotted on normal probability paper.

It can be seen from the graph (figure 4.12) that a 1% unit around 95% probability is of about four times the size range as that around the 50% probability. This tends to aggravate the errors discussed under the theory of compensating errors, hence it is usual to draw a best straight line through the central points.

If all particles greater or smaller than a certain particle size have been removed, the curve becomes asymptotic towards these sizes. It is also possible to determine whether the distributions are homogeneous or heterogeneous since, in the latter case, points of inflection occur. These and other cases are discussed in detail by Irani and Callis [39, p. 47].

4.8.2 The log-normal distribution

According to the normal law, it is differences of equal amounts in excess or deficit from a mean value which are equally likely. With the log-normal law, it is ratios of equal amounts which are equally likely. In order to obtain a symmetrical curve of the same shape as the normal curve, it is therefore necessary to plot the relative frequency against log size (see figure 4.10).

The equation of the log-normal distribution is obtained by replacing x with $z = \ln x$, in equation (4.49). Then:

$$y = \frac{1}{\sigma_z\sqrt{2\pi}} \quad \exp\left[-\frac{(z-\bar{z})^2}{2\sigma_z^2}\right]$$

(4.51)

where $\quad y = \dfrac{d\phi}{d \ln x}$, $\quad \sigma_z$ is the standard deviation of z

and $\quad \bar{z} = \dfrac{\Sigma z d\phi}{\Sigma d\phi} \quad (\phi = N, S \text{ or } W)$

$$\bar{z} = \frac{\Sigma z d\phi}{\phi}$$

or $\quad \ln x_g = \dfrac{\Sigma \ln x \, d\phi}{\phi}$

Therefore $x_g = [\sqrt{\Pi x^{d\phi}}]^{1/\phi}$

(4.52)

$\Pi x^{d\phi}$ is the product of the group data in which the frequency of particles of size x is $d\phi$, that is, the mean of a log-normal distribution is the geometric mean, i.e. the arithmetic mean of the logarithms:

$$x_g^\phi = x_1^{d\phi_1} \, x_2^{d\phi_2} \ldots x_r^{d\phi_r} \ldots x_n^{d\phi_n}$$

Since the particle size is plotted on a logarithmic scale, the presentation of data on a log-probability graph is particularly useful when the range of sizes is large.

As before, the median particle size of the data presented on the graph (figure 4.13) is the 50% median size (20 μm) and this is equal to the geometric mean size x_g. The geometric standard deviation is:

$$\log \sigma_g = \log x_{84} - \log x_{50}$$

$$= \log x_{50} - \log x_{16}$$

or $\quad 2 \log \sigma_g = \log x_{84} - \log_{16}$

$$= \log \frac{x_{84}}{x_{16}}$$

(4.53)

From figure 4.13

$$\log \sigma_g = \tfrac{1}{2} \log \frac{40}{10}$$

$$= \log 2$$

Therefore:

$$\sigma_g = 2.$$

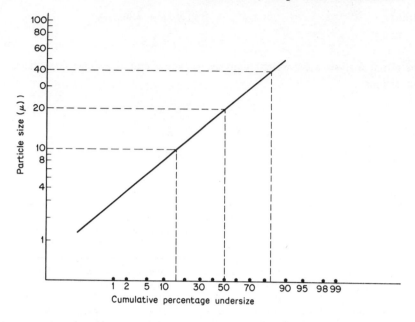

Fig. 4.13 A log-normal distribution plotted on log-probability paper.

As a rule, if the number distribution of a given variable obeys a certain distribution law, the weight distribution does not and vice versa. This is not true for the log-normal distribution. If the number distribution is log-normal, the surface and weight distributions are also log-normal with the same standard deviation [38, p. 85]. Conversion from one distribution to another is easy using the following equations (see section 4.12).

(a) $$\ln x_{NL} = \ln x_{gN} + 0.5 \ln^2 \sigma_g$$

(b) $$\ln x_{NS} = \ln x_{gN} + 1.0 \ln^2 \sigma_g$$

(c) $$\ln x_{NV} = \ln x_{gN} + 1.5 \ln^2 \sigma_g$$

(d) $$\ln x_{NM} = \ln x_{gN} + 2.0 \ln^2 \sigma_g \qquad (4.54)$$

where x_{gN} is the geometric mean (median) of the number distribution and:

$$x_{NL} = \frac{\Sigma x\,dN}{\Sigma\,dN}; \quad x_{NS}^2 = \frac{\Sigma x^2\,dN}{\Sigma\,dN}; \quad x_{NV}^3 = \frac{\Sigma x^3\,dN}{\Sigma\,dN}; \quad x_{NM}^4 = \frac{\Sigma x^4\,dN}{\Sigma\,dN}$$

Derived sizes are obtained in the following manner:

Surface-volume $$x_{sv} = \frac{\Sigma x^3\,dN}{\Sigma x^2\,dN}$$

$$= x_{NV}^3 / x_{NS}^2$$

Therefore:
$$\ln x_{sv} = 3 \ln x_{NV} - 2 \ln x_{NS}$$
$$= \ln x_{gN} + 2.5 \ln^2 \sigma_g \tag{4.55}$$

If the initial analysis was a weight analysis, the above equations may be utilized using the conversions:

(a)
$$\ln x_{gS} = \ln x_{gN} + 2 \ln^2 \sigma_g$$

(b)
$$\ln x_{gV} = \ln x_{gN} + 3 \ln^2 \sigma_g \tag{4.56}$$

where x_{gN}, x_{gS} and x_{gV} are the number, surface and volume geometric mean diameters.

Table 4.15 Log-normal distribution

Particle size range (μm)	Average size (x)	$\log x$	Cumulative % oversize ϕ	Percentage in range $d\phi$	$\dfrac{d\phi}{d \log x}$	$\log x \, d\phi$
$\sqrt{2} - 2$	1.68	0.225	0.4	0.4	2.7	0.09
$2 - 2\sqrt{2}$	2.38	0.376	3.5	3.1	20.5	1.17
$2\sqrt{2} - 4$	3.36	0.526	14.5	11.0	72.8	5.79
$4 - 4\sqrt{2}$	5.76	0.677	36.3	21.8	144.2	14.77
$4\sqrt{2} - 8$	6.72	0.827	63.6	27.3	180.8	22.58
$8 - 8\sqrt{2}$	9.52	0.978	85.6	22.0	145.7	21.52
$8\sqrt{2} - 16$	13.4	1.029	95.7	10.1	66.9	10.40
$16 - 16\sqrt{2}$	19.0	1.179	99.6	3.9	25.8	4.60
$16\sqrt{2} - 32$	26.9	1.430	100.0	0.4	2.7	0.57
						81.49

Assuming a weight distribution ($d\phi = dW = x^3 \, dN$):

$$\log x_{gV} = \frac{\Sigma \log x d\phi}{d\phi}$$

$$= 0.815$$

$$x_{gV} = 6.53 \; \mu m$$

σ_g may also be obtained from the table, but both these values may be obtained more readily from a graph giving:

$$x_{gV} = x_{median} = 6.6 \; \mu m$$
$$\sigma_g = 1.64 \; (\log \sigma_g = 0.215)$$

The number distribution will have a median:

$$\log x_{gN} = \log x_{gV} - 6.9 \log^2 \sigma_g$$
$$= 0.815 - 6.9 \times 0.215^2$$
$$x_{gN} = 3.28 \; \mu m$$

Similarly:
$$\log x_{gS} = \log x_{gV} - 2.3 \log^2 \sigma_g$$
$$x_{gS} = 5.11 \; \mu m$$

Mode:
$$\ln x_m = \ln x_{gV} - \ln^2 \sigma_g$$
$$x_m = 5.11.$$

In each case the slope of the log-normal line, hence the standard deviation σ_g, will be the same; from equations (4.54):

$$X_{LS} = 3.67$$
$$X_{VS} = 4.10$$
$$X_{VM} = 4.55$$
$$X_{NL} = 3.30$$
$$X_{NS} = 3.48$$
$$X_{NM} = 3.87$$

4.8.3 The Rosin–Rammler distribution

For broken coal, a distribution function has been developed which has since been found to apply to many other materials [40]. For example, the particle size distribution of moon dust is found to closely follow a Rosin–Rammler distribution, hence it is assumed that the lunar surface was formed as a result of crushing forces due to impact [78].

Let the size distribution of broken coal be obtained by sieving and let the weight percentage retained on the sieve of aperture x be denoted by R; a plot of R against x gives the cumulative percentage oversize curve.

From the probability considerations the authors obtain:

$$\frac{dF(x)}{dx} = 100nbx^{n-1} \exp(-bx^n) \qquad (4.57)$$

where n and b are constants, b being a measure of the range of particle size present and n being characteristic of the substance being analysed. Integrating gives:

$$R = 100 \exp(-bx^n) \qquad (4.58)$$

This reduces to:
$$\text{log-log } \frac{100}{R} = \text{constant} + n \log x \qquad (4.59)$$

If log-log $100/R$ is plotted against $\log x$, a straight line results. The peak of the distribution curve for $n = 1$ is at $100/e = 36.8\%$, and denoting the mode of the distribution curve by x_m equation (4.58) gives $b = 1/x_m$.

The sieve opening for $R = 36.8\%$ is used to characterize the degree of comminution of the material, and since the slope of the line on the Rosin–Rammler graph depends on the particle size range, the ratio of $\tan^{-1}(n)$ and x_m is a form of variance.

This treatment is useful for monitoring grinding operations for highly skewed distributions, but should be used with caution since the device of taking logs always reduces scatter, hence taking logs twice is not to be recommended.

4.8.4 Mean particle sizes and specific surface evaluation for Rosin–Rammler distributions

The moment-volume mean diameter is given by:

$$x_{vm} = \frac{\Sigma x \Delta W}{\Sigma \Delta W} \qquad (4.60)$$

Since $\Delta W = \Delta F(x)$, defining $F(x)$ as 100 gives from equations (4.57) and (4.60):

$$x_{vm} = \frac{1}{100} \int_0^\infty 100 nbx^n \exp(-bx^n)\, dx$$

$$= \frac{1}{n\sqrt{b}} \ \Gamma\left(\frac{1}{n} + 1\right)$$

The surface-volume mean diameter may be similarly evaluated as:

$$x_{sv} = \frac{1}{n\sqrt{b}\Gamma(1 - \frac{1}{n})}, \quad n > 1$$

These can be evaluated from tables of gamma functions for experimental values of n and the specific surface determined.

4.8.5 Other particle size distribution equations

Various other size distribution functions have been proposed. These are usually in the form of two-parameter (b and n) potential distribution functions such as:

(a)   Gates–Gaudin–Schumann [60, 61]

$$\phi_{GGS} = (bx)^n \qquad (4.61)$$

(b)   Gaudin–Meloy [62]

$$\phi_{GM} = [1 - (1-bx)^n] \qquad (4.62)$$

where ϕ is the undersize fraction.

Three- and four-parameter functions with more accuracy have also been proposed [63–65].

4.8.6 Simplification of two-parameter equations

Tarjan [66] converted a two-parameter size distribution function from the form $\phi = f(x)$ to the form $\phi = f(\frac{x}{x_{0,5}})$ where $x_{0.5}$ is the median size. This results in an easy-to-handle

function with a high degree of correspondence to the more complicated logarithmic function (ϕ_K) below [67].

$$\phi_K = \Phi \, \ln(bx)^n \qquad (4.63)$$

where

$$\Phi = \frac{1}{\sqrt{2\pi}} \int_{-\infty^-}^{u} \exp\left(-\frac{u^2}{2}\right) . \, du \qquad (4.64)$$

Let the parameter b of equations (4.58), (4.61), (4.62) and (4.63) be expressed in terms of $x_{0.5}$ when $\phi = 0.50$ ($R = 50$ in equation (4.55)).

$$CGS: \qquad 0.50 = (bx_{0.5})^n$$

$$RR: \qquad 0.50 = \exp(-bx_{0.5}^n)$$

$$GM: \qquad 0.50 = (1 - bx_{0.5})^n$$

Substituting back for b gives:

equation (4.58)

$$\phi_{RR} = 1 - \exp\left[-\left(\frac{x}{x_{0.5}}\right)^n \ln 2\right] \qquad (4.65)$$

equation (4.61)

$$\phi_{GGS} = 0.5 \left(\frac{x}{x_{0.5}}\right)^n \qquad (4.66)$$

equation (4.62)

$$\phi_{GM} = 1 - \left[1 - \left(\frac{x}{x_{0.5}}\right)(1 - \sqrt[n]{0.5})\right]^n \qquad (4.67)$$

4.8.7 Evaluation of non-linear distributions on log-normal paper

A bimodal distribution is detectable on log-probability paper by a change in slope of the line. It is also possible to deduce further features of the distribution. Figure 4.14 shows a bimodal distribution in which the parent distributions do not intersect on a log-probability plot. These distributions are asymptotic to the parent distributions. The geometric means of the parent distributions may be obtained by plotting relative percentage frequency against particle size on log-linear paper (figure 4.15). The area under the two quite distinct curves gives the proportions of the two constituents. From the modes to the 34% levels in areas gives the two standard deviations.

Figure 4.16 shows a bimodal distribution in which the parent distributions intersect on a log-probability plot. These distributions are asymptotic to the parent distribution having the widest size range (i.e. high standard deviation). The point of inflection passes through both distributions. If the separation of means is large, these may be obtained from a plot of relative percentage frequency against particle size on log-linear paper. If the separation of means is small, it is difficult to resolve these distributions (figure 4.17).

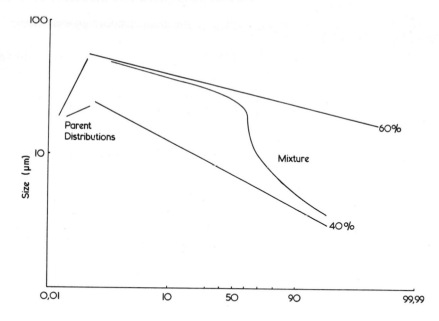

Fig. 4.14 Bimodal non-intersecting distributions.

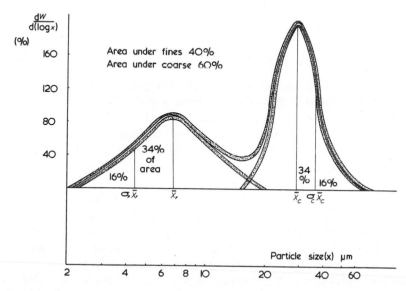

Fig. 4.15 Relative percentage per log-micrometre of a bimodal distribution with little overlap.

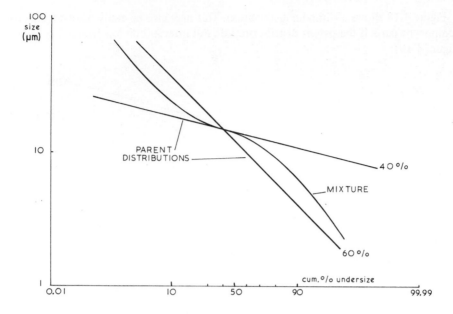

Fig. 4.16 Bimodal intersecting distributions.

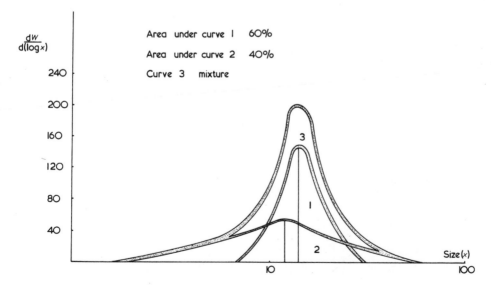

Fig. 4.17 Relative percentage per log-micrometre of a bimodal distribution with small separation of means.

Figure 4.18 shows a trimodal distribution. This may also be easily resolved into its component parts if the parent distributions do not intersect on log-probability paper (figure 4.19).

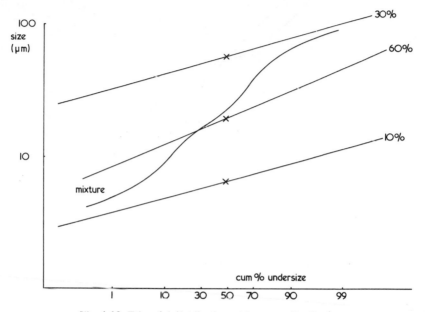

Fig. 4.18 Trimodal distribution with parent distributions.

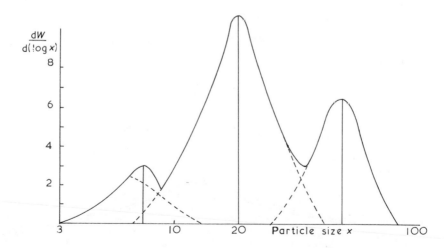

Fig. 4.19 Relative percentage per log-micrometre of a trimodal distribution with little overlap.

4.8.8 Derivation of shape factors from parallel log-normal curves

The two curves on figure 4.20 are analyses of the same powder using gravimetric (X-ray and pipette sedimentation) and volumetric (Coulter) techniques.

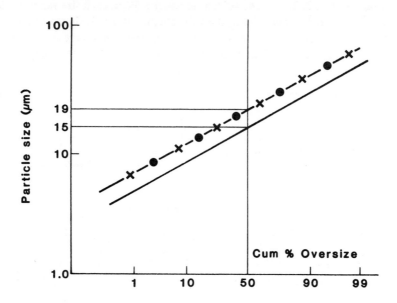

Fig. 4.20 Comparison between Coulter and X-ray (×) and pipette methods (•).

Multiplying the diameter by the latter technique by a factor of 1.27 brings the two curves into coincidence, hence:

$$1.27 \, d_{Stk} = d_v$$

Since:

$$d_{Stk}^2 \doteq \frac{d_v^3}{d_s}$$

this gives:

$$\frac{d_v}{d_{Stk}} = 1.27; \quad \frac{d_s}{d_v} \doteq 1.27^2$$

and, from equation (4.17):

$$\psi_w = 0.384$$

For a spherical particle, these ratios are unity, increasing with increasing divergence from sphericity.

4.9 The law of compensating errors

In any method of size analysis, it is always possible to assign the wrong size to some of the particles. If this error is without bias, the possibility of assigning too great a size is equally as probable as assigning too small a size. This will modify the extremes of the distribution, but will have little effect on the central region. An illustration with particles of mean size $1, \sqrt{2}, 2, 2\sqrt{2}, 4, 4\sqrt{2}, 8$, in which 25% in each size range are placed in the size category below and 25% in the size category above, is shown in figure 4.21. An illustration with bias is also shown with particles having the same size ranges as in the first example, but with 25% in each category being placed in the category above.

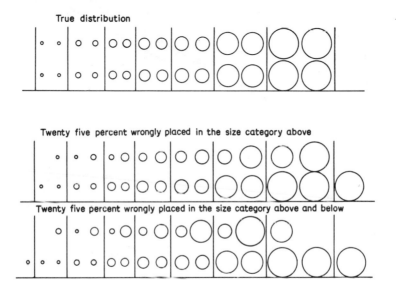

Fig. 4.21 Law of compensating variables.

It can be seen that there is the same number of particles in each of the central-size categories, irrespective of whether bias is present or not. If the number distribution is converted into a weight distribution, this is still true, but wrongly assigning a size of $8\sqrt{2}$ to 25% by number of the particles in the top-size category gives an apparently coarser distribution. For measurement sizes in arithmetic progression of sizes, the effect is small provided sizing is carried out at 10 or more size intervals and, for a log-normal distribution plotted as a relative frequency curve against the logarithm of particle size, the position of the mode is only slightly affected.

Table 4.16 is for a log-normal distribution having a mean size of 8.6 μm and a standard deviation of 0.320. Wrongly placing 25% of each category by weight in the size category above and 25% in the size category below, gives a mean size of 8.6 μm and a standard deviation of 0.284.

Table 4.16

Upper size limit (μm)	Mean size (μm)	Cumulative percentage undersize		Percentage in range	
		true	error	true	error
1	1.2	0	0.08	0.3	0.08
$\sqrt{2}$	1.7	0.3	0.50	1.1	0.42
2	2.4	1.4	2.15	4.1	1.65
$2\sqrt{2}$	3.4	5.5	6.85	9.5	4.70
4	4.8	15.0	17.38	19.0	10.53
$4\sqrt{2}$	6.8	34.0	35.25	24.0	17.87
8	9.6	58.0	57.00	20.0	21.75
$8\sqrt{2}$	13.4	78.0	76.28	13.0	19.25
16	19.0	91.0	89.385	6.4	13.10
$16\sqrt{2}$	28.9	97.4	96.30	2.0	6.95
32	38.0	99.4	99.03	0.5	2.73
$32\sqrt{2}$	53.8	99.9	99.80	0.1	0.77
64	76.0	100	99.98		0.18
			100		0.02

Figure. 4.22 shows a log-normal distribution of geometric mean size 10 μm and geometric standard deviation 2. This distribution is deficient in sub-6 μm particles and is asymptotic to this size. If $(x - 6)$ is taken as the particle size, a straight line results to give the original distribution. A similar sort of plot occurs when there is a deficiency of coarse particles and this may be similarly resolved.

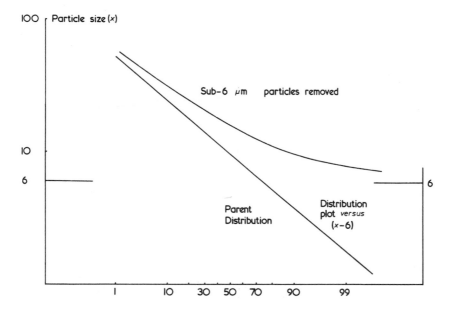

Fig. 4.22 Log-normal distribution with deficiency of sub-6 μm particles.

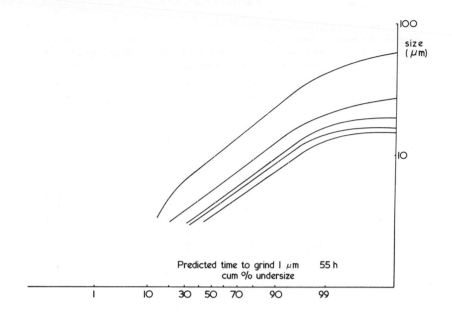

Fig. 4.23 Andreasen analysis monitoring a grinding operation (1).

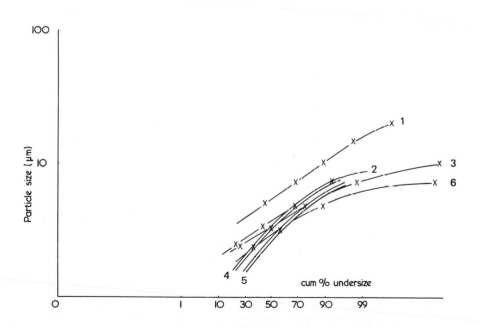

Fig. 4.24 Andreasen analysis monitoring a grinding operation (2).

Figure 4.23 shows Andreasen analysis monitoring a grinding operation. Since, in this case, the new surface created is proportional to grinding time, it is possible to predict future performance. Similarly, less accurately, maximum size is inversely proportional to grinding time (see table below).

Grinding time t (hours)	Mean particle size (μm)	Maximum particle size x_g (μm)	$x_g t$	$x_m t$
9	5.3	27.5	47.7	239
13	4.1	19	53.3	247
15	3.75	16	56.3	240
16	3.42	14	54.7	224

Hence, the predicted time to grind to 1 μm mean size is 55 hours; the predicted time to produce sub-10 μm particles is 24 hours.

Figure 4.24 gives six analyses from a grinding operation. For samples 4 and 5, the grinding variables have been altered.

4.10 Alternative notation for frequency distribution

The notation used in this chapter is widely used but an alternative notation has been developed in Germany [41–47]. Although elegant, the German notation requires some memorization and is probably most suitable for frequent usage and computer application. It is included here, in a shortened form, for the sake of completeness.

4.10.1 Notation

Let the fractional number smaller than size x be:

$$Q_0(x) = \int_{x_{min}}^{x} q_0(x)\, dx \qquad (4.68)$$

or

$$q_0(x) = \frac{d\,Q_0(x)}{dx} \qquad (4.69)$$

Hence $q_0(x)$ is the fractional number in the size range x to $x + dx$.
Further:

$$Q_0(x)_{max} = \int_{x_{min}}^{x_{max}} q_0(x)\, dx = 1 \qquad (4.70)$$

The subscript may be varied to accommodate other distributions, viz: q_r and Q_r where:

$r = 0$ for a number distribution
$r = 1$ for a length distribution
$r = 2$ for an area distribution
$r = 3$ for a volume distribution

4.10.2 Moment of a distribution

The moment of a distribution is written as:

$$M_{k,\,r} = \int_{x_{min}}^{x_{max}} x^k q_r(x)\, dx \tag{4.71}$$

[Note $M_{0,r} = 1$.]

For an incomplete distribution:

$$M_{k,\,r}(x_e, x_u) = \int_{x_e}^{x_u} x^k q_r(x)\, dx \tag{4.72}$$

4.10.3 Transformation from $q_t(x)$ to $q_r(x)$

If $q_t(x)$ is known $q_r(x)$ may be determined using the following:

$$q_r(x) = \frac{x^{r-t} q_t(x)}{\int_{x_{min}}^{x_{max}} x^{r-t} q_t(x)\, dx}$$

$$= \frac{x^{r-t} q_t(x)}{M_{r-t,\,T}} \tag{4.73}$$

The denominator is necessary to normalize the distribution function.
Examples
To convert from a number to a weight distribution put $t = 0, r = 3$. To convert from a surface to a weight distribution put $t = 2, r = 3$.

Effect of particle shape
This transformation is derived with the assumption that particle shape does not change with particle size. More correctly:

$$q_r(x) = \frac{\alpha(x) x^{r-t} q_t(x)}{\int_{x_{min}}^{x_{max}} \alpha(x) x^{r-t} q_t(x)\, dx} \tag{4.74}$$

where α is a shape coefficient.

4.10.4 Relation between moments

Putting $t = 0$ in equation (4.73) and substituting $q_0(x)$ for $q_r(x)$ in equation (4.71) gives:

$$M_{k,r} = \frac{\displaystyle\int_{x_{min}}^{x_{max}} x^k x^r q_0(x)\, dx}{M_{r,0}}$$

$$M_{k,r} = \frac{M_{k+r,0}}{M_{r,0}} \tag{4.75}$$

More generally, substituting $q_r(x)$ from equation (4.73) in equation (4.71) gives:

$$M_{k,r} = \frac{M_{k+r-t,\,t}}{M_{r-t,\,t}} \tag{4.76}$$

Examples

To determine the surface-volume mean diameter from a number distribution put $t = 0, r = 2, k = 1$:

$$M_{1,2} = \frac{M_{3,0}}{M_{2,0}} = x_{sv}$$

To determine the surface-volume mean diameter from a surface distribution put $t = 2, r = 2, k = 1$:

$$M_{1,2} = \frac{M_{1,2}}{M_{0,2}}$$

4.10.5 Means of distributions

(a) Distribution means are given by the equation:

$$\bar{x}_{1,r} = \frac{\displaystyle\int_{x_{min}}^{x_{max}} x q_r(x)\, dx}{\displaystyle\int_{x_{min}}^{x_{max}} q_r(x)\, dx}$$

$$\bar{x}_{1,r} = \frac{M_{1,r}}{M_{0,r}} = \frac{M_{r+1,0}}{M_{r,0}}$$

$$= M_{1,r} \tag{4.77}$$

(b) Arithmetic means are given by the equation:

$$\bar{x}_{k,0} = \left[\frac{\displaystyle\int_{x_{min}}^{x_{max}} x^k q_0(x)\, dx}{\displaystyle\int_{x_{min}}^{x_{max}} q_0(x)\, dx} \right]^{1/k}$$

$$\bar{x}_{k,0} = k \sqrt{(M_{k,0})} \tag{4.78}$$

(c) More generally:

$$(\bar{x}_{k,r})^k = \frac{\int_{x_{min}}^{x_{max}} x^k q_r(x)\,dx}{\int_{x_{min}}^{x_{max}} q_r(x)\,dx}$$

$$(\bar{x}_{k,r})^k = M_{k,r} \tag{4.79}$$

4.10.6 Standard deviations

For a number distribution, the variance is defined by:

$$\sigma_0^2 = \int_{x_{min}}^{x_{max}} (x - \bar{x}_{1,0}) q_0(x)\,dx$$

$$= \int_{x_{min}}^{x_{max}} x^2 q_0(x)\,dx - \bar{x}_{1,0}^2 \int_{x_{min}}^{x_{max}} q_0(x)\,dx$$

$$\sigma_0^2 = M_{2,0} - (M_{1,0})^2 \tag{4.80}$$

More generally:

$$\sigma_r^2 = \int_{x_{min}}^{x_{max}} (x - \bar{x}_{1,r})^2 q_r(x)\,dx$$

$$\sigma_r^2 = M_{2,r} - (M_{1,r})^2 \tag{4.81}$$

Alternatively:

$$\sigma_r^2 = M_{1,r}\left[\frac{M_{2,r}}{M_{1,r}} - M_{1,r}\right]$$

Putting $t = r + 1$ and $k = 1$ in equation (4.76) gives:

$$\sigma_r^2 = M_{1,r}\,[M_{1,r+1} - M_{1,r}] \tag{4.82}$$

From equation (4.79):

$$\sigma_r^2 = \bar{x}_{1,r}\,[\bar{x}_{1,r+1} - \bar{x}_{1,r}] \tag{4.83}$$

4.10.7 Coefficient of variance

$$C_r'^2 = \frac{\sigma_r^2}{(\bar{x}_{1,r})^2} \tag{4.84}$$

$$= \frac{\bar{x}_{1,r+1}}{\bar{x}_{1,r}} - 1 \tag{4.85}$$

4.10.8 Applications

(a)   Calculation of volume-specific surface

$$S_v = \frac{surface\ area}{volume}$$

$$= \frac{\int_{x_{min}}^{x_{max}} \alpha_s(x)\, x^2 q_0(x)\, \mathrm{d}x}{\int_{x_{min}}^{x_{max}} \alpha_v(x)\, x^3 q_0(x)\, \mathrm{d}x}$$

where $\alpha_s(x)$ and $\alpha_v(x)$ are the surface and volume shape coefficients. Assuming that these are independent of particle size and defining the ratio of α_s to α_v as α_{sv}, the surface-volume shape coefficient is:

$$S_v = \alpha_{sv} \frac{M_{2,0}}{M_{3,0}} = \frac{\alpha_{sv}}{M_{1,2}} = \alpha_{sv}\, M_{1,3} \tag{4.86}$$

Also, since $M_{1,2} = \bar{x}_{1,2}$

$$S_v = \frac{\alpha_{sv}}{\bar{x}_{1,2}} \tag{4.87}$$

(b)   Calculation of the surface area of a size increment

$$S_v(x_e, x_u) = \frac{\int_{x_e}^{x_u} \alpha_s x^2 q_0(x)\, \mathrm{d}x}{\int_{x_e}^{x_u} \alpha_v x^3 q_0(x)\, \mathrm{d}x}$$

$$= \alpha_{sv} \frac{M_{2,0}(x_e, x_u)}{M_{3,0}(x_e, x_u)}$$

$$= \alpha_{sv} \left[\frac{M_{2,0}(x_e, x_u)}{M_{3,0}} \right] \times \left[\frac{M_{3,0}}{M_{3,0}(x_e, x_u)} \right]$$

Now:

$$Q_r(x_i) = \int_{x_{min}}^{x_i} q_r(x)\, \mathrm{d}x = \frac{\int_{x_{min}}^{x_i} x^r q_0(x)\, \mathrm{d}x}{M_{r,0}}$$

$$= \frac{M_{r,0}(x_{min}, x_i)}{M_{r,0}}$$

so: $$Q_3(x_u) - Q_3(x_e) = \frac{M_{3,0}(x_e, x_u)}{M_{3,0}}$$

hence:

$$S_v = \alpha_{sv} \left[\frac{M_{2,0}(x_e, x_u)}{M_{3,0}} \right] \times \left[\frac{1}{Q_3(x_u) - Q_3(x_e)} \right] \tag{4.88}$$

The application of this equation enables a surface area to be calculated from a summation of increments, i.e.

$$S_v = \alpha_{sv} \frac{M_{2,0}}{M_{3,0}}$$

$$= \frac{\alpha_{sv}}{M_{3,0}} \left[\int_{x_{min}}^{x_{u_1}} x^2 q_0(x)\, dx + \int_{x_{u_1}}^{x_{u_2}} x^2 q_0(x)\, dx + \ldots \right]$$

$$= \frac{\alpha_{sv}}{M_{3,0}} \sum_{i=1}^{i=n} M_{2,0}(x_{e_i}, x_{u_i}) \tag{4.89}$$

$e_i = $ min, $u_n = $ max.

4.10.9 Transformation of abscissa

Suppose, in an analysis, ξ, which is a function of x, is measured. Since the amount of material between sizes x and $x + dx$ is constant, there must be a simple relationship between $q(\xi)$ and $q(x)$. Let

$$x = f(\xi) \quad \text{so that} \quad \xi = \phi(x)$$

then

$$q_r^*(\xi) = q_r(x) \cdot \frac{dx}{d\xi} \tag{4.90}$$

For example, suppose the following relationship holds:

$$\xi = x^k$$

then

$$\frac{d\xi}{dx} = kx^{k-1}$$

and

$$q_r^*(\xi) = \frac{1}{kx^{k-1}}\, q_r(x) \tag{4.91}$$

In general, suppose we have the $q_u(x)$ distribution and wish to find the $q_r(x^k)$ distribution. Now from equation (4.73):

$$q_r(x) = \frac{x^{r-u} q_u(x)}{M_{r-u,\, u}}$$

Substituting in equation (4.91) gives:

$$q_r^*(\xi) = \frac{x^{r-u}q_u(x)}{M_{r-u,\,u}} \cdot \frac{1}{kx^{k-1}}$$

(4.92)

Example

In a Coulter counter analysis, the pulse height V is proportional to particle volume, i.e.

$$\xi = V = p^3 x^3$$

Calculation of $M_{r,\,0}$ $(= [\bar{x}_{r,\,0}]^r)$

By definition:

$$M_{r,\,0} = \int_{x_{min}}^{x_{max}} x^r q_0(x)\, dx$$

$$= \frac{1}{p^r} \int_{q_{min}}^{q_{max}} \xi^{r/3} q_0^*(\xi)\, d\xi$$

$$= \frac{M_{r/3,\,0}(\xi)}{p^r}$$

(4.93)

Also, since $q_r(x) = \dfrac{x^r q_0(x)}{M_{r,\,0}}$

substituting in equations (4.90) and (4.94):

$$q_r(x) = x^r q_r^*(\xi) \frac{d\xi}{dx} \cdot \frac{p^r}{M_{r/3,\,0}(\xi)}$$

$$q_r(x) = \frac{3pq^{(r+2)/3} q_0^*(\xi)}{M_{r/3,\,0}(\xi)}$$

(4.94)

To calculate $M_{k,\,r}$

$$M_{k,\,r} = \int_{x_{min}}^{x_{max}} x^k q_r(x)\, dx$$

$$= \frac{1}{p^k} \int_{x_{min}}^{x_{max}} \frac{\xi^{(k+r)/3} q_0^*(\xi)\, d\xi}{M_{r/3,\,0}(\xi)}$$

$$= \frac{M_{(k+r)/3,\,0}(q)}{M_{r/3,\,0}(q)}$$

(4.95)

To calculate volume-specific surface:

$$S_v = \alpha_{sv} \frac{M_{2,0}}{M_{3,0}}$$

which, using equation (4.95), may be written:

$$S_v = \alpha_{sv} \cdot p \frac{M_{2/3,0}(\xi)}{M_{1,0}(\xi)} \qquad (4.96)$$

Thus, specific surface can be determined from moments calculated directly from Coulter counter data.

Calculation of mean size $\bar{x}_{k,r}$

$$\bar{x}_{k,r} = k\sqrt{(M_{k,r})}$$

$$= \frac{1}{p} \; k\sqrt{\left(\frac{M_{(k+r)/3,0}(\xi)}{M_{r/3,0}(\xi)} \right)} \qquad (4.97)$$

4.11 Phi-notation

In geological literature dealing with particle size distribution, a very advantageous transformation of particle size is commonly used. Because it is a logarithmic transformation, it simplifies granulometric computations in the same way as logarithms in mathematical operations. This transformation replaces ratio scale numbers, based on millimetre values of particle size, by interval scale numbers, based on the logarithms of those values. Although several transformations based on decadic logarithms were also suggested (zeta-transformation [55] and gamma-transformation [48]) more than thirty years ago, it has been broadly adopted, particularly in the USA. After the redefinition by McManus [49] and the comments of Krumbein [11], the transformation is:

$$\phi = -\log_2 X_i \quad \text{or} \quad X_i = 2^{-\phi} \qquad (4.98)$$

where X_i is a dimensionless ratio of a given particle size, in millimetres, to the standard particle size of 1 mm.

Phi-values can be found if the common decadic logarithms of X_i are multiplied by $(\log 10^2)^{-1} = 3.3219282$. Conversely, phi-values can be converted into their millimetre (or more precisely X_i) equivalents if their decadic antilogarithms are multiplied by $\log 10^2 = 0.30103$. For easy manipulation, a conversion chart [50, p. 244] or a conversion table [51, 52] can be used.

By using the phi-notation, the statistical measurements acquire a great simplicity. The standard deviation, σ_ϕ, used in this notation refers to its quartile estimate:

$$\sigma_\phi = 0.5(\phi_{84} - \phi_{16}) \qquad (4.99)$$

Similarly the ϕ skewness:

$$\phi = \frac{\phi_{16} + \phi_{84} - 2\phi_{50}}{\phi_{84} - \phi_{16}} \qquad (4.100)$$

Other statistical measurements used in geology for particle size distribution characterization (moment, quartile and others) have been reviewed [53, 54].

4.12 Manipulation of the log-probability equation

Consider the log-normal equation:

$$\frac{d\phi}{d \ln x} = \frac{1}{\ln \sigma_g \sqrt{(2\pi)}} \exp\left[-\left(\frac{\ln x - \ln x_g}{\sqrt{2} \ln \sigma_g}\right)^2\right] \tag{4.101}$$

ϕ being a general term for the frequency, being number, length, surface or volume (weight) (i.e. $\phi = N, L, S$ or V).

Let

$$X = \frac{(\ln x - \ln x_g)}{\sqrt{2} \ln \sigma_g} \tag{4.102}$$

then

$$\sqrt{2} \ln \sigma_g \, . \, dX = d \ln x \tag{4.103}$$

and

$$\int d\phi = \frac{1}{\sqrt{\pi}} \int \exp(-X^2) \, dX \tag{4.104}$$

The fraction undersize the geometric mean size x_g is obtained by inserting the limits $x = 0, x = x_g$, i.e. $X = -\infty, X = 0$.

$$\phi_{x_g} = \frac{1}{\sqrt{\pi}} \int_{-\infty}^{0} \exp(-X^2) \, dX$$

$$= \tfrac{1}{2} \tag{4.105}$$

Therefore the geometric mean size is the median size.

The fraction lying within one standard deviation of the mean is obtained by inserting the limits $x = x_g, x = \sigma_g x_g$, i.e. $X = 0, X = 1/\sqrt{2}$.

$$\phi_{(x_g - \sigma_g x_g)} = \frac{1}{\sqrt{\pi}} \int_{0}^{1/\sqrt{2}} \exp(-X^2) \, dX$$

$$= \frac{1}{\sqrt{\pi}} \left[X - \frac{X^3}{3} + \frac{X^5}{10} - \frac{X^7}{42} + \ldots\right]^*$$

$$= 0.3413 \tag{4.106}$$

$$* \ \exp(-X^2) = \underset{n \to \infty}{\text{Limit}} \left[1 + \left(\frac{-X^2}{n}\right)\right]^n$$

$$= 1 + n\left(\frac{-X^2}{n}\right) + \frac{n(n-1)}{2!} \left(\frac{-X^2}{n}\right)^2 + \ldots \frac{n!}{(n-r)!r!} \left(\frac{-X^2}{n}\right)^r + \ldots$$

$$\int \exp(-X^2) \, dX = X - \frac{X^3}{3} + \frac{1}{2!}\frac{X^5}{5} - \frac{1}{3!}\frac{X^7}{7} + \ldots$$

$$+ \ldots \frac{(-1)^{r-1}}{(r-1)!} \frac{X^{2r-1}}{2r-1} \ldots$$

Hence 34.13% of the distribution lies between the median size x_g and size $\sigma_g x_g$. (Compare this to the normal-probability curve which contains 34.13% of the distribution between the sizes $\bar{x}$ and $\bar{x} + \sigma$.) The geometric standard deviation is therefore: the ratio of the 84.13% and 50% sizes and the ratio of the 50% and 15.87% sizes:

$$\log \sigma_g x_g - \log x_g = \log x_{84.13} - \log x_{50}$$

$$\sigma_g = x_{84.13}/x_{50} \tag{4.107}$$

4.12.1 Average sizes

Consider a log-normal distribution by number, such that:

$$\int_{-\infty}^{+\infty} d\phi = \sum_{r=0}^{r=\infty} dN_r = \frac{1}{\ln \sigma_g \sqrt{(2\pi)}} \int_{-\infty}^{+\infty} \exp\left(-\left[\frac{\ln x - \ln x_{gN}}{\sqrt{2} \ln \sigma_g}\right]^2\right) d \ln x \tag{4.108}$$

$$= 1$$

i.e. the distribution is normalized.

dN_r is the number of particles, in a narrow size range, of mean size x_r; x_0 and x_∞ are the smallest and largest particles present in the distribution. x_{gN} is the geometric mean (median) of the number distribution and σ_g is the geometric standard deviation (which is the same for number, length, surface and weight).

$$x_{NL} = \frac{\sum_{r=0}^{r=\infty} x_r \, dN_r}{\sum_{r=0}^{r=\infty} dN_r}$$

$$= \sum_{r=0}^{r=\infty} x_r \, dN_r = \frac{1}{\ln \sigma_g \sqrt{(2\pi)}} \int_{-\infty}^{+\infty} \exp\left[-\left(\frac{\ln x - \ln x_{gN}}{\sqrt{2} \ln \sigma_g}\right)^2\right] x \, d \ln x \tag{4.109}$$

$$x^2_{NS} = \sum_{r=0}^{r=\infty} x_r^2 \, dN_r = \frac{1}{\ln \sigma_g \sqrt{(2\pi)}} \int_{-\infty}^{+\infty} \exp\left[-\left(\frac{\ln x - \ln x_{gN}}{\sqrt{2} \ln \sigma_g}\right)^2\right] x^2 \, d \ln x \tag{4.110}$$

$$x^3_{NV} = \sum_{r=0}^{r=\infty} x_r^3 \, dN_r = \frac{1}{\ln \sigma_g \sqrt{(2\pi)}} \int_{-\infty}^{+\infty} \exp\left[-\left(\frac{\ln x - \ln x_{gN}}{\sqrt{2} \ln \sigma_g}\right)^2\right] x^3 \, d \ln x \tag{4.111}$$

Substituting from equations (4.102) and (4.103) gives:

$$x_{NL} = \frac{x_{gN}}{\sqrt{\pi}} \int_{-\infty}^{+\infty} \exp\left(\sqrt{2} \ln \sigma_g \, X - X^2\right) dx \tag{4.112}$$

$$x^2_{NS} = \frac{x_{gN}^2}{\sqrt{\pi}} \int_{-\infty}^{+\infty} \exp\left(2\sqrt{2} \ln \sigma_g \, X - X^2\right) dx \tag{4.113}$$

$$x^3_{NV} = \frac{x_{gN}^3}{\sqrt{\pi}} \int_{-\infty}^{+\infty} \exp\left(3\sqrt{2} \ln \sigma_g \, X - X^2\right) dx \tag{4.114}$$

Making the transformations:

$$Y_1 = X - \frac{\sqrt{2}}{2} \ln \sigma_g \quad \text{in equation (4.112)}$$

$$Y_2 = X - \sqrt{2} \ln \sigma_g \quad \text{in equation (4.113)} \tag{4.115}$$

$$Y_3 = X - \frac{3\sqrt{2}}{2} \ln \sigma_g \quad \text{in equation (4.114)} \tag{4.116}$$

$$x_{NL} = \frac{x_{gN}}{\sqrt{\pi}} \exp(\tfrac{1}{2}\ln^2 \sigma_g) \int_{-\infty}^{+\infty} \exp(-Y_1^2)\, dY_1 \tag{4.117}$$

$$x^2{}_{NS} = \frac{x_{gN}^2}{\sqrt{\pi}} \exp(2\ln^2 \sigma_g) \int_{-\infty}^{+\infty} \exp(-Y_2^2)\, dY_2 \tag{4.118}$$

$$x^3{}_{NV} = \frac{x_{gN}^3}{\sqrt{\pi}} \exp[(9/2)\ln^2 \sigma_g] \int_{-\infty}^{+\infty} \exp(-Y_3^2)\, dY_3 \tag{4.119}$$

The integration yields a value $I = \sqrt{\pi}$ giving:

$$\ln x_{NL} = \ln x_{gN} + 0.5 \ln^2 \sigma_g \tag{4.120}$$

$$2\ln x_{NS} = 2\ln x_{gN} + 2\ln^2 \sigma_g \tag{4.121}$$

$$3\ln x_{NV} = 3\ln x_{gN} + 4.5\ln^2 \sigma_g \tag{4.122}$$

Similarly:

$$4\ln x_{NM} = 4\ln x_{gN} + 8\ln^2 \sigma_g \tag{4.123}$$

4.12.2 Derived average sizes

If the number-size distribution of a particulate system has been determined and found to be log-normal, equations (4.109) to (4.112) may be used to determine other average sizes.

For example, the mean size of a weight distribution is given by:

$$x_{VM} = \frac{\displaystyle\sum_{r=0}^{r=\infty} x_r^4\, dN_r}{\displaystyle\sum_{r=0}^{r=\infty} x_r^3\, dN_r}$$

$$= \frac{x_{NM}^4}{x_{NV}^3}$$

Therefore $\qquad \ln x_{VM} = 4\ln x_{NM} - 3\ln x_{NV} \tag{4.124}$

Substituting from equations (4.122) and (4.123):

$$\ln x_{VM} = \ln x_{gN} + 3.5 \ln^2 \sigma_g \qquad (4.125)$$

Similarly, the mean size of a surface distribution is given by:

$$\ln x_{SV} = \ln x_{gN} + 2.5 \ln^2 \sigma_g \qquad (4.126)$$

Using this equation, the specific surface of the particulate system may be determined since:

$$S_V = 6/x_{SV} \qquad (4.127)$$

4.12.3 Transformation of the log-normal distribution by count into one by weight

If a number distribution is log-normal, the weight distribution is also log-normal with the same geometric standard deviation. Using the same treatment as was used to derive equation (4.109) gives, for a weight analysis:

$$\ln x_{VM} = \ln x_{gV} + \tfrac{1}{2} \ln^2 \sigma_g \qquad (4.128)$$

Comparing with equation (4.114) gives:

$$\ln x_{gV} = \ln x_{gN} + 3.0 \ln^2 \sigma_g \qquad (4.129)$$

Since the relations between the number-average sizes and the number-geometric mean are known (equations (4.120) to (4.123)), these can now be expressed as relationships between number-average sizes and the weight (volume) geometric mean x_{gV}.

$$\ln x_{NL} = \ln x_{gV} - 2.5 \ln^2 \sigma_g \qquad (4.130)$$

$$\ln x_{NS} = \ln x_{gV} - 2.0 \operatorname{in}^2 \sigma_g \qquad (4.131)$$

$$\ln x_{NV} = \ln x_{gV} - 1.5 \ln^2 \sigma_g \qquad (4.132)$$

$$\ln x_{NM} = \ln x_{gV} - 1.0 \ln^2 \sigma_g \qquad (4.133)$$

Other average sizes may be derived from the above, using a similar procedure to that used to derive equations (4.125) and (4.126) to give:

$$\ln x_{SV} = \ln x_{gV} - 0.5 \ln^2 \sigma_g \qquad (4.134)$$

Similarly, for a surface distribution, the equivalent equation to equation (4.129) is:

$$\ln x_{gS} = \ln x_{gN} + 2.0 \ln^2 \sigma_g \qquad (4.135)$$

Substituting this relationship into equations (4.130) to (4.133) yields the equivalent relationships relating surface-average sizes with the surface-geometric mean diameter.

4.13 Relationship between median and mode of a log-normal distribution

The log-normal equation may be written:

$$\frac{d\phi}{dx} = \frac{1}{x \ln \sigma_g \sqrt{(2\pi)}} \exp(-x^2)$$

where

$$2 \ln^2 \sigma_g X^2 = (\ln x - \ln x_g)^2$$

$$\sqrt{2} \ln \sigma_g \frac{dX}{dx} = \frac{1}{x}$$

At the mode:

$$\frac{d^2\phi}{dx^2} = 0 = -\sqrt{2} X - \ln \sigma_g$$

i.e.

$$\ln x_m = \ln x_g - \ln^2 \sigma_g \tag{4.135}$$

where x_m is the mode.

4.14 An improved equation and graph paper for log-normal evaluations [56]

Using the relationship:

$$\frac{1}{x} = \exp(-\ln x) \tag{4.137}$$

equation (4.101) may be transformed into the following form:

$$\frac{d\phi}{dx} = \frac{1}{\ln \sigma_g \sqrt{(2\pi)}} \exp(-\ln x) \exp\left[-\frac{1}{2 \ln^2 \sigma_g} \left(\ln \frac{x}{x_g} \right)^2 \right]$$

$$= \frac{1}{\ln \sigma_g \sqrt{(2\pi)}} \exp\left[-\frac{1}{2 \ln^2 \sigma_g} \{2 \ln^2 \sigma_g \ln x + (\ln x - \ln x_g)^2\} \right]$$

$$\left\{ \begin{array}{c} \\ \end{array} \right\} = \ln^2 x - 2 \ln x (\ln^2 x_g - \ln^2 \sigma_g) + \ln^2 x_g$$

Replacing x_g by x_m (equation (4.135)):

$$\left\{ \begin{array}{c} \\ \end{array} \right\} = 2 \ln x_m \ln^2 \sigma_g + \ln^4 \sigma_g + (\ln x - \ln x_m)^2$$

Therefore

$$\left[\begin{array}{c} \\ \end{array} \right] = -\ln x_m - \frac{\ln^2 \sigma_g}{2} - \frac{(\ln x - \ln x_m)^2}{2 \ln^2 \sigma_g}$$

Hence:

$$\frac{d\phi}{dx} = \frac{1}{x_m \ln \sigma_g \sqrt{(2\pi)}} \exp\left(-\tfrac{1}{2} \ln^2 \sigma_g \right) \exp\left[-\frac{\ln^2 x/x_m}{2 \ln^2 \sigma_g} \right]$$

This form of the log-normal equation is more convenient for use since the variable only appears once. It may be simplified further:

$$\frac{d\phi}{dx} = A \exp(-b \ln^2 x/x_m) \tag{4.138}$$

where

$$b = \frac{1}{2 \ln^2 \sigma_g}$$

and

$$A = \sqrt{\frac{b}{\pi}} \cdot \frac{\exp(-1/4b)}{x_m}$$

The relationship between geometric mean and mode (equation (4.135)) takes the form:

$$C = \frac{x_m}{x_g} = \exp\left(-\frac{1}{2b}\right)$$

This modified form of the log-normal equation simplifies parameter determination from log-probability plots of experimental data. The graph paper may be furnished with additional scales of b and C both being determined by drawing a line parallel to the distribution through the pole (0.25 μm, 50%).

4.14.1 Applications

Consider a log-normal distribution with a geometric mean

$$x_g = 6.75 \ \mu m$$

and

$$\sigma_g = \frac{x_{84\%}}{x_{50\%}} = \frac{11.1}{6.75} = 1.64$$

The mode, according to equation (4.135), is:

$$x_m = 5.27 \ \mu m$$

making:

$$\frac{d\phi}{dx} = 0.1344 \exp\left[-2.02 \ln^2 \frac{x}{5.27}\right]$$

This form is particularly useful when further mathematical computation is envisaged such as for grade efficiency, $G_c(x)$, calculation since:

$$G_c(x) = E_T \frac{dF_c(x)}{dF(x)}$$

$$= E_T \frac{d\phi_c}{dx}$$

where $d\phi_c/dx$ is the relative frequency of the coarse product, $d\phi/dx$ is the relative frequency of the feed and E_T is the total efficiency.

References

1 Tsubaki, J. and Jimbo, G. (1979), *Powder Technol.*, **22**, 2, 161–70.
2 Heywood, H. (1947), Symposium on *Particle Size Analysis*, Inst. Chem. Engrs, Suppl. 25, 14.
3 Heywood, H. (1963), *J. Pharm. Pharmac.*, Suppl. 15, 56T.
4 Heywood, H. (1973), *Harold Heywood Memorial Lectures*, Loughborough University, England.
5 Wadell, H. (1932), *J. Geol.*, **40**, 243–53.
6 Wadell, H. (1935), *ibid.*, **43**, 250–80.
7 Wadell, H. (1934), *J. Franklin Inst.*, **217**, 459.
8 Wadell, H. (1934), *Physics*, 5, 281–91.
9 Krumbein, W.C. (1934), *J. Sediment. Petrol.*, **4**, 65.
10 Krumbein, W.C. (1941), *ibid.*, **11**, 2, 64–72.
11 Krumbein, W.C. (1964), *ibid.*, **34**, 195.
12 Hausner, H.H. (1966), *Planseeber Pulvermetall.*, **14**, 2, 74–84.
13 Medalia, A.I. (1970/71), *Powder Technol.*, **4**, 117–38.
14 Church, T. (1968/69), *ibid.*, **2**, 27–31.
15 Cole, M. (June 1971), *Am. Lab.*, 19–28.
16 Pahl, M.H., Schädel, G. and Rumpf, H. (1973), *Aufbereit, Tech.*, 5, 257–64; 10, 672–83; 11, 759–64.
17 Davies, R. (1975), *Powder Technol.*, **12**, 111–24.
18 Beddow, J.K. (1974), Report A390-CLME-74-007, The University of Iowa, 52242.
19 Laird, W.E. (1971), *Particle Technology*, Proc. Seminar, Indian Inst. Technol., Madras (eds D. Venkateswarlu and A. Prabhakdra Rao), 67–82.
20 Cartwright, J. (1962), *Ann. Occup. Hyg.*, 5, 163.
21 Fair, G.L. and Hatch, L.P. (1933), *J. Am. Wat. Wks. Ass.*, **25**, 1551.
22 Crowl, V.T. (1963), Paint Research Station Report, No. 325, Teddington, London.
23 Ellison, J.McK. (1954), *Nature*, **173**, 948.
24 Hodkinson, J.R. (1962), PhD Thesis, London University.
25 Tsubaki, J. and Jimbo, G. (1979), *Powder Technol.*, 161–78.
26 Tsubaki, J., Jimbo, G. and Wade, R. (1975), *J. Soc. Mat. Sci. Japan*, **24**, 262, 622–6.
27 Schwarcz, H.P. and Shane, K.C. (1969), *Sedimentology*, **13**, 213–31.
28 Meloy, T.P. (1969), *Screening*, AIME, Washington, D.C.
29 Meloy, T.P. (1971), *Eng. Found.* Conference, Deerfield.
30 Meloy, T.P. (1977), *Powder Technol.*, **16**, 2, 233–54.
31 Meloy, T.P. (1977), *ibid.*, **17**, 1, 27–36.
32 Gotoh, K. and Finney, J.L. (1975), *ibid.*, **12**, 2, 125–30.
33 Mandelbrot, B.P. (1977), *Fractal Form Chance and Dimension*, W.H. Freeman & Co., San Francisco.
34 Kaye, B.H. (1977), *Proc. Particle Size Analysis Conference*, Salford, Chem. Soc., Analyt. Div., London.
35 Flook, A.G. (1978), *Powder Technol.*, **21**, 2, 295–8.
36 Medalia, A.J. (1970/71), *ibid.*, **4**, 117–38.
37 Morony, M.J., *Facts from Figures*, Pelican.
38 Herdan, G. (1960), *Small Particle Statistics*, Butterworths.
39 Irani, R.R. and Callis, C.F. (1963), *Particle Size: Interpretation and Applications*, Wiley, NY.
40 Rosin, P. and Rammler, E. (1933), *J. Inst. Fuel*, 7, 29; (1927), *Zemast*, **16**, 820, 840, 871, 897; (1931), *Zemast*, **20**, 210, 240, 311, 343.
41 Rumpf, H. and Ebert, K.F. (1964), *Chem. Ingr. Tech.*, **36**, 523–37.
42 Rumpf, H. and Debbas, S. (1966), *Chem. Eng. Sci*, **21**, 583–607.
43 Rumpf, H., Debass, S. and Schönert, K. (1967), *Chem. Ingr. Tech.*, **39**, 3, 116–25.
44 Rumpf, H. (1961), *ibid.*, **33**, 7, 502–8.
45 Leschonski, K., Alex, W. and Koglin, B. (1974), *ibid.*, **46**, 3, 23–6.
46 D1N 66141, February 1974.

47 Leschonski, K., Alex, W. and Koglin, B. (1974), *Chem. Ingr. Tech.*, **46**, 3, 101–6.
48 Baturin, V.P. (1943), *Reports Acad. Sci. USSR* (Moscow), **38**, 7 (in Russian).
49 McManus, D.A. (1963), *J. Sediment. Petrol.*, **33**, 670.
50 Krumbein, W.C. and Pettijohn, F.J. (1938), *Manual of Sedimentary Petrolography*, Appleton-Century-Crofts, New York.
51 Page, H.G. (1955), *J. Sediment. Petrol.*, **25**, 285.
52 Griffiths, J.C. and McIntyre, D.D. (1958), *A Table for the Conversion of Millimetres to Phi Units*, Mineral Ind. Expl. Sta., Pennsylvania State University.
53 Folk, R.L. (1966), *Sedimentology*, **6**, 73.
54 Griffiths, J.C. (1962), In *Sedimentary Petrography* (ed. H.B. Milnes), **1**, Macmillan, New York, Ch. 16.
55 Krumbein, W.C. (1937), *Neues Jahrb. Mineral Geol.* Beil-Bd, **73A**, 137–50.
56 Svarovsky, L. (1973), *Powder Technol.*, **7**, 6, 351–2.
57 Cartwright, J. (1962), *Ann. Occup. Hyg.*, **5**, 163–71.
58 Johnston, J.E. and Rosen, L.J. (1976), *Powder Technol.*, **14**, 195–201.
59 Schurt, G.A. and Johnston, J.E. (1975), *Proc. 13th AEC Air Cleaning Conference*, CONF-740807, **2**, UC-70, 1039–44.
60 Gates, A.O. (1915), *Trans. AIME*, **52**, 875–909.
61 Schumann, R. (1940), *ibid.*, Tech. publ. 1189.
62 Gaudin, A.M. and Meloy, T.P. (1962), *ibid.*, 43–50.
63 Austin, L.R. and Klimpel, R.R. (1968), *Trans. Soc. Mech. Engrs*, 219–24.
64 Harris, C.C. (1969), *Trans. AIME*, **244**, 187–90.
65 Svensson, J. (1955), *Acta Polytech. Scand.*, **167**, 53.
66 Tarjan, G. (1974), *Powder Technol.*, **10**, 73–6.
67 Kolmogoroff, A.N. (1941), *Dokl. Akad. Nauk SSSR*, Novaja Ser., **31**, 2.
68 Stein, F. and Corn, M. (1976), *Powder Technol.*, **13**, 133–41.
69 Narva, O.M. *et al.* (1977), *Zavod. Lab.*, **43**, 4, 477–8.
70 Naylor, A.G. and Kaye, B.H. (1972), *J. Colloid. Interfac. Sci.*, **39**, 1, 103–8.
71 White, E.T. *et al.* (1972), *Powder Technol.*, **5**, 2, 127–30.
72 Kaye, B.H. and Treasure, C.R.G. (1966), *Br. Chem. Engng*, **11**, 1220.
73 Seidel, H. (1966), *Staub*, **26**, 329.
74 Wise, M.E. (1954), *Philips Res. Rep.*, 9, 231.
75 Petersen, E.E., Walker, P.L. and Wright, C.C. (1952), ASTM Bull. TP 116.
76 *Reporting Particle Size Characteristics of Pigments* (1955), ASTM D1366-55T.
77 Scott, K.J. and Mumford, D. (1971), *CSIR Spec. Rep. Chem.* 155, Pretoria, South Africa.
78 Martin, P.M. and Mills, A.A. (1977), *Moon* (Netherlands), **16**, 2, 215–19.
79 Beddow, J.K., Sisson, K. and Vetter, Q.F. (1976), *Powder Metall. Int.*, **2**, 69–76; **3**, 107–9.
80 Tracey, V.A. and Llewelyn, D.M. (1976), *ibid.*, **8**, 3, 126–8.
81 Beddow, J.K., *Powtech 77*, Heyden.
82 Beddow, J.K., Philip, G.C. and Nasta, M.D. (1975), *Planseeber. Pulvermetall.*, **23**, 3–14.
83 Beddow, J.K. *et al.* (1976), 8th Ann. Mtg. Fine Particle Soc., Chicago.
84 Mandelbrot, B.B. (1977), *Fractals: Form, Chance and Dimension*, W.H. Freeman, San Francisco, California, USA; Reading, UK.

5 Sieving

5.1 Introduction

Sieving is an obvious means of classification and it has been used since early Egyptian times for the preparation of foodstuffs. The simplest sieves would be made of some woven material but punched plate sieves are recorded in early Egyptian drawings and by 1556 Agricola is illustrating woven wire sieves [1].

Such sieves were used for powder classification and the inception of test sieving did not arise until sieve aperture sizes were standardized. Standard apertures were first proposed in 1867 by Rittinger who suggested a $\sqrt{2}$ progression of sizes based on 75 μm [2]. Modern standards are based on a fourth root of two progression, apart from the French AFNOR series which is based on a tenth root of ten. This range has also been extended downwards but the tolerances are rather liberal which limits their acceptance.

Sieves are often referred to by their mesh size which is the number of wires per linear inch. The American ASTM range is from 400 mesh to 4.24 in. The apertures for the 400 mesh are 37 μm hence the wire thickness is 26.5 μm and the percentage open area is 34.

Standards are in the process of modification to match the international ISO series which is based on a root-two progression starting at 45 μm.

Sieve analysis is one of the simplest, most widely used methods of particle size analysis, that covers the approximate size range 20 μm to 125 mm using standard woven wire sieves. Micromesh sieves extend the range down to 5 μm or less and punched plate sieves extend the upper range.

The sieve size d_A is the minimum square aperture through which the particles can pass. Fractionation by sieving is a function of two dimensions only, maximum breadth and maximum thickness for, unless the particles are excessively elongated, the length does not hinder the passage of particles through the sieve apertures (figure 5.1).

This definition only applies to sieves having square apertures.

Sieve analysis results can be highly reproducible even when using different sets of sieves. Although most of the problems encountered in sieving have been known for many years and solutions proposed, reproducibility is rarely achieved in practice due to the failure to take cognisance of these problems [3].

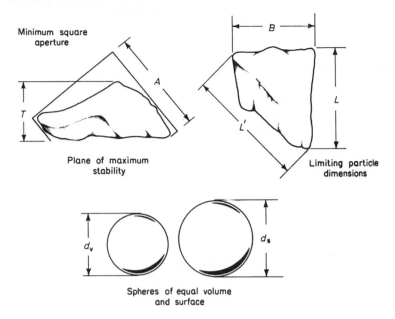

Fig. 5.1 Equivalent particle diameters (after Heywood).

5.2 Woven-wire and punched plate sieves

Sieve cloth is woven from wire and the cloth is soldered and clamped to the bottom of open cylindrical containers. Although the apertures are usually described as square they deviate from this shape due to the three-dimensional structure of the weave. Heavy-duty sieves are often made of perforated plate giving rise to circular holes. Various other aperture shapes, such as slots for sieving asbestos fibres, are also available. Fine sieves are usually woven with phosphor bronze wire, medium with brass and coarse with mild steel. Special purpose sieves are also available in stainless steel and the flour industry frequently uses nylon and silk.

A variety of sieve aperture ranges are currently used and these may be classified as coarse (4 to 100 mm), medium (0.2 to 4 mm) and fine (less than 0.2 mm). Large-scale sieving machines are used for the first range and these require a charge of 50 to 100 kg of powder [4]. There are a range of commercial sieve shakers available for medium-sized particles and these usually classify the powder into five or six fractions: the loading is 50 to 100 g. The fine range extends down to 37 μm with woven wire mesh and below this for micromesh sieves.

Some of the standard specifications are listed below: Great Britain, BS 410 (1969): Woven-wire test sieves; USA, ASTM E11-70: Woven-wire test sieves; E161-607: Micromesh (electroformed) sieves; Germany, DIN 4187: Perforated plate sieves; DIN 4188: Woven-wire test sieves; DIN 4195: Textile sieve cloths; France, AFNOR NFX 11-501: Woven-wire test sieves; International, ISO R-565-1972(E): Woven-wire and perforated plate test sieves. As well as these, several industries have their own specifications.

Wahl [5] describes the production of highly wear-resistant sieves in soft annealed plates of chromium-steel by punching, plasma cutting or mechanical working followed by heating and strain hardening.

A method of preparing a metal sieve cylinder has also been described in which parallel grooves are cut on a metal mandrel [6]. These are filled with an insulating material to produce a network of conducting and non-conducting lines. After passivating the mandrel, copper wire is wound perpendicular to the linear net to give the pattern and the whole is immersed in a nickel-plating bath; 2500 holes cm^{-2} are produced with a nickel deposition of 25 to 30 μm.

5.3 Electroformed micromesh sieves

Basically, the photo-etching process is as follows: a fully degreased metal sheet is covered on both sides with a photosensitive coating. The desired pattern is applied photographically to both sides of the sheet. Subsequently the sheet is passed through an etching machine and the unexposed metal is etched away. Finally the photosensitive coating is removed. A supporting grid is made by printing a coarse line pattern on both sides of a sheet of copper foil coated with photosensitive enamel. The foil is developed and the material between the lines is etched away. The mesh is drawn tautly over the grid and nickel-plated on to it. The precision of this method gives a tolerance of ±2 μm for apertures up to 500 μm. Some typical sieve meshes are illustrated in figure 5.2.

For square mesh sieves the pattern is ruled on to a wax-coated glass plate, with up to 8000 lines per inch with each line 0.0001 in wide and the grooves are etched and filled.

Micromesh sieves were first described by Daescher et al. [7]. They are available from Buckbee Mears in aperture sizes from 5 to 150 μm with apertures in a fourth root of two progression for plus 25 μm apertures. Other aperture sizes are available and Zwicker reports using them down to 1 μm in size [8].

The percentage open area decreases with decreasing aperture size ranging from 2.4% for 5 μm aperture sieves to 31.5% for 40 μm aperture sieves and this leads to greatly extended sieving times for the smaller aperture sieves.

Endecotts make them in the size range 5.5 to 31 μm in a root-two progression. These sieves have about twice the open area of the Daescher sieves and are therefore more fragile. Burt [9] examined samples of British sieve cloth and found that the average width of metal between openings for each grade of cloth was one-third to one-half the minimum recommended for sieves of American manufacture [10].

Veco manufacture round and square aperture sieves. The apertures for the former are in the shape of truncated cones with the small circle uppermost [11]. This reduces blinding but also reduces the percentage open area and therefore prolongs the sieving time. Where thicker sieves are required, the 'Veco' type sieves are subjected to further electrodeposition on both sides to produce the 'Vero' type sieves with biconical apertures.

Micromesh sieves are also manufactured in the German Democratic Republic [12] using a process which is said to give better bonding of the mesh on to the support grid.

Fig. 5.2 Electroformed micromesh sieves. (a) 75 μm, (b) 45 μm, (c) 30 μm, (d) 20 μm.

The tolerances with these sieves are much better than those for woven-wire sieves, the aperture being guaranteed to within 2 μm of nominal size. Each type of sieve has different advantages for specific problems [13]. When ultrasonics are not necessary for cleaning the sieves and the sample retained in the apertures can be dissolved out, Endecotts are to be preferred because of their large percentage open area. For insoluble materials Buckbee Mears are recommended for sub-15 μm and Veco for plus-15 μm.

Dry sieving is often possible with the coarser mesh sieve, and this may be speeded up and the lower limit extended to about 15 μm with the air-jet technique. Aggregation of the particles may sometimes be reduced by drying or adding about 1% of a dispersant, stearic acid and fine silica being two possibilities. If this is unsuccessful, wet methods have to be used [13, 14]. Ultrasonic vibrations are often used as an aid to sieving or for cleaning blocked sieves; the danger of breakage of the delicate mesh becomes very likely under these conditions, readily occurring at frequencies of 50 Hz and sometimes at frequencies as high as 20 kHz. Endecotts sieves should never be subjected to ultrasonics [13], and a recommended safe frequency for other sieves, according to Colon [11], is 40 kHz for blockages which cannot be cleared by other means, though Crawley [15] reported damage to an 11 μm sieve at a frequency of 80 kHz and a power level of 40 W, and subsequently recommended a frequency of 800 kHz and power level of 20 W. Daescher [14] confirmed Rosenberg's [16] earlier findings that the rate of cavitation erosion is less in hydrocarbons than in alcohol and about six times greater in water than in alcohol. By saturating alcohols with carbon dioxide, the rate is less than in degassed alcohol. He tested three ultrasonic cleaners, 40 kHz, 60 W; 40 kHz, 100 W; 90 kHz, 40 W, and found that the high-frequency cleaner produced the least amount of erosion.

Veco recommend 15 to 20 s at a time in a low-power 40 kHz ultrasonic bath containing an equivolume mixture of isopropyl or ethyl alcohol and water with the sieve in a vertical position and this is in general accord with ASTM Standard E161–70 [10].

Alpine recommend a cleaning time varying from 10 to 20 min for 10 μm sieves down to 2 to 4 min for sieves in the size range 50 to 100 μm.

While very expensive, these sieves find a ready market in specialized fields [17].

5.4 British Standard specification sieves [18]

The Standard is divided into three parts. Part 1 covers a range of woven-wire fine mesh test sieves of aperture width in a fourth root of two progression from a maximum of 3.35 mm to a minimum of 0.045 mm. The series incorporates a root-two progression proposed as ISO sieves for international co-ordination. Two grades are provided of equal nominal dimensions but different tolerances. Also included is a range of woven-wire medium test sieves of aperture width from a maximum of ½ in to a minimum of $\frac{1}{32}$ in. In addition, requirements are given for the manufacture of the sieves, including the frame mounting cover and finish and the wire cloth, together with requirements for marking, and the method of examination for acceptance.

Parts 2 and 3 of the Standard deal with coarse test sieves (perforated plates) having square apertures of width from 4 in to $\frac{3}{16}$ in for general purposes, and also a range of

coarse heavy-duty test sieves for single-hole gauges specifically for use with blast-furnace coke, having square apertures of width from 8 in to $\frac{1}{8}$ in.

It is a matter of some interest that, of the 26 sizes in the BS fine mesh series, 11 are now identical with the corresponding values in the American series, and a further 4 are within one micrometre of the corresponding American values; the appendices include tabular summaries of the USA Standard series, the German Standard and the French.

The mesh for the sieves is woven wire of circular cross-section, the most commonly used materials being phosphor bronze for fine sieves (aperture $< 250\ \mu m$), brass for coarser sieves and mild steel for apertures greater than 1 mm. The mesh number, a common method of designating sieves, is the number of wires per linear inch of sieve cloth.

From the table reprinted from BS 410 (1962) (Table 5.1) it can be seen that the 200-mesh sieve has a nominal aperture of 75 μm and a nominal wire diameter of 52 μm.

Table 5.1

Aperture width (μm)	Mesh number	Aperture tolerances			Percentage open area
		Average	Intermediate	Maximum	
75	200	4.6	19	33	35
45	350	3.8	17	30	38

Tolerances
   Average: The average aperture width must not deviate from the nominal aperture width by more than this amount.
   Intermediate: Not more than 6% of the apertures may deviate from the nominal aperture width by more than this amount.
   Maximum: No aperture may deviate from the average aperture width by more than this amount.

The average aperture width must lie between 70.4 and 79.6 μm with not more than 6% of the apertures outside the size range 56 to 94 μm and all the apertures in the size range 42 to 108 μm. For the 350-mesh sieve of nominal aperture 45 μm all apertures must be in the range 15 to 75 μm.

Since these requirements apply to the mesh and not to the sieve, individual sieves may have tolerances outside these limits. In the author's experience examination of individual sieves, according to BS 410, suffices to show that in nearly every case the quality is far higher than required by the Standard, although this cannot be generally assumed.

Leschonski [19] examined eight 50 μm sieves to DIN 4188 and found that the number distribution of apertures was log-normal with one exception. The median varied between 47.3 and 63.2 μm and the standard deviations ranged from 1.8 to 8.6 μm, thus greatly exceeding the standard specification.

Ilantzis [20] examined thirteen sieves to AFNOR NFX 11–501 with nominal aperture widths from 40 to 630 μm and found that only four lay within the specification.

5.5 Methods for the use of fine sieves

The sieving operation may be carried out wet or dry, by machine or by hand, for a fixed time or until powder passes through the sieve at a constant low rate.

5.5.1 Machine sieving

Machine sieving is carried out by stacking the sieves in ascending order of aperture size and placing the powder on the top sieve. A closed pan, a receiver, is placed at the bottom of the stack to collect the fines and a lid is placed at the top to prevent loss of powder. A stack usually consists of five or six sieves in a root-two progression of aperture size. The stack is vibrated for a fixed time and the residual weight of powder on each sieve determined. Results are usually expressed in the form of a cumulative percentage in terms of the nominal sieve aperture.

For routine control purposes it is usual to machine sieve for 20 min after which time the sieving operation is deemed to be complete. In BS 1796 [21] it is recommended that sieving be continued until less than 0.2% of the sample passes through in any 5-min sieving period. In ASTM D452 a 20-min initial sieving period is recommended followed by 10-min periods during which the amount passing should be less than 0.5% of the total feed. For coarse aggregates sieving is deemed complete when the rate falls below 1% per minute (ASTM C136).

It is generally recommended that if losses during sieving exceed 0.5% of the total feed the test should be discarded. Preliminary hand sieving on the finest sieve should be carried out for the removal of dust; this dust would otherwise pass through the whole nest of sieves and greatly prolong the sieving time; it would also percolate between sieves in the nest and increase powder loss.

The sieving action of many commercial machines is highly suspect and frequently subsequent hand sieving will produce a sieving rate far greater than is produced on the machine. The best type of sieving action is found with the types of shakers exemplified by the Pascal Inclyno and the Tyler Ro-tap which combine a gyratory and a jolting movement although the simpler vibratory sieves may be suitable in specific cases. ASTM B214 suggests 270 to 300 rotations per minute for granular metals combined with 140 to 160 taps to reduce blinding of sieve apertures.

In BS 1796, which applies to the sieving of material from 3350 to 53 μm in size, it is suggested that the sieving operation be carried out in 5-min stages at the end of which the sieves should be emptied and brushed in order to reduce blocking of apertures. Test sieving procedures are also described in ISO 2591 (Zurich) (1973).

Modifications to the methods may be necessary for materials that are not free-flowing, are highly hygroscopic, very fragile, have abnormal particle shapes or have other properties which cause difficulty in sieving. For example, in ASTM C92 the fines are first removed by washing through the finest sieve; the residue is then dried and analysed in the dry state.

5.5.2 *Wet sieving*

These techniques are useful for material originally suspended in a liquid and necessary for powders which form aggregates when dry-sieved and a full description of the techniques is given in BS 1796.

Automated wet sieving has been proposed by several authors [17, 67]. In most of these methods a stack of sieves is filled with a liquid and the sample is fed into the top sieve. Sieving is accomplished by rinsing, vibration, reciprocating action, vacuum, ultrasonics or a combination [13]. A successful method is a combination of elutriation and sieving. Commercial equipment is available in which the sample is placed in the top sieve of a nest of sieves and sprayed with water while the nest is being vibrated.

Several methods of wet sieving using micromesh sieves have been described. Mullin [66] describes a technique of wet sieving using an 80 kHz ultrasonic bath with an output of 40 W in which the microsieve rests on a support which in turn rests on a beaker in the bath. Sieving intervals are 5 min with an initial load of 1 g; sieving continues using other beakers until no further powder can be seen passing through the sieve.

Colon [13] rinses the fines through the sieve aperture with a suitable liquid after 0.5 to 1 g has been dispersed in a small volume of the liquid. Sieving is then continued by moving the sieve up and down in a glass beaker filled with the sieving liquid, so that the direction of flow of the liquid through the sieve openings is continually reversed; this helps to clean blocked apertures and disperse agglomerates. Depending on the type of micromesh sieve and the resistance to breakage of the particles during sieving, ultrasonics may be used. After some time, which should preferably be standardized, the sieve is transferred to a glass beaker with fresh sieving liquid and sieving continued. Sieving is deemed complete when the number of particles passing through the sieve is negligible compared with the total number of particles.

If about 1 g of powder is dispersed in 1 litre of liquid, this may be poured through a sieve supported in a retort stand. For woven-wire and coarse micromesh sieves, the sieves may be mechanically rapped to facilitate sieving. The powder is then rinsed off the sieve and weighed or the sieve is dried and weighed [22]. With fine sieves, an ultrasonic probe may be necessary and the sieving liquid must have a low surface tension, e.g. acetone, otherwise it will not flow through the sieve apertures. The liquid may be recovered by distillation. Similar methods have been described by other investigators [27, 28].

A wet screening device for the size range 10 to 100 μm, which includes a sieve vibrator of variable amplitude, is commercially available (figure 5.3).

Daescher [14] describes a method using a set of tared sieves mounted on a special funnel held in a filter flask. One to three grams of the powder are placed on the top sieve and washed through each sieve in turn with a suitable polar liquid or hydrocarbon containing a trace of dispersant. At the same time alternate pulses of pressure and suction are applied to the filter flask. This pulsating action orientates the particles in such a way as to speed up the sieving action. After the sample has been washed through each sieve they are dried and weighed. A full analysis can be completed in less than one hour.

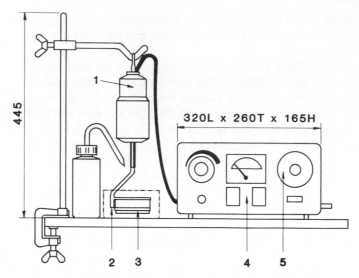

320L x 260T x 165H

Fig. 5.3 The Alpine wet sieving device. The device consists of a variable-amplitude sieve vibrator (1) with support (2), microsieve (3), regulator with voltmeter for adjusting amplitude (4) and timer (5).

Ioos [23] describes a device which makes it possible to carry out particle size analysis from 60 μm down to 5 μm by means of sieves arranged vertically above one another and subjecting the nest of microsieves to ultrasonics.

The Gallie-Porritt apparatus (BS 4398) consists of a metal funnel terminating in a short cylindrical outlet in which a wire sieve cloth is soldered. Water, at a pressure greater than 2 bar, is supplied by a nozzle designed to discharge a spreading jet through the sieve. A similar arrangement is provided for another tube to give a gentle stream of water to keep the level of the water in the funnel constant throughout the test. About 25 g of powder are slurried and introduced into the funnel at the commencement of the test which continues until the water issuing from the apparatus is clear. The residual mass is determined in order to find the undersize percentage.

A method for wet sieving clays is described in ASTM C325–56 in which the fines are washed out first and the rest is washed through a nest of sieves. In ASTM D313 and C117 washing through a 200-mesh sieve is suggested for the removal of fines and in D185–72 the use of a camel-hair brush is recommended to facilitate passage through a 325-mesh sieve.

5.5.3 Hand sieving

Hand sieving is time-consuming but necessary for dependable dry sieving data. Consider a powder in the approximate size range 100 μm to 1 mm. A representative sample of approximately 50 g is first obtained and the whole of the sample used in the analysis.

The preferred method of sampling is with a spinning riffler or, failing that, a chute splitter. Even though hand quartering is recommended in some standards (e.g. ASTM D346) this should never be used.

A 90-μm sieve should be nested on to a catch pan and the tared sample placed on the mesh and the whole sealed with a lid. The sieve should be slightly inclined to the horizontal and rapped with a cylindrical piece of wood about 8 in long and of 1 in diameter. (The heel of the hand is recommended in ASTM C136 and is an acceptable alternative if your hand can take it!) The rate of rapping should be about 150 per min and the sieve should be rotated one-eighth of a turn every 25 raps. After about 10 min the residue on the sieve is transferred to a 1-mm sieve which is nested in a second catch pan and the process is repeated. The fines are carefully removed from the first catch pan for subsequent weighing. It is suggested that white card be placed on the bench (approximately 60 cm square) so that any accidental spillage can be recovered.

After about 10 min the sieve and lid are transferred to the empty catch pan and sieving is continued for 5 min. The amount passing through is weighed and the sieving continued until the rate of passage is less than 0.1% of the total feed per min. The sieve residue is removed for weighing; any material adhering to the underside of the sieve is transferred to the undersize residue. The sieve may then be placed upside down on a sheet of paper and the particles wedged in apertures removed by brushing and added to the sieve residue.

This process is repeated with the 710-, 500-, 355-, 250-, 180-, 125- and finally the 90-μm sieve, the final residue being added to the dust collected initially. Brushing is not recommended for sieves with apertures less than 200 μm due to the possibility of damage [3].

Sieves should be washed and dried after use. Ultrasonics should be used to remove particles clogging the apertures or these may be leached out if this can be done without damage to the sieve.

The results may be expressed in terms of the nominal size although it would be preferable to use sieves calibrated against a standard powder. In ASTM C429 and D2772-69 (reapproved 1973) the use of a reference set and a matching set is proposed. The reference set should be used on every fiftieth analysis for comparison and when sieves wear out they should be replaced with ones from the reference set which in turn are replaced by new sieves.

5.5.4 Air-jet sieving

The principle of operation of this instrument (figure 5.4) is that air is drawn upwards, through a sieve, from a rotating slit so that material on the sieve is fluidized. At the same time a negative pressure is applied to the bottom of the sieve which removes fine particles to a collecting device (a filter paper). With this technique, there is a reduced tendency to blind the apertures, and the action is very gentle, making it suitable for brittle and fragile powders. Sieving is possible with some powders down to 10 μm in size but with others balling occurs [24, 25]. The reproducibility is much better than by hand or machine sieving. Size analyses are performed by removal of particles from the fine end of the size distribution by using single sieves consecutively.

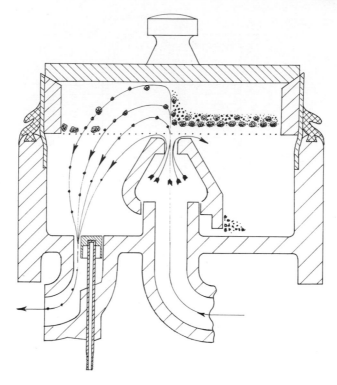

Fig. 5.4 Mode of action of the Alpine air-jet sieve.

The end-point of sieving can be determined by microscopic examination of the cleanness of the sample [26], or by adopting the same criteria as are used in conventional dry sieving. Sieving is usually completed in 3 to 5 min using a 5 g to 10 g sample on 8-in diameter sub-75-μm sieves. Sieving time is more protracted with finer mesh sieves, and a sieving time of 20 min is usual with a 1 g load on a 3-in diameter 20-μm sieve.

A discussion of the techniques has been presented by Jones [27], and Lauer [4, 65] has appraised it by microscopic examination of the powder fractions.

5.5.5 *The sonic sifter [28]*

The Allen–Bradley sonic sifter [29] is produced as a laboratory (LP3) and industrial (P60) model. It is claimed to be able to separate particles in the 2000 to 20 μm range for most materials and 5660 to 10 μm in some cases. It combines two motions to provide particle separation, a vertical oscillating column of air and a repetitive mechanical pulse. The sonic sifter moves the air in the sieve stack (figure 5.5). The oscillating air sets the sample in a periodic vertical motion which reduces sieve blinding and breaks down aggregates and yet produces very little abrasion, thus reducing sieve wear and particle breakage.

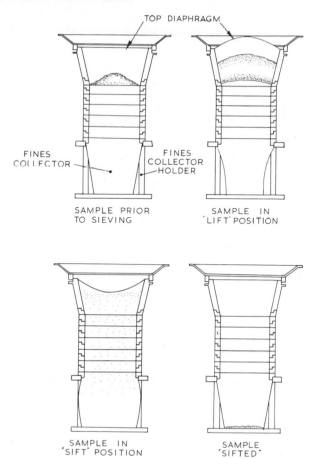

TOP DIAPHRAGM

FINES
COLLECTOR

FINES
COLLECTOR
HOLDER

SAMPLE PRIOR
TO SIEVING

SAMPLE IN
"LIFT" POSITION

SAMPLE IN
"SIFT" POSITION

SAMPLE
"SIFTED"

Fig. 5.5 Diagram of four conditions during sieving operation.

5.5.6 Felvation

The term *felvation* has been applied to a technique for grading powder using an elutriation process, with the sieves acting as stops for the coarse powder in suspension [30]. The apparatus used by Burt [9] is shown in figure 5.6.

The powder to be classified is dispersed in the predispersion unit A, which is connected to the bottom of the first felvation column B. The needle valve C is opened to allow liquid from the header tank to carry suspended particles into the conical base of the column, where they are fluidized. The flow rate is gradually increased until the finer particles are elutriated up the column to D, which consists of $\frac{1}{2}$-in square micromesh, or $\frac{3}{4}$-in diameter circular woven-wire, mesh cloth. The particles continue into the next felvation column E where they meet, and if fine enough pass through sieving surface F. The flow rate is increased in steps until larger and larger particles are elutriated and continued until the ascending particles are too large to pass through the sieves. The end-

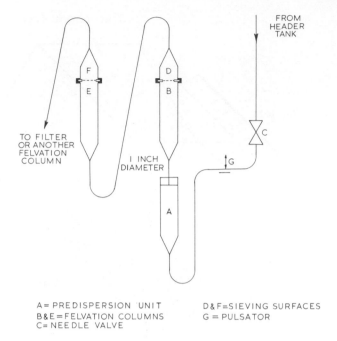

FROM
HEADER
TANK

F
E

D
B

C

G

TO FILTER
OR ANOTHER
FELVATION
COLUMN

I INCH
DIAMETER

A

A = PREDISPERSION UNIT   D & F = SIEVING SURFACES
B & E = FELVATION COLUMNS   G = PULSATOR
C = NEEDLE VALVE

Fig. 5.6 Felvation apparatus.

point is when the liquid above the sieves becomes clear. The powder passing the finest sieve is collected on a suction filter and the various fractions are contained in the bodies of the appropriate columns. Burt used 0.5 to 1.0 g for the micromesh sieves and found that the efficiency of separation increased with decreasing sample size and when only three felvation units were used. Separation efficiency increased if the fluid flow was pulsed two or three times per second, so a pulsator G was added to the equipment. British sieve cloth was used and found to be too fragile for this purpose. The technique was used more successfully with 2-g samples with woven-wire sieves.

The technique is not proposed as an alternative to standard sieving methods, but may well be useful if only small samples are available or, with hazardous materials, where small samples are desirable for safety reasons. The technique has also been used [9] to grade 5000-g samples in the size range 64 to 45 μm.

5.5.7 Self-organized sieve (SORSI)

This system was developed by Kaye in an attempt to eliminate some of the problems associated with nest sieving and to automate the process [31, 32]. A hexagonal sieving chamber is supported on discs and rotated as shown in the diagram (figure 5.7). A steady stream of powder is fed into the entry port and as the sieve moves the powder is gently moved across the sieve; as the sieve is rotated further the powder is picked up by a vane and dumped out of the other side of the sieve.

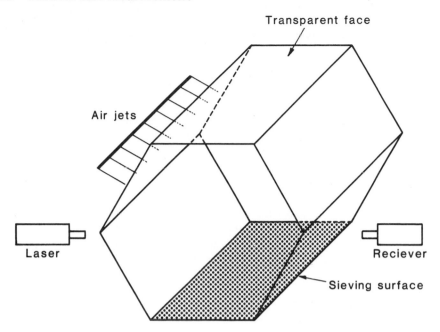

Fig. 5.7 SORSI (Self-Organized Sieve).

The mesh is automatically cleaned every cycle by an air jet to eliminate blinding and, while in a vertical position, it is interrogated by a laser beam to see if there are any damaged, distorted or blinded apertures.

A very low rate of feed and a series of mesh sizes can be used in sequence and the fractions passing the sieves can be collected and weighed automatically to give a continuously generated sieve analysis of the powder.

Kaye [33] also postulated that partitioning of sieves will improve performance. For new sieves it will isolate the effects of extremes of aperture size and the cut size will lie nearer the nominal aperture than it would with a conventional sieve. The better mechanical support will reduce wear and tear. For old sieves, partitioning will reduce the effect of faults due to wear. Kaye suggests the reduction of an 8-in diameter sieve to a honeycomb of $\frac{1}{2}$-in square partitions.

5.6 Sieving errors

The apertures of a sieve may be regarded as a series of gauges which reject or pass particles as they are presented at the aperture. The probability that a particle will present itself at an aperture depends on the following factors:

(1) The particle size distribution of the powder.
(2) The number of particles on the sieve (load).
(3) The physical properties of the particles (e.g. surface).

(4) The method of shaking the sieve.
(5) The dimension and shape of the particles.
(6) The geometry of the sieving surface (e.g. open area/total area).

Whether or not the particle will pass the sieve when it is presented at the sieving surface
will then depend upon its dimension and the angle at which it is presented.

   The size distribution given by a sieving operation depends also on the following
variables:

(1) Duration of sieving.
(2) Variation of sieve aperture.
(3) Wear.
(4) Errors of observation and experiment.
(5) Errors of sampling.
(6) Effect of different equipment and operation.

The effects of sieving time and sieving load have been investigated by Shergold [34]. The
tests were carried out with 14-, 52- and 200-mesh sieves using samples of sand specially
prepared so that 50% by weight of each sample could pass through the appropriate
sieve. Some of the results using the 200-mesh sieve are given in Table 5.2.

Table 5.2

Sample weight (g)	Sieving time (min)			
	5	10	20	40
	Percentage retained on sieve (P)			
500	83.6	80.7	76.5	73.8
250	67.2	64.3	61.6	57.8
125	58.6	58.0	55.2	53.2
62.5	56.6	55.0	53.2	52.3

These results are of the form $P = k \log_e t$. Shergold's results showed that the smaller
the sieve aperture, the greater the effect of overloading; and the greater the discrepancies between the results for different loadings. He also showed that although, in general,
there is no end-point for sieving, the approach to the true percentage is quicker for
small sieve apertures. Since it is evident that a reduction in sample size is more effective
than prolonging the sieving, he recommends that the sample should be as small as is
compatible with convenient handling, 100–150 g for coarse sand and 40–60 g for fine
sand with a sieving time of 9 min.

   Heywood [35] also investigated the effect of sample weight on sieving time using 20-,
50- and 100-g samples of coal dust on 60-, 100-, 150- and 200-mesh sieves. He found
that neither the time required to attain the end-point of 0.1% per minute nor the residual percentage on the sieve was affected by the weight of sample.

   Other variables that have been investigated include sieve motion [36, 37], percentage
open area [38], calibration [39–41], static [42], humidity [40] and accuracy for a

particular application [35, 43–46]. Heywood [45], for example, describes experiments carried out at seven laboratories, in which 50-g samples of coal dust were sieved on BS sieves of 72, 100, 150 and 200 mesh. The average percentage retained on a specified sieve in all the trials was taken to correspond to the quoted or nominal sieve aperture, and the percentage retained on that sieve in a single trial was taken to correspond to the effective sieve aperture. From a graph of the cumulative percentage passing through the sieve against the nominal sieve aperture, the effective sieve aperture may be read. For example, Heywood found that the average percentage passing through the 52-mesh (295 μm) sieve was 76.9%. In a particular analysis 74% passed through this sieve, the 76.9% on the graph being at an effective aperture of 280 μm, the aperture error being − 15 μm. Heywood, by averaging over all the sieves, arrives at the following values for the standard deviations of aperture errors expressed as a percentage of the nominal aperture (Table 5.3).

Table 5.3

Different sieves, different methods	8.30
Different sieves, same method	3.71
Same sieves, same method:	
machine sieving	0.61
hand sieving	0.80

Although differences between analyses are inevitable, standardization of procedure more than doubles the reproducibility of a sieving operation. A useful statistical analysis of Heywood's data is to be found in Herdan [47, p. 121].

5.7 Mathematical analysis of the sieving process

The mechanism of sieving can be divided into different regions with a transition region in between [48], as illustrated in figures 5.8 and 5.9. The first region relates to the passage of particles much finer than the mesh openings and the law

$$P = at^b \tag{5.1}$$

is obeyed where

a = fraction passing sieve per unit time or fraction per tap for hand sieving,
b = a constant very nearly equal to 1,
t = sieving time or number of taps for hand sieving,
P = cumulative weight fraction through the sieve.

Whitby assumed a to be a function of the variables, total load on sieve (W), particle density (ρ_s), mesh opening (S), sieve open area (A_o), sieve area (A), particle size (d), and bed depth on sieve (T).

This reduces to $a = f\left(\dfrac{W}{\rho_s S A_o}, \dfrac{S}{d}, \dfrac{A_o}{A}, \dfrac{T}{d}, \dfrac{A}{S^2}\right)$, an identity with seven variables and two dimensions; hence a is a function of five dimensionless groups.

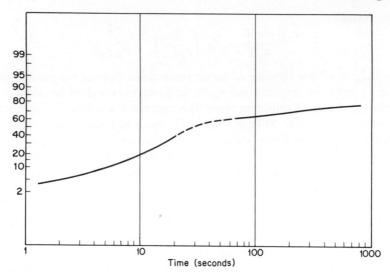

Fig. 5.8 The rate at which particles pass through a sieve plotted on log-probability paper (after Whitby [48]).

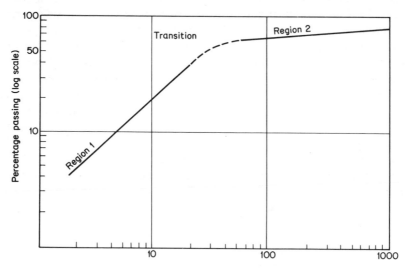

Fig. 5.9 The rate at which particles pass through a sieve plotted on log-log paper (after Whitby [48]).

Since A_o/A is constant for any sieve and A/S^2 is so large that it is unlikely to have any appreciable effect, and the effect of variation in T/d is negligible, the equation becomes:

$$a = f\left(\frac{W}{\rho_s S A_o}, \frac{S}{d}\right) \tag{5.2}$$

Whitby found:

$$\frac{a\,W}{A_o S} = C_1 \left(\frac{S}{k_s \bar{d}_m}\right) \frac{h}{\log \sigma_{gP}} \tag{5.3}$$

where $k_s \bar{d}_m$ is a linear function of the geometric mass mean of the particle size distribution. C_1 and h are constants and σ_{gP} is the geometric standard deviation at a particular size on the size distribution curve. This expression was found to hold for wheat products, crushed quartz, St Peter's sand, glass beads and other similar materials.

Whitby suggests that a good procedure would be to select as end-point a time at the beginning of region 2. This can be done by plotting the time–weight curve on log-probability paper (figure 5.8) and selecting as an end-point a time at the beginning of region 2. It is difficult to determine this point accurately in practice, and an alternative procedure used by the author is to transfer the straight line for region 2 on to a log-log scale (figure 5.9) and define as the end-point the intersection of the resulting curve with the straight line for region 1.

Using the rate test, the sieving operation is terminated some time during the second region. The true end-point, when every particle capable of passing through the sieve has done so, is not reached unless the sieving time is unduly protracted.

The second region relates to the passage of 'near mesh' particles. These are defined as particles which will pass through the sieve openings in only a limited number of ways relative to the many possible particle orientations with respect to the sieve surface.

The sieving of such particles is a statistical process, that is, there is always an element of chance as to whether a particular particle will or will not pass through the sieve. In the limit the sieving process is controlled by the largest aperture through which the ultimate particle will pass in only one particular orientation. In practice there is no definite 'end-point' to a sieving operation so this is usually defined in an arbitrary manner.

The rate method is fundamentally more accurate than the time method but is more tedious to apply in practice and for most routine purposes a specified sieving time is adequate.

Several authors have derived equations for the rate of sieving during region 2 when the residual particles are near mesh [37, 48–53]. The general relationship derived is of the form:

$$-\frac{dW}{dt} = k(W_t - W_r)^m \tag{5.4}$$

where W_t is the residual weight at time t and W_r is the weight of particles too large to pass through the sieve. The weight of 'near mesh' particles is therefore equal to $W_n = W_t - W_r$.

Equation (5.4) may be written in terms of percentages:

$$\frac{dR}{dt} = k(R_t - R_r)^m \tag{5.5}$$

Kaye [37] and Jansen and Glastonbury [53] postulate that $m = 1$ and Kaye suggests that a plot of $\log(R_t - R_r)$ against time will yield a value for the ultimate end-point R_r.

Even if Kaye's argument is accepted, the value of R_r is questionable since it cannot apply to the nominal aperture of the sieve. As sieving progresses the smaller apertures become ineffective since all particles finer than these apertures will have passed through the sieve. The sieving operation is therefore controlled by the largest aperture in the sieve and the final particle to pass through the sieve will be able to do so in only one preferred orientation.

Conventional dry sieving is not recommended for brittle materials since attrition takes place and an end-point is difficult to define. An investigation of this problem has been carried out by Gupta *et al.* [54]. If the rate of passage of particles does not decrease with sieving time it may be due to either particle attrition or a damaged sieve.

5.8 Calibration of sieves

A procedure recommended for the optical examination of test sieves is described in BS 410 (1962). Extensive researches by MacCalman [55] failed to determine any method by which the effective aperture of a sieve could be calculated from optical measurements. It is clear, however, that oversize apertures are more undesirable than undersize apertures, since the latter are merely ineffective whilst, with prolonged sieving, the former result in an increase in the size of particles passed by the sieve. The effective aperture is thus likely to be greater than the mean aperture determined by optical examination, and the system of intermediate apertures was introduced in order to reduce this difference [45, 56].

According to Leschonski [3] no manufacturer of sieve cloth has available a fast testing machine that records aperture width distributions although such an instrument is available [57]. This instrument, developed at Carl Zeiss, is capable of measuring 50 mesh per second using a travelling photometric system with an average error of ±0.3 μm. Leschonski points out that measurements of this kind yield two aperture size distributions, one for the warp and one for the weft, and that these differ. In his experience the median of the aperture width distribution between warp wires not only matches the nominal aperture width better than the median of the weft but the size distribution also has a smaller spread. Since the smaller dimension controls the cut one should therefore measure the smaller opening of each individual aperture and use this for calibration purposes.

Kaye *et al.* [58] suggest optical processing using amplitude spatial filtering to speed up the process of assessing the quality of a sieve. This technique operates on a large number of apertures simultaneously and not sequentially as with the normal optical methods and it enhances the damaged or distorted area of the sieve. Damaged regions of the sieve are located with ease and quantitative measurements of the damage can be made.

The aperture width distribution may also be analysed using spherical particles [59]. These are fed on to the sieve which is shaken a few times and then the excess removed. Many spheres will have jammed into the sieve cloth and may be removed for microscopic examination. This technique has the advantage that for apertures which are not pre-

cisely square, the smallest dimension, which to a first approximation controls the cut, is analysed.

A simple method of dealing with single sets of sieves is due to Stairmand [60]. It is known that the products of comminution are often log-normally distributed. Hence, if the sieve analysis is recorded on log-probability paper, a best straight line can be drawn through the points, while a zig-zag is obtained by joining them. In this way the effective aperture of the sieves may be determined. If the same material is used for different sets of sieves, these can be corrected to a standard. Kiff [61] states that commercial glass beads are also usually of a log-normal distribution and may be used similarly.

ASTM E11–70 includes reference to standard materials available from the National Bureau of Standards in Washington. These consist of glass spheres in the various mesh ranges 8 to 35, 20 to 70, 50 to 170 and 140 to 400.

Sieves may be calibrated by a technique first suggested by Andreasen [62] and later by Andersen [63]. This is a counting and weighing technique applied to the fraction of particles passing the sieve immediately prior to the end of an analysis. These will have a very narrow size distribution and the average particle size may be taken as the cut size of the sieve. A minimum number (n) of particles needs to be weighed to obtain accurate volume diameters (d_v); let this weight be m_s and let the particle density be ρ_s, then:

$$d_v = \sqrt[3]{\left(\frac{6m_s}{\pi \rho_s n}\right)}$$

Particles larger than 250 μm can be easily counted by hand and, if weighed in batches of 100, d_v is found to be reproducible to three significant figures. For particles between 100 and 250 μm in size it is necessary to count in batches of 1000 using a magnifier. For sizes smaller than this the Coulter counter may be used. Sieve analyses are then plotted against the volume diameter in preference to the nominal sieve diameter.

The method is tedious and time-consuming and Sub-Committee BCR 4.13.1 (Community Bureau of Reference, Brussels) has prepared several quartz samples, by this method, which may be used as calibration material. These are available from the National Physical Laboratory, London [64].

The calibration material may be fed to a stack of sieves and the analytical cut size read off the cumulative distribution curve of the calibration material.

References

1 Heywood, H. (1970), *Proc. Particle Size Analysis Conf.* (eds M.J. Groves and J.L. Wyatt-Sargent), Soc. Analyt. Chem.
2 von Rittinger, P.R. (1887), *Aufbereit.*, 222 and 243.
3 Leschonski, K. (1977), *Proc. Particle Size Analysis Conf.* Chem. Soc. Analyt. Div., London.
4 Lauer, O. (1966), *Grain Size Measurements on Commercial Powders,* Alpine, A.G. Augsburg.
5 Wahl, W. (1976), Ger. Offen. 2 413 521.
6 Stork Brabant, B.V. (1975), Neth. Appl., 74 07348.
7 Daescher, M.W., Seibert, E.E. and Peters, E.D. (1958), *Symp. Particle Size Measurement,* ASTM Sp. Publ. 234, 26–56.
8 Zwicker, J.D. (1966), Report No. AW-FR-2-66, Aluminium Company of Canada Ltd, Arvida, Canada.

9 Burt, M.W.G. (1970), *Proc. Soc. Analyt. Chem.*, 7, 9, 165–8.
10 ASTM Specification for *Precision Micromesh Sieves*, Designation E161–70.
11 Colon, F.J. (1965), *Chemy Ind.*, Feb. 263.
12 Heidenreich, E. (1977), *Proc. Particle Size Analysis Conf.* Chem. Soc. Analyt. Div., London.
13 Colon, F.J. (1970), *Proc. Soc. Analyt. Chem.*, 7, 9, 163–4.
14 Daescher, M.W. (1969), *Powder Technol.*, 2, 6, 349–55.
15 Crawley, D.F.C. (1968), *J. Scient. Instrum.* 1, series 2, 576–8.
16 Rosenberg, L.D. (1960), *Ultrasonic News*, 4, 16.
17 Irani, R.R. and Callis, C.F. (1963), *Particle Size Measurement, Interpretation and Application*, Wiley, NY.
18 BS 410 (1962): *Specification for Test Sieves.*
19 Leschonski, K. (1970), *Proc. Particle Size Analysis Conf.* (eds M.J. Groves and J.L. Wyatt-Sargent), Soc. Analyt. Chem.
20 Ilantzis, M.A. (1961), *Annls Inst. Tech. Bat. Trav. Pub.*, 14, 161, 484–512.
21 BS 1796 (1952): *Methods for the Use of British Standard Fine Mesh Sieves.*
22 Niedick, E.A. (1969), *Z. Zuckerind.*, 19, 9, 495–506.
23 Ioos, E. (1965), *Staub Reinhalt. Luft*, 25, 12, 540–3.
24 Brown, O.E., Bobrowski, G.S. and Kovall, G.E. (1970), ASTM Sp. Tech. Publ. 473, 82–97.
25 Malhetra, V.M. and Zalderns, N.G. (1970), ASTM Sp. Tech. Publ. 473, 98–105.
26 Lauer, O. (1958), *Staub*, 18, 306.
27 Jones, T.M. (1970), *Proc. Soc. Analyt. Chem.*, 7, 9, 159–63.
28 Suhm, H.O. (1969), *Powder Technol.*, 2, 6, 356–62.
29 US Patent 3 045 817.
30 Kaye, B.H. (1966), *Symp. Particle Size Analysis*, Soc. Analyt. Chem., London.
31 Kaye, B.H. (1977), Dechema Monogram 79 (1589–1615), Part B, 1–19.
32 Kaye, B.H. (1977), *Proc. Powder Technology Conf.*, Powder Advisory Centre, Chicago.
33 Kaye, B.H. (1978), *Powder Technol.*, 19, 121–3.
34 Shergold, F.A. (1946), *Trans. Soc. Chem. Ind.*, 65, 245.
35 Heywood, H. (1945/6), *Trans. Inst. Min. Metall.*, 55, 373.
36 Fahrenwald, A.W. and Stockdale, S.W. (1929), US Bureau Mines Report, Investigation 2933.
37 Kaye, B.H. (1962), *Powder Metall.*, 10, 199–217.
38 Weber, M. and Moran, R.F. (1938), *Ind. Engng Chem., Analyt. Edn*, 19, 180.
39 Carpenter, F.G. and Deitz, V.K. (1951), *J. Res. Natn. Bur. Stand.*, 47, 139.
40 Moltini, E. (1956), *Ind. Mineraria* (Rome), 7, 771; *Appl. Mech. Rev.*, 11, 345.
41 Carpenter, F.G. and Deitz, V.K. (1950), *J. Res. Natn. Bur. Stand.*, 45, 328.
42 Allen, M. (1958), *Chem. Engng*, 65, 19, 176.
43 Fritts, S.S. (1937), *Ind. Engng Chem.*, 9, 180.
44 MacCalman, D. (1937), *Ind. Chem.*, 13, 464; (1938), *ibid.*, 14, 64; (1939), *ibid.*, 15, 161.
45 Heywood, H. (1956), *Inst. Min. Metall. Bull.*, 477.
46 Ackerman, L. (1948), *Chem. Engng Mining Rev.*, 41, 211.
47 Herdan, G. (1960), *Small Particle Statistics*, Butterworths.
48 Whitby, K.T. (1958), *Symp. Particle Size Measurement*, ASTM Sp. Publ. 234, 3–25.
49 Fagerhalt, G. (1945), GEC, Gads Forlag, Copenhagen.
50 Hukki, R.T. (1943), PhD Thesis, Massachusetts Inst. of Technol.
51 Bodziony, J. (1960), *Bull. Acad. Polon. Sci., Ser. Sci. Tech.*, 8, 99–106.
52 Mitzutani, S. (1963), *J. Earth Sci.*, Nagoya Univ., 11, 1–27.
53 Jansen, M.L. and Glastonbury, J.R. (1968), *Powder Technol.*, 1, 334–43.
54 Gupta, V.S., Fuerstenan, D.W. and Mika, T.S. (1975), *ibid.*, 11, 3, 257–72.
55 MacCalman, D., *Ind. Chem.* (1937), V13, 464, 507; (1938), V14, 64, 101, 143, 197, 231, 306, 363, 386, 498; (1939), V15, 161, 184, 247, 290.
56 Heywood, H. (1938), *Proc. Instn. Mech. Engrs*, 140, 257.
57 Meister, G., Thiemer, R. and Lenski, G. (1964), *Jenaer Rundschau*, 9, 3, 143–7.
58 Konowalchuk, H., Naylor, A.G. and Kaye, B.H. (1976), *Powder Technol.*, 13, 97–101.

59 Leschonski, K. (1960), *Getreide Mehl*, **8**, 12, 140–4.
60 Stairmand, C.J. (1947), *Symp. Particle Size Analysis, Instn. Chem. Engrs*, Suppl. **25**, 18.
61 Kiff, P.R. (1973), *Proc. Soc. Analyt. Chem.*, **10**, 5, 114–15.
62 Andreasen, A.H.M. (1927), *Sprechsaal*, **60**, 515.
63 Andersen, J. (1931), *Zement*, **20**, 224–6, 242–5.
64 Wilson, R. (1977), *Proc. Particle Size Analysis Conf.* Chem., Soc. Analyt. Div., London.
65 Lauer, O. (1960), *Staub*, **20**, 69–71.
66 Mullin, J.W. (1971), *Chemy. Ind.*, **50**, 1435–6.
67 Peterson, J.L. (1969), US Pat. 3 438 490, Method and apparatus for wet-sizing finely divided solid materials.

⑥ Microscopy

6.1 Introduction

Microscopy is often used as an absolute method of particle size analysis since it is the only method in which the individual particles are observed and measured. It also permits examination of the shape and composition of particles with a sensitivity far greater than for any other technique. The representativeness of the sample under analysis is critical since measurements are carried out on such minute quantities. Sampling techniques and sample preparation should, therefore, be carefully considered and the statistical factors governing accuracy should be well known.

The great advances in microscopy made during the past few years have led to a flood of papers and numerous new instruments, many of them having similarities. The difficulty of selecting the most suitable microscope technique has now reached similar proportions to the difficulty of selecting the most suitable size analysis technique from the many that are available. To aid in this choice there are certain guide-lines, such as the range of sizes under consideration, cost, the number and frequency of analyses required, and so on. It is hoped that this chapter will help the reader make a wise decision and also to produce accurate and meaningful analyses.

6.2 Optical microscopy

The optical microscope is used for the examination of particles from about 150 to 0.8 μm in size. Above 150 μm a simple magnifying glass is suitable, while for smaller particles it is necessary to use electron microscopy.

Its most severe limitation is its small depth of focus, which is about 10 μm at a magnification of 100 x and about 0.5 μm at 1000 x. The surface of particles larger than about 5 μm can be studied by reflected light, but only transmission microscopy, with which silhouettes are seen, can be used for sub-5 μm particles. The edges of the images seen in a microscope are blurred due to diffraction effects. Charman [68], in an investigation into the accuracy of sizing by optical microscopy, showed that for particles greater than about 1 μm in diameter the estimated size under ideal conditions was about 0.13 μm too high; a 0.5 μm particle gave a visual estimate of 0.68 μm and all particles smaller than about 0.2 μm appeared to have a diameter of 0.5 μm. Rowe [72] showed that wide differences may occur between operators in particle sizing because of this effect.

The limit of resolution of an optical microscope is expressed by the fundamental formula:

$$d = \frac{f\lambda}{2NA} \qquad (6.1)$$

where d is the limit of resolution, λ is the wavelength of the illuminant, NA is the numerical aperture of the objective and f is a factor of about 1.3 to allow for the inefficiency of the system [27].

For $\lambda = 0.6\ \mu m$ the resolving power is a maximum with $NA = 0.95$ (dry) and $NA = 1.40$ (wet) giving $d_{min} = 0.41\ \mu m$ and $0.28\ \mu m$ respectively. Particles having a separation less than this merge to form a single image. Particles smaller than the limit appear as diffuse circles and image broadening occurs, even for particles coarser than d_{min}, and this results in the oversizing found by Charman. Some operators routinely size down to these levels but the British Standard 3406 [31] is probably correct in stipulating a minimum size of 0.8 μm and limited accuracy from 0.8 up to 2.3 μm.

The images produced may be viewed directly or by projection. Binocular eyepieces are preferable for particle examination but monoculars for carrying out size analyses since, by using a single eyepiece, the tube length can be varied to give stepwise magnification. Most experienced operators prefer direct viewing, but projection viewing, less tiring to the eye, is often used for prolonged counting. Projection may be front or back. With the former, the operation is carried out in a darkened room due to the poor contrast attainable. Back projection gives better illumination, but image definition is poor; this can be rectified by using a system whereby two ground-glass screens are placed with their faces in contact and one is moved slowly relative to the other [73]. Some automatic and semi-automatic counting and sizing devices work from negatives or positives. The principal criticism that can be levelled against photographic methods, in conjunction with optical microscopy, is that only particles in good focus can be measured and this can lead to serious bias. Although photographic methods are often convenient and provide a permanent record, the processing time may well offset any advantage obtained by using a high-speed counting device. This is particularly true when a weight count is required since, on statistical considerations, a large number of fields of view are required for accurate results.

6.2.1 Sample preparation

The most difficult problem facing the microscopist is the preparation of a slide containing a uniformly dispersed, representative sample of the powder. Many methods of carrying out this procedure have been suggested, but the final result often depends more on the skill of the operator than the procedure itself. Some acceptable procedures for easily dispersed powders are described by Green [1] and Dunn [2]. The method of Orr and Dallevalle [3] for the production of permanent slides is to place a small representative sample of the powder to be analysed in a 10-ml beaker, add 2 or 3 ml of a solution containing about 2% collodion in butyl acetate, stir vigorously and place a drop of suspension on the still surface of distilled water in a large beaker. The film pro-

duced as this drop spreads and evaporates may then be picked up on a clean microscope slide and completely dried. A dispersing agent may be added to prevent flocculation.

Alternative techniques are to produce a mounting medium of: collodion in amyl
acetate, Canada balsam in xylol, polystyrene in xylene, dammar in turpentine, gelatin
in water, gum arabic in glycerine, styrex in xylene, or rubber in xylene [111]. With a
1% solution, this may be formed by dropping on to the surface of distilled water; with
a 0.5% solution, it may be cast directly on to a microscope slide. The particle suspension may then be sprayed on, or a droplet can be allowed to evaporate, leaving the particles [4].

Dullien [120] uses Cargill's series H compound, having a refractive index of 2.0, as a
mounting medium for salt particles. It gives a transparent yellow background for the
particle and since it has a higher refractive index than salt, the particles appear as dark
spots. A range of systems is necessary in order to select one where the difference in refractive index is easily detectable.

If a permanent slide is not required, an effective procedure is to place a small sample
of the powder on a microscope slide and add a few drops of the dispersing fluid. Some
operators work the powder into the fluid with a flexible spatula, others roll it with a
glass rod. Both these procedures may produce fracture of the particles and a preferable
alternative is to use a small camel-hair brush. Further dispersing fluid is then added
until the concentration is satisfactory. A drop of the suspension is then placed on
another slide with the brush and a cover slip put carefully in place so as to exclude air
bubbles. For a semi-permanent slide, the cover slip may be sealed with amyl acetate or
glue. Cedar oil and glycerol are two satisfactory dispersing fluids; a dispersing agent
may be added to eliminate flocculation.

Harwood [5] describes two methods for dispersing difficult powders. One involves
the use of electrical charges to repel the particles and fixing the aqueous suspension
with a gelatin-coated slide to overcome Brownian movement. The other, for magnetic
materials, involves heating the sample to a temperature above the Curie point, dispersing and fixing it on a slide to cool. Rosinski et al. [6] have investigated several techniques in order to find which gave the best reproducibility. Allen [7] mounted the
powder directly into clear cement, dispersing it by using sweeping strokes of a needle
and spreading a film on a microscope slide to dry. Lenz [8] embedded particles in a
solid medium and examined slices of the medium.

A method has also been described in which particles were suspended in filtered agar
solution which was poured on to a microscope slide; this set within seconds [100].

Variations in analyses may occur with operators and techniques due to orientation
of particles on the slide. Ellison [9], for example, showed that if particles are allowed
to fall out of suspension on to a microscope slide, they will do so with a preferred orientation. Also, if the dispersing is not complete, the presence of flocculates will give the
appearance of coarseness (Green [10]).

Pidgeon and Dodd [91], who were interested in measuring particle surface area using
a microscope, developed methods for preparing slides of particles in random orientation.
For sieve-size particles, a thin film of Canada balsam was spread on the slide and

heated until the material was sufficiently viscid, determined by scratching with a fine wire until there was no tendency for the troughs to fill in. Particles sprinkled on the slide at this stage were held in random orientation. After a suitable hardening time, a cover glass coated with glycerol or warm glycerol jelly, was placed carefully on the slide. Sub-sieve powders were dispersed in a small amount of melted glycerol jelly. When the mixture started to gel a small amount was spread on a dry slide. After the mount had set, it was protected with a cover slip coated with glycerol jelly. With this technique, it was necessary to refocus for each particle since they do not lie in a single plane.

The sizing of fibrous particles by microscopy presents serious problems including overlapping. Timbrell [74, 75] found that certain fibres show preferred orientation in a magnetic field, e.g. carbon and amphibole asbestos. He dispersed the fibres in a 0.5% solution of collodion in amyl acetate and applied a drop to a microscope slide, keeping the slide in a magnetic field until the film had dried. For SEM examination an aqueous film may be drained through a membrane filter held in a magnetic field.

Various means of particle identification are possible with optical microscopy. These include dispersion staining for identification of asbestos particles [92] and the use of various mounting media for particle identification [93].

Proctor *et al.* [102, 103] 'froze' particles in solidifying medium of perspex monomer and hardener. This was poured into a plastic mould which was slowly rotated to ensure good mixing. Microscope analyses were carried out on thick sections using the Zeiss-Endter; a lower limit of 5 μm was due to contamination.

Zeiss [97] described a method for measuring sections of milled ferrite powder. The powder is mixed in a 40 : 60 volume ratio with epoxy resin using a homogenizing head rotating at 25 000 rpm. The mixture is then poured into a ½-in diameter mould and the casting is cured at 60°C and 1000 psi to eliminate air bubbles. The casting is then polished in a vibratory polisher using 0.3 and 0.05 μm alumina in water. A photomicrograph of the polished section is used for subsequent analysis.

Millipore recommend filtering a suspension of the particles through a 0.2 μm PTFE membrane filter which is then placed on a dry microscope slide. The slide is inverted over a watch glass half-filled with acetone, the vapours of which render the filter transparent after two to five minutes. The filter is allowed to dry for two to five minutes and is then ready for analysis (see also section 3.3).

6.2.2 *Particle size distributions from measurements on plane sections through packed beds*

When the size distribution of particles embedded in a continuous solid phase is required, the general approach is to deduce the distribution from the size distribution of particle cross-sections in a plane cut through the particle bed. This problem has exercised workers in many branches of science who have tended to work in isolation and this has led to much duplication of effort.

The historical development of this technique is reviewed by Eckhoff and Enstad [101] and the relevant theory, that of Scheil, is reviewed by Dullien *et al.* [119]. A more recent theoretical analysis [98] has been criticized on several grounds [99].

Dullien *et al.* [120–122] examined salt particles embedded in a matrix of Wood's metal using the principles of quantitative stereology. They then leached out the salt particles and examined the matrix using mercury porosimetry. Poor agreement was obtained and this they attributed to the mercury porosimetry results being controlled by neck diameter.

Nicholson [21] considered the circular intersections of a Poisson distribution of spherical particles to estimate the particle size distribution.

6.3 Particle size

The image of a particle seen in a microscope is two-dimensional. From this image an estimate of particle size has to be made. Accepted diameters are (figure 6.1):

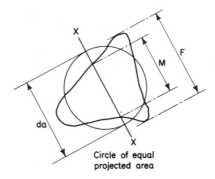

Fig. 6.1 Representative two-dimensional diameters.

(1) Martin's diameter (*M*) is the length of the line which bisects the particle image. The lines may be drawn in any direction which must be maintained constant for all the image measurements (Martin [11], Heywood [12]).
(2) Feret's diameter (*F*) is the distance between two tangents on opposite sides of the particle, parallel to some fixed direction (Feret [13]).
(3) Longest dimension. A measured diameter equal to the maximum value of Feret's diameter.
(4) Maximum chord. A diameter equal to the maximum length of a line parallel to some fixed direction and limited by the contour of the particle.
(5) Perimeter diameter. The diameter of a circle having the same circumference as the perimeter of the particle.
(6) The projected area diameter (d_a) is the diameter of a circle having the same area as the particle viewed normally to a plane surface on which the particle is at rest in a stable position.

Mean diameters from these measurements are all independent of particle orientation since they are either taken in some preferred direction or, in the last two cases, are two-dimensional measurements.

It has been shown [14, 15] that the relationship between specific surface, S_v, and Martin's diameter is:

$$M = \frac{4}{S_v}$$

Herdan [16, p. 46] points out that the volume surface diameter is:

$$d_{vs} = \frac{6}{S_v}$$

the proportionality factor being a minimum of six for spherical particles. Thus Martin's diameter is systematically different from the volume surface diameter. Experiments confirm that, on the whole, $M < d_a < F$. The ratios of these three diameters remain fairly constant for a given material and may be expressed as a shape function. For example, $F/M = 1.2$ for Portland cement and 1.3 for ground glass [17].

Heywood [12] measured crushed sandstone which had passed a $1\frac{1}{8}$-in square sieve aperture and been retained on a 1-in square aperture. He determined the projected area of the particles using a planimeter and calculated the mean projected diameter; he next estimated the diameter using both the opaque and transparent circles on a globe and circle graticule and also determined Martin's and Feret's diameters. His conclusion, based on an examination of 142 particles, was that Feret's diameter was grossly in error for elongated particles, but that the Martin and projected area diameters are sufficiently in agreement for all practical purposes. Walton [20] disputed this and showed that Feret's diameter, averaged over all particle orientations is equal to particle perimeter/π. (See also [14].)

Herdan [16, p. 140] examined Heywood's data more rigorously. His results were:

(a) Feret's diameter was significantly different from the other four diameters;
(b) Martin's diameter shows significant difference from that obtained using the globe
      and circle graticule, if the planimeter results are accepted as standard.

He concluded that there was no definite advantage to be gained by laboriously measuring profiles.

As one might expect, the projected area diameters gave the best estimate of the true cross-sectional areas of the particles. This does not rule out the use of the other diameters if they are conveniently measured, since the cross-sectional area diameter of a particle is not necessarily its optimum dimension.

6.4 Transmission electron microscopy (TEM)

TEM is often used for the direct examination of particles in the size range 0.001 to 5 μm [56]. The TEM produces an image on a fluorescent screen or a photographic plate by means of an electron beam. Although attempts have been made to perform analyses directly on the images visible on the fluorescent screen, this involves tying down

the instrument for long periods of time; hence it is more usual to carry out the analyses on the images recorded photographically.

Many particle size studies can be carried out at magnifications of less than 4000x and several relatively cheap instruments are now available giving magnifications up to 10 000x.

Calibration is usually effected with narrowly classified polystyrene latices available from Dow Chemicals but for high-accuracy work diffraction gratings are required.

6.4.1 Specimen preparation

Specimens for electron microscopy are usually deposited on or in a thin (10 to 20 nm) membrane which rests on a grid. These grids are usually made of copper and form a support for the film which is usually self-supporting only over a very small area (figure 6.2). Since most materials are opaque to the electron beam, even when only a few Ångstroms thick, special problems exist in the production of suitable mounted specimens. Specimen support films are usually made of plastic or carbon, though other materials have also been used. Suitable film solutions may be made up of 2% w/v collodion in amyl acetate or 2% w/v formvar (polyvinyl formal) in ethylene dichloride or chloroform.

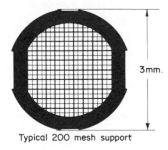

Typical 200 mesh support

Fig. 6.2 Electron microscope grid.

Films may be produced in the following manner [57, 118]. A dish about 20 cm in diameter is filled with distilled water and a large circle of wire gauze (200-mesh) is placed at the bottom of the dish. A number of grids are placed on the wire gauze, then two drops of the film solution are dropped on to the surface of the water and the film that results after the solvent has evaporated is removed with a needle; this ensures that the water surface is clean. A second film, formed in the same way, is removed by raising the wire gauze containing the grids. The wire gauze and grids are then allowed to dry. A pre-examination of the film-covered grids in the electron microscope is desirable as this enables dirty films to be rejected and the film polymerizes in the electron beam, thus greatly increasing its strength.

An alternative procedure is to clean a microscope slide with a detergent solution and polish with a soft cloth without rinsing away the detergent, so as to form on the sur-

face a hydrophilic layer to facilitate subsequent stripping of the membrane. The slide is dipped into a solution of formvar in ethylene dichloride (0.3 to 0.7% w/v depending on the thickness of film required) and allowed to drain dry. The film may then be floated on to a water surface and mounted on grids as before. If individual grids are required, the film may be cut into small squares with a needle or razor blade. The mounting operation is easier if a special jig is used [64]. The jig is a brass cylinder about 1 in long and of ¼ in diameter with a hole of the same diameter as the specimen grid drilled and tapped through its axis. A set screw in the threaded hole is adjusted so that its end is flush with the face of the plug and then withdrawn slightly to leave a shallow recess. The specimen grid is held in this recess and the membrane is lifted from the water surface with a wire loop of slightly larger diameter than the jig, surplus water being carefully removed with blotting paper. The wire loop is then lowered over the jig and when the membrane is dry the grid is raised by means of the screw, surplus membrane being removed by scoring round the grid with a needle. Special apparatus has also been described for producing plastic films of uniform thickness suitable for the preparation of replicas [58].

Carbon films are prepared under vacuum (10^{-3} mmHg) by electrical discharge [59] from two pointed hard graphite rods. Films are best deposited on microscope slides cleaned with detergent and placed about 10 to 15 cm from the source. A thickness indicator, consisting of a drop of vacuum oil on a piece of white glazed porcelain, is placed beside the slide. During the discharge, the porcelain not covered with oil takes on a brownish colour changing to a light chocolate shade as the film thickness increases from 5 to 10 nm, the latter being a suitable thickness for general use. Evaporation is completed in about half a second with a current of about 50 A. The film may then be floated on to the surface of distilled water and picked up as a whole or scored into small squares. A simple method of producing a suitable specimen is to place a few milligrams of the powder on a microscope slide, add a drop of 1 to 2% w/v suspension of formvar in a suitable solvent and rub out on the slide with a glass rod. Further solvent is added if required and the dispersion is spread out over the slide and allowed to dry. The film is removed from the slide and mounted on the grid as before.

Alternatively the sample may be dispersed in linseed oil which is then thinned with white spirit. The dispersion is next spread out on a microscope slide which is immersed in white spirit for a few minutes to remove the oil. After drying the slide a thin layer of carbon is deposited on the specimen to form a supporting film. Finally, this is floated off on water, as before, and picked up on a grid for examination [65].

A dispersed sample may also be obtained by means of one of the aerosol sampling devices described in Chapter 3. A suitable technique is to form a sandwich of plastic film particles and 20 nm thick carbon. The underlying plastic may then be washed away with solvent and the specimen examined after shadowing [50, 57, 59].

A suspension of the powder may be made up and a drop placed on a grid by means of a pipette or hypodermic syringe. However, this often produces an uneven deposit. A more uniform deposit is often produced by spraying the suspension on to the grid; several suitable spray guns have been described [56, 60].

Timbrell [76] modifies the method of Hamilton and Phelps described in section 6.10, for the preparation of transparent profiles to facilitate electron microscopy. The metal film is floated on to a water surface and picked up on a grid. In cases of difficulty the slide is first dipped into 1% hydrofluoric acid to release the edge of the film and the process is completed in water. Although the metal film is strong enough to be floated off whole, it is usual to score it into small squares as described earlier and then remove the separate pieces.

In order to obtain reliable results for particle size analysis, as many separate grids as possible must be prepared and a large number of electron micrographs taken from each.

In an analysis of the errors involved in electron microscopy, Cartwright and Skidmore [64] examined the optical microscopy specimens produced by thermal precipitation. The specimens were then stripped from the microscope slides and the fines were counted, using an electron microscope. Good agreement was obtained by examining 1000 particles in the electron microscope using about sixty fields of view (60 micrographs) and almost 4000 particles in the optical microscope. To obtain accurate magnification calibration, four overlapping micrographs along the bar of a readily identifiable grid square were taken and the total length of the image of the grid bar was measured from the micrographs and directly, using an optical microscope.

The surface areas of dust samples as determined by electron and optical microscopes have also been compared by Joffe [66]. Pore size distributions of thin films of Al_2O_3, as measured by TEM, have been compared with those determined by gas adsorption/desorption [130]. It has also been suggested that electron microscopy gives a truer estimate of surface area than adsorption techniques [131].

6.4.2 Replica and shadowing techniques

Replicas are thin films of electron-transparent material which are cast on opaque specimens in order that their surface structure may be studied. The basic procedure is to form a film on the substance to be examined, separate the two and examine the film. If a reverse of the original is unsatisfactory, a positive replica may be obtained by repeating the process. One method is to deposit the specimen on to a formvar-covered grid, vacuum-deposit about 10 nm of carbon, remove the formvar by rinsing with chloroform and finally remove the specimen with a suitable solvent [61]. Instead of a backing film, it is sometimes possible to prepare a carbon replica of a dried suspension deposited on a microscope slide, the replica being washed off the slide in a water bath or a bath of hydrofluoric acid. The carbon film is then transferred to a bath containing a solvent for the specimen. The film may be strengthened immediately after being deposited by dipping the slide in a 2% w/v solution of Bedacryl 122X in benzene which is removed by a suitable solvent after the film has been deposited on a grid [56]. Numerous variations of these techniques have been used [3, 16, 50, 56, 62, 94, 95].

In order to determine surface characteristics and particle thickness, it is usual to deposit obliquely a film of heavy metal on to the specimen or its replica. The metal is applied by deposition in a hard vacuum by a small source in order that a nearly parallel beam may reach the specimen. The technique was originated by Williams and Wyckoff in 1946 [63] and has been used extensively since.

6.4.3 Chemical analysis

When particles are bombarded with electrons they emit radiation which depends upon their chemical composition. The Auger [69, 70] process can be studied by using mono-chromatic or polychromatic radiation or electron beams. It is a secondary electron process which follows the ejection of an electron from an inner-shell level. The hole is filled by an electron falling to the vacant level which provides energy for another electron to be emitted. The energy of this Auger electron is characteristic of the molecule involved. There have been many studies on metal surfaces using vacuum ultra-violet techniques and the energy distribution curves (EDCs) obtained give information on the band structure of the metals. The use of soft X-rays is known as electron spectroscopy for chemical analysis (ESCA). In addition to ejecting electrons from the valence shell orbits, the X-rays have sufficient energy to eject electrons from some of the inner shells. These are essentially atomic in nature and the spectra produced are characteristic of the atom concerned rather than the molecule of which it forms a part [71]. An introduction to these techniques has been written by Szalkowski [123].

These, and other techniques, may be applied to electron microscopy to permit chemical assays and particle size analyses to be carried out concurrently.

6.5 Scanning electron microscopy (SEM)

In SEM a fine beam of electrons of medium energy (5–50 keV) is caused to scan across the sample in a series of parallel tracks. These electrons interact with the sample, producing secondary electron emission (SEE), back-scattered electrons (BSE), light or cathodoluminescence and X-rays. Each of these signals can be detected and displayed on the screen of a cathode ray tube like a television picture. Examinations are generally made on photographic records of the screen. The SEM is considerably faster and gives more three-dimensional details than the TEM; its operation is described by, for example, Hay and Sandberg [104].

Samples as large as 25 mm x 25 mm can be accommodated and parts viewed at magnifications varying from 20 to 100 000x at resolutions of 15–20 nm as compared to 0.3–0.5 nm for the TEM. Because of its great depth of focus the SEM can give considerable information about the form of a particle and its surface morphology. Its depth of focus is nearly 300 times that of the optical microscope.

In both the SEE and BSE modes, the particles appear as being viewed from above. In the SEE mode, where the particles appear to be diffusely illuminated, particle size can be measured and aggregation behaviour can be studied but there is little indication of height. The BSE mode in which the particles appear to be illuminated from a point source gives a good impression of height due to the shadows. Several of the current methods of particle size analysis can be adapted for the measurement of images in SEM photographic records. There is also active interest in the development of analysis techniques that will make more use of the three-dimensional image presentation.

The Le Mont Scientific B-10 system features an energy-dispersive X-ray detector. Particles are located and interrogated to find size and shape; various software options

are available. The Bausch and Lomb system (see section 6.9) has also been applied to electron beam microscopy [113, 114].

Various methods of sample preparation have been described. Krinsley and Margolis [105] mounted sand grains in Duco cement on the SEM target stub. This technique is not suitable for size analysis and cannot be used for fine particles [107]. Willard and Hjelmstad [106] tried to improve this technique, in order to mount fine coal, by using a double-backed adhesive tape. They had limited success with particles smaller than 30 μm and found that the method was unsuitable for larger particles. White *et al.* [108] prepared specimens of alumina by aspirating droplets of alumina suspension in NH_4OH solution on to aluminized glass slides and then fixing the slides to the SEM stub. This method has been criticized as being unsuitable for particles which readily agglomerate or which swell in aqueous solutions. Furthermore, the method is very slow [107].

Turner *et al.* [107] prepared an adhesive layer by immersing about three feet of 'Scotch tape' in 150 cm^3 of carbon tetrachloride or chloroform, agitating long enough for the adhesive layer to be dissolved and then removing the tape. The thickness of the adhesive layer on the SEM stub was controlled by adjusting the concentration and the size and number of drops of solution applied. A suitable amount of powder was dispersed in acetone (0.5% w/v) and a drop taken on a glass rod and dropped on to the stub. The mounted specimen was then coated with a 15 nm layer of aluminium. They also describe a method for studying agglomerates using freeze drying.

6.6 Manual methods of sizing particles

A frequently described method of measuring particle size is to use a microscope fitted with an optical micrometer, a movable cross-hair being built into the ocular. The cross-hair is moved by a calibrated micrometer drum until it coincides with one edge of the particle. It is then moved to the other end of the particle, or some other conventionally adopted limit, precautions being taken to eliminate the effect of backlash. The difference in readings is a measure of the size of the particle. This technique is very time-consuming and has been largely superseded by the use of calibrated ocular scales.

The simplest form consists of a glass disc which is fitted on to the field stop of the ocular. Engraved upon the disc is a scale which is calibrated against a stage graticule. The stage graticule consists of a microscope slide on which is engraved a linear scale. The image of this scale is brought into coincidence with the ocular scale by focussing. With a single tube microscope the magnification may be varied somewhat by racking the tube in or out. The stage graticule is then replaced by a microscope slide containing the particles to be examined.

The slide is placed on a microscope stage which is capable of movement in perpendicular directions and is examined in strips (figure 6.3). As a particle image passes over the scale it is sized and recorded. For statistical accuracy, particles overlapping one end of the scale are neglected and particles overlapping the other end counted. Strips are selected so as to give complete coverage of the slide. The analyses are best carried out by two operators, one sizing and one recording, the operators alternating their duties at regular intervals to reduce eye strain. An alternative is to use a counting array with five

or more size intervals and with a cumulative register so that the total count is also recorded [18].

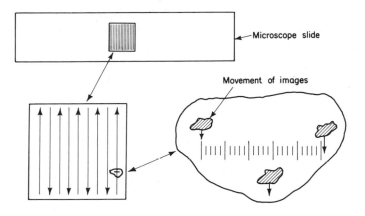

Fig. 6.3 Strip scanning using a calibrated scale.

6.6.1 Graticules

Ocular graticules having a linear scale are satisfactory for the measurement of linear dimensions of particles. Particle sizes obtained with a linear eyepiece are best classified arithmetically, hence it is most suited to particles having a narrow size range.

This type of eyepiece has been criticized on the grounds that the diameters so measured are greater than those derived by other methods. To overcome this objection, grids inscribed with opaque and transparent circles have been developed. These permit a direct comparison between the projected area of a particle and the area of the circles. According to Cauchy the projected area is a quarter of the surface area for a random dispersion of convex particles [24], hence this dimension may be related to other properties of the powder.

The first of this type, the globe and circle graticule, described by Patterson and Cawood (figure 6.4) [25], had 10 globes and circles ranging in diameter from 0.6 to 2.5 μm when used with a +2 mm × 100 objective eyepiece combination and was recommended by Watson [26] for thermal-precipitator work. Fairs [27] designed a graticule using reference circles with a root-two progression in diameter except at the smallest size. He considered this to be superior to the Patterson–Cawood where the series is much closer. May [28] (figure 6.5) also describes a graticule with a root-two progression. Watson (figure 6.6) [29] has developed a graticule designed specifically to measure particles in the 0.5 to 5 μm range. This was compared with a line graticule and the Patterson–Cawood globe and circle graticule by Hamilton and Holdsworth [30]. As the sizing of particles by visual comparison with reference circles some distance from the particle may be subject to appreciable operator errors, the line graticule was included in order to determine whether more consistent results were obtained by this method. With this graticule, particles are sized as they cross the reference lines and the diameter so

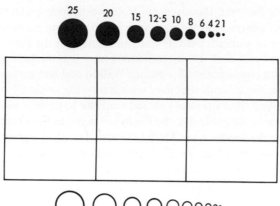

Fig. 6.4 Patterson–Cawood graticule.

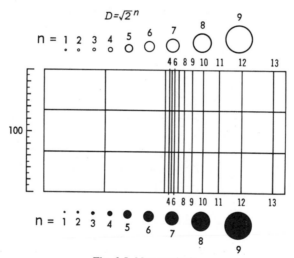

Fig. 6.5 May graticule.

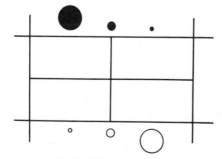

Fig. 6.6 Watson graticule.

measured is Feret's diameter. Hamilton and Holdsworth found systematic differences in the mean counts of operators in the size range 1 to 5 μm, but no evidence that this was altered by the type of graticule used. It was also found that the Feret diameter over-estimated the size of coal dust. Their conclusions were that all three graticules gave equally accurate and consistent results but the Watson and line graticules were preferred by the operators on the grounds that they were less trying to use.

Fairs [27] modified the Patterson–Cawood graticule to cover a wider size range (128 : 1) on three separate graticules, the diameters being in a constant root-two progression except for the smaller sizes. Fairs [77] also described a graticule for use with the projection microscope which was incorporated in the projection screen instead of being in the eyepiece; the nine circles were in a root-two progression. The method was adopted as the British Standard graticule [31] (figure 6.7).

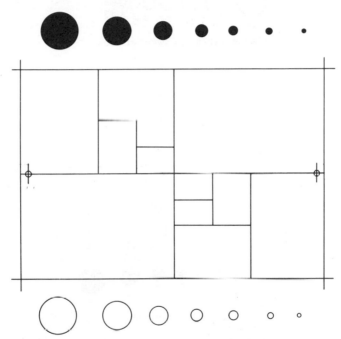

Fig. 6.7 The British Standard graticule (Graticules Ltd). Diagram from BS 3406 (1961): *Methods for the Determination of the Particle Size of Powders*, Part IV, *Optical Microscope Method*. (Reproduced by permission of the British Standards Institution, 2 Park Street, London W1, from whom copies of the complete standard may be obtained.)

Guruswamy [78] has designed a graticule in which the diameters of 11 circles are arranged in a constant ratio of 1.2589 to use the fact that $\log_{10} 1.2589 = 1$. This choice of the ratio and the novel system of marking the circles are claimed to facilitate the rapid calculation of size parameters from the data.

Particle thickness may be determined by a stereophotogrammetric procedure proposed by Aschenbrenner [19] (see also [3, p. 19]).

6.6.2 Training of operators

Although the use of linear diameters such as Martin's and Feret's gives the most reproducible analyses, the projected area diameter is more representative of particle size;
hence the globe and circle graticules are the most popular. When comparing an irregular
profile with a circle, untrained operators have a tendency to oversize the profile. A
method of correcting this is to compare the analysis of a trained operator with that of
the trainee. When the trainee recognizes the bias, he readily corrects it. Heywood [33]
produced a set of hand-held test cards. The trainee is required to compare each of the
profiles with the reference circles and assign it to a size group. The area of each profile
is previously determined by counting squares, hence the trainee can recognize and correct bias. Watson and Mulford [34] extended the technique by inscribing a number
next to each profile and reducing them photographically so that they could be examined
under a reversed telescope, giving more realistic conditions. In a series of tests with nine
operators, it was found that five were underestimating and four overestimating, seven
of the nine being badly biased. The nine operators were trained microscopists who were
aware of the natural tendency to oversize and had overcorrected. It was also noticed
that all nine observers were consistent in their bias but reduced it only slightly on a
second reading.

Fairs [35] used a projection microscope for training purposes. A trained operator
and a trainee can then examine given areas together and compare their results. This
technique is also used for sizing (ASTM (1951) E20-SIT, 1539) but is not recommended for particle diameters less than 2 μm.

Holdsworth et al. (loc. cit. [76]) demonstrated the necessity for training operators
and showed that gross count differences on the same samples at different laboratories
were much reduced after interlaboratory checks.

Nathan et al. [96] compared three commonly used microscopic measurement
techniques and confirmed the oversizing by untrained operators. They suggested that
unskilled operators produce the best results with a line graticule and that experienced
microscopists perform best using an image-splitting device.

6.7 Semi-automatic aids to microscopy

Semi-automatic aids to counting and sizing have been developed to speed up analyses
and reduce the tedium of wholly manual methods. The advantage of these aids over
automatic microscopy, apart from their relative cheapness, is that human judgement is
retained. The operator can select or reject particles, separate out aggregates and discri-
minate over the choice of fields of view. Many such aids have been developed and these
differ widely in degree of sophistication, price, ease, mode and speed of operation.

The Zeiss–Endter particle size analyser (figure 6.8) [32, 109] has been developed so
that a direct comparison may be made between the projected area of a particle and the
area of a reference circle. The reference circle is a spot of light adjustable in size by an
iris diaphragm (I) and centred on a particle. After the spot of light has been made equal
in area to the projected area of the particle, it is recorded in a preset size category
Z (1–8) by the depression of a foot switch (F). The instrument has been designed to

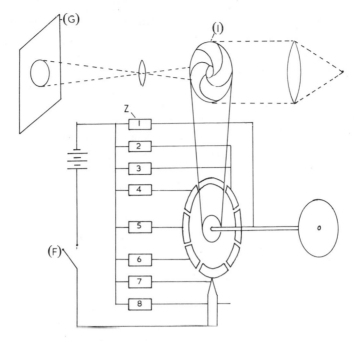

Fig. 6.8 Zeiss–Endter particle size analyser.

work with a photomicrograph (G) which may be obtained from an electron microscope, thus its range of applicability is extended to around 0.01 μm.

Although this gives the size distribution very conveniently, the data have to be transferred manually if further calculations have to be carried out, e.g. to evaluate a mean size or other statistics of the distribution. The instrument can be modified quite inexpensively to give a dc voltage output which is proportional to spot size. This allows particle size to be logged automatically and subsequent calculation to be performed without operator transcription. Exner *et al.* [112] apply the instrument to size and shape determination of lead powder, defining shape as $f = 4\pi a/p^2$ where a is projected area and p is particle perimeter. A modification of Endter's instrument which is rugged and simpler, but not as versatile, has been described by Becher [55].

Electron micrograph counts are greatly facilitated by the use of a projector, a transparent screen and a large transparent graticule connected to a counting device [65]. The micrograph is projected on to the screen and the area of each image compared directly with the areas of a series of circles, in a root-two progression, on the graticule. To record the size, an electrical contact by the appropriate circle is touched with a movable contact. This activates the appropriate counter as well as a counter to sum all the particles classified. A counting speed of over 1000 particles per hour is possible. The Spri particle analyser is a device on a similar principle to the above using a V-notch to give the size gradings. It may be interfaced with a computer or electronic counter unit. Particles may also be measured using specially constructed callipers [67].

The basic module of the DIGIPLAN [132] is an electronic planimeter with a built-in microprocessor. The image structure under analysis is traced out and stored in different count channels. Measured parameters are area and length which in the model AM 02 are extended to other functions: maximum diameter, angle of orientation, Feret's diameter, form factor and centre of gravity coordinates.

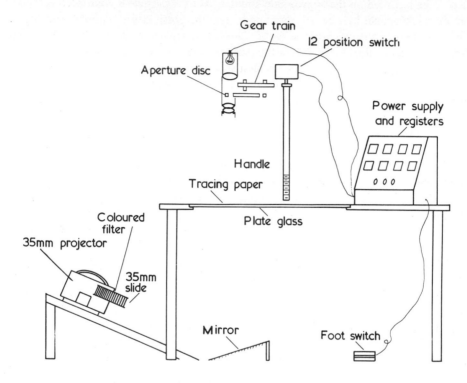

Fig. 6.9 Chatfield particle size analyser.

The Chatfield particle size comparator [79] (figure 6.9) was devised for the size classification of particles in the size range below 10 μm. The system operates on 35-mm film records and is based on the projection of a photograph on to a translucent screen and the comparison of the particle sizes with a superimposed spot of light projected from the other side. The photograph is projected by a standard projector via a surface-silvered mirror and the screen is a sheet of tracing paper put on a glass plate set into a bench top. The particle images can thus be seen from above by the operator seated at the bench. A second projector is used to project a spot of light downwards on to the sheet of paper. The diameter of this spot can be changed by rotation of a disc which has a number of apertures of different diameters. The aperture selected is illuminated by parallel light and its image is focussed on the tracing paper at a magnification of about 4 when using a lens of 2 in focal length. Change of aperture is controlled through a gear train by rotating a handle at bench level. The whole spot projection system is sup-

ported at constant height above the bench by a pantograph arrangement. This allows
the projected spot to be moved over the screen by moving the handle. Also coupled to
the gear train is a 12-position rotary switch connected to a series of electromagnetic
counters. When an individual particle has been sized by selecting the aperture with the
area which best matches that of the particle image, use of a foot switch causes the particle to be recorded on the appropriate counter. As each particle is counted it is marked
on the tracing paper by a pencil, thus eliminating the possibility of double counting.
The spots are in root-two progression of diameter, starting with the smallest at 0.14 cm.
With a total magnification of 7000x the particle size range extends from 0.2 to 1.0 μm in
11 logarithmically equal steps. The smallest spot size complies with the requirement
of BS 3406 in that the smallest size should be 20 times the limit of resolution of the
unaided eye. A performance trial showed that both experienced and untrained operators
were able to size particles at a speed in excess of 30 per minute.

In the Lark particle counter [80] designed for measuring particles in loose powders
the record of analysis consists of a series of pinholes in a chart 3½ in wide. An electro-
mechanical counter gives the total number of particles measured. The system consists
of an adjustable slotted eyepiece which is fitted to a microscope of the stage focussing
type. This eyepiece contains in its focal plane a fixed hairline and a second movable
hairline. A particle to be measured is moved across the field of view by a square mechanical stage until its image reaches the stationary hairline. The second hairline is then
brought up to the image by adjusting the free end of a lever the other end of which is
mechanically linked to this hairline. The free end of the lever has a spring-loaded knob
which can be depressed to make a pinpoint mark on the chart. As this knob rises from
the chart it activates a bar and ratchet which advances the charge by $\frac{1}{16}$ in and also operates the counter. The chart is calibrated by a stage micrometer. By ruling lines on the
chart corresponding to a scale of sizes and counting the pinholes between these lines,
the size frequency distribution in terms of Feret's diameter can be determined. About
1000 particles per hour can be counted.

In its original form the Humphries micrometer eyepiece (*loc. cit.* [76]) made use of a
fixed and a movable hairline. The modified form, developed in collaboration with
Malies Instruments Ltd can be attached to the microscope in which it replaces the
customary eyepiece, a 10x ocular forming an integral part of the instrument. The two
vertical hairlines seen in the field of view are attached to frames and move towards or
away from each other as a milled head to the side of the eyepiece is turned. Grain sizing
is achieved by adjusting the position of the hairlines so that they bracket the image sym-
metrically and then turning the milled head on the eyepiece so that they touch the
edges of the grain. When this has been effected a push-button at the centre of the
milled head is pressed. This closes a contact and operates one of a bank of 16 electromagnetic counters; a seventeenth records the total number of grains examined. Classification and counting are carried out according to a scale of sizes with a ratio of $4\sqrt{2} : 1$
between classes. This scale is directly related to Krumbein's [81] phi-scale and by adjustment of the overall tube length of the microscope, the class limits can usually be
made to fit these scales exactly. By the selection of suitable objectives, particles from
2 μm to 5 μm can be classified. The rate at which particles can be classified depends

largely on the dimension chosen for the analysis. The measurement of the long axis involves rotation of the eyepiece in the body tube. It is claimed that the longest dimension and Feret's diameter can be measured at about 400 and 1000 grains per hour respectively. The counting can be made on slow-moving objects such as particles in liquid suspension without loss of accuracy.

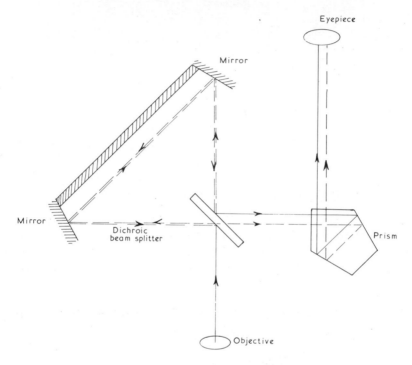

Fig. 6.10 The Watson image-shearing eyepiece.

If the maximum width parallel to a fixed direction (shear size) is an acceptable measure of particle size, the use of a semi-automatic device is recommended. The Watson image-shearing eyepiece (figure 6.10), developed by Dyson [52, 54, 82], employs an image-shearing principle which is obtained by splitting the beam and separating the images by the rotation of two mirrors using a micrometer drum. The principle may be easily understood by referring to figure 6.11. With zero shear, the two images coincide; (a) with a small amount of shear the images overlap; (b) with greater shear a bright gap appears between the images. Since it is possible to change the direction of shear, selected diameters may be measured, a useful feature for particles with widely differing perpendicular diameters.

The eyepiece is first calibrated and preset. It can be seen at a glance which particles are greater than the preset size. When a representative number of fields have been examined at this setting, a new setting is made and the particles oversize recorded. The number of fields examined at any size level may be chosen so as to give a statistically

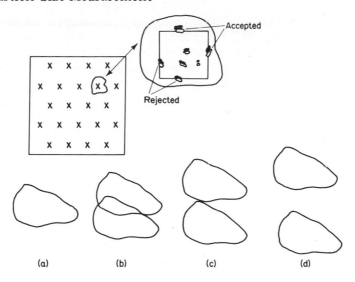

Fig. 6.11 Particle sizing using the image-shearing principle.

acceptable count. An improved optical and mechanical system is claimed for the Vickers-AEI image-splitting eyepiece [53] in which two rotating prisms are used to produce shear (figure 6.12). Each prism consists of a rhomboidal and right-angled prism

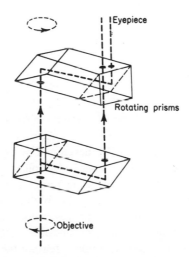

Fig. 6.12 The Vickers image-splitting eyepiece. The operation of the eyepiece is based on the image-splitting principle. A single ray of light is split into two at the partially reflecting surface of the lower prism block. One ray is reflected along the prism block and then upwards into the eyepiece; the other ray is transmitted through the lower prism block and reflected along the upper prism where it is finally reflected upwards into the eyepiece.

cemented together with a substantially neutral semi-reflecting interface. These are mounted so as to rotate about a vertical axis and are constrained by means of a simple linkage to rotate by equal amounts in opposite directions, the rotation being produced by using a micrometer head. Timbrell [22] modified a normal microscope optical system by the introduction of a small mirror inclined at 45° to the vertical to reflect the light beam from the objective into a horizontal eyepiece. The mirror is mounted on to the diaphragm of a loudspeaker vibrating at about 50 Hz. The source of illumination is arranged to flash when the mirror is at the extremities of its vibration, the duration of the flash being short compared with the period of the vibration. Because of the persistence of vision, two images of each particle are seen. The amplitude of vibration of the mirror is controlled by varying the energizing current which is therefore proportional to the separation of the images.

Under ideal conditions, the accuracy of these instruments is 0.025 μm. The apparatus can be preset to chosen values of 10 size limits. A manually operated switch enables the amount of shear to be increased in stepwise fashion to correspond to the preset limits in turn. Pressure on a foot switch registers the particle on the appropriate one of 10 electromagnetic counters which display the numbers together with the total. The apparatus can be connected to digital read-out and print-out on tape facilities for processing the results by computer. Barnett and Timbrell [23] claim extremely high reproducibility for the instrument down to 2 μm.

This device was modified so that it could be inserted in an ordinary microscope tube and in this form is available from Microsizers. Fleming Instruments in collaboration with NRDC market a similar instrument. An electrometer incorporating similar facilities but electronic rather than electrical, has been developed by Coulters, Hialeh, Florida. This instrument features a television display of the particles under examination. A particular feature of Timbrell's device is that particles can be made to vibrate in any direction and the two images from any one particle can be made to rotate about each other. This permits the measurement of maximum and minimum diameters as well as the Feret and shear size. The images can also be made to vibrate at two different amplitudes so that the overlap varies between, say, 10 μm and 12 μm; hence particles can be classified into narrow size categories [76]. An application of the Timbrell/Coulter Shearicon to difficult particulates has been presented by Perry *et al.* [124].

6.8 Automatic counting and sizing

The need to count and size large numbers of samples of airborne dust stimulated the development of automatic microscopes. In these instruments, the particles are passed through a light beam which then falls on a photocell. Changes in the intensity of the light beam due to the presence of particles are monitored and recorded. The instrument types may be categorized as spot scanners and slit scanners [36].

The spot scan methods [37–39] are based on the flying spot microscope of Roberts and Young [40]. In this, the scan is produced by a moving spot of light on a cathode-ray tube which is projected through a microscope on to the specimen. When a particle interrupts the light beam a photocell is activated and the particle is recorded. Special

memory devices in the electronics prevent the same particle from being counted twice [41]. This technique forms the basis of the Rank Cintel instrument. Causley and Young [44], Furmidge [45] and Phillips [46], in discussing this technique, show that certain designs of apparatus are suspect and that there is always the probability that re-entrant particles will be counted twice.

The MR particle size distribution analyser utilizes the scanning spot of a television camera for determining particle size distribution. The instrument records the number of intercepts of the spot by particles and the size distribution may be derived by successive scans of the field at different discriminator levels. A conventional microscope is used with a beam splitter below the eyepiece, so that the image is passed both into the eyepiece and into a television camera. The output from the television camera is displayed on a screen, thus enabling the operator to select and focus on a field without the fatigue of peering through a microscope. The output is also fed to a selector unit which responds to the changes in intensity produced when the scanning spot passes over the particle. The selector unit output is displayed on a screen, so that the operator can check that the unit is operating at the correct level of discrimination and then passed to an indicating unit which displays on a meter the number of intercepts of the scanning spot with particles larger than the selected size.

In the Mullard film-scanning particle analyser [42], the principle of scanning a sample directly has been abandoned in favour of using an intermediate photographic image. This extends the range of the instrument, since electron micrographs may be used, and it also provides a permanent record. Against these advantages must be weighed the disadvantage of the extra operation of photographing and developing transparencies of the slide. Crowl [43] states that this innovation extends the lower range of the instrument to 0.08 μm and that improved performance is obtained by using negative images, i.e. white particles on black background. He concludes that the counting speed of 1000 particles in half an hour is little better than the counting speeds obtainable by a well-trained operator, and although its reproducibility is high, its inability to distinguish between single particles and aggregates produces errors. Crowl substantiates Phillips by stating that there is evidence that the instrument performs more satisfactorily with regularly shaped particles.

The theory of slit scanning has been covered by Hawksley [47] and the technique is used in the Casella counter [48]. The method of operation is to project the image of a slide on to a slit using a conventional microscope. The slide is mechanically scanned and the signals produced as the particle images pass over the slip are electronically recorded. The effects of coincidence and overlap are eliminated by varying the width of the slit. The reproducibility and accuracy are found to be very good down to 2.0 μm using spherical opaque particles [49]. In recent years these instruments have been superseded by image analysers which have many advantages over the earlier systems.

6.9 Quantitative image analysers

Quantitative image analysers accept samples in a variety of forms; photographs, electron micrographs and direct viewing. The particles are scanned by a television camera and

displayed on a console and electrical information describing the sample is passed to a detector. Parameters which can be measured include count, size distribution by area or some statistical diameter, projected length and intersect count and chord-size distribution; from these many secondary parameters may be evaluated. At the detector the features to be measured are selected and passed to an analog computer which counts the chosen data and stores or presents them in one of a variety of ways.

A great number of basically similar instruments have been described and several of these are commercially available.

The πMC is distributed by Millipore for clinical, biological, contamination and powder applications. The QMS was distributed by Bausch and Lomb for metallurgical and photogrammetric work. Both these instruments were developed by Bausch and Lomb. The Omnicon Pattern Analysis System now replaces the QMS [114, 125]; with it one can measure area, longest dimension (L), breadth (perpendicular to L), d_F, perimeter, convex perimeter plus derived measurements and shape factors; a light pen is provided for selective features.

The Quantimet 720 is not based on standard television scanning, but employs a digitally controlled scan which divides the field into 650 000 picture points and allows the use of more sophisticated logic [87]. Applications include inhalation aerosols [110].

Other instruments include the Leitz Classimat [126], the Ameda and the Optomax.

The Classimat [88] has been modified by the addition of software analysis of the binary image to measure area, perimeter and maximum chord length simultaneously (*loc. cit.* [127]).

The Histotrak is especially designed to work interactively. By using a light pen, desired features can be analysed.

A series of papers have been written by Gahm [115–117] who discusses image analysis using the Zeiss Microvideomat. This 'programmable linear analyser' is described as a multipurpose instrument for image analysis and a survey is presented of the stereological field, specifically for mineralogists.

The Epiquant is a new instrument manufactured by VEB Zeiss Jena.

The Quantimet, Classimat and Omnicon have also been compared and gave excellent agreement [127]; comparison has also been made with electrical resistivity techniques (Coulter and Celloscope).

Detailed descriptions of these instruments are given by Davies [83]. Quantitative image analysis is reviewed by Jesse [84] and discussed in several papers in *Microscope* [85, 86, 129].

6.10 Specimen improvement techniques

Automatic and quantitative microscopes tend to give erroneous results for transparent particles. To overcome this problem, Amor and Block [89] have used the silver-staining technique to make the particles opaque. The particles are dry-mounted on to a thin film of tacky collodion on a microscope slide. Silver is then deposited from solution using the silver mirror reaction. Preliminary sensitizing of the crystalline surface ensures that much more silver is deposited on the particles than on the collodion. They

also give details for a method of staining particles in aqueous solution prior to deposition on a membrane filter for analysis.

Hamilton and Phelps [90] adapted the metal-shadowing technique for the preparation of transparent profiles of dust particles. The process consists of evaporating *in vacuo* a thin metal film in a direction normal to a slide containing particles. The particles are then removed by a jet of air or water, leaving sharp transparent profiles.

6.11 Statistical considerations governing the determination of size distributions by microscope count [51]

Since it is impracticable to size all the particles deposited on the microscope slide or electron microscope grid, a representative sample must be chosen. As with all sampling techniques, the measurement sample should be taken at random, or according to some predetermined pattern, from the analysis sample. The simplest procedure is to examine a number of fields spread uniformly over the slide or grid. Since it is necessary to size about 600 particles in order to obtain a statistically accurate count, and it is inadvisable to size more than six particles in any one field of view, it is necessary to examine at least 100 fields. The adopted procedure depends upon whether a number or a weight distribution is required, the former being the simpler of the two analyses.

6.11.1 Frequency distribution determination

The percentage standard error of the mean size in a number distribution is $100/\sqrt{n}$, where n is the number of particles sized. Thus, if only a mean size is required, one may estimate the accuracy of the determination. For powders containing a narrow range of sizes, the necessary count may be substantially lower than that given by the above expression, and Irani and Callis [4] suggest that the mean size should be calculated as the count progresses. This will tend towards some fixed value and the count may be ended when this limit can be estimated.

For a full number distribution, the number to be counted in any size range to achieve a given accuracy is given by:

$$\sigma(F_1) = \left\{ \frac{F_1(100 - F_1)}{n_t} \right\}^{1/2}$$

where $\sigma(F_1)$ = standard deviation expressed as a percentage of the total by frequency,
F_1 = frequency of particles in a given size range,
n_t = total number of particles of all size ranges counted.

The function $F_1(100 - F_1)$ has a maximum value for a size range containing 50% of the particles. Hence, if the required accuracy for $\sigma(F_1)$ is obtained in the size range containing most of the particles, it will hold for all other size ranges. In the case of a size analysis on a frequency basis, the same area should be examined for all classes.

6.11.2 Weight distribution determination

The number of particles to be sized to achieve a given accuracy is given by:

$$\sigma(M_q) = M_q/\sqrt{n_d}$$

where $\sigma(M_q)$ = standard deviation expressed as a percentage of the total by weight,
$\quad\quad M_q$ = percentage by weight in the given size range,
$\quad\quad n_d$ = number of particles counted in the size range.

If, in a particular case, 10% by weight of the particles lie in the top range and an accuracy $\sigma(M_q)$ = 2% is required, the number of particles (n_d) to be counted in that size range is equal to 25.

In order to maintain this accuracy for all other size ranges, the control factor

$$\Omega = \left\{ \frac{nd^6}{(ka)^2} \right\}^{1/2}$$

calculated for the top range exceeds that counted for all other ranges.

d = volume mean diameter of the size range,
a = area of one field,
k = number of areas examined.

In optical microscopy it is often necessary to examine the whole of the slide in strips in order to find 25 particles in the top size range and sometimes necessary to examine more than one slide. It is advisable to make a preliminary scan at the lowest magnification in order that an estimate of Ω may be made and the areas to be scanned for the next two size ranges estimated. Derivations of the above formulae and full operating instructions are given in BS 3406 [31].

6.12 Conclusion

The microscope is an invaluable tool to the particle size analyst. If there are relatively few particles present that are smaller than 1 μm in size, the optical microscope should be used. This should also be used to examine every sub-sieve (greater than 1 μm) powder awaiting analysis in order to reduce the possibilities of errors. For example, the Coulter counter will give a size analysis for a sub-micrometre powder although the size of the individual pulses generated by the particles is below the discriminator range of the instrument, since the counting system for this instrument will count doublets and triplets as single particles if the concentration is high enough.

Number counts can be carried out quickly and accurately, but it is not widely understood that weight counts can also be accurately carried out on as few as 600 particles. The method to be adopted however, is entirely different; hence BS 3406 should be carefully followed.

Image-shearing eyepieces are very useful for size-grading narrowly classified powders and are relatively inexpensive. The Timbrell system is more expensive, more rapid and easier to use. It is, therefore, useful if many analyses are required on a routine basis.

The image-analysing computers are even more sophisticated and expensive but provide far more information.

For sub-micrometre particles it is necessary to use a TEM or SEM, the latter being extremely useful for studies of surface topography.

References

1 Green, M. (1921), *J. Franklin Inst.*, **192**, 657.
2 Dunn, E.J. (1930), *Ind. Engng. Chem., analyt. Edn*, **2**, 59.
3 Orr, C. and Dallevalle, J.M. (1959), *Fine Particle Measurement*, Macmillan, NY.
4 Irani, R.R. and Callis, C.F. (1963), *Particle Size*, Wiley, NY.
5 Harwood, M.G. (1954), *B.J. appl. Phys.* suppl. 3, S193.
6 Rosinski, J., Glaess, H.E. and McCulley, C.R. (1956), *Analyt. Chem.*, **28**, 486.
7 Allen, R.P. (1942), *Ind. Engng. Chem., analyt. Edn*, **14**, 92.
8 Lenz, F. (1954), *Optik*, **11**, 524.
9 Ellison, J. McK. (1954), *Nature*, **179**, 948.
10 Green, M. (1946), *Ind. Engng. Chem.*, **38**, 679.
11 Martin G. *et al.* (1923), *Trans. Ceram. Soc.*, **23**, 61; (1926), **25**, 51; (1928), **27**, 285.
12 Heywood, H. (1946), *Trans. Inst. Min. Metall.*, **55**, 391.
13 Feret, R.L. (1931), *Assoc. Int. pour l'essai des Mnt.* **2**, Group D, Zurich.
14 Tomkieff, S.L. (1945), *Nature*, **155**, 24.
15 Moran, P.A.P. (1944), *Nature*, **154**, 490.
16 Herdan, G. (1960), *Small Particle Statistics*, Butterworths.
17 Steinheitz, A.R. (1946), *Trans. Soc. Chem. Ind.*, **65**, 314.
18 Crowl, V.T. (1961), *Paint Res. Station, Teddington, Memorandum No.* 291, **12**, 24.
19 Aschenbrenner, B.C. (1955), *Photogrammetric Engng.*, **21**, 376.
20 Walton, W.H. (1948), *Nature*, **162**, 329.
21 Nicholson, W.L. (1976), *J. Microsc.*, **107**, 3, 323-4.
22 Timbrell, V. (1952), *Nature*, **170**, 318-9.
23 Barnett, M.I. and Timbrell, V. (Oct., 1962), *Pharm. J.*, 379.
24 Cauchy, A. (1840), *C.R. Acad. Sci., Paris*, **13**, 1060.
25 Patterson, H.S. and Cawood, W. (1936), *Trans. Faraday Soc.*, **32**, 1084.
26 Watson, H.H. (1936), *Trans. Inst. Min. Metall.*, **46**, 176.
27 Fairs, G.L. (1943), *Chemy Ind.*, **62**, 374-8.
28 May, K.R. (1965), *J. scient. Instrum.*, **22**, 187.
29 Watson, H.H. (1952), *B.J. Ind. Med.*, **19**, 80.
30 Hamilton, R.J. and Holdsworth, J.F. (1954), *B.J. appl. Phys.*, suppl. 3, S101.
31 BS 3406 (1963) Part 4.
32 Endter, F. and Gebauer, H. (1956), *Optik*, **13**, 87.
33 Heywood, H. (1946), *Bull. Inst. Min. Metall.*, nos. 477, 478.
34 Watson, H.H. and Mulford, D.F. (1954), A particle profile test strip for assessing the accuracy of sizing irregularly shaped particles with a microscope. *B.J. appl. Phys.*, suppl. 3, S105.
35 Fairs, G.L., Discussion. *ibid.*, S108.
36 Walton, W.H., Survey of the automatic counting and sizing of particles. *ibid.*, S121.
37 Vick, F.A. (1956), *Sci. Prog.*, **94**, 176, 655.
38 Morgan, B.B. (1957), Automatic particle counting and sizing. *Research (Lond.)*, **10**, 271.
39 Taylor, W.K. (1954), An automatic system for obtaining particle size distributions with the aid of the flying spot microscope. *B.J. appl. Phys.*, suppl. 3, S173.
40 Roberts, F. and Young, J.Z. (1952), *Nature*, **169**, 962.
41 Bell, H.A. (1954), Stages in the development of an arrested scan type microscope particle counter. *B.J. appl. Phys.*, suppl. 3, S156.
42 *Mullard Film Scanning Particle Analyser*, L. 188, Mullard Ltd, Technical Leaflet.

43 Crowl, V.T. (1960). The use of the Mullard film scanning particle size distribution counting from electron micrographs. *Res. Mem. No.* 284, Research Association of British Paint, Colour and Varnish Manufacturers.
44 Causley, D. and Young, J. (1955), *Z. Res.* **8**, 430.
45 Furmidge, C.G.L. (1961), *B.J. appl. Phys.*, **12**, 268.
46 Phillips, J.W. (1954), Some fundamental aspects of particle counting and sizing by linear scans. *B.J. appl. Phys.*, suppl. 3, S133-7.
47 Hawksley, P.G.W., Theory of particle sizing and counting by track scanning. *ibid.*, S125-32.
48 *Casella Automatic Particle Counter and Sizer* (Booklet 906A), Cooke, Troughton & Simms Ltd.
49 Allen, T. and Kaye, B.H. (1965), *Analyst*, **90**, 1068, 147.
50 Walton, W.M. (1947), The application of the electron microscope to particle size measurement. Symp. Particle Size Analysis, *Inst. Chem. Eng.*, **25**, 64-76.
51 Fairs, G.L. (1951), *J.R. microsc. Soc.*, **71**, 209.
52 Dyson, J. (1960), *J. opt. Soc. Am.*, **50**, 754-7.
53 Payne, B.O. (1964), *Microscope*, **14**, 6, 217.
54 Dyson, J. (1961), *AEI Engng.*, **1**, 13.
55 Becher, P. (1964), *J. Colloid Sci.*, **19**, 468.
56 Kay, D.H. (1965), *Techniques for Electron Microscopy*, 2nd edn. Blackwell Scientific Publications, Oxford.
57 Drummond, D.G. (ed.) (1950), *The Practice of Electron Microscopy*, Royal Microscopical Society, London.
58 Revell, R.S.M. and Agar, A.W. (1955), *B.J. appl. Phys.*, **6**, 23.
59 Bradley, D.E. (1954), *B.J. appl. Phys.*, **5**, 65.
60 Backus, R.C. and Williams, R.C. (1950), *J. appl. Phys.*, **21**, 11.
61 Bradley, D.E. and Williams, D.J. (1957), *J. gen. Microbiol.*, **17**, 75.
62 Bailey, G.W. and Ellis, J.R. (1954), *Microscope*, **14**, 6, 217.
63 Williams, R.C. and Wyckoff, R.W.G. (1946), *J. appl. Phys.*, **17**, 23.
64 Cartwright, J. and Skidmore, J.W. (1953), Report No. 79, SMRE Sheffield.
65 Crowl, V.T. (1961), Report No. 291, Paint Research Station, Teddington.
66 Joffe, A.D. (1963), *B.J. appl. Phys.*, **14**, 7, 429.
67 Maclay, W.N. and Grindter, E.M. (1963), *J. Colloid Sci.*, **18**, 343.
68 Charman, W.N. (1961), Ph.D. Thesis, London Univ.
69 Taylor, N.J. (1969), *Vacuum*, **19**, 575; *J. Vacuum Sci. Tech.* (1969), **6**, 241.
70 Chang, C.C. (1971), *Surface Sci.*, **25**, 23.
71 Brundle, C.R. (1972), *Surface and Defect Properties of Solids*, **6**, Ch. 6, Chem. Soc., London.
72 Rowe, S.H. (1966), *Microscope*, **15**, 216.
73 Welford, G.A. (1960), *Optics in Metrology*, **85.**
74 Timbrell, V. (1972), *J. appl. Phys.*, **43**, 11, 4839.
75 Timbrell, V. (1972), *Microscope*, **20**, 365.
76 Timbrell, V. (1973), *Harold Heywood Memorial Symposium*, Loughborough Univ., England.
77 Fairs, G.L. (1951), *J.R. microsc., Soc.*, **71**, 209.
78 Guruswamy, S. (1967), *Particle Size Analysis*, Soc. Analyt. Chem., 29-31.
79 Chatfield, E.J. (1967), *J. scient. Instrum.* **44**, 615.
80 Lark, P.D. (1965), *Microscope*, **15**, 1-6.
81 Krumbein, W.C. (1934), *J. Sediment. Petrol*, **4**, 65-7.
82 Dyson, J. (1959), *Nature*, **184**, 1561.
83 Davies, R. (1970), Illinois State Microscopical Society Seminar.
84 Jesse, A. (1971), *Microscope*, **19**, 1, 21-30.
85 Cole, M., *ibid.*, 87-103.
86 Huna, W., *ibid.*, 2, 205-18.
87 Williams, G. (1971), *Bull. Soc. fr. Ceram.* **90**, 59-63.
88 Stutzer, M., *ibid.*, 65-68.
89 Amor, A.F. and Block, M. (1968), *J.R. microsc. Soc.*, **88**, 4, 601-5.

 90 Hamilton, R.J. and Phelps, B.A. (1956), *B.J. appl. Phys.* **7**, 186.
 91 Pidgeon, F.D. and Dodd, C.G. (1954), *Analyt. Chem.*, **26**, 1823–8.
 92 McCrone, W.C. (1970), *Microscope,* **18**, 1, 1.
 93 Delly, J.G. (1969), *ibid.*, **17**, 205–11.
 94 Corcoran, J.F. (1970), *Fuel,* **49**, 3, 331–4.
 95 Eckert, J.J.D. and Caveney, R.J. (1970), *J. Phys. E.,* 413–14.
 96 Nathan, I.F., Barnett, M.I. and Turner, T.D. (1972), *Powder Technol.,* **5**, 2, 105–10.
 97 Anon. (1967), *Ceramic Age,* December.
 98 Barbery, G. (1974), *Powder Technol.,* **9**, 5/6, 231–40.
 99 Sahu, B.K. (1976), *ibid.,* **13**, 295–6.
100 Ellison, J. McK. (1954), *Nature,* **173**, 948.
101 Eckhoff, R.K. and Enstad, G. (1975), *Powder Technol.,* **11**, 1–10.
102 Proctor, T.D. and Harris, G.W. (1974), *J. Aerosol. Sci.,* **5**, 1, 81–90.
103 Proctor, T.D. and Barker, D., *ibid.,* 91–9.
104 Hay, W. and Sandberg, P. (1967), *Micropaleontology,* **13**, 407–18.
105 Krinsley, D. and Margolis, S. (1969), *Trans. N.Y. Acad. Sci.,* **31**, 457–77.
106 Willard, R.J. and Hjelmstad, K.E. (1969/70), *Powder Technol.* **3**, 311–13.
107 Turner, G.A., Fayed, E. and Zackariah, K. (1972), *ibid.,* **6**, 33–37.
108 White, E.W. *et al.,* (1970), *Proc. 3rd Ann. Scanning Electron Microscope Symp.,* Illinois Institute of Technology Res. Inst., Chicago, pp. 57–64.
109 Duncan, A.A. (1974), Report MH SMP-74-19-F. Dept. NTIS, USA 7 pp.
110 Hallworth, G.W. and Barnes, P. (1974), *J. Pharm. Pharmac.,* Suppl. 26, 78–79.
111 ASTM (1976), *Annual Book of Standards,* Part 41, Particle Size Measurement, Microscopy, E20–68 (Reapproved 1974).
112 Exner, H.E. and Linck, E. (1977), *Powder Metall. Int.,* **9**, 3, 131–3.
113 Morton, R.R.A., *Measurement and Analysis of Electron Beam Microscope Images,* Form 7050, Bausch and Lomb.
114 Morton, R.R.A. and Martens, A.E. (1972), *Res. Develop.,* **23**, 1, 24–26, 28.
115 Gahm, J. (1975), *Spec. Iss. Pract. Metall.,* 5, 29–46, Dr Riederer, Stuttgart.
116 Gahm, J. (1975), *Spec. Iss. Res. Film,* **8**, 6, 553–68.
117 Gahm, J. (1975), *Fortschr. Minerol.* **53**, 1, 79–128.
118 Walton, W.H. (1947), *Trans. Inst. Chem. Engrs,* Suppl. 25, 64.
119 Dullien, F.A.L., Rhodes, E. and Schroeter, S.R. (1969/70), *Powder Technol.,* **3**, 124–35.
120 Dullien, F.A.L. and Mehta, P.N. (1972), *ibid.,* **5**, 179–94.
121 Dullien, F.A.L. (1973), *Am. Chem. Soc. Div. Org. Coat. Plast. Chem.* Pap. 33, 2, 516–24.
122 Dullien, F.A.L. and Dhawan, G.K. (1974), *J. Colloid Interfac. Sci.,* **47**, 2, 337–49.
123 Szalkowski, F.J. (1977), *ibid.,* **58**, 2, 199–215.
124 Perry, R.W., Harris, J.E.C. and Scullion, H.J. (1977), In: *Particle Size Analysis Conf.* (ed. M.J. Groves) Chem. Soc. Anal. Div., (1978), Heyden.
125 Morton, R.R.A. and McCarthy, C. (1975), *Microscope,* **23**, 4, 239–60.
126 Davies, R. (1972), *Leitz-Nutteilung. Wiss.,* U. Techn. Suppl. **1**, 3, 65–74.
127 Alliet, D.F., Tietjen, T.A. and Wood, D.H. (1977), In: *Particle Size Analysis Conf.* (ed. M.J. Groves) Chem. Soc. Anal. Div., (1978), Heyden.
128 Anon. (1975), *Microscope,* **23**, 2, VI and VII.
129 Jesse, A. (1976), Automatic image analysis; Bibliography (1973–1975); 676 References, *ibid.,* **24**, 1, 1–95.
130 Chu, Y.F. and Ruckenstein, E. (1976), *J. Catalysis,* **41**, 3, 373–83.
131 Ruzek, J. and Zbuzek, B. (1975), *Silikaty,* **19**, 1, 49–66.
132 Anon. (1978), *Powder Metall. Int.,* **10**, 2, 95.

7 Interaction between particles and fluids in a gravitational field

7.1 Introduction

The settling behaviour of particles in a fluid is widely used for particle size determination. The simplest case to consider is the settling velocity, under gravity, of a single sphere in a fluid of infinite extent. Many experiments have been carried out to determine the relationship between settling velocity and particle size. A unique relationship between drag factor and Reynolds number has been found, and this relationship reduces to a simple equation, the Stokes equation, which applies at low Reynolds number, relating settling velocity and particle size. In this chapter this equation is developed and its limits explored. It is shown that, for the purpose of particle size measurement, the time for a particle to reach a steady velocity (the acceleration time) is negligible.

If the concentration is monitored at a fixed depth below the surface for an initially homogeneous suspension of spheres, this will remain constant until the largest particle present in the suspension has fallen from the surface to the measurement zone. The concentration will then fall, being at all times proportional to the concentration of particles smaller than the diameter given by the Stokes equation for that particular time and depth of fall.

Any sample removed from this depth should not contain particles with diameters larger than the Stokes diameter. In practice this is not true, due to particle–particle interaction, and this is investigated here in some detail. In general, pairs of equally sized spheres in close proximity will fall with a greater terminal velocity than for a single sphere. For unequally sized particles the situation is more complex; the larger particle may actually pick up the smaller one so that it revolves round the large one as a satellite. Assemblies of spheres tend to diverge due to the rotation effect caused by the greater velocity of the streamlines on the envelope of the assembly. A cluster of particles acts as a single large particle of appropriate density and reduced rigidity and has a much greater velocity than the settling velocities of the particles of which it is composed.

At volume concentrations as low as 1%, the suspension settles *en masse* and the rate of fall of the interface gives an average size for the particles with a modified Stokes equation. This equation is very similar to the permeametry equation for a fluid passing through a fixed bed of powder. The settling velocities of particles are also reduced in the proximity of container walls though this effect is usually quite small.

The volume concentrations recommended in various Standards are: 2–4% (BS 1377

(1961), for hydrometers); up to 1% (BS 3406 (1963) Part 2); 0.2% (DIN 66111 (1973)).

It is generally recommended that the concentration should be kept as low as is compatible with weighing requirements. A suitable test to determine the highest applicable concentration is to carry out one analysis at 0.5% concentration and one at 0.2%. If there are significant differences between the two results, which will be manifested as a finer analysis where interaction is important, i.e. at the higher concentration, one should work at the lower concentration and if available the even lower concentrations possible with the photosedimentation technique or the sedimentation balance.

The data from sedimentation analyses are interpreted as equivalent Stokes diameters. Stokes' equation applies only at low Reynolds number when inertial terms may be neglected. Under these conditions the Stokes diameter for a sphere is the same as its physical diameter. As particles become larger, turbulence sets in and particles settle more slowly than Stokes' law predicts, the Stokes diameters becoming smaller than the physical diameters.

The difficulties that arise are even more pronounced with irregularly shaped particles since the Stokes diameter is limited to laminar settling. Under laminar flow conditions particles tend to settle in random orientation; once the boundary condition is exceeded particles take up a preferred orientation to give maximum resistance to drag. The measured size distribution becomes finer with increasing Reynolds number in a manner difficult to predict.

As particle size decreases the particles are acted upon by the fluid molecules to give a variable settling rate to particles of the same size. Indeed, a proportion of the particles will actually rise during a time interval, although the concentration at a fixed depth, on the average, will fall. Impressed upon this effect there are convection currents which may be set up by surface evaporation or temperature differences. All settling suspensions appear to be basically unstable due to preferred paths up the sedimentation tank for the fluid displaced by the settling particles. The fluid tends to rise up the walls of the containing vessel and to dissipate itself as convection currents at the top of the sedimentation column and carry with it some of the finer particles. This leads to an excess of fines at the top of the container and an overestimation of the fines percentage in an analysis. For these reasons, for accurate results below about 2 μm, centrifugal techniques should be used.

Reproducible data are possible at high Reynolds number, high concentrations and with sub-micron particles. These data may be highly inaccurate and in general precise values of erroneous sizes and concentrations are of limited worth.

7.2 Relationship between drag coefficient and Reynolds number for a sphere settling in a liquid

When a particle falls under gravity in a viscous fluid, it is acted upon by three forces: a gravitational force W acting downwards; a buoyant force U acting upwards and a drag force F_D acting upwards. The resulting equation of motion is:

$$mg - m'g - F_D = m \frac{du}{dt} \tag{7.1}$$

where m is the mass of the particle, m' is the mass of the same volume of fluid, u is the particle velocity and g is the acceleration due to gravity.

When the terminal velocity is reached, the drag force is equal to the motive force on the particle, that is, the difference between the gravitational attraction and the Archimedes upthrust.

For a sphere of diameter D and density ρ_s falling in a fluid of density ρ_f, the equation of motion becomes:

$$F_D = (m - m')g$$

$$F_D = \frac{\pi}{6}(\rho_s - \rho_f)gD^3 \tag{7.2}$$

Dimensional analysis of the general problem of particle motion under conditions of equilibrium [34] shows that there is a unique relationship between two dimensionless groups, the Reynolds number, Re, and the drag coefficient C_D where:

$$Re = \frac{\rho_f u D}{\eta} \tag{7.3}$$

where η is the viscosity of the fluid:

$$C_D = \frac{\text{drag force}}{\text{cross-sectional area of particle} \times \text{dynamic pressure on particle}}$$

$$C_D = \frac{F_D}{\pi \dfrac{D^2}{4} \times \rho_f \dfrac{u^2}{2}} \tag{7.4}$$

The experimental relationship between the Reynolds number and the drag coefficient is shown graphically in figure 7.1. This graph is divided into three regions, a laminar flow or Stokes region, an intermediate region and a turbulent flow or Newton region.

7.3 The laminar flow region

Stokes [35] assumed that when the terminal velocity is reached, the drag on a spherical particle falling in a viscous fluid of infinite extent is due entirely to viscous forces within the fluid and deduced the expression:

$$F_D = 3\pi D \eta u_{Stk} \tag{7.5}$$

where u_{Stk} is the terminal velocity in the Stokes region. Alternatively u_{Stk} is the terminal velocity as given by Stokes' equation, the difference between this and the free-falling velocity increasing with increasing Reynolds number.

Substituting in equation (7.2) gives, for a sphere:

$$u_{Stk} = \frac{(\rho_s - \rho_f)g\,D^2}{18\eta} \tag{7.6}$$

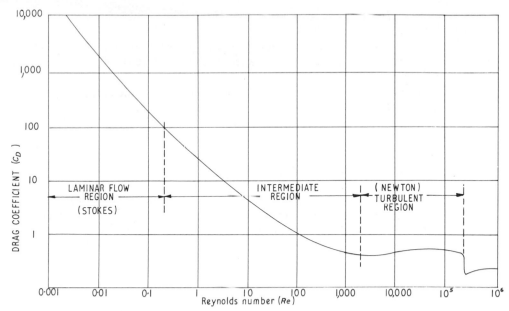

Fig. 7.1 Experimental relationship between drag coefficient and Reynolds number for a sphere settling in a liquid.

The assumptions made in the derivation of Stokes' law of settling velocities are:

(1) The particle must be spherical, smooth and rigid and there must be no slip between it and the fluid.
(2) The particle must move as it would in a fluid of infinite extent.
(3) The terminal velocity must have been reached.
(4) The settling velocity must be low so that all inertia effects are negligible.
(5) The fluid must be homogeneous compared with the size of the particle.

The relationship between C_D and Re in the laminar flow region is, from equations (7.3), (7.4) and (7.5):

$$C_D = \frac{24}{Re} \,. \tag{7.7}$$

C_D tends to this value at low Reynolds numbers (see figure 7.1). Equation (7.6), the Stokes equation, may therefore be used for low Reynolds numbers; the terminal velocities calculated thereby will be about 5% too great at $Re = 0.2$ and the derived diameters will be about 2.5% smaller than the physical diameter.

7.4 Critical diameter for laminar flow settling

Writing the diameter as given by the Stokes equation as D_{Stk}, the critical diameter above which Stokes' equation should not be used is given by making $Re = 0.2$ and eliminating

u from equations (7.3) and (7.6) giving:

$$\text{critical value of } D^3_{Stk} = \frac{3.6\eta^2}{(\rho_s - \rho_f)\rho_f g} \tag{7.8}$$

For quartz particles settling in water:

$$\rho_s = 2650 \text{ kg m}^{-3}$$

$$\rho_f = 1000 \text{ kg m}^{-3}$$

$$g = 9.81 \text{ m s}^{-2} \qquad \underline{D_{Stk} = 60.6 \ \mu m}$$

$$\eta = 0.001 \text{ N s m}^{-2}$$

The critical value for quartz particles settling in air, $\eta = 18 \times 10^{-6}$ N s m^{-2}, $\rho_f = 1.39$ $\times 10^{-4}$ kg m^{-3}, is 30 μm. Very many problems concerned with particle motion are beyond the validity of Stokes' equation and a rapid method of calculating the terminal velocity in such cases is required.

7.5 Particle acceleration

In sedimentation analyses it is assumed that the terminal velocity is reached instantaneously. The usual procedure is to agitate the suspension and assume that when this ceases the particles are all falling with their terminal velocities. In actual fact, the particles are in random motion, though a visual examination suffices to show that all the particles are falling within a few seconds. If it is assumed that the particles are initially at rest in a still fluid, the equation of motion, in the laminar flow region, becomes:

$$\frac{\pi}{6} (\rho_s - \rho_f)gD^3 - 3\pi D\eta u = \frac{\pi}{6} \rho_s D^3 \frac{du}{dt} \tag{7.9}$$

Simplifying by putting

$$\rho = \frac{\rho_s - \rho_f}{\rho_s} \quad \text{and} \quad X = \frac{18\eta}{\rho_s D^2}$$

$$\frac{du}{dt} = \rho g - Xu \tag{7.10}$$

$$\int_0^u \frac{du}{\rho g - Xu} = \int_0^t dt$$

$$u = \frac{\rho g}{X} [1 - \exp(-Xt)] \tag{7.11}$$

As t approaches infinity, u approaches the Stokes velocity u_{Stk} as given in equation (7.6).

Theoretically a particle never reaches its terminal velocity but, for practical purposes, it can be assumed that the velocity is sufficiently near the terminal velocity for the error to be neglected.

From equation (7.11), the velocity is 0.99 times the terminal velocity when:

$$u = 0.99\, u_{Stk}$$

$$1 - \exp(-Xt) = 0.99$$

$$Xt = \log_e 100$$

$$t = \frac{4.6D^2\, \rho_s}{18\eta}$$

For spheres of density 2650 kg m^{-3} in water, $\eta = 0.001$ N s m^{-2} and in air, $\eta = 18 \times 10^{-6}$ N s m^{-2}, the times taken to reach 99% of the terminal velocity for particles of different diameters are shown in Table 7.1.

Table 7.1

Diameter (μm)	Times (ms) Water	Air
5	0.017	0.95
10	0.068	3.70
50	1.70	95.0

From Table 7.1 it can be seen that the assumption that a particle falling from rest reaches its terminal velocity instantaneously does not introduce any appreciable errors in the Stokes region.

The distance covered, h, during the acceleration time is given by integrating equation (7.11):

$$\int_0^h \mathrm{d}h = u_{Stk} \int_0^t [1 - \exp(-Xt)]\; \mathrm{d}t$$

$$h = u_{Stk}t \left\{ 1 - \frac{1}{Xt}\,[1 - \exp(-Xt)] \right\} . \qquad (7.12)$$

At 99% of the terminal velocity $1 - \exp(-Xt) - 0.99$, as before, and $Xt = 4.6$:

$$h = 0.785 u_{Stk} t.$$

Since t is very small, the distance fallen in achieving a velocity equal to 0.99 times the settling velocity is also very small at low Reynolds number.

7.6 Errors due to the finite extent of the fluid

When the fluid is of finite extent, there are two effects: the fluid streamlines about the particle impinge on the walls of the containing vessel and are reflected back on the particle causing increased drag; also, since the fluid is stationary at a finite distance from the particle there is distortion of the flow pattern which reacts back on the particle.

Both effects increase the drag on the particle leading to too low an estimate of particle size in sedimentation. Thus the drag on a sphere is given by:

$$F = 3\pi D\eta u \left(1 + k\frac{D}{L}\right)$$
(7.13)

where L is the distance from the centre of the particle to the walls of the containing vessel. The numerical constant k has been obtained theoretically for a sphere [1]. For a single wall $k = 0.563$, for two walls $k = 1.004$ and for a circular cylinder $k = 2.104$. Each extra wall increases the drag by an approximately equal amount, thus the settling velocity of a sphere in a cylinder is much the same over a large part of the central area, the increase in drag due to displacement towards one side being offset by the decrease due to displacement from the other. As long as the inertial terms are negligible, there is no force tending to move the sphere towards or away from the wall.

A sphere moving near a single plane wall will, however, rotate, as if it were rolling on the wall, at an angular velocity given by [2, p. 327]:

$$\frac{3u}{2D}\left(\frac{D}{L}\right)^4 \left\{1 - \frac{3}{4}\frac{D}{L}\right\}.$$

Between two parallel walls with $D \ll L$, where the sphere is located such that its distance from one wall is three times its distance from the other, rotation is in the opposite direction to rolling on the near wall with angular velocity:

$$\frac{1}{80}\frac{D}{L^2} \bigg/ \left(1 - 0.326\frac{D}{L}\right)$$

where L is the distance to the nearer wall.

Inserting equation (7.13) in Stokes' equation gives:

$$D = D_{Stk}\left(1 + \frac{1}{2}k\frac{D}{L}\right)$$
(7.14)

neglecting second- and higher-order terms, where D is the true diameter and D_{Stk} the diameter obtained using the unmodified Stokes equation.

For a 100 μm sphere settling in a 0.5 cm diameter cylinder, $L = 0.25$ cm, the error in particle size is given by:

$$D = D_{Stk}\left(1 + \frac{1}{2}\frac{2.104}{0.25}\frac{10^{-4}}{10^{-2}}\right)$$

$$= 1.041 D_{Stk}, \quad \text{an error of about 4\%.}$$

The effect of the bottom of the container has been evaluated by Lorentz [1] who modified equation (7.5) in the following manner:

$$F_D = 3\pi D\eta u \left(1 + \frac{9}{16}\frac{D}{L}\right)$$
(7.15)

The correction term is negligible if the sampling is carried out at a distance greater than 1000 diameters from the ends of the suspension and is very small for a distance as small as 50 diameters.

7.7 Errors due to discontinuity of the fluid

The Stokes drag needs modification for the molecular nature of real fluids. At one extreme, when the pressure is very low and the mean free path length of the gas molecules (λ) is much larger than the particle size, the resistance to particle motion is due to bombardment by individual molecules acting independently and is very much less than the Stokes drag, leading to an increased settling velocity:

$$u = u_{Stk}\ \frac{4.49\lambda}{BD} \tag{7.16}$$

where B depends on the nature of the molecular reflections and lies between 1 and 1.4 [3].

At the other extreme, when the pressure is higher and the particles are much larger than the mean free path length, the discontinuity effect gives rise to 'slip' between the particle and the medium, leading to the following modification to Stokes' equation:

$$u = u_{Stk}\ \left[1 + \frac{2\lambda}{D}\left(\frac{2-f}{f}\right)\right] \tag{7.17}$$

where f is the fraction of molecules undergoing diffuse reflection at the particle surface and is experimentally found to be of the order of 0.90. In the intermediate region, when the mean free path length and particle size are of the same order, neither treatment is applicable.

Cunningham [4] introduced a correction term to Stokes' law of the form:

$$u = u_{Stk}\ \left(1 + \frac{2A\lambda}{D}\right) \tag{7.18}$$

where A is a constant approximately equal to unity. Experimental data, analysed by Davies [5], gave the more accurate empirical equation:

$$u = u_{Stk}\left\{1 + \frac{\lambda}{D}\left[2.514 + 0.800\exp\left(-0.55\frac{D}{\lambda}\right)\right]\right\} \tag{7.19}$$

For air at pressure P cm Hg and $20°$ C, λ may be replaced by $(5.0 \times 10^{-4}/P)$. For air at 76 cm Hg and $20°$ C and D in micrometres, the equation becomes:

$$u = u_{Stk}\left\{1 + \frac{1}{D}\left[0.1663 + 0.0529\exp\left(-8.32D\right)\right]\right\}.$$

The mean free path is here defined as:

$$\lambda = \frac{2\eta}{P}\sqrt{\pi\ \frac{RT}{8M}}$$

where R is the molar gas constant, T the absolute temperature and M the molecular weight of the gas.

Lapple [6] gives the following typical values for spherical particles in air at 70° F:

$$D(\mu m) \quad 0.1 \quad 0.25 \quad 0.50 \quad 1.0 \quad 10.0$$

$$\frac{2A\lambda}{D} \quad 1.88 \quad 0.68 \quad 0.33 \quad 0.16 \quad 0.016.$$

The correction may be important for aerosols but not for particles in liquids.

7.8 Brownian motion

When a particle is sufficiently small, collisions with individual fluid molecules may displace the particle by a measurable amount. This results in a random motion of the particles in addition to any net motion in a given direction due to the action of external forces such as gravity [7, 8]. Quantitatively, this random motion may be expressed as follows [2, p. 412]:

$$\bar{X}^2 = \frac{4RTK_m t}{3\pi^2 \eta ND} \tag{7.20}$$

where $\bar{X}$ is the statistical average linear displacement in a given direction in time t, R is the gas constant, T the absolute temperature, N is Avogadro's number and K_m the correction for discontinuity of the fluid discussed above.

A comparison of Brownian movement displacement and gravitational settling displacement is given in Table 7.2 [9].

Table 7.2 Comparison of Brownian movement displacement and gravitational settling displacement.

	Displacement in 1.0 second (μm)					
	in air at 70° F (1 atm)		in water at 70° F		in water at 70° F	
Particle diameter (μm)	due to Brownian movement*	due to gravitational settling†	due to Brownian movement*	due to gravitational settling†	γ	$k = \dfrac{100\,\gamma}{1+\gamma}$ (%)
0.10	29.4	1.73	2.36	0.005	31.1	96.9
0.25	14.2	6.3	1.49	0.0346	3.15	75.9
0.50	8.92	19.9	1.052	0.1384	0.556	35.7
1.0	5.91	69.6	0.745	0.554	0.0983	5.0
2.5	3.58	400	0.334	13.84	0.00995	1.0
10.0	1.75	1550	0.236	55.4	0.00031	0.03

* Mean displacement given by equation (7.20).

† Distance settled by a sphere of density 2000 kg m^{-3}, including Cunningham's correction.

γ is defined in equation (7.23).

This effect limits the use of gravitational sedimentation in water for particle size analysis to particles greater than about 1 μm where, for a specific gravity of 2, the Brownian motion exceeds the gravitational motion. Even for specific gravities of 5, the magnitudes of the two motions are similar, i.e. 0.528 m s^{-1} and 2.22 m s^{-1} respectively.

Calculating γ and percentage Brownian motion k, for an analysis time t_f = 20 min, gives columns 6 and 7 in Table 7.2. According to Moore and Orr [36] the effect of Brownian motion may be neglected provided k is less than 20% (i.e. D = 0.69 μm for the data in Table 7.2). This size may be reduced by adjustment of such parameters as liquid density, viscosity, temperature, time and height of analysis.

Brownian motion is normally negligible [8, p.16] when compared with even the most feeble convection currents and this further limits the smallest size at which gravitational sedimentation gives meaningful results.

Muta and Watanabe [37] examined samples extracted during Andreasen analyses and detected particles of 1 μm, for example, when Stokes' equation predicted a maximum size of 0.89 μm. They concluded that the conventional sedimentation analysis of powder samples of wide size range containing sub-micron particles yielded highly erroneous results, and that this might be attributed to the redistribution of fine particles caused by disturbance of the suspension in the course of long-term sedimentation.

Allen [38] reported the presence of convection currents which appeared in sedimenting suspensions. He concluded that these were caused by a basic instability in sedimenting suspensions which determines the lower size for which gravity sedimentation could be used. The liquid displaced by sedimenting particles has a tendency to flow towards the surface up the sides of walls. This he attributed to the lower velocity of particles settling in the vicinity of walls and the pressure gradient across these particles which cause them to rotate in such a manner as to cause such a flow. The effect is most apparent when there are large particles together with fines, i.e. a wide size distribution. The large particles initiate the flow at a velocity of about 15 μm s^{-1}, and particles in the vicinity of the walls, having a settling velocity smaller than this, are entrained and carried towards the surface.

Allen and Baudet [39] also compared several centrifugal techniques with the Andreasen and the Sedigraph X-ray sedimentometer. As an example, for kaolinite, the centrifugal techniques gave the percentage smaller than 0.2 μm as 31 to 36.6%; the Andreasen gave 48%; the Sedigraph gave 80%. The high percentage of fines as given by the Andreasen technique was attributed to the prolonged settling time required. In the Sedigraph, high concentrations were also required since the particles were poor absorbers of X-rays and this caused retarded settling. Other cases are known where the Andreasen results are comparable with centrifugal results. For example, for quartz particles, size range 0.2 to 0.8 μm, the author has found that Andreasen results, using a volume concentration of 0.2%, agree with pipette centrifuge results.

For settling in a centrifugal field Brugger [76] compared the mean square displacement in a radial direction by Brownian motion in time Δt with the centrifugal settling. He concluded that the lower size limit is several hundredths of a micron. With the assumption that the effect can be neglected if the radial distance covered by Brownian motion is one-tenth of the distance covered by centrifugal motion he develops the fol-

lowing equation:

$$D_{min}^3 = \frac{12}{\pi} \frac{RT}{N} \frac{1}{(\rho_s - \rho_f)\omega^2} \frac{\ln r/s}{(r - s)^2} \qquad (7.21)$$

For example, using typical values: $(R/N) = 1.37 \times 10^{-23}$ J K^{-1}, $T = 300$ K, $r = 0.07$ m, $s = 0.04$ m, $(\rho_s - \rho_s) = 1000$ kg m^{-3}, $\omega = 25$ rad s^{-1}, gives $D_{min} = 0.01$ μm.

Moore and Orr [36] developed a parameter (γ) defined as the mean ratio of Brownian to settling motion and used this to guide the choice of system parameters to minimize Brownian diffusion effects. The equation developed was :

$$\alpha = \left(\frac{2tRT}{3\pi D\eta N}\right)^{1/2} \left(\frac{18\eta}{\Delta\rho g D^2 t}\right) \qquad (7.22)$$

The mean value of α, defined as γ, for a system is calculated by integrating equation (7.22) over the time required for a sedimentation analysis (t_f).

$$\gamma = \frac{36}{\Delta\rho g D^{5/2}} \left(\frac{2RT\eta}{3\pi t_f}\right)^{1/2} \qquad (7.23)$$

Comparison of equations (7.20) and (7.22), neglecting K_m, gives:

$$v_f = \sqrt{\frac{\pi}{2}} \frac{X}{t_f} \qquad (7.24)$$

where v_f is the Brownian velocity.

7.9 Viscosity of a suspension

When discrete solid particles are present in a fluid, they themselves cannot take part in any deformation the fluid may undergo, and the result is an increased resistance to shear. Thus a suspension exhibits a greater resistance to shear than a pure fluid. This effect is expressed as an equivalent viscosity of the suspension. As the proportion of solids increases, so the viscosity increases. Einstein [10] deduced the equations:

$$\eta_T = \eta_0 (1 + kc) \qquad (7.25)$$

where η_0 is the viscosity of the fluid, η_T the viscosity of the suspension, c the volume concentration of solids and k a constant which equals 2.5 for rigid, inertialess spherical particles.

The formula is found to hold for very dilute suspension, but requires some modification for c greater than 1% [1]. However, in the dilute suspensions used in sedimentation analysis, the effect is usually smaller than errors inherent in the determination of η by conventional methods.

7.10 Calculation of terminal velocities in the transition region

The general equation relating the size of a sphere with its settling velocity may be written:

$$C_D \frac{\pi D^2}{4} \rho_f \frac{u^2}{2} = \frac{\pi}{6} (\rho_s - \rho_f) g D^3$$

(equations (7.2) and (7.4)).

For Reynolds number $Re < 0.2$, equation (7.7) may be inserted in the above to yield Stokes' equation. At higher Reynolds numbers the equation is only soluble using experimental values of C_D (figure 7.1).

Attempts at theoretical solutions of the relationship between C_D and Re have been made. Oseen [12] partially allowed for inertial effects and obtained:

$$C_D = \frac{24}{Re} \left(1 + \frac{3}{16} Re\right). \tag{7.26}$$

This equation is more complicated than Stokes', and in practice is found to be equally inaccurate, since the second term overcorrects Stokes' equation and gives a value of C_D as much in excess as Stokes' is too low.

Proudman and Pearson [13] pointed out that Oseen's solution could only be used to justify Stokes' law and not as a first-order correction. They obtain as a first-order correction:

$$C_D = \frac{24}{Re} \left(1 + \frac{3}{16} Re + \frac{9}{20} Re^2 \ln Re\right) \tag{7.27}$$

Goldstein [14] solved the equation without approximation and obtained:

$$C_D = \frac{24}{Re} (1 + 1.88 \times 10^{-1} Re - 1.48 \times 10^{-2} Re^2$$

$$+ 3.46 \times 10^{-3} Re^3 - 8.75 \times 10^{-4} Re^4). \tag{7.28}$$

Schillar and Nauman [15] fitted an empirical equation to the experimental values and obtained for $Re < 700$:

$$C_D = \frac{24}{Re} (1 + 0.15 Re^{0.687}). \tag{7.29}$$

The data, as expressed above, are most inconvenient for computation since C_D and Re contain both the velocity and diameter. Davies [16] has given a statistical analysis of the published data and expressed the result in the form of Re as a function of $C_D Re^2$, which is suitable for calculating the velocity when the diameter is known:

For $Re < 0.4$ or $C_D Re^2 < 130$

$$Re = \frac{C_D Re^2}{24} - 2.34 \times 10^{-4} (C_D Re^2)^2 + 2.02 \times 10^{-6} (C_D Re^2)^3$$

$$- 6.91 \times 10^{-9} (C_D Re^2)^4. \tag{7.30}$$

For $3 < Re < 10^4$, $100 < C_D Re^2 < 4.5 \times 10^7$

$$\log Re = -1.29536 + 0.986 \log (C_D Re^2) - 4.6677$$

$$\times 10^{-2} \log (C_D Re^2)^2 + 1.1235 \times 10^{-3} \log (C_D Re^2)^3. \quad (7.31)$$

A method for simplifying the calculation was proposed by Heywood [17], who presented the experimental data in tabular form. These data were embodied in tables presenting $C_D Re^2$ in terms of $Re C_D^{-1}$ and vice versa. Since the former expression is independent of velocity and the latter is independent of particle diameter, the velocity may be determined for a particle of known diameter and the diameter determined for a known settling velocity.

Heywood also presented data for non-spherical particles in the form of correction tables for four values of the volume-shape coefficient $(\alpha_{v,a})$ from microscopic measurement of particle projected areas.

The particle volume V is related to the projected area diameter d_a by the equation:

$$V = \alpha_{v,a} d_a^3 \quad (7.32)$$

$$\left(\alpha_{v,a} = \frac{\pi}{6} \text{ for spherical particles}\right).$$

It is more convenient to present Heywood's data in the form shown in Table 7.3. If settling velocities are required for spheres of known diameter, $C_D Re^2$ is plotted against Re on logarithmic paper and the velocities are derived from the graph (figure 7.2). If the converse is required, a graph of $C_D Re^{-1}$ against Re is plotted on logarithmic paper.

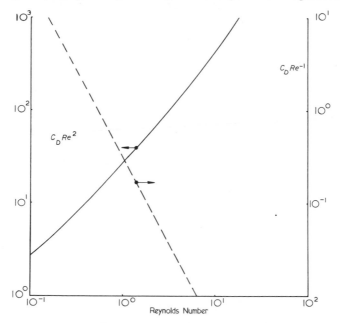

Fig. 7.2 Graphs of $C_D Re^2$ against Re (full line) and $C_D Re^{-1}$ against Re (broken line).

Table 7.3 Table of drag coefficients

1	2	3	4	5	6	7	8	9	10	11	12	13	14	15	16	17	18	19
				C_D for various α_L,a values					log C_D Re² for various α_v,a values					log C_D Re⁻¹ for various α_v,a values				
Re	C_D	$C_D Re^2$	$C_D Re^{-1}$	0.4	0.3	0.2	0.1	log Re	π/6	0.4	0.3	0.2	0.1	π/6	0.4	0.3	0.2	0.1
10^{-2}	2400	2.4×10^{-1}	2.4×10^{5}	3400	4400	6060	9300	$\bar{2}.0000$	$\bar{1}.3802$	$\bar{1}.5315$	$\bar{1}.6435$	$\bar{1}.7825$	$\bar{1}.9685$	5.3802	5.5315	5.6435	5.7825	5.9685
	1200	4.8×10^{-1}	6.0×10^{4}	1700	2200	3030	4650	$\bar{2}.3010$	$\bar{1}.6812$	$\bar{1}.8325$	$\bar{1}.9445$	0.0830	0.2695	4.7782	4.9294	5.0414	5.0414	5.3665
	484	1.21×10^{0}	9.78×10^{3}	680	880	1212	1860	$\bar{2}.699$	0.0828	0.2304	0.3424	0.4814	0.6675	3.9903	4.1271	4.2455	4.3845	4.5705
10^{-1}	244	2.44×10^{0}	2.44×10^{3}	340	440	606	930	$\bar{1}.000$	0.3874	0.5315	0.6435	0.7825	0.9685	3.3874	3.5315	3.6435	3.7825	3.9685
	123	4.92×10^{0}	6.15×10^{2}	175	218	300	460	$\bar{1}.301$	0.6920	0.8325	0.9405	1.0792	1.2672	2.7889	2.9420	3.0374	3.1761	3.3617
	51.4	1.29×10^{1}	1.03×10^{2}	70	90	127	176	$\bar{1}.699$	1.1106	1.2430	1.3522	1.5017	1.6435	2.0128	2.1461	2.2553	2.4048	2.9445
10^{0}	27.2	2.72×10^{1}	2.72×10^{1}	36.3	50	69	93	0.000	1.4346	1.5599	1.6990	1.8388	1.9685	1.4346	1.5599	1.6990	1.8388	1.9685
	15.0	6.0×10^{1}	7.5×10^{0}	20.3	27.5	39	55	0.301	1.7782	1.9096	2.0414	2.1931	2.3424	0.8751	1.0065	1.1383	1.2900	1.4393
	7.12	1.78×10^{2}	1.424×10^{0}	10.4	13	18	28	0.699	2.2504	2.4150	2.5119	2.6532	2.8241	0.1535	0.3181	0.4150	0.5563	0.7482
10^{1}	4.35	4.35×10^{2}	4.35×10^{-1}	6.3	8.0	10.8	17.4	1.000	2.6385	2.7993	2.9031	3.0334	3.2405	$\bar{1}.6385$	$\bar{1}.7993$	$\bar{1}.9031$	0.0334	$\bar{1}.2405$
	2.74	1.10×10^{3}	1.37×10^{-1}	3.97	5.12	7.3	12.0	1.301	3.0414	3.2009	3.3114	3.4654	3.6812	$\bar{1}.1367$	$\bar{1}.2978$	$\bar{1}.4082$	$\bar{1}.5623$	$\bar{1}.7782$
	1.56	3.9×10^{3}	3.12×10^{-2}	2.40	3.2	4.45	8.6	1.699	3.5911	3.7782	3.9031	4.0882	4.3324	$\bar{2}.4942$	$\bar{2}.6812$	$\bar{2}.8062$	$\bar{2}.9494$	$\bar{1}.2355$
10^{2}	1.10	1.1×10^{4}	1.1×10^{-2}	1.70	2.35	3.7	7.7	2.000	4.0414	4.2304	4.3711	4.5682	4.8803	$\bar{2}.0414$	$\bar{2}.2304$	$\bar{2}.3711$	$\bar{2}.5682$	$\bar{2}.8865$
	0.808	3.23×10^{4}	4.04×10^{-3}	1.37	1.9	3.17	7.2	2.301	4.5092	4.7388	4.8573	5.1031	5.4594	$\bar{3}.6064$	$\bar{3}.8357$	$\bar{3}.9777$	$\bar{2}.0354$	$\bar{2}.5563$
	0.568	1.42×10^{5}	1.14×10^{-3}	1.12	1.48	2.68	7.1	2.699	5.1523	5.4472	5.5682	5.8261	5.5492	$\bar{3}.0569$	$\bar{3}.3502$	$\bar{3}.4713$	$\bar{3}.7292$	$\bar{2}.1523$
10^{3}	0.460	4.6×10^{5}	4.6×10^{-4}	1.00	1.42	2.40	7.0	3.000	5.6628	6.0000	6.1523	6.3802	6.8451	$\bar{4}.6629$	$\bar{4}.3000$	$\bar{3}.1523$	$\bar{3}.3802$	$\bar{3}.8451$
	0.420	1.68×10^{6}	2.1×10^{-4}															
	0.410	1.03×10^{7}	8.2×10^{-5}															
10^{4}	0.42	4.2×10^{7}	4.2×10^{-5}															
	0.45	1.8×10^{8}	2.25×10^{-5}															
	0.48	1.2×10^{9}	9.6×10^{-6}															
10^{5}	0.48	4.8×10^{9}	4.8×10^{-6}															
	0.44	1.76×10^{10}	2.2×10^{-6}															
	0.20	5×10^{10}	4.0×10^{-7}															
10^{6}	0.22	2.2×10^{11}	2.2×10^{-7}															

$$C_D Re^2 = \frac{4(\rho_s - \rho_f)\rho_f g}{3\eta^2}\, D^3 \; ; \quad C_D Re^{-1} = \frac{4(\rho_s - \rho_f)g}{3\rho_f^2 u^3}\,\eta \; ; \quad V = \alpha_{v,a} d_a^3$$

$$= P^3 D^3$$

$$= Q^3/u^3$$

Example

Calculate the terminal velocity of a spherical particle of diameter 300 μm, density 2650 kg m^{-3} falling in water at 150° C, from equations (7.2), (7.3) and (7.4).

$$C_D Re^2 = \frac{4}{3} \frac{\rho_f (\rho_s - \rho_f)}{\eta^2} gD^3 :$$

$$C_D Re^2 = 448.$$

From figure 7.2, $Re = 10.2$

Hence:

$$u = \frac{\eta}{\rho_f D} \times 10.2$$

$$u = 3.88 \text{ cm s}^{-1}$$

Data

$D = 3 \times 10^{-4}$ m

$\rho_s = 2650$ kg m^{-3}

$\rho_f = 1000$ kg m^{-3}

$\eta = 0.00114$ N s m^{-2}

$g = 9.81$ m s^{-2}

An alternative to plotting the data is to use the tabulated values of log $C_D Re^2$ and log Re (Table 7.3, columns 10 and 9):

when
$$C_D Re^2 = 448$$
$$\log C_D Re^2 = 2.651$$

From Table 7.4:

Table 7.4

log Re	log C_D Re^2
1.301	3.0414
1.000	2.6385

Interpolating, when log $C_D Re^2 = 2.651$, log $Re = 1.0093$.

Hence $Re = 10.22$, giving $u = 3.89$ cm s^{-1}.

7.11 The turbulent flow region

For $Re > 500$, $C_D = 0.44$ and is roughly constant. For low Reynolds number, the drag is due mainly to viscous forces and the streamlines about a settling particle are all smooth curves. With increasing Reynolds number, the boundary layer begins to detach itself from the rear of the particle and form vortices, figure 7.3(a). Further increase in Re causes the vortices to increase in size and move further downstream, figure 7.3(b); at very high Reynolds numbers, the wake becomes fully turbulent and the vortices break up and new vortices are formed, figure 7.3(c).

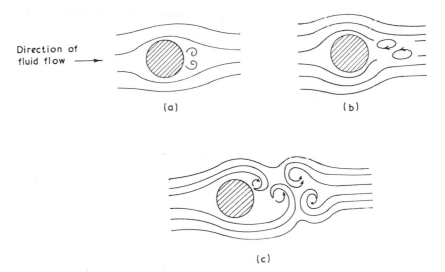

Direction of
fluid flow ⟶

(a)   (b)

(c)

Fig. 7.3 Formation of vortices behind a spherical particle as the relative velocity
between the fluid and the particle increases.

This effect may be demonstrated as follows: A table-tennis ball held just under the surface of a container of water will, on release, rise to a much greater height than a ball
initially held several inches below the surface. Since the momentum carried by the ball
on breaking the surface is much greater in the latter case, one would expect this ball to
rise higher. The explanation is that the mass of the ball is much greater in the latter
case since it includes its associated vortices. Due to the low velocity the vortices have
not formed in the first case.

7.12 Non-rigid spheres

Non-rigid particles, e.g. liquid droplets, will deform in such a way that the drag is reduced [2, p. 129]. The diameter calculated from the terminal velocity D_{Stk} will, therefore, be less than the true diameter D. It has been shown that the drag is

$$F_D = 3\pi D \eta_2 \, u \left\{ \frac{3\eta_1 + 2\eta_2}{3\eta_1 + 3\eta_2} \right\} \tag{7.33}$$

where η_1 and η_2 are the viscosities of the drop and fluid respectively.

For the case of a gaseous bubble rising slowly through a liquid, at a **Reynolds**
number in the Stokes region, $\eta_1 \ll \eta_2$, and thus:

$$F_D = 2\pi D \eta u.$$

This result is, in fact, identical to that for a solid sphere at whose surface perfect slip
occurs.

Comparing with Stokes' equation gives, for a sphere of diameter D,

$$D_{Stk}^2 = D^2 \left\{ \frac{3\eta_1 + 3\eta_2}{3\eta_1 + 2\eta_2} \right\} \tag{7.34}$$

For a raindrop falling in air; $\eta_1 = 1000 \times 10^{-6}$ N s m^{-2}

$$\eta_2 = 180 \times 10^{-6} \text{ N s m}^{-2}$$

$$D_{Stk} = 1.04 \, D.$$

Experimentally, small bubbles behave like solid spheres having terminal velocities closely approaching Stokes' which, according to Levich [18] may be attributed to impurities at the interface.

7.13 Non-spherical particles

7.13.1 Stokes' region

Homogeneous symmetrical particles can take up any orientation as they settle slowly in a fluid of infinite extent.

Spin-free terminal states in all orientations are attainable for ellipsoids of uniform density and bodies of revolution with fore- and aft-symmetry, but the terminal velocities of such particles will depend on their orientation. A set of identical particles can, therefore, have a range of terminal velocities. This range is, however, fairly limited. Heiss and Coull [19], for example, found that the ratio of maximum to minimum velocities for discs and cylinders was less than 2 : 1, even with a length : diameter ratio of 10 : 1.

Particles which are symmetrical in the sense that the form of the body is similarly related to each of three mutually perpendicular coordinate planes, as for example, a sphere or cube, not only fall stably in any orientation, but fall vertically with the same velocity in any orientation. An orienting force exists for less symmetrical particles.

Asymmetric particles, such as ellipsoids and discs, do not generally fall vertically, unless they are dropped with a principal axis of symmetry parallel to a gravity field, but tend to drift to the side. Thin, flat, triangular laminae fall edgeways unless equilateral. Few particles possess high symmetry and small, local features exert an orienting influence.

Not all bodies are capable of attaining steady motion; with unsymmetrical bodies spiralling and wobbling may occur. For an oblate spheroid [8, p. 144] of eccentricity ϵ and equatorial diameter a, the maximum drag force is:

$$F_D = 3\pi a \eta (1 - \tfrac{1}{3}\epsilon)\, u \text{ for } \epsilon \to 0. \tag{7.35}$$

For a sphere of equal volume to the above oblate spheroid:

$$F_D = 3\pi a \eta (1 - \tfrac{1}{3}\epsilon)\, u.$$

Hence a sphere of equal volume has a smaller resistance than the spheroid and the same can be shown to hold for a sphere of equal surface.

Happel and Brenner [2, p. 156] also determined the correction to Stokes' law for a prolate spheroid settling with its axis of revolution parallel to the direction of motion, and showed that when the major diameter a greatly exceeds the equatorial diameter b, the spheroid behaves as a long thin rod.

For this limiting case:

$$F_D = \frac{2\pi a \eta u}{(\ln a/b + 0.1935)}. \tag{7.36}$$

Because of the logarithmic term the resistance changes but slowly with the ratio (a/b).

A thin circular disc of diameter a, thickness b [2. p. 204], will have components of

velocity, horizontally $u_h = \dfrac{\pi abg}{128} \dfrac{\Delta \rho}{\eta} \sin 2\phi$

and vertically $u_v = \dfrac{\pi abg}{128} \dfrac{\Delta \rho}{\eta} (5 - \cos 2\phi)$ $\qquad$ (7.37)

where ϕ is the angle between the normal to the plane of the disc and the vertical (figure 7.4).

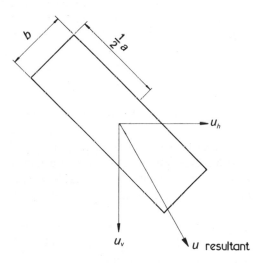

Fig. 7.4 Direction of motion of a disc settling in a fluid.

If β is the angle between the downward vertical and the direction of motion of the disc, then:

$$\tan \beta = \frac{\sin 2\phi}{5 - \cos 2\phi}$$

The angle is a maximum when the disc orientation is:

$$\phi = 39.2°$$

$$\beta = 11.5°$$

corresponding to a maximum ratio of horizontal to vertical velocities of 0.204.
    When $\phi = 0$, the disc will fall with its face horizontal.

$$F_D = 8\eta a u \tag{7.38}$$

When $\phi = 90°$ the disc will fall edge on and

$$F_D = \tfrac{16}{3}\eta a u \tag{7.39}$$

Averaging over all orientations:

$$F_D = 6\eta a u \tag{7.40}$$

With $b = 0.1a$, the values of the constants change only slightly (e.g. the constant in equation (7.38) increases by about 0.4%).
    These equations are similar to the Stokes equation ($F_D = 3\pi d_a \eta u$) with the drag diameter, d_a, of the same order as the disc diameter.
    Pettyjohn and Christiansen [20] made an extensive experimental study of isometric particles and proposed the following relationship:

$$d_{Stk}^2 = d_v^2 \times 0.843 \log \frac{\psi}{0.065} \tag{7.41}$$

where ψ, the sphericity, is the ratio of the surface area of a sphere of the same volume as the particle to the surface area of the particle.

$$\psi^2 = d_v^2/d_s^2 \tag{7.42}$$

    Hawksley [3] showed that this was equivalent to stating that the drag diameter closely approximates to the surface diameter.
    For the Stokes region, the modified form of equation (7.6) is:

$$3\pi d_a\, \eta u_{Stk} = \frac{\pi}{6} (\rho_s - \rho_f) g d_v^3$$

giving

$$d_{Stk} = \sqrt{\frac{18\eta u_{Stk}}{(\rho_s - \rho_f)g}} \tag{7.43}$$

where

$$d_{Stk}^2 = \frac{d_v^3}{d_a} \tag{7.44}$$

d_f,   the free-falling diameter, is defined as the diameter of a sphere having the same free-falling speed as the particle in a fluid of the same density and viscosity.

d_{Stk},   the Stokes diameter, is defined as the free-falling diameter in the laminar flow region ($Re < 0.2$).

d_v,   the volume diameter, is defined as the diameter of the sphere having the same volume as the particle.

Since the drag diameter is otherwise indeterminable, it is usual in practice to assume that in the Stokes region $d_d = d_s$. This is found to hold at very low Reynolds number but as the Reynolds number increases $d_d > d_s$.

7.13.2 The transition region

In the transition region particles fall with their largest cross-sectional area horizontal according to a report by Davies [5].

Hawksley [3] proposed that the C_D-Re relation can be used provided the following definitions are employed.

$$C_D = \frac{4}{3} \psi \left(\frac{\rho_s - \rho_f}{\rho_f} \right) \frac{d_v g}{u^2} \tag{7.45}$$

$$Re = \frac{1}{\sqrt{\psi}} \frac{u d_v \rho_f}{\eta}. \tag{7.46}$$

In figure 7.5 the drag coefficient C_D is plotted against Reynolds number for different sphericities.

It is more convenient to extend Heywood's technique to non-spherical particles. Using data from figure 7.4, $C_D Re^2$ may be evaluated in terms of the volume diameter d_v, and if this is known the free-falling velocity, u_f may be determined. Heywood determined u for particles of different projected area diameters d_a, as seen with a microscope.

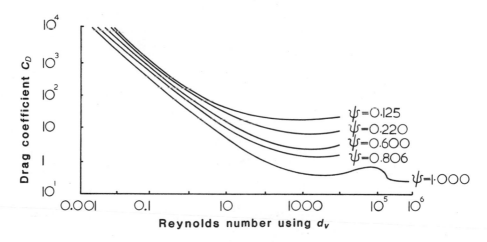

Fig. 7.5 Drag coefficient *versus* Reynolds number for particles of different sphericities.

The volume of such a particle is given by equation (7.32) as:

$$V = \alpha_{v,\,a}d_a^3.$$

Hence, modified values of Pd_a and u/Q were determined for $\alpha_{v,\,a} = 0.1, 0.2, 0.3,$ 0.4. These data are plotted in Table 7.3 in the form of $\log C_D Re^2$ and $\log C_D Re^{-1}$ which are applied as in the following worked example.

Worked example
In the earlier example the terminal velocity was found for a sphere of diameter 300 μm. Consider now a particle of projected area diameter 300 μm and volume-shape coefficient $\alpha_{v,a} = 0.25$.

As before, $\log C_D Re^2 = 2.651$.

A section of Table 7.3 is reproduced below (Table 7.5).

Table 7.5

Column	9	11	12	Interpolating $\alpha_{v,\,a}$
	$\log Re$	$\log C_D Re^2$		
Line		$\alpha_{v,a} = 0.3$	0.2	0.25
9	0.699	2.5119	2.6532	2.5826
10	1.000	2.9031	3.0334	2.9683
Interpolating	0.706 ←————————————			2.651

Hence Re = 5.082, giving

u = 1.94 cm s^{-1}, about half the velocity of a sphere with the same projected area diameter.

7.14 Concentration effects

Most of the theoretical work on particle–particle interaction has been limited to the study of pairs of spheres. If the particles are close together, they may be considered as a single particle and a correction factor applied, provided their centre-to-centre distance is small compared to the distance from the container walls. As they move farther apart, their separate effects must be considered, and these must include the field reflections from the container walls. The net effect in the first case is a reduction of the drag on the individual particles so that they fall with a greater terminal velocity than for a single sphere. Happel [2, p. 27] plots the ratio of the drag force exerted on either sphere to that exerted on a single sphere, against L/D, the ratio of interparticle separation (spheres touching when $L = D$), and against particle diameter for the cases of

spheres falling (*a*) parallel, and (*b*) perpendicular to their line of centres. The results are shown in Table 7.6:

Table 7.6   Ratios of drag force for pairs of spheres and single spheres for spheres falling (*a*) parallel, and (*b*) perpendicular to their line of centres, where L is the centre-to-centre separation of spheres of diameter D.

L/D	1	2	3	4	5	6	7
a	0.65	0.73	0.80	0.83	0.86	0.80	0.91
b	0.70	0.83	0.87	0.91	0.93	0.94	0.95

The particles will also rotate as shown in figure 7.6.

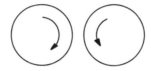

Fig. 7.6   Direction of rotation of two spheres falling close together.

For all practical purposes, this interaction becomes negligible for separations greater than about 10 diameters. For more than two spheres, interaction is more complex; assemblies of spheres will tend to diverge slowly, that is repel each other, due to the rotation effect. For three spheres in a vertical line, with the top two closer together so they fall more rapidly, the centre sphere will join forces with the lower sphere and, leaving its original companion behind, fall as a doublet (Kynch [21]). A large sphere falling in a vertical line close to a small sphere can pick it up so that it revolves as a satellite. Oseen [12] shows that, for two identical spheres falling in the same vertical line, the retardation on the trailing sphere is smaller than on the leading sphere so that they will move towards each other.

It is important to distinguish between two cases: an assembly of particles which completely fills the fluid, and a cluster of particles. The descent of a given particle creates a velocity field which tends to increase the velocity of nearby particles. In opposition to this the downward motion of each particle must be compensated for by an equal-volume upflow. If the particles are not uniformly distributed, the overall effect is a net increase in settling velocity, since the return flow will predominate in particle-sparse regions. On the other hand, a system of uniformly distributed particles will be retarded to much the same extent.

A cluster of particles in an infinite fluid can be treated as a single large particle of appropriate density and reduced rigidity, that is as a liquid drop [3, p. 131]. A very large increase in settling rate could arise which is of importance at low concentrations.

For a dilute uniform assembly of uniform spheres of diameter D, the settling velocity is reduced by the factor:

$$\frac{u_h}{u} = \frac{1}{1 + 1.3 \left(\dfrac{D}{L}\right)} \tag{7.47}$$

where u_h is the settling velocity of a particle in the presence of other particles, u is the free-falling velocity and L is the interparticle separation.

In a cubic assembly the concentration of particles by volume is

$$c = \frac{\pi}{6} \left(\frac{D}{L}\right)^3,$$

hence:

$$\frac{u_h}{u} = \frac{1}{1 + 1.61c^{1/3}}. \tag{7.48}$$

Famularo [22] investigated the problem further using a digital computer and found values of the constant for different assemblies as follows: cubic, 1.91; rhombohedral, 1.79; random, 1.30.

Burgers [23] considered a random assembly of particles and arrived at the expression:

$$\frac{u_h}{u} = \frac{1}{1 + 6.88c}. \tag{7.49}$$

The numerical constant has been questioned by Hawksley [3, p. 131], who suggests that in practice the particles will accelerate to an equilibrium arrangement with a reduced constant of 4.5. The form of the expression has also been criticized by Happel and Brenner [2, p. 376].

Maude and Whitmore [24] suggest that:

$$\frac{u_h}{u} = (1 - c)^\beta \tag{7.50}$$

where $4.67 > \beta > 4.2$ for $Re < 1$ and β is a function of particle shape and size distribution. Richardson and Zaki [25] also propose a relationship similar to equation (7.50) with $\beta = 4.65$.

Brinkman [26] proposes:

$$\frac{u_h}{u} = 1 + \tfrac{3}{4}(1 - \epsilon)\left\{1 - \sqrt{\left(\frac{8}{1 - \epsilon} - 3\right)}\right\} \tag{7.51}$$

for $0.6 < \epsilon < 0.95$, where the porosity $\epsilon = 1 - c$.

Steinour [27] developed, on theoretical grounds, a modification of Stokes' law which resulted in the equation:

$$\frac{u_h}{u} = \left(\frac{\epsilon^3}{1-\epsilon}\right) \theta\,(\epsilon) \tag{7.52}$$

Steinour, working with tapioca in oil and spherical glass beads in 0.1% aqueous sodium hexametaphosphate, reported that $\theta\,(\epsilon)$ was sensibly constant over the range $0.3 < \epsilon < 0.7$ giving:

$$\frac{u_h}{u} = 0.123 \left(\frac{\epsilon^3}{1-\epsilon}\right) \tag{7.53}$$

He also derived a general expression for $\theta\,(\epsilon)$:

$$\theta\,(\epsilon) = \left(\frac{1-\epsilon}{\epsilon}\right) 10^{-A\,(1-\epsilon)} \tag{7.54}$$

Substituting into equation (7.52) gives:

$$\frac{u_h}{u} = \epsilon^2\, 10^{-A\,(1-\epsilon)} \tag{7.55}$$

A is equal to the slope of the plot of log (u_h/u) against ϵ, and for Steinour's results $A = 1.82$.

Powers [28] found that, in order to represent his experimental data, he had to modify the above equation by introducing a factor w_i, where w_i was included to compensate for the liquid dragged down by the settling particles.

Powers' final equation was:

$$u_h = \frac{(\rho_s - \rho_f)g}{5\eta S_v^2}\,\frac{(\epsilon - w_i)^3}{(1-\epsilon)}\,.$$

For spherical particles $S_v = 6/D$, hence:

$$\frac{u_h}{u_{Stk}} = 0.10\,\frac{(\epsilon - w_i)^3}{(1-\epsilon)}\,. \tag{7.56}$$

Steinour defined the immobile liquid per unit of total volume as $a(1-\epsilon)$, where $a = w_i/(1-w_i)$. To correct equation (7.53), $a(1-\epsilon)$ must be subtracted from each value of ϵ giving:

$$\frac{u_h}{u} = 0.123\,\frac{[\epsilon - a\,(1-\epsilon)]^3}{(1+a)\,(1-\epsilon)} \tag{7.57}$$

$$= 0.123\,\frac{(1+a)^2}{(1-\epsilon)}\left(\epsilon - \frac{a}{1+a}\right)^3\,.$$

In terms of w_i, this becomes:

$$\frac{u_h}{u} = \frac{0.123\,(\epsilon - w_i)^3}{(1 - \epsilon)\,(1 - w_i)^2}\,.$$

(7.58)

Similarly substitution into equation (7.55) gives:

$$\frac{u_h}{u} = \frac{(\epsilon - w_i)^2}{(1 - w_i)^2}\,10^{-1.82}\,\left\{\frac{1 - \epsilon}{1 - w_i}\right\}$$

(7.59)

Bed expansion in particulate fluidization may be described by the Richardson–Zaki equation [79].

$$u_h = u_i \epsilon^n$$

(7.60)

where u_h is the superficial gas velocity and n is an exponential parameter which is given by:

$$n = \left(4.45 + 18\,\frac{d}{D_B}\right) Re^{-0.1}$$

(7.61)

for $1 < Re < 200$.

u_i is the extrapolated value of u_h at $\epsilon = 1$ and is related to particle terminal velocity u_{Stk} by:

$$\log u_{Stk} = \log u_i + \frac{d}{D_B}$$

(7.62)

where d = particle diameter and D_B = bed diameter (see [60]).

These relationships have also been applied to sedimentation where the problem of dealing with particles which are aggregated into sedimentation units with properties similar to flocs has been recognized [64, 80]. Such units effectively immobilize a relatively large volume of fluid, thus reducing the apparent porosity. In this situation Scott [80] replaces ϵ in equation (7.60) by an 'effective' porosity

$$u_h = u_{Stk}(1 - kc)^{4.65}$$

(7.63)

where c is solids concentration and k is the volume of aggregates per unit weight of solid.

$$k = \frac{1}{(1 - w_i)}$$

Equation (7.63) is also applicable to rough irregular particles, as treated by Whitmore [59], in which a layer of fluid may be considered to be immobilized in the surface irregularities.

The linear rate of settling of the interface (u_h) increases with increasing initial liquid volume fraction (ϵ). Additionally, since $(1 - \epsilon)$ decreases to zero as ϵ increases, a plot of $u_h(1 - \epsilon)$ against ϵ will pass through a maximum at a porosity ϵ_1. The quantity $u_h(1 - \epsilon)\rho_s$ is known as 'solid flux' and indicates the mass transfer of solid per unit cross-section per unit time down the sedimentation column.

Applying this concept to equation (7.60) a graph of 'solid flux' against porosity has a slope with $d \ll D_B$ so that $u_i = u_{Stk}$

$$\frac{\mathrm{d}}{\mathrm{d}\epsilon} (u_h (1 - \epsilon)\rho_s) = \frac{\mathrm{d}}{\mathrm{d}\epsilon} [u_{Stk}(1 - \epsilon)\rho_s \epsilon^n]$$

For maximum mass transfer

$$\frac{\mathrm{d}}{\mathrm{d}\epsilon} (u_h(1 - \epsilon)\rho_s) = 0$$

Defining the porosity at this point as ϵ_1, gives

$$\epsilon_1 = \frac{n}{n + 1}.$$

Hence, the Richardson–Zaki equation takes the form

$$u_h = u_{Stk}\epsilon \left(\frac{\epsilon_1}{1 - \epsilon_1} \right) \tag{7.64}$$

This physical significance of n, as a function of the porosity at which solids flux is a maximum, was pointed out by Davies et al. [55]. In a later paper [53], they propose a relationship of the form

$$u_h = u_{Stk} \, 10^{-bc}$$

$$u_h = u_{Stk} 10^{-b\rho_s(1 - \epsilon)} \tag{7.65}$$

where ρ_s is the solids density.

For ϵ tending to unity this is equivalent to Steinour's equation (equation (7.55)).

By differentiating equation (7.65) one obtains [53, 61]

$$\epsilon_1 = 1 - \frac{1}{2.303 \, b \, \rho_s} \tag{7.66}$$

This is equivalent to equation (7.64) and ϵ_1 may be evaluated by plotting $\log u_h$ against c.

Equation (7.65) is criticized by Dixon et al. [57] as being equivalent to equation (7.60) at low concentrations and incorrect at high concentrations. Dixon et al. also criticize the use of equation (7.66) to evaluate ϵ_1. In a reply to these criticisms Davies and Dollimore [58], defend their approach.

7.15 Hindered settling

At very low concentrations particles settle singly and, under laminar flow conditions, Stokes' law is valid. At higher concentrations particle–particle interaction occurs which, on average, should lead to reduced settling rates. At very high concentrations particles tend to settle en masse, and the rate of fall of the interface (u_h) is given by equations of

the form (7.47) to (7.57). With closely graded powders the interface is sharp becoming more diffuse for powders with a wide size range which, in some cases, form more than one interface. The formation of more than one interface is due to fines being swept out as the bulk of the suspension settles, forming a suspension of fines over the suspension of coarse; the fines supernatant being, in its turn, subject to hindered settling. These multiple interfaces persist until, at some critical concentration, mechanical interlocking occurs [40]. The equations developed to represent such systems have great similarities to those developed to explain the flow of fluid through a fixed bed of powder (permeametry).

7.15.1 Low-concentration effects

Several investigators have shown that the settling rate increases with concentration, when the concentration is low, and this is probably due to cluster formation. Kaye and Boardman [29] used a cylindrical 62-mm diameter tube filled with liquid paraffin and visually determined the settling rate of 900-μm red marker spheres in the presence of glass spheres of size 850, 400 and 100 μm. A maximum settling rate of 1.6 times the value found in very dilute suspensions was reported at volume concentrations around 1%. Twenty measurements were carried out for each individual point and considerable scatter was found around the maxima. It was also found that this enhancement of settling rate was greatest for powders having a narrow size range. Johne [30] used 200-μm glass spheres in a 35-mm diameter tube filled with 148 cP motor oil. The sedimentation velocity was determined by following single 200-μm diameter radio-active marker spheres. A maximum settling rate of 2.4 times the values found in very dilute suspensions was reported at volume concentrations around 1%.

Koglin [41] continued Johne's work with an emphasis on wall effects. He attributed the differences between the results of Kaye and Boardman and those of Johne to these effects. Since Boardman visually determined the settling time between two marker lines on the sedimentation tube he selected particles in the centre of the tube whereas Johne's method was non-selective.

The increasing rate of sedimentation at increasing concentrations was also demonstrated by Jovanovic [43] who showed that the weight-mean diameter of aluminium powder, as determined with a Sartorius sedimentation balance, increased as the powder concentration was increased from 0.02 to 0.05 volume percentage. This he attributed to particle agglomeration.

Jayaweera et al. [44], in experiments with collections of 2- to 7-μm spheres, also found settling rates greater than Stokes' velocity for a single sphere.

Barford [45] carried out similar experiments to those of Jovanovic with polishing powders in the 15 to 30 μm range. The maximum settling rate found was 1.12 to 1.40 times the Stokes rate at a concentration by volume between 0.1 to 0.2% and this he attributed to cluster formation.

Davies and Kaye [42] used a Cahn sedimentation balance to investigate cluster formation and found that, for a powder having a narrow size range, cluster formation was not eliminated at volume concentrations as low as 0.47%; for a powder having a

wide size range cluster formation was not important at concentrations as high as 0.414%.

The laws of probability predict that, even in dilute suspensions, the unequal distribution of particle separation will result in cluster formation giving rise to enhanced settling. It is therefore desirable to use as low a concentration as possible when carrying out sedimentation size analyses.

7.15.2 High-concentration effects

Several investigations have been carried out on the settling behaviour of spherical particles in a concentrated suspension [2, p. 413] with best agreement with equations of the form (7.50). Equations of the form (7.51) indicate too high a concentration dependence of the rate of settling.

The apparatus required for hindered settling experiments was described in 1960 [46], modified apparatus was described in 1970 [47] and the experimental procedure has also been described [48].

There are various ways of dealing with experimental data. Dollimore and McBride [49, 77] discuss three:

(1) Apply equation (7.53) combined with Stokes' equation to find an 'uncorrected particle radius'.
(2) Apply equation (7.58) and plot $[u_h (1 - \epsilon)]^{1/3}$ against ϵ to find w_i and equation (7.52) to find $\theta (\epsilon)$ and a 'corrected particle radius'.
(3) Apply equation (7.55) and plot $\log (u_h/\epsilon^2)$ against ϵ.
(4) Plot $\log u_h$ against c and extrapolate to zero concentration.

Dollimore and Heal [31] applied method 2 to determine the thickness of the film on the outside of the particle. Ramakrishna and Rao [50] criticized this approach and prefer method 3. In a later discussion [51] they stated that their criticism was because equation (7.58) was being used outside the range of validity for this equation (i.e. at ϵ greater than 0.8) and in this region equation (7.55) must be used. Pierce, in the same discussion [52] suggested the use of equation (7.63).

Dollimore and co-workers [54–56], in later investigations, found wide variations in A and in $\theta(\epsilon)$. Dollimore and Horridge [54] found $A = 20.2$ for china clay; Thompson [cited in 55] found $24.3 < A < 36.6$ for calcium carbonate in aqueous sodium chloride; Dollimore and Owens [56] found $A = 81$ and $A = 206$ for non-porous silica in water and n-heptane respectively.

Theoretical expressions relate hindered settling with concentration. General conclusions from the published literature are that at low concentrations hindered settling is more likely the greater the polarity of the system in general and the solid in particular [55], is more likely with dense solids in viscous liquids [61] and that the stability of some systems is due to electrical double layers [62]. Davies and Dollimore [62] conclude that:

$$\text{Hindrance} \propto f \left\{ \frac{\text{Surface density of charge on particle, particle density, liquid viscosity}}{\text{Difference in viscosity between suspension and bulk liquid, the cation–liquid stability constant}} \right\}$$

Sarmiento and Uhlherr [63] have also considered temperature effect on settling behaviour and suggest that at low concentrations the change in settling velocity with temperature change can be predicted solely from the change in viscosity of the suspending liquid; at high concentrations this is no longer possible since the floc characteristics undergo change. Change in sediment volume has also been used to predict particle size [64, 76].

The general conclusions that can be made are that settling is extremely complex in the high-concentration regions and several equations may apply according to the range of porosity considered and the presence or absence of flocculation. The determined particle size decreases with the addition of dispersing agents and it is suggested that the size so determined is floc size [33]. The technique is a useful and simple one for determining average floc size.

7.16 Electro-viscosity

Charged particles in weak electrolytes have associated with them an electrical double layer. When these particles settle under gravity the double layer is distorted with the result that an electrical field is set up which opposes motion. This effect was first noted by Dorn [65] and was studied extensively by Elton *et al.* [66–71] and later by Booth [72, 73].

For a spherical particle of diameter d, the electrical force, F_E, is equal to the product of the electrical field E and the charge q. For a surface charge per unit area σ, the force on the sphere is $d^2 \sigma E$. If the sum of this force and the viscous force is set equal to the gravitational force equation (7.6) is modified to:

$$u'_{Stk} = \frac{(\rho_s - \rho_f)gd^2}{18\eta} - \frac{d\sigma E}{3\eta} \tag{7.67}$$

that is, Stokes' velocity is diminished by the term $(d\sigma E/3\eta)$.

The specific conductivity of a solution (k) is equal to the current density divided by the electrical field. For N spheres of diameter d, per unit volume of suspension:

$$k = \frac{\pi d^2 \sigma N}{E} u'_{Stk} \tag{7.68}$$

But N is equal to the total weight of particles in suspension (m) divided by the volume of solution (V) and the weight of a single particle.

$$N = m/V \bigg/ \frac{\pi}{6} \rho_s d^3$$

Substituting in equations (7.67) and (7.68) and rearranging yields

$$u'_{Stk} = u_{Stk}/(1 + (2m\sigma^2/\rho_s \eta Vk)) \tag{7.69}$$

Pavlik and Sansone [74] found that the size distribution of spherical particles, in the size range 5 to 40 μm, obtained by sedimentation in double-distilled de-ionized water plus a wetting agent was significantly different from that obtained in 0.1 N KCl plus a

wetting agent. Coulter counter data agreed with the latter data. These results were confirmed by Sansone and Civic [75].

The electro-viscous effect should be eliminated, for Stokes' law to apply, by the use of non-ionic sedimentation liquids.

The magnitude of this effect was discussed by Siano [78] for a 0.1% suspension of 1.101 μm polystyrene settling in water. For a zeta potential of 50 mV the electric field strength is 1 mV cm^{-1} making the correction term negligible at 0.04%.

References

1 Lorentz, H. (1906), *Abh. u. Th. Phys.*, **82**, 541.
2 Happel, J. and Brenner, H. (1965), *Low Reynolds Number Hydrodynamics*, Prentice Hall (1951).
3 Hawksley, P.G.W. (1951), *BCURA Bull.* **15**, 4.
4 Cunningham, E. (1910), *Proc. R. Soc.,* **A83**, 357.
5 Davies, C.N. (1954), *Proc. Phys. Soc.,* **57**, 259.
6 Lapple, C.E. (1950), *Chemical Engineering Handbook* (ed. J.M. Perry), McGraw-Hill, NY.
7 Green, M.L. and Lane, W.R. (1957), *Particulate Clouds: Dusts, Smokes and Mists*, Spon, p. 58.
8 Boothroyd, R.G. (1971), *Flowing Gas Solids Suspensions,* Chapman & Hall, London.
9 Fuchs, N.A. (1964), *Mechanics of Aerosols, Trans.* (ed. C.N. Davies), Pergamon, Oxford.
10 Einstein, A. (1906), *Ann. Phys. Leipzig,* **19**, 289; (1911), **34**, 591.
11 Kynch, G.J. (1954), *B. J. appl. Phys.*, suppl. 3.
12 Oseen, C.W. (1927), *Neuere Methoden und Ergebrisse in der Hydrodynamik,* Leipzig Akademische Verlag.
13 Proudman, I. and Pearson, J.R.A. (1957), *J. Fluid. Mech.*, **2**, 237.
14 Goldstein, S. (1938), *Modern Developments in Fluid Dynamics,* Clarendon Press.
15 Schillar, L. and Nauman, A.Z. (1933), *Ver. dt. Ing.*, **77**, 318.
16 Davies, C.N. (1947), *Trans. Inst. Chem. Engrs,* suppl. **25**, 39.
17 Heywood, H. (1962), *Proc. Symp. on the Interaction between Fluids and Particles,* Inst. Chem. Engrs, London.
18 Levich, V.G. (1962), *Physiochemical Hydrodynamics,* Prentice Hall, Englewood Cliffs, NJ.
19 Heiss, F. and Coull, J. (1952), *Chem. Eng. Prog.,* **48**, 3, 133–40.
20 Pettyjohn, E.A. and Christiansen, E.B. (1968), *ibid.,* **44**, 157.
21 Kynch, G.J. (1959), *J. Fluid. Mech.,* **5**, 193.
22 Famularo, J. (1962), Engng. Sci. D. Thesis, New York Univ.
23 Burgers, J.M. (1941), *Proc. K. ned. Akad. Wet.,* **44**, 1045; (1942), **45**, 9.
24 Maude, A.D. and Whitmore, R.L. (1958), *R. J. appl. Phys.,* **4**, 477.
25 Richardson, J.F. and Zaki, W.N. (1954), *Chem. Eng. Sci.,* **3**, 65.
26 Brinkman, H.C. (1947), *Appl. Sci. Res.,* **A1**, 27; (1948), **A1**, 81; (1949), **A2**, 190.
27 Steinour, H.H. (1944), *Ind. Engng. Chem.* **36**, 618, 840, 901.
28 Powers, T.C. (1939), *Proc. Am. Concr. Inst.,* **35**, 465.
29 Kaye, B.H. and Boardman, R.P. (1962), *Proc. Symp. Interaction between Fluids and Particles,* Inst. Chem. Engrs., London.
30 Johne, R. (1966), Diss. Karlsruhe (1965); also *Chemie-Ing-Tech.* (1966), **38**, 428–30.
31 Dollimore, D. and Heal, G.R. (1962), *J. appl. Chem.,* **12**, 445.
32 Ramakrishna, V. and Rao, S.R. (1965), *ibid.,* **15**, 473.
33 Dollimore, D. (1972), *J. Powder Technol.,* **8**, 207–120.
34 Rayleigh, Lord (1892), *Phil. Mag.,* **5**, B4, 59.
35 Stokes, Sir G.G. (1891), *Mathematical and Physical Paper III,* Cambridge University Press.
36 Moore, D.W. and Orr, C.Jr. (1973), *Powder Technol.,* **8**, 13–18.

37 Muta, A. and Watanabe, S. (1970), *Proc. Conf. Particle Size Analysis* (eds. M.J. Groves and J.L. Wyatt-Sargent), Soc. Anal. Chem., Bradford, pp. 178–93, 196, 197.
38 Allen, T. (1970), *ibid., Discussion*, pp. 194, 195, 198.
39 Allen, T. and Baudet, M.G. (1977), *Powder Technol.*, **18**, 131–8.
40 Davies, R. and Kaye, B.H. (1971/2), *ibid.*, **5**, 61–68.
41 Koglin, B. (1970), *Proc. Conf. Particle Size Analysis* (eds. M.J. Groves and J.L. Wyatt-Sargent), Soc. Anal. Chem., Bradford, pp. 223–35.
42 Davies, R. and Kaye, B.H. (1970), *ibid.*, pp. 207–22.
43 Jovanovic, D.S. (1965), *Kolloid Z., Polymere*, **203**, 1, 42–56.
44 Jayaweera, K.O.L.F., Mason, B.J. and Slack, G.W. (1964), *J. Fluid Mech.*, **20**, 121–8.
45 Barford, N. (1972), *Powder Technol.*, **6**, 1, 39–44.
46 Dollimore, D. and Griffiths, D.L. (1960), *Proc. 3rd Int. Cong. Surface Activity,* Cologne, **V2**, Section B, 674.
47 Christian, J.R., Dollimore, D. and Horridge, T.A. (1970), *J. Phys. E*, **3**, 744.
48 Sarmiento, G. and Uhlherr, P.H.T. (1977), *Proc. 5th Austral. Chem. Eng. Conf.,* Canberra, 296.
49 Dollimore, D. and McBride, G.B. (1973), *Powder Technol.*, **8**, 207–12.
50 Ramakrishna, V. and Rao, S.R. (1966), *J. appl. Chem.*, **15**, 473.
51 Ramakrishna, V. (1970), *Proc. Conf. Particle Size Analysis* (eds. M.J. Groves and J.L. Wyatt-Sargent), Soc. Anal. Chem., Bradford, p. 206.
52 Pierce, T.J. (1970), *ibid.*, p. 205.
53 Davies, L., Dollimore, D. and McBride, G.B. (1977), *Powder Technol.*, **16**, 45–49.
54 Dollimore, D. and Horridge, T.A. (1971), *Trans. Br. Ceramic Soc.*, **70**, 191.
55 Davies, L., Dollimore, D. and Sharpe, J.H. (1976), *Powder Technol.*, **13**, 123–32.
56 Dollimore, D. and Owens, N.F. (1972), *Proc. 6th Int. Cong. Surface Activity,* Zurich, (cit. [55]).
57 Dixon, D.C., Buchanon, J.E. and Souter, P. (1977), *Powder Technol.*, **18**, 283–4.
58 Davies, L. and Dollimore, D. (1977), *ibid.*, 285–7.
59 Whitmore, R.L. (1957), *J. Inst. Fuel*, **30**, 238.
60 Capes, C.E. (1974), *Powder Technol.*, **10**, 303–6.
61 Davies, L. and Dollimore, D. (1977), *ibid.*, **16**, 59–61.
62 Davies, L. and Dollimore, D. (1978), *ibid.*, **19**, 1–6.
63 Sarmiento, G. and Uhlherr, P.H.T. (1979), *ibid.*, **22**, 139–42.
64 Michaels, A.S. and Bolger, J.C. (1962), *Ind. Engng Fund.*, **1**, 24.
65 Dorn, E. (1880), *Wied, Ann.* **10**, 46 (cited by Abramson, H.A. (1934), *Electrokinetic Phenomena,* Chemical Catal. Co., NY).
66 Elton, G.A.H. (1948), Electroviscosity 1, *Proc. R. Soc.*, **A194**, 259.
67 Elton, G.A.H. (1948), Electroviscosity 2, *ibid.*, **A194**, 275.
68 Elton, G.A.H. (1949), Electroviscosity 3, *ibid.*, **A197**, 568.
69 Dulin, C.I. and Elton, G.A.H. (1952), *J. Chem. Soc.*, 286.
70 Elton, G.A.H. and Hirschler, F.G. (1954), *B. J. appl. Phys.*, suppl. 3, S60.
71 Elton, G.A.H. and Hirschler, F.G. (1962), *J. Chem. Soc.*, 2953.
72 Booth, F. (1950), *Proc. R. Soc.*, **A203**, 533.
73 Booth, F. (1954), *J. Chem. Phys.*, **22**, 1956.
74 Pavlik, R.E. and Sansone, E.B. (1973), *Powder Technol.*, **8**, 159–64.
75 Sansone, E.B. and Civic, T.M. (1975), *ibid.*, **12**, 1, 11–18.
76 Brugger, K. (1976), *ibid.*, **14**, 187–8.
77 Dollimore, D. and McBride, G.B. (1970), *Proc. Conf. Particle Size Analysis* (eds. M.J. Groves and J.L. Wyatt-Sargent), Soc. Anal. Chem., Bradford, pp. 199–206.
78 Siano, D.B. (1979), *J. Colloid Interfac. Sci.*, **68**, 1, 111–27.
79 Richardson, J.F. and Zaki, W.N. (1954), *Trans. Inst. Chem. Eng.*, **32**, 35.
80 Scott, K.J. (1968), *Ind. Chem. Fund.*, **7**, 484.

⑧ Dispersion of powders

8.1 Discussion

In many size analysis methods it is necessary to incorporate a powder into a liquid medium such that the particles are evenly dispersed. If the powder surface is lyophobic the powder is difficult to disperse, if it is lyophilic the powder disperses easily (for water dispersions the terms are hydrophobic and hydrophilic respectively). Dispersing agents are, therefore, added to wet the surface of lyophobic materials to make them lyophilic [16].

Starting with a dry solid and a liquid medium, several separate stages in the dispersion process may be distinguished. Firstly, there is the process of wetting which is defined as the replacement of the solid–air interface by a solid–liquid interface. Secondly, there is the process of disaggregation of clusters of particles, and thirdly the process of dispersion stabilization [43].

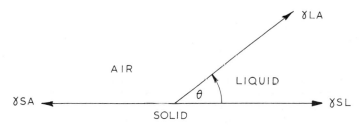

Fig. 8.1 The spreading of a liquid on the surface of a solid.

If a liquid is placed on a solid, it will spread if (see figure 8.1):

$$\gamma_{SA} \geqslant \gamma_{SL} + \gamma_{LA} \cos \theta \tag{8.1}$$

where $\gamma_{SL}, \gamma_{SA}, \gamma_{LA}$ are the interfacial tensions between the solid and the liquid, solid and air and liquid and air and θ is the angle of contact between the solid and the liquid (γ_{SA} and γ_{LA} will rapidly fall to γ_{SV} and γ_{LV} as the solid surface becomes saturated with vapour, but for simplicity the former suffixes will be retained).

The spreading coefficient is defined as:

$$S_{LS} = \gamma_{SA} - \gamma_{SL} - \gamma_{LA}. \tag{8.2}$$

The liquid will spread on the solid if the spreading coefficient is positive (see figure 8.2).

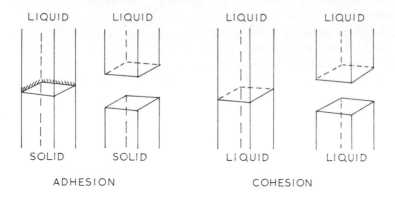

LIQUID   LIQUID   LIQUID   LIQUID

SOLID   SOLID   LIQUID   LIQUID

ADHESION   COHESION

Fig. 8.2 Work of adhesion and work of cohesion.

The work of adhesion is defined as the work necessary to separate one square centimetre of interface between two phases:

$$W_A = \gamma_{SA} + \gamma_{LA} - \gamma_{SL}. \qquad (8.3)$$

The work of cohesion is defined as the work necessary to separate one square centimetre of one phase:

$$W_C = 2\gamma_{LA}. \qquad (8.4)$$

Hence, the spreading coefficient may be defined as:

$$S_{LS} = W_A - W_C \qquad (8.5)$$

and, if positive, infers a greater affinity between the liquid and the solid than between the liquid's own molecules.

The energy of immersion is defined as the surface energy loss per unit area of surface on immersion:

$$W_i = \gamma_{SA} - \gamma_{SL}. \qquad (8.6)$$

Equations (8.1) and (8.3) may be combined to give:

$$W_A \geqslant \gamma_{LA} (1 + \cos \theta). \qquad (8.7)$$

The ease of displacement of air from the surface of a powder is thus enhanced if W_A is increased. This is frequently accomplished in aqueous systems by the use of surface-active agents to reduce the contact angle θ to zero if possible. At equilibrium, from equation (8.1):

$$\cos \theta = \frac{\gamma_{SA} - \gamma_{SL}}{\gamma_{LA}} \qquad (8.8)$$

The addition of a surface-active agent usually causes a reduction in γ_{LA}, and if absorbed a reduction in γ_{SL}. Both effects lead to better wetting. The change in γ_{SA} is probably negligible in most cases so the dominating factor is γ_{LA}, the surface tension of the liquid phase.

The difficulty in the use of these equations in practice is the experimental one of determining θ for fine powders. Bartell and co-workers [5] developed a method in which a pressure was applied to prevent liquid from penetrating a plug of powder. The required pressure to prevent penetration is given by the Laplace equation:

$$\Delta P = \frac{2\gamma_{LA}\cos\theta}{r}.$$
(8.9)

For a liquid which wets the solid, one has:

$$\Delta P_0 = \frac{2\gamma_{LA}{}^0}{r}$$

so that:

$$\cos\theta = \frac{\Delta P}{\Delta P_0}\left\{\frac{\gamma_{LA}{}^0}{\gamma_{LA}}\right\}.$$
(8.10)

The principle of the method is to obtain the effective capillary radius r using a non-wetting liquid and repeat the measurement with a wetting liquid.

This method is difficult to use and the following simpler method has been described [3]. The distance of penetration L in time t into a horizontal capillary, or in general when gravity may be neglected, is given by the Washburn equation [2]:

$$\frac{L^2}{t} = r\frac{\gamma_{LA}\cos\theta}{2\eta}$$
(8.11)

where η is the viscosity of the liquid.

For a packed bed of powder this equation becomes:

$$\frac{L^2}{t} = \frac{r}{(K^2)}\frac{\gamma_{LA}\cos\theta}{2\eta}$$
(8.12)

where the bracketed term is an unknown factor dependent on the packing. If several liquids are used with the same powder, uniform high values of $(r/K^2)\cos\theta$ are found and it is assumed that these correspond to $\cos\theta = 1$. This factor is then used to obtain values of θ with other liquids.

Heertjes and Kossen [38] present a full discussion of their techniques together with descriptions of the required apparatus and experimental procedures. They consider both methods unsuitable for the determination of $\cos\theta$ and propose a new method, the h-ϵ method. Briefly, the method consists of determining the height of a drop of liquid placed on the top of a cake of compressed powder which has been saturated with the liquid. The theory is presented in a previous paper [39].

An analysis has been presented [28] of the wetting of a fine powder by aqueous solutions of anionic wetting agents in terms of adhesion tension, spreading coefficient and capillary forces involved in the displacement of air from externally wetted aggregates. This study reveals how the film preceding the bulk liquid can retard or markedly accelerate the submersion of powder.

The next stage in the dispersion process is the breakdown of aggregates and agglomerates after wetting. For easily wetted material penetration of liquid into the voids between particles may provide sufficient force to bring about disintegration. Often, however, mechanical energy is required and this is usually introduced by spatulation or stirring, though the use of ultrasonics is now quite widely practised.

The stability of a wetted, dispersed system depends upon the forces between particles. The random motion of the particles brings them into close contact and under certain circumstances causes them to flocculate. The frequency of collision depends upon the concentration, viscosity and temperature. Whether two approaching particles will combine or not depends on the potential barrier between them. The potential energy can be considered to consist of two terms, the attractive, due to London–van der Waals' forces and the repulsive, due to the electrical double layers which exist around particles; this double layer consists of an inner layer of ions at the surface of the particle and a cloud of counter-ions surrounding it. The interaction curve (figure 8.3) follows an exponential decay pattern; the sum for the attraction A and repulsion R is the total energy curve B and if the maximum has an energy of more than $15\,kT$, where k is Boltzmann's constant and T the absolute temperature, the system is stable [4]. With

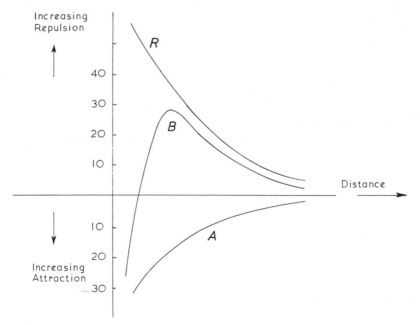

Fig. 8.3 Potential energy (net) curve against distance from particle: A, force of attraction, R, force of repulsion, B, resultant force.

large particles, the net potential energy curve may show a minimum at appreciable distances of separation. This concept cannot account for the action shown by many non-ionic, surface-active agents, and in these cases it has been suggested that the size of the agent could lead to a repulsion due to the steric nature of the absorbed layer, i.e. the molecules extend so far out into the media that two approaching particles do not get close enough to flocculate.

In non-aqueous media of low dielectric constant ionic charge stabilization is unlikely to be very important. In such cases stabilization depends on steric or entropic repulsion and polymeric agents are preferred against the long-chain paraffinic types which are so successful in aqueous media.

In practice it is necessary to decide whether an agent is required to wet out the solid so that it will disperse in the liquid concerned or whether the problem is one of stabilization. The minimum of wetting agent to ensure adequate dispersion is added. Any combination which causes foaming should be rejected since this may cause separation of the finer particles. If this is not feasible, an anti-foaming agent may be used. The present state of the art is such that dispersing agents are chosen almost at random, but with some knowledge of the surface chemistry involved [14].

A simple test for wetting efficiency is to make up suspensions using the same concentration of powder but different agents and allow the suspension to settle out. About a gram of powder is mixed with the dispersing liquid in a 10 ml container. Slow settling, a clear interface between the clear liquid and turbid lower layers and small depth of sediment indicate the best agent or, if several concentrations of the same agent are being tested, the best concentration [6, 7]. It is important that no vibration be imparted to the containers during settlement which may take several hours. Another test for the degree of flocculation of a paste is to measure the difference between the smear and flow points. The test is made by adding known quantities of dispersing medium to a known weight of the powder and working it in with a spatula. The difference is noted between the amount required for the powder to smear and to flow; the more dispersed the sample the smaller the difference [9].

It has been found that for pigments in solvents a high dielectric constant leads to a more dispersed system. In general, polar liquids disperse polar solids and non-polar liquid disperse non-polar solids. For polar solids dispersed in non-polar liquids, it is possible to use the polarity of the surface as a means of anchoring a stabilizing molecule to the surface. The effectiveness is characterized by the heat of wetting which may be determined by calorimetry.

A guide to the choice of dispersing medium and dispersing agents is given in tables 8.1 and 8.2. It is also necessary to adjust the density and viscosity of the medium to suit the density and particle size of the powder so that settling is within the Stokes region. These values are best taken from a comprehensive standard book [18, 48, 49].

Effectiveness of deflocculants in dilute suspensions may be studied by light absorption using the photosedimentometer [10]. If the optical density of an agitated suspension falls with time this is an indication of either flocculation or dissolution, whereas an increasing optical density indicates breakdown of flocs. Neither of these situations is tolerable.

Dilute suspensions in electrolytes may be studied with a Coulter counter. It is frequently found using the instrument that the count level at the lower sizes decreases with time, and this may usually be attributed to flocculation or dissolution (this loss is usually not marked enough to significantly affect the resulting particle size analysis). A problem that arises with this technique is that the dispersing liquid is an electrolyte and one would expect that this would reduce the energy barriers and decrease the stability of the system [15].

The final criterion is the practical test; if size analyses are carried out with two systems, the one showing the finer distribution will be more accurate than the other. This, however, is not conclusive since both systems may have some flocculation.

One test which is often suggested is to examine, by transmission microscopy, a drop of suspension placed on a microscope slide. Although a visual examination of particles in order to determine their approximate size is always recommended, a false impression may be obtained as to their state of dispersion due to the wide difference in environment between this testing situation and the actual analysis.

Since many wetting agents are hygroscopic, where the concentration of extracted suspension is to be determined by drying and weighing, the dried residue should be cooled in a desiccator and contact with the atmosphere during weighing reduced to a minimum. It is suggested, though there is very little evidence in support, that some wetting agents are preferentially absorbed on to finer powders leading to errors in powder concentration derived from the weight of dried residue.

Dispersion of magnetic suspensions may be effected by subjecting the suspension to a high-frequency magnetic field [40]. For the Autometrics PSM 200, 400 Hz was selected at a high peak magnetic field strength of 800 öersted [41].

A range of non-ionic fluorochemical surfactants (FC–170C, FC–433, FC–430, FC–431) manufactured by 3M Corporation, has been found to be particularly useful for dispersing sub-micron powders. The addition of 0.01% of any of these surfactants to distilled water reduces its surface tension from 72 to less than 25 dyn cm⁻¹. FC–430 is a 100% active, viscous liquid, FC–431 is a solution of 50% active solids in ethyl acetate. These surfactants exhibit good solubility in most solvents.

The demarcation between individual particles and groups of particles is often very blurred. In some systems there are aggregates which are tightly bound together with some chemical bonding or sintering and strong mechanical forces are required to break the bonding, flocculates which are loosely bound together and are readily separated by simply wetting, and primary particles.

Koglin [42] has reviewed the methods of assessing the degree of agglomeration in suspension. He states that the only direct method is microscopy. Estimates of degree of flocculation are possible by rheological properties or sediment volume. The Coulter principle is only suitable for aggregates since flocs disintegrate due to the shear experienced as the suspension passes through the aperture. Koglin states that the degree of aggregation can be determined by comparing the size distribution of the agglomerated suspension with the size distribution of a completely dispersed system. The methods he uses are photoextinction and the sedimentation balance. His review contains 57 references.

Dispersion of pigments is often effected by prolonged milling [44-46]. It is argued that pigment is made up of agglomerates which are easily reduced to their constituent aggregates (1-20 μm in size). These aggregates may be further subdivided to primary particles (0.3-1 μm) with the application of considerably more force. Further reduction to crystallites (0.01-0.6 μm) is effected by crystal fracture. Milling times of around twenty hours are required to attain stable conditions.

8.2 The use of glidants to improve flowability of dry powders

Glidants are frequently added to powders, which are to be analysed in the dry state, in order to improve their flow properties. Various mechanisms have been proposed to explain their mode of action [31, 32], which include:

(1) Reduction of interparticle friction by coating particles.
(2) Reduction of surface roughness.
(3) Reduction of interparticle attractive forces by creating a physical barrier between particles.
(4) Reduction of static electrical charge on particles.

An optimum concentration of glidant exists for most systems [31, 33] and amongst the techniques used to estimate this percentage are angle of repose [34], flow through an orifice [34], vibrating funnel [35] and bulk density.

These additives are particularly useful with fine powders for analysis by micromesh sieving and classification. In the absence of other data a weight concentration of about 1% is used. Fatty acids (usually stearic), fine silica ('Aerosil 2000' from Degussa, Frankfurt, W. Germany), purified talc and magnesium stearate are often used. York [36] examined three typical glidants with lactose mixtures using the Jenike shear cell [37] to assess flowability and found that fine silica produced the greatest increase in flowability, from 1.9 for lactose to 10.4 for a 1% concentration; magnesium stearate and purified talc gave smaller increases, up to a maximum of 8.2 and 5.8 respectively. All three curves indicated an optimum concentration (1.75, 2.25 and 0.85% respectively). He also concluded that the angle of internal friction is not a useful indicator of flowability.

8.3 Density determination

One of the parameters often required for particle size analysis is particle density. Density can be determined by measuring the mass and the volume of a sample of particles. The volume of a body with no well-defined geometry can be determined with a pyknometer. The body displaces a liquid or a gas and its volume can be determined from this displacement either directly or, with the use of gas pyknometers, from physical laws.

The usual method for determining this physical constant has been by liquid pyknometer and full details of relevant techniques may be found in BS 733 (1952) and BS 3483, Part B8 (1978). If liquids other than water are used they should have a low evaporation rate under vacuum and a high boiling point ($>$ 170 $^\circ$C); aromatic or aliphatic

compounds are suitable. The use of oil presents a particular problem in that it is difficult to ensure that the outside of the bottle is completely oil-free.

The following method has been found suitable, using distilled water plus dispersing
agent, for materials which are not soluble in water.

The pyknometer is carefully cleaned, dried and weighed. It is then half-filled with dry
powder using a thistle funnel in order that particles do not adhere to the neck of the
bottle.

It is then weighed, placed in a vacuum desiccator and evacuated. The pyknometer is
filled to approximately the three-quarters full level while under vacuum, with dispersant
(0.1% w/v Calgon in distilled water). This is to ensure that no air bubbles are trapped in
the powder. The suspension is then left to settle for several hours.

Air is next admitted to the desiccator, and the pyknometer is carefully topped up
using a squeezy bottle. The stopper is carefully and gently inserted so that no air bubbles
are trapped. The pyknometer is placed in a water bath, set at one or two degrees above
ambient temperature, with the water up to the neck of the bottle. When no more liquid
is being extruded through the capillary the bottle is removed, dried and weighed. The
stopper is then removed, the bottle topped up, the stopper re-inserted and the pyknometer replaced in the water bath. When temperature equilibrium has been reached the
bottle is removed, dried and weighed as before. This exercise is repeated six times.

The bottle is then emptied and filled with dispersing liquid. It is replaced in the water
bath and the above procedure carried out again, six times.

The bottle is emptied and the exercise repeated (six times) with distilled water as the
filling liquid.

The density is determined using the formula below:

M_1 = mass of pyknometer (empty)
M_2 = mass of pyknometer + dispersant
M_3 = mass of pyknometer + powder
M_4 = mass of pyknometer + powder + dispersant to fill
ρ = density of the solution
ρ_w = density of water at water bath temperature
ρ_d = density of dispersant at water bath temperature
T = water bath temperature
M_5 = mass of pyknometer + distilled water
ρ_p = density of powder
ρ_a = density of air at ambient temperature

$$\rho_p - \rho_a = \frac{M_3 - M_1}{(M_2 - M_1) - (M_4 - M_3)} (\rho - \rho_a) \qquad (8.13)$$

where

$$\rho - \rho_a = \frac{M_2 - M_1}{M_5 - M_1} (\rho_w - \rho_a) \qquad (8.14)$$

Specimen result

M_1	=	21.1287	±	0.0002 g
M_3	=	29.9503	±	0.0002 g

M_4	=	51.2696
		51.2697
		51.2686
		51.2690
		51.2686
		51.2686

Repeat analyses

$\overline{M}_4$	=	51.2690	±	0.0007 g

M_5	=	45.7942
		45.7945
		45.7950
		45.7944
		45.7947
		45.7941

		Run 1	Run 2
M_1	=	31.7351	31.1351
M_2	=	82.9351	82.0516
M_3	=	48.8142	48.6568
M_4	=	93.5016	92.8897

$\overline{M}_5$	=	45.7945	±	0.0005 g

M_2	=	45.8150
		45.8160
		45.8152
		45.8155
		45.8157
		45.8151

$\overline{M}_2$	=	45.8154	±	0.0006 g
T	=	24.4 °C		
ρ_w	=	0.9972 g cm^{-3}		
ρ_a	=	0.0012 g cm^{-3}		

Using data from the specimen result the density obtained is ρ_p = 2.6121 g cm^{-3} with a standard deviation of 0.0013 g cm^{-3}.

Repeat analyses: ρ_p = 2.614 and 2.613.

A special apparatus has been described which is claimed to be simpler to operate than most other methods for the determination of the apparent density of porous carbons [12]. The apparatus is designed so that the solid can be thoroughly outgassed and de-gassed dilatometric liquid brought into contact with it without exposing the latter to the atmosphere. Other techniques have been described earlier [20]; the most commonly used liquid being mercury but volatile liquids have frequently been employed as molecular probes for the investigation of pore structure [21].

Burt *et al.* [17] consider the liquid pyknometer method to be unsuitable for powders which are predominantly below 5 μm. For these fine powders there is a greater chance of strongly adsorbed gas being present on the surface. In order to remove this gas the powder may often need treatment at high temperature or under high vacuum. For particles with rough surfaces also, it is possible that air contained within surface cracks and pits may not easily be removed. They propose the use of a centrifugal pyknometer in which a suspension, prepared in the usual way for sedimentation analysis, is placed; the suspension is centrifuged and the density determined in the usual manner. The density so obtained is usually slightly lower than that obtained by other methods due to the presence of adsorbed and trapped gas.

The method of May and Marienko [29] for determining the density of small amounts of material using a micropyknometer is described by Stein *et al.* [30] who filled, under vacuum, a 1-cm^3 micropyknometer containing the dust, using ethylene glycol as the fill liquid. Stein *et al.* found the method time-consuming and difficult and developed a new method using air as the fill 'liquid'. Their micropyknometer had a 2-mm bore stem which was calibrated at two volume levels, 0.3 and 0.5 cm^3. These were determined accurately using mercury. A known weight of powder was placed in the pyknometer and the neck sealed with a mercury plug. This was forced down to the 0.5 cm^3 level in a pressure chamber at a pressure P_1 and then to the 0.3 cm^3 level at a pressure P_2. The volume of powder (v) is given by:

$$v = \frac{P_2(0.3) - P_1(0.5)}{P_2 - P_1}$$

(8.15)

The volume of the dust sample could be determined to ± 0.5 mm^3 corresponding to an accuracy of ± 3% for a 25-mg sample of density 1.5 g cm^{-3}.

A discussion of available air pyknometers is presented by Thudium [22]. He criticizes instruments in which the measurement consists of an absolute volume change [23, 24] or a variable containing the absolute volume change [25] since there is no linear relationship between the measured parameter and sample volume. In instruments containing only one chamber [23, 24] or having one chamber as pressure reference only [25, 27] every fluctuation in temperature gives an error of measurement. In systems having two chambers [26] only differences between the two chambers give an error.

He criticizes the Beckmann [26], a two-chamber instrument, on the grounds that the large volume change (half initial volume) could be reduced by 50% if pressure was reduced rather than increased; moreover, sorption effects would be reduced. An instrument embodying these recommendations has been described [47]. Thudium was particularly interested in micropyknometers for measuring volumes less than 5 mm^3 and considered that the best available micropyknometer was that described by Hänel [27] which would measure a volume of 20 to 40 mm^3 with an accuracy of ±10%. He then described a micropyknometer similar to the Beckmann with expansion using a micrometer syringe, thus reducing the pressure change to less than 10 mbar.

8.4 Viscosity

The viscosity of the suspending medium should have a value that will fulfil the following conditions.

(1) The largest particle in the suspension should settle under laminar flow conditions, i.e. the Reynolds number should be less than 0.2.
(2) The free-falling velocity of the largest particle should be restricted so that it takes at least one minute for it to reach the measurement zone.

For the first condition, the relationship between Stokes diameter and viscosity is given by inserting a value of 0.2 for the Reynolds number.

$$Re \ = \ \frac{\rho_f v d}{\eta} \quad \text{in Stokes' equation:}$$

$$d_{Stk}^3 \ = \ \frac{3.6 \, \eta^2}{(\rho_s - \rho_f) \rho_f g}. \tag{8.16}$$

For a particle of density ρ_s = 2700 kg m^{-3} in a suspending medium of density ρ_f = 1000 kg m^{-3}, the required viscosity for a particle of size 75 μm is therefore 0.0014 N s m^{-2}.

Stokes' law does not apply when the Reynolds number exceeds 0.2, and the diameters calculated using Stokes' equation are smaller than the correct values. This is due to flow being no longer laminar; eddies are set up behind the particle which retard its motion. Further, it is generally accepted that under streamline conditions particles fall in random orientation, whereas particles orientate themselves for maximum drag under turbulent motion conditions. Corrections may be applied using Heywood's table. Alternatively, for comparison purposes, a frequency may be plotted or tabulated against free-falling velocity.

8.5 Sedimentation systems

The usual concentration of wetting agent used is 0.1% weight by volume.

Table 8.1 List of wetting and dispersing agents

A   sodium linoleate
B   potassium silicate
C   sodium hexametaphosphate (Calgon)
D   Dispersal T
E   potassium citrate
F   calcium chloride (0.05 g l^{-1})
G   trisodium phosphate
H   aerosol OT 1%
I   gallotannic acid
J   sodium tartrate
L   sodium silicate
M   trinatrium phosphate
N   perminal BX
O   sodium oxalate

Table 8.1 (Cont.)

P sodium pyrophosphate
R oleic acid
S sodium citrate
T tannic acid
U sodium carbonate
V potassium chloride 0.001 M
W sodium hydroxide
X xylene
Y Daxad 23 0.02%
α Cetrimide B.P. (cetyl trimethyl ammonium bromide) cationic [17]
β Dispersal T (sodium methylene dinaphthalene sulphonate) anionic
γ Nonidet P40 (alkylphenol/ethylene oxide condensate) non-ionic
θ Lissapol N.X. (alkylphenol/ethylene oxide condensate) non-ionic
ω Nonidet P42

Table 8.2 Dispersing solutions for powders

Material	Liquid	Wetting agent
Alkali salts	cyclohexanol	–
Alumina	n-butanal, n-butylamine, linseed oil	xylene
Aluminium	cyclohexanol	–
	carbon tetrachloride	–
	water	C, J or O
	dilute hydrochloric acid (pH adjusted to 3)	
Aluminium oxide	water	P
Anthracite	water	M or N
Antimony trioxide	water	P or C
Arsenates	water	P
Arsenious oxide	octyl alcohol	–
	cyclohexanol	–
	liquid paraffin	2% fatty acid
Ash	water	P
Barium carbonate	methyl alcohol	–
	methyl alcohol	–
Barium strontium carbonate	water–ethyl alcohol mixture	–
Barium sulphate	water	C or θ or α
	water–methanol mixture	–
Barytes	water	C or P
Beryl	water	L or C
Blast-furnace slag	water	C
Bronze powder	cyclohexanol	
Brown coal	cyclohexanol + 10% methanol phthalsaurediaethylester	–
Cadmium sulphide	water	P
	ethylene glycol	–
Calcium arsenate	1 : 1, ethyl alcohol : water	–
Calcium carbonate	water	P, D or X
	xylene	–

Table 8.2 (Cont.)

Material	Liquid	Wetting agent
Calcium compounds	water	C
Calcium oxide	ethylene glycol	–
Calcium phosphate	water	P
Calomel	cyclohexanol	–
Carbon black	water	H, I or T
Carborundum	water	C
Cellulose powder	benzene	M
Cement	methyl alcohol saturated with glycerol	–
	ethylene glycol	F
	ethyl alcohol	F
	kerosene	R
	benzene	–
	isopropanol	–
	absolute alcohol	F (anhydrous)
	methyl alcohol	P
	butyl alcohol	–
Ceramic grog	water	C
Cerussite	water	C
Chalk	water	L
	water	E
	acetone	–
	petroleum	–
Chalk (precipitated)	isopropanol	–
Charcoal	water	O, A or P
	aqueous ammonia ($\rho = 0.91$, 5.8 v/v%)	–
Chromium oxide	water	P
Chromium pigment	water	P
Chromium powder	isobutyl alcohol	–
Clay china	water	G
Coal	water	F
	ethyl alcohol	–
	cyclohexanol	–
	1 : 1, cyclohexanol : methyl alcohol	–
Cobalt	95% ethyl alcohol in water	–
	isobutyl alcohol	–
	diethylester of phthalic acid	–
Coke	water	N or A
	isobutyl alcohol	–
	isopropyl alcohol	F
	1 : 1, ethylene glycol : ethyl alcohol	F
Copper powder	water	θ
Glass powder	water	θ
Haematite	water	–
Hydrated lime	ethyl alcohol	–
	isopropanol	–

Table 8.2 (Cont.)

Material	Liquid	Wetting agent
Ilmenite	water	–
Iron and iron alloys	rape oil and acetone	–
	1 : 1, soya bean oil : acetone	–
Kaolin	water	L or P
	water + few drops of ammonia	
Kieselguhr	water	–
Lead	acetone	–
	water	–
	cyclohexanol	–
	isoamylalcohol	–
Lead cyanamide	water	P
Lead monoxide	xylene	–
Lead oxide	water	P
Lead pigments	water	P
Lignite	cyclohexanol + 10% methanol	–
	isobutyl alcohol	–
	diethyl ester or phthalic acid	–
Limestone	water	(L + P) or (P + S)
	25% aqueous glycerol	–
Lithopone	water	D
	33% aqueous glycerol	–
Magnesite	ethylene glycol	–
Magnetite	water/ethyl alcohol/methyl alcohol/nitrobenzene	–
Manganese	isobutyl alcohol	–
Manganese dioxide	water	P
Methyl methacrylate	water	–
Molybdenum	ethyl alcohol	–
	acetone	–
	glycerol	–
	aqueous glycerol	–
Moulding sand	water	W
Nickel	rape oil + acetone	–
	aqueous glycerol	–
	cyclohexanol + 10% acetone	–
Organic powders	isobutyl alcohol and diethyl phthalate mixtures	–
	octyl alcohol	–
	isoamyl alcohol	–
Phosphate ores	water	C
Phosphorus	water	B or B + Y
Pigments	water	P
	isopropanol	–
Plaster	water	E
	alcohol–glycol	E
Polyvinyl acetate	water	C
Polyvinyl chloride	water	T
Pulp	water	L
Pumicite	water	–
Quartz	water	–

Table 8.2 (Cont.)

Material	Liquid	Wetting agent
Red lead	paint prepared in linseed oil and dispersed in white spirit (aluminium stearate)	−
Red phosphor	4% conc. in methylated spirits	−
Sand	water	L
	butyl phthalate + alcohol	−
Shales	alcohol	F
Silica	water	C, O or P
	1 : 1, water : xylene	−
	water + alcohol	−
Silicates	water	P
Sillimanite	1 : 1, water : ethyl alcohol	−
	1 : 1, water : ethyl alcohol	−
	water	P
Slag (cement)	isopropanol	−
	water	−
Soils and clays	water	O
	butyl phthalate + alcohol	−
Starch	isobutyl alcohol	−
	diethyl ester of phthalic acid	−
	isobutyl alcohol + diethyl phthalate	−
Steel powder	water	θ
Sugar	isobutyl alcohol	−
	diethyl ester of phthalic acid	−
	isoamyl alcohol	−
Sulphides	ethylene glycol	−
Talcum	water	C
Thorium	33% aqueous glycerol	−
Tin	butyl alcohol	P or C
Titanium dioxide	water	C or α or θ
	xylene, linseed oil	−
Tricalcium phosphate	water	−
Tungsten	water	0.01% C + W [19]
	glycerol	−
	acetone + rape oil	−
	ethyl alcohol	−
Tungsten ores	water	P
Tungsten carbide	ethylene glycol	−
	oil	−
Uranium oxides	aqueous glycerol	−
	isobutyl alcohol	−
	water	C
Zinc	ethyl alcohol	−
	butyl alcohol	−
	acetone	−
Zinc oxide	water	C or P
Zirconium	0.0001 N hydrochloric acid in methyl alcohol	−
	isobutyl alcohol	−
Zirconium oxide	water	C or R

8.6 Densities and viscosities of some aqueous solutions

Aqueous glycerol solutions

Some relationships between temperature T (°C), density ρ (g ml^{-1}), viscosity η (centipoise) and fractional volume concentration of glycerol C are given below:

$$T = 20°C \qquad \rho = 1 + 0.26C; \text{ for all } C. \qquad\qquad (8.17)$$

$$\eta = 1.0 + 4.4C; 0 < C < 0.25. \qquad\qquad (8.18)$$

$$\eta = 24.0 - 139.4\,C + 207.2\,C^2; 0.25 < C < 0.75. \qquad\qquad (8.19)$$

$$C = 0.50 \qquad \eta = 29.1 - 1.86\,T + 0.036\,T^2$$
$$20 < T < 25. \qquad\qquad (8.20)$$

$$C = 0.75 \qquad \eta = 99.5 - 5\,T + 0.08\,T^2; 15 < T < 25. \qquad\qquad (8.21)$$

Change in density with temperature may be neglected over limited temperature ranges. For example, $\rho = 1.195 \pm 0.002$ for $T = 20 \pm 5$,

$$C = 0.75 \qquad\qquad (8.22)$$

From equation (8.19) it can be seen that

$$\frac{\Delta\eta}{\eta} = \frac{(414C - 139)\Delta C}{24 - 139.4C + 207.2C^2}.$$

For $C = 0.75$

$$\frac{\Delta\eta}{\eta} = \frac{0.171\Delta C}{36} = \frac{4.7\Delta C\%}{}.$$

Therefore, at $C = 0.75$ an error of $\Delta C = 1\%$ leads to a 4.7% error in viscosity, hence a 2.2% error in Stokes diameter.

Alcohol–water

$$0.30 < C < 0.80 \qquad\qquad 30\rho = 31.2 - 8.3C + C^2$$
$$\eta = 1.2 + 7.4C - 8C^2.$$

Ethylene glycol–water

$$0.25 < C < 0.75 \qquad\qquad \rho = 1 + 0.12C$$
$$\eta = 2 - 4.2C + 13.6C^2.$$

Ethanol

$$T = 20°C$$

% by weight	34	36	38	40	42	44	46	48
ρ_f (g cm^{-3})	0.9468	0.9431	0.9392	0.9352	0.9311	0.9269	0.9227	0.9183
η (cP)		2.767	2.803	2.829	2.846	2.852	2.850	2.843	2.832

By using a solution of 42% by weight of ethanol the rate of change of viscosity with temperature is a minimum.

Water

$$20 < T < 100°C$$

$$\log_{10}\left(\frac{\eta_T}{\eta_{20}}\right) = \frac{1.3272(20-T) - 0.001053(20-T)^2}{T + 105}$$

$$T = 20°C$$

$$\log_{10}\eta_T = \frac{1301}{998.333 + 8.1855(T-20) + 0.00585(T-20)^2} - 3.30233$$

where $\eta_{20} = 1.002$ cP.

Further data are available in [48] and [49].

8.7 Standard powders

Narrowly classified powders are required to calibrate sensing zone instruments. Some of the light-scattering instruments are precalibrated by the manufacturers but with the electrical sensing zone method calibration is carried out, on a regular basis, by the user.

Early calibration materials were usually pollens since these had a very narrow size range. Arguments still persist as to whether these swell in electrolytes and age with storing. Polymer latices are also available as calibration standards.

A method of preparing narrowly classified powder has been described by Muta [50]. A fine copper powder of approximately the right size was mixed with 20 times as much calcium carbonate and heated for 10 minutes at 1100°C in a hydrogen atmosphere. The calcium carbonate was then dissolved out using a solution of 10 parts water to 1 part concentrated nitric acid by volume. The copper spheres so produced had a mean diameter of 6 μm and 80% by weight lay in the size range 1.4–10.5 μm.

This method was used by Colon [51] for the production of glass spheres. Several methods are available for the preparation of narrowly classified fractions. The simplest method is separation between consecutive sieves. Glass and metal spheres of diameter greater than 50 μm and standard deviation of 10% to 15% produced by this method are commercially available in quantity. A much smaller standard deviation is obtained by separation between two sieves of the same nominal diameter but the rate of production is lower. Even narrower fractions are obtainable by collecting the particles wedged in a sieve mesh after a sieving operation but the quantity produced is very small. Colon used a combination of sedimentation and sifting to produce glass spheres in commercially viable quantities (see Table 8.3 Central Technical Institute).

Metal spheres have been available for some time in commercial quantities, particularly between sieve sizes. Glass spheres are also available and these are often preferred due to their lower densities [57]. Glass beads in the size range 1 to 30 μm (Minnesota Mining and Manufacturing Co.) were carefully treated and examined to provide NBS Standard 1003 [54, 58].

Narrowly classified latices have been made available for some time from Dow Chemicals [56] but considerable doubt has been cast on the accuracy of the sizing. A 3.49-

μm polystyrene latex was independently sized using electron microscopy and found to have a mean diameter of 3.40 μm [52]. This sample was used by Coulter Electronics as a standard to test other Dow latices [53]; these are available from Coulter in the form of 10 ml suspensions.

The National Physical Laboratory (NPL) act as distributing agents for the Community Bureau of Reference (BCR) standards which have been financed by the Commission of

Table 8.3 Narrowly classified powders

Material	Size (μm)		Source
Polystyrene latex	0.557		Dow Chemicals
	0.796		
	1.099	1.15*	
	1.3		
	1.857	1.87*	
	2.0		
	2.68	2.64*	
	3.49	3.40†	
	Nominal size (μm)		
Polystyrene latex	1		Coulter Electronics
	10		
	13		
	20		
	40		
	80		
	175		
Polyvinyl toluene	2		
	3		
Particle suspension	5		
Puff ball spheres	3.5		
Paper mulberry pollen	15		
Ragweed pollen	20		
Hazel pollen	25		
Michaelmas daisy pollen	27		
Lycopodium powder	27		
Rye grass	34		
Pecan pollen	40		
Corn pollen	80		
Glass spheres	4.7	0.4	Central Technical
	8.6	0.6	Institute
	11.1	0.7	
Calibration spheres for sensing zone methods			Particle Information Service

† Size determined independently by electron microscopy [52].
* Size determined by Coulter Electronics using the 3.40 μm latex as standard [53].

the European Communities. BCR has had five quartz powders prepared as reference materials for the calibration of apparatus. These powders are certified by their Stokes diameters except for RM 68 which is certified by volume diameter. The powders were analysed by gravity sedimentation except for RM 68 which was analysed by sieving. Analyses were carried out at five laboratories and the results were compared to give a measure of the quality of the standards.

NPL have also started a feasibility study on surface area and pore size distribution standards which should be available at the end of 1980. AC fine test dust is available from General Motors for calibrating classifiers and ASME also produce standard powders for this purpose. A list of addresses is given in Appendix 2.

Table 8.4 Standard powders

Material	Size range (μm)	Source
Quartz (BCR reference material)		National Physical Laboratory (NPL)
RM 66	0.35–2.5	
RM 67	3–20	
RM 68	140–650	
RM 69	12–90	
RM 70	0.5–12	
Glass beads [54, 55] standard reference material 1003 based on Stokes' settling	5–30	National Bureau of Standards (NBS)
Glass spheres	50	
	8–35 mesh	
	20–70 mesh	
	50–170 mesh	
	140–400 mesh	
AC fine test dust		General Motors
Standard powders under development		
Titania	0.1–3	
Quartz	0.1–3	
	3–40	
	40–1000	
	10–100	
	1–10	
Zirconia, monodisperse	1–2	
	5	
	15	
	30	
	60	

Table 8.5 Surface area standards

The following certified standard powders are available from the National Physical Laboratory (NPL).

Material	Surface area ($m^2 g^{-1}$)
α –Alumina	2.1
	0.3
	0.1
Sterling FT–G (2700)	11
(graphitized carbon black)	
Vulcan 3–G (2700)	71
(graphitized carbon black)	
Silica TK800	165
(non-porous silica)	
Meso-porous Gasil silica [1], data on pore size distribution available on request	286

References

1 Adamson, A.W. (1963), *Physical Chemistry of Surfaces*, Wiley, NY.
2 Washburn, E.D. (1921), *Phys. Rev.*, 17, 374.
3 Crowl, V.T. and Wooldridge, W.D.S. (1967), *Wetting*, SCI Monograph No. 25, 200.
4 Crowl, V.T. (1967), In: *Agriments* (ed. D. Patterson), Elsevier, p. 192.
5 Bartell, F.E. and Walton, C.W. (1934), *J. Phys. Chem.*, 38, 503.
6 Rossi, C. and Baldocci, R. (1951), *J. Appl. Chem.*, 1, 446.
7 von Buzagh, A. (1937), *Colloid Systems*.
8 Herdan, G. (1960), *Small Particle Statistics*, Butterworths, p. 347.
9 BS 3406 (1963): Part 2, p. 39.
10 Koglin, B. (1969), Turbidimetric investigation of the activity of surface active agents in dispersing suspension. Private communication: Lehrstuhl für Mechanische Verfahrenstechnik, Universität Karlsruhe (TH), W. Germany.
11 Galatchi, G.L. (1969), *Stud. Cercet. Fiz.*, 21, 7.
12 Dollimore D. *et al.* (1970), *J. Phys. E.*, 3, 465–6.
13 Keng, E.Y.H. (1970), *Powder Technol.*, 3, 3, 179–80.
14 Bryant, D.P. (1968), *Proc. Soc. Analyt. Chem.*, 5, 8, 165–6.
15 Groves, M.J. *ibid.*, 166–8.
16 Parfitt, G. (1973), *Dispersion of Powders in Liquids*, Applied Science Publishers.
17 Burt, M.W.G., Fewtrell, C.A. and Wharton, R.A. (1973), *Powder Technol.*, 8, 223–30.
18 Perry, J.H. (1963), *Chemical Engineers' Handbook*, McGraw-Hill.
19 Kellie, J.L.F. (1970), In: *Particle Size Analysis* (eds. M.J. Groves and J.L. Wyatt-Sargent), Soc. Anal. Chem., p. 176.
20 Bond, R.L. and Spencer, D.H.T. (1957), Proc. 1st Conf. *Industrial Carbon and Graphite*, Soc. Chem. Ind. London, pp. 231–51.
21 Spencer, D.H.T. (1967), *Porous Carbon Solids* (ed. R.L. Bond), Academic Press, NY, pp. 87–154.
22 Thudium, J. (1976), *J. Aerosol Sci.*, 7, 2, 167–74.
23 Baranowski, J. (1973), *Ochrona Powietrza*, 2, 38.
24 Juda, J. (1966), *Staub*, 26, 197.
25 Krutzch, J. (1954), *Chemiker-Zeitung*, 78, 49.

26 Müller, G. (1964), *Methoden der Sedimentuntersuchung,* E. Schweltzerbartsche Verlagbuch-handlung, Stuttgart.
27 Hänel, G. (1972), Bestimmung Physikalischer Eigenschaften Atmosphärischer Schwebetulchen als Funktion der Relativen Luftfeuchtigkeit, Diss. Universität, Mainz.
28 Carino, L. and Mollet, H. (1975), *Powder Technol.,* 11, 189–94.
29 May, I. and Marienko, J. (1966), *Am. Mineralog.,* 51, 931–4.
30 Stein, F. Penkala, S. and Buchino, J. (1971/2), *Powder Technol.,* 5, 317–19.
31 Jones, T.M. (1969), Symp. *Powders,* Soc. Cosmetic Chemists, Dublin.
32 Peleg, M. and Mannheim, C.H. (1973), *Powder Technol.,* 7, 45.
33 Pilpel, N. (1970), *Mfg Chem. Aerosol News,* 4, 19.
34 Gold, G. *et al.* (1966), *J. Pharm. Sci.,* 55, 1291.
35 Leoveanu, O., Zaharia, N. and Pilea, V. (1966), *Rev. Chim.,* 17, 112.
36 York, P. (1975), *Powder Technol.,* 11, 197–8.
37 Jenike, A.W. (1961), Bull. 108, Eng. Expl. Sta., Utah State Univ.
38 Heertjes, P.M. and Kossen, N.W.F. (1966), *Powder Technol.* 1, 33–42.
39 Kossen, N.W.F. and Heertjes, P.M. (1965), *Chem. Eng. Sci.,* 20, 593.
40 Hartig, H.E., Oristad, N.I. and Foot, N.J. (1951), Univ. Minnesota Mines Experimental Station Information Circular, No. 7.
41 Hathaway, R.E. and Guttrals, D.L. (1976), *Can. Min. Metall. Bull.,* 69, 766, 64–71.
42 Koglin, B. (1977), *Powder Technol.,* 17, 2, 219–27.
43 Parfitt, G.D. *ibid.,* 157–62.
44 Karpenko, I., Dye, R.W. and Engel, W.H. (1962), *TAPPI,* 45, 1, 65–69.
45 Carr, W. (1970), *JOCCA,* 53, 884.
46 Carr, W. (1977), *Powder Technol.,* 17, 2, 183–90.
47 Keng, E.Y.H. (1969/70), *Powder Technol.,* 3, 179–80.
48 DIN 66111 (1973): *Particle Size Analysis; Sedimentation Analysis in a Gravitational Field; Fundamentals.*
49 West, R.C. (ed.) (1973), *Handbook of Chemistry and Physics.* The Chemical Rubber Company.
50 Muta, A., Saito, N. and Uchara, Y. (1967), *Particle Size Analysis,* Proc. Conf. Soc. Analyt. Chem., London, p. 215.
51 Colon, F.J. *et al.* (1973), *Powder Technol.* 8, 307–10.
52 Matthews, B.A. and Rhodes, C.T. (1970), *J. Colloid Interfac. Sci.,* 32, 339.
53 Harfield, J.G. and Wood, W.M. (1970), *Particle Size Analysis,* Proc. Conf. Soc. Analyt. Chem., Bradford, 1971, pp. 293–300.
54 Hunt, C.M. and Woolf, A.R. (1969), *Powder Technol.* 3, 9.
55 US Department Commerce, Washington, D.C. (1965).
56 Bradford, E.B. and Vanderhoff, J.W. (1966), Symp. *Particle Size Distribution,* Pittsburgh, PA.
57 Gaudin, A.M. and Bowdish, F.W. (1944), *Mining Tech.,* 8, 3, 1–6.
58 Carpenter, F.G. and Dietz, V.R. (1951), *J. Res. nat. Bur. Stand.,* 47, 3, 139–47.

⑨ Incremental methods of sedimentation size analysis

9.1 Basic theory

The particle size distribution of a fine powder may be determined by examining a sedimenting suspension of the powder. The powder may be introduced as a thin layer on top of a column of clear liquid, the two-layer technique; or it may be uniformly dispersed in the liquid, the homogeneous suspension technique. In the incremental method of size analysis by sedimentation, changes with time in the concentration or density of the suspension at known depths are determined and from these the size distribution may be found. In the cumulative method, the rate at which the powder is settling out of suspension is determined and, from a knowledge of this, the size distribution may be found. Incremental methods may be divided into fixed time and fixed depth methods, the latter being more popular, although a combination is sometimes used.

9.1.1 Variation in concentration within a settling suspension

Let a mass $m_s = \rho_s v_s$ of a solid be dispersed in a mass $m_f = \rho_f v_f$ of fluid, ρ and v being density and volume respectively.

Initially, the concentration throughout the suspension will be uniform and equal to:

$$C(h, 0) = \frac{\text{mass of solids}}{\text{volume of solids} + \text{volume of fluid}}$$

$$C(h, 0) = \frac{m_s}{v_s + v_f} \tag{9.1}$$

where $C(h, 0)$ is the concentration at depth h, time $t = 0$.

Consider a small horizontal element in the suspension at a depth h. At the commencement of sedimentation the particles leaving the element are exactly balanced by the particles entering it from above. When the largest particles initially present at the surface of the suspension leave the element, there are no similar particles entering to replace them. Hence the concentration within the element falls and becomes equal to the concentration of particles smaller than D in the suspended phase, where D is the size of the particle which falls with a velocity h/t. The concentration of the suspension at depth h at time t may be written:

$$C(h, t) = \frac{m'_s}{v'_s + v_f} = \int_{D_{min}}^{D} F(D)\,dD \qquad (9.2)$$

where m'_s is the mass and v'_s the volume of solids in a volume v_f of fluid at time t from commencement of sedimentation and at a depth h from the surface of the suspension. Hence:

$$C(h, 0) = \frac{m_s}{v_s + v_f} = \int_{D_{min}}^{D_{max}} F(D)\,dD \qquad (9.3)$$

From these three equations:

$$\frac{C(h, t)}{C(h, 0)} = \frac{m'_s}{m_s} = \frac{\int_{D_{min}}^{D} F(D)\,dD}{\int_{D_{min}}^{D_{max}} F(D)\,dD} \qquad (9.4)$$

It has been assumed that the difference between v'_s and v_s is negligible compared with v_f.

Thus, if a graph of $100\,C(h, t)/C(h, 0)$ is plotted against D, the resulting curve shows the cumulative percentage undersize by weight.

9.1.2 Relationship between density gradient and concentration

Let $\phi(h, t)$ be the density of the suspension at a depth h and at time t. Then:

$$\phi(h, t) = \frac{m'_s + m_f}{v'_s + v_f}$$

$$= \frac{\rho_s v'_s + \rho_f v_f}{v'_s + v_f}$$

$$= \frac{\rho_f(v'_s + v_f) + (\rho_s - \rho_f)v'_s}{v'_s + v_f}$$

$$= \rho_f + \frac{\rho_s - \rho_f}{\rho_s} \cdot \frac{m'_s}{v'_s + v_f}$$

$$= \rho_f + \frac{\rho_s - \rho_f}{\rho_s} \cdot C(h, t)$$

Also

$$\phi(h, 0) = \rho_f + \frac{\rho_s - \rho_f}{\rho_s} C(h, 0)$$

Therefore

$$\frac{C(h, t)}{C(h, 0)} = \phi = \frac{\phi(h, t) - \rho_f}{\phi(h, 0) - \rho_f} \tag{9.5}$$

where ϕ is the fraction undersize D.

Thus a plot of $100 \times \dfrac{\phi(h, t) - \rho_f}{\phi(h, 0) - \rho_f}$ against D gives a cumulative percentage curve.

9.2 Resolution for incremental methods

Assuming Stokes' law to apply:

$$D^2 = k(h/t)$$

Differentiating with respect to h:

$$2D(dD/dh) = k/t$$
$$= D^2/h$$
$$dD/dh = D/2h$$

or

$$dD/D = dh/2h \tag{9.6}$$

The variation in size within an element of thickness dh, from the calculated Stokes diameter is therefore $\pm \frac{1}{2} dD$ which equals:

$$\left[\pm \frac{D}{4} \cdot \frac{dh}{h} \right].$$

States [1] defines the resolution for incremental techniques as the reciprocal of the variation in particle size in the sampling zone, i.e.

Resolution $1/dD = 2h/aD$ where $a = dh$ is the width of the sampling zone.

Heywood [2] points out that this is misleading since the relative size variation is the important factor. This is greatest for the hydrometer method, although the effect depends on the type of size distribution concerned.

Consider the following cases, in Table 9.1.

Table 9.1

h (cm)	dh (cm)	$\dfrac{dD}{D} = \dfrac{dh}{2h}$
10	10	0.5
20	2	0.05
5	0.2	0.02

The first case applies to the hydrometer method of analysis and, at a nominal Stokes diameter of 50 μm, the size variation within the sampling zone is 25 μm. If the weight frequency of particles in the 50 to 62.5 μm range is equal to that in the 37.5 μm range, the effect is balanced out, otherwise a bias results. The general effect of such a bias is to mask peaks in a multimodal distribution.

Heywood carried out experimental work on the relationship between width of sampling zone and resolution and concluded that if the sampling zone is less than one-sixth of the mean depth, the practical error introduced is negligible. The hydrometer, in spite of the low resolution, gives reasonably accurate results for clay particles, but is not, in general, recommended for other types of powdered materials.

9.3 The pipette method

In the pipette method of particle size analysis, the concentration changes occurring within a settling suspension are followed by drawing off definite volumes by means of a pipette.

The method was described first in 1922 by Robinson [3], who used a normal laboratory pipette, and Jennings *et al.* [50], who used a pipette taken out downwards from the sedimentation vessel with a suction system consisting of nine tubes of which eight were bent outwards, four on each of two different radii. In 1923 Krauss [51] suggested a variable-height system of three suction tubes closed at the bottom, each having six to eight horizontal bores. Koettgen [52] used a normal pipette with a bent top and recommended that the pipette should be rotated during the withdrawal of the sample. Köhn [53] used a pipette closed at the tip and with several horizontal holes. Lehmann [54] and Lorenz [55] in 1932 reported a pipette closed at the tip with four or six horizontal apertures.

Andreasen [4] in 1928 was the first to leave the pipette in the sedimentation vessel for the duration of the analysis. The apparatus described in 1930 by Andreasen and Lundberg [7] is the one in general use today [5, 6].

The question that arises is whether some of these complicated constructions are really necessary. By making certain assumptions, an estimate may be made of the magnitude of the various errors and the modifications required to reduce them to acceptable limits [53, 9]. Before discussing these a description of the basic equipment is given. The Andreasen equipment (figure 9.1) consists of a graduated sedimentation vessel (0 to 20 cm) which holds between 500 and 600 ml when filled to the 20-cm mark. The stem of the pipette is fused to a ground-glass socket which fits the neck of the sedimentation vessel, so that the stem is centrally positioned in the sedimentation vessel and its tip is level with the zero fiduciary mark. Above the socket is a two-way tap so that suspension may be drawn into a 10-ml container which may then be emptied into a 25-ml beaker or centrifuge tube.

The recommended procedure is to take representative samples of powder to make up a volume concentration of between 0.2 and 1.0% by volume. The powder is made into a paste and then a slurry by the slow addition of dispersing liquid while the powder is being mixed with a spatula. At this stage, dispersing agent may be added, e.g. two or

three drops of Nonidet P42. Other systems may be made up in bulk beforehand, e.g. 0.1% sodium hexametaphosphate in distilled water, and the mixing is done with this liquid. Further dispersing may be carried out in an ultrasonic bath. The suspension is washed into the sedimentation vessel and the level made up to the top fiduciary mark. The analysis is preceded by violent agitation, preferably not with a stirrer, since this imparts a centrifugal motion to the suspension. A recommended method is continually to invert the container by hand for about a minute. Zero time is the time at which agitation ceases. Since the particles are not at rest when $t = 0$, an error is introduced. Because of this, it is not advisable to withdraw samples for times shorter than 1 min. Acceptable time scales are 2 : 1 progression of t-values to give a root-two progression in particle size D as calculated from Stokes' equation.

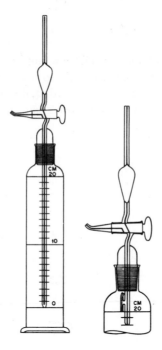

Fig. 9.1 Andreasen's fixed-position apparatus (Gallenkamp).

Ten seconds before withdrawal time, suction is applied to the withdrawal pipette, either orally or with a suction bulb. Rate of withdrawal should be such that the 10-ml sample is withdrawn in 20 s. The two-way tap is turned and excess pressure applied to empty the pipette into a tared container (a 25-ml capacity beaker). About 5 ml of fill liquid, without dispersing agent, are then drawn into the pipette from a beaker and the beaker is then removed, so that air is drawn into the 10-ml container. This is also blown into the sample container, leaving the pipette clean in preparation for the next sample.

The concentration of solids in the samples may be determined by centrifuging, drying and weighing or simply drying and weighing. The drying temperature should not be too high or 'spitting' and subsequent loss of powder may occur. With hygroscopic dispersing agents, special care must be taken to eliminate uptake of moisture from the atmosphere as the containers cool. They are, therefore, cooled in a desiccator and some desiccant is placed in the balance so that the air near the pan is dry. The containers are removed from the desiccator singly and weighed as quickly as possible. Alternatively, the dispersing agent may be removed by filtering off the liquid instead of evaporating it [9, 61].

The equation yielding the percentage undersize D_{Stk} is:

$$P = 100 \; \frac{m_t}{v} . \frac{v_s}{m_s}$$

$$= K m_t \tag{9.7}$$

where m_t is the weight of powder in the aliquot sample of volume v, and m_s/v_s is the solid concentration in the initial homogeneous suspension. The largest size present in each sample d_{Stk} is calculated from Stokes' equation, due allowance being made for the fall in height of the suspension due to withdrawing samples.

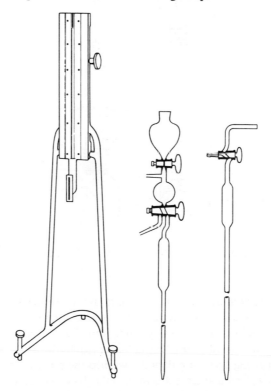

Fig. 9.2 The fixed-depth pipette.

If $C(h, 0)$ is too large, i.e. greater than 1.0% by volume, hindered settling may occur, leading to erroneous results, whereas low values of $C(h, 0)$ lead to inaccuracies in weighing the small samples withdrawn. It is therefore advisable to carry out duplicate analyses, one at about 1% by volume and one at about 0.5% by volume. If the former gives a significantly coarser distribution than the latter, further analyses are required to determine the optimum concentration, which is usually between these two limits. Fixed-position pipettes of varying stem length are available and the final sample may be taken with one of these in order to complete the analyses in a shorter time.

A modification to this technique is to use a fixed-depth pipette [6, 11] (figure 9.2). This pipette is inserted immediately before the sample is required. After the sampling procedure has been carried out, the pipette is withdrawn so that the collected sample may be emptied into a container. If several suspensions have been prepared, each sample may be taken from a different suspension which has previously been undisturbed. This makes the analysis more complex, but gives a marginal increase in accuracy.

An alternative apparatus [7] consists of a sedimentation vessel (figure 9.3) with a horizontal sampling tube and a loose pipette. When the pipette is connected to the sampling tube, the pressure inside the vessel will make the suspension rise into the pipette beyond the two-way stopcock which is then closed and the pipette removed, the sampling tube being closed with a glass rod. The suspension above the stopcock is then blown out through the side tube of the stopcock and the pipette is emptied into a weighed vessel, which is evaporated and weighed.

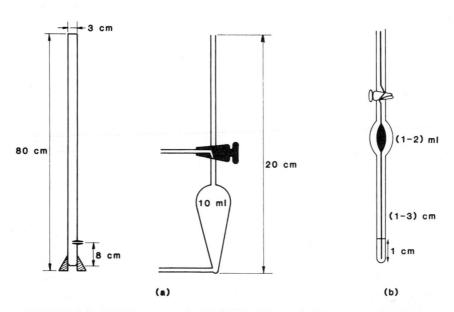

Fig. 9.3 (a) The side-arm pipette (Berg [20]). (b) Berg's pipette for fine material [15]. The pipette is immersed in the suspension immediately prior to sampling with the bottom of the pipette resting on the bottom of a container holding the suspension.

If a dispersing agent is used, the weight in the aliquot sample withdrawn is deduced from the weight added to the whole sample and a correction applied. Errors can arise with hygroscopic dispersing agents or if the dispersing agent is adsorbed by the solid. These errors may be eliminated by washing out the dispersing agent [9].

9.3.1 Experimental errors

Since the sample is not withdrawn instantaneously, the size of the largest particle present in the sampling zone will vary throughout the sampling time. The error due to this will be greater at the commencement of the analysis but soon becomes negligible. Rapid withdrawal of the sample is not recommended since preferential sampling always occurs when sampling is not isokinetic. Small particles beyond the sampling zone are drawn into it due to their low inertia, thus suggesting a finer distribution. This has been confirmed experimentally by Johnson [8], who varied the sampling time between 12 and 140 s. In BS 3406 a compromise time of 20 s is recommended. For reproducible results, this time should be strictly adhered to.

In the theory it is assumed that the sample is withdrawn from a narrow cylindrical element, whereas with the Andreasen pipette the sampling zone is, to a first approximation, a sphere. Leschonski [9] showed that the error introduced is less than 1%, provided the ratio of sphere radius to sedimentation height remains below 0.1.

With a vertically downward positioned pipette, such as the Andreasen, there is a liquid volume below the tip of the pipette into which no solid particles can sediment from the layers lying above. This situation creates an instability which results in the setting up of density difference convection currents around the lower end of the pipette. Johnson [8] found that a horizontally directed pipette gave a coarser analysis than one directed downwards and attributed it to this effect. Similar experiments conducted by Allen [25] showed no significant difference between these two types. Kast [56] used pipettes with open tips and different external diameters of the capillary and found that with large external diameters there was a displacement of the percentage undersize curves towards smaller values.

Leschonski [9] designed a pipette to conform with theory (figure 9.4). The pipette is extended to the bottom of the vessel and the sample withdrawn through a series of holes around its circumference at a fixed depth. Leschonski pointed out that this idea was not new since Andreasen [4] employed a pipette extending to the bottom of the vessel for the sake of sturdiness and for accurate height adjustment. Lorenz [55] also reported on a similar system which was not incorporated in the resulting commercial apparatus, and Berg's pipette for fine material [15] also rests on the bottom of the container holding the suspension.

With the Andreasen pipette, a small amount of liquid is retained in the capillary tube after each withdrawal and gives rise to a systematic positive error, i.e. an over-estimation of the percentage undersize. Leschonski [9] showed theoretically that this error could be substantial for narrow distributions. This error can be reduced by shortening the capillary volume. In the MSO-Maschinen und Schleifmittelwerke [57], the capillary runs out of the side of an 80-cm long sedimentation column. **Kuncewicz and Krzyzewski**

[58] prefer the pipette by Esenwein on account of its shorter capillary which, again, projects laterally from the sedimentation cylinder.

If the error is to be completely avoided, the sample residue must be removed from the capillary shortly before taking the next sample. This measure was recommended by Andreasen in 1928 [4], but in the apparatus described in 1930 [7] this instruction is missing.

The pipette described by Leschonski [9] incorporates a subsidiary bulb with a volume matched to that of the capillary (figure 9.4). The capillary extends to the bottom of the vessel and has four withdrawal apertures staggered by 90°, located about 30 mm above the bottom of the vessel. The shape of the bulb is also unconventional with a gradual transition from the bulb to the stem. This design reduces the possibility of over-running the upper fiduciary mark. As a further precaution an additional volume scale is provided in the vicinity of this mark to eliminate the practice of allowing excess suspension to run back into the sedimentation cylinder, one which is to be deprecated since more solids are flushed back than were originally present in the excess volume.

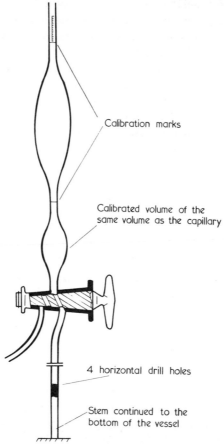

Calibration marks

Calibrated volume of the same volume as the capillary

4 horizontal drill holes

Stem continued to the bottom of the vessel

Fig. 9.4 Improved pipette design (Leschonski [9]).

Pavlik and Sansone [64] state that particles can adhere to the ground-glass assembly of the stopcock. They therefore eliminated the stopcock, replacing it with a separate 10-cm^3 pipette which was pressed down on a seal during extraction.

Agitation prior to analysis causes convection currents which may introduce considerable errors in the early readings [59]. With apparatus which on account of its size could not be shaken by hand, Andreasen [4] used a long hollow tube through which air was blown, a method also used in [57] but rejected by Leschonski due to the possibility of flotation occurring. His preferred method is with a perforated plate plunger-type stirrer.

Convection currents may also be set up due to temperature variations and it is advisable to keep the temperature constant to $\pm 0.01^\circ C \, min^{-1}$. Oden [60] states that for colloidal gold suspensions the upper limit is $0.2^\circ C$ per day. For extended analyses it is usual to immerse the container in a thermostatic bath taking care that no vibrations are imparted to the system. An acceptable alternative method of lagging the sedimentation bottle has been suggested by Heywood [10].

If the sedimentation vessel is not set up absolutely vertically, convection currents are detectable, whereas these are not seen when the vessel is set up correctly. Further errors are caused by disturbance to the suspension when samples are withdrawn, and, with the variable-depth pipette, when the pipette is inserted and withdrawn. These errors could be reduced by using several sedimentation vessels and taking each sample from a different vessel, but this more elaborate procedure is not recommended for routine work.

This technique was recently reviewed by Alex [21] and compared with other methods by Grandillo [23]. It has also been the subject for an American Standard for cement [22] and used for pharmaceutical powders [24]. It is a highly reproducible method, the cost of the equipment is low and several analyses may be carried out by semi-skilled labour per day.

9.4 The photosedimentation technique

9.4.1 Introduction

The photosedimentometer combines gravitational settling with photoelectric measurement. The principle of the technique is that a narrow horizontal beam of parallel light is projected through a suspension at a known depth h on to a photo-cell. Assuming an initially homogeneous suspension, the concentration of particles in the light beam will be the same as the concentration in the suspension. If the particles are allowed to settle, the number of particles leaving the light beam will be balanced by the number entering it from above. However, after the largest particle present in the suspension, d_m, has fallen from the surface to the measurement zone the emergent light flux will begin to increase since there will be no more particles of this size entering the measurement zone from above. Hence, the concentration of particles in the light beam at any time, t, will be the concentration of particles smaller than d_{Stk}, where d_{Stk} is given by Stokes' equation (equation (7.43)):

$$d_{Stk} = \sqrt{\frac{18\eta h}{(\rho_s - \rho_f)gt}} \qquad (9.8)$$

the symbols having their usual meaning.

It can be shown that the attenuation of the beam of light is related to the projected surface area of the particles in the light beam and from this relationship the particle-size distribution may be determined.

9.4.2 Theory

Consider a sedimentation tank of length L cm, measured in the direction of the light beam, containing the suspension of powder under analysis, the concentration of the powder being C g cm^{-3} of suspension.

Let the incident light intensity on an element of thickness δL be I and the emergent light intensity be $I - \delta I$. If the area of the light beam is A, the reduction in flux due to the presence of particles may be attributed to a fall in the overall intensity of the light beam or a reduction in the area of the light beam.

The emergent flux may be written:

$$(I - \delta I)A = (A - \delta A)I$$

$$\frac{\delta I}{I} = \frac{\delta A}{A} \qquad (9.9)$$

where δA is the effective cross-sectional area of particles in the beam perpendicular to the direction of propagation. This equation holds, provided the beam of light becomes homogeneous again between adjacent particles.

If there are n_r particles of size d_r in 1 g of powder,

$$\delta A = -AC\delta L \sum_{r=0}^{r=st} K_r k_r n_r d_r^2 \qquad (9.10)$$

where k is a constant depending on the shape of the particles ($k = \pi/4$ for spheres); d_0 and d_{st} are the smallest and largest particles present in the light beam.

The extinction coefficient K_r is defined as:

$$K_r = \frac{\text{light obscured by particles of size } d_r}{\text{light which would be obscured by this particle if the laws of geometric optics held}}$$

From equations (9.9) and (9.10):

$$\int_{I_0}^{I_t} \frac{dI}{I} = -\int_0^L C dL \sum_{r=0}^{r=st} K_r k_r n_r d_r^2$$

Integrating gives:

$$\ln \frac{I}{I_0} = - CL \sum_{r=0}^{r=st} K_r k_r n_r d_r^2$$

Taking logarithms to base 10 and defining the optical density as:

$$D = \log_{10} I/I_0$$

gives:
$$D = - CL (\log_{10} e) \sum_{r=0}^{r=st} K_r k_r n_r d_r^2 \qquad (9.11)$$

Consider a small change in optical density, ΔD_r, as the maximum size of particle in the beam changes from d_r' to d_r'' where $d_r = \frac{1}{2}(d_r' + d_r'')$.

$$\frac{\Delta D_r}{K_r} = - k_r CL (\log_{10} e) n_r d_r^2 \qquad (9.12)$$

The cumulative distribution undersize by surface, assuming that K is constant for the restricted size range of any powder under analysis, is given by:

$$\frac{\sum\limits_{r=0}^{r=st} n_r d_r^2}{\sum\limits_{r=0}^{r=max^m} n_r d_r^2} = \frac{\sum\limits_{r=0}^{r=st} \frac{\Delta D_r}{K_r}}{\sum\limits_{r=0}^{r=max^m} \frac{\Delta D_r}{K_r}} \qquad (9.13)$$

The cumulative distribution undersize by weight is given by:

$$\frac{\sum\limits_{r=0}^{r=st} n_r d_r^3}{\sum\limits_{r=0}^{r=max^m} n_r d_r^3} = \frac{\sum\limits_{r=0}^{r=st} \frac{\Delta D_r}{K_r} d_r}{\sum\limits_{r=0}^{r=max^m} \frac{\Delta D_r}{K_r} d_r} \qquad (9.14)$$

The weight-specific surface of the powder is derivable from the initial concentration of the suspension and the maximum optical density (D_M).

$$S = \sum_{r=0}^{r=max^m} \alpha_{s,r} n_r d_r^2 \qquad (9.15)$$

$$W = \rho_s \sum_{r=0}^{r=max^m} \alpha_{v,r} n_r d_r^2 \qquad (9.16)$$

where α_s and α_v the surface and volume shape coefficients, may be assumed constant for the restricted size range of any powder under analysis. Further, since n_r is the number of particles of size d_r in 1 g of powder, $W = 1$. Hence:

$$S = \alpha_s \sum_{r=0}^{r=max} {}^{m} n_r d_r^2 \qquad (9.17)$$

Combining with equation (9.12) gives:

$$S_W = \frac{\alpha_s}{kCL(\log_{10} e)} \sum_{r=0}^{r=max} {}^{m} \frac{\Delta D_r}{K_r} \qquad (9.18)$$

This may be written:

$$S_W = \frac{\alpha_s}{kL(\log_{10} e)} \frac{1}{K_m} \frac{D_M}{C} \qquad (9.19)$$

where K_m is a mean value for the extinction coefficient. For non-re-entrant particles, the ratio of the surface and projected area shape coefficients (α_s/K) is equal to 4. For re-entrant particles, the surface area obtained by making this assumption is the envelope surface area. Equation (9.19) simplifies to:

$$S_W = \frac{9.2}{K_m L} \frac{D_M}{C} \qquad (9.20)$$

9.4.3 The extinction coefficient

The photosedimentation technique has many advantages over most other particle size analysis methods. The attenuation of a beam of light can be measured accurately; the suspension is not disturbed by the insertion of a probe or other measuring device; the sample of powder required is small; a test can be carried out rapidly; the required concentration is low which reduces the possibility of particle–particle interaction and experimental results are obtained in a form which lends itself to automatic recording and remote-control techniques. The method has not yet become popular due to the breakdown in the laws of geometric optics which occurs as the particle size approaches the wavelength of the incident radiation.

For small particles (around 80 μm with light of wavelength 0.60 μm), an amount of light flux, equal in magnitude to that incident upon the particle, is bent away from the forward direction. As the particle size decreases the scattered light is contained in an increasing solid angle; however, no matter how small the receiver, some of the light is accepted until the particle size is similar to the wavelength of light when the light is predominantly back-scattered. Thus large opaque particles will obstruct an amount of light proportional to their cross-sectional area. With a narrow-angle receiver, this amount will double as the particle size decreases from about 80 to 6 μm.

With a partially transparent particle, interference due to the light transmitted by the particle will also occur. If a 6-μm particle attenuates a fraction F of the light incident

upon it, a fraction $1 + F$ will not be collected by a small-angle receiver, hence the extinction coefficient will more than double. It cannot, therefore, be assumed that each particle obstructs the beam with its own geometric cross-section, and complex diffraction, scattering, interference and absorption effects have to be considered. These effects are compensated for in the equations by the insertion of an extinction coefficient K.

Early experimenters were either unaware of, or neglected, the parameter K, and this attitude has persisted because of the difficulty in determining it [30, 31]. Some research workers have used monochromatic light and determined K theoretically [32, 33]. Rose and Lloyd [34–36] attempted to derive a universal calibration curve for K against d for white light. This curve has been challenged by Allen [37], who shows that an alternative interpretation may be placed on Rose's experimental data. In later papers [26, 27] Allen showed that a correction curve for diffraction effects could be derived from Fraunhofer diffraction theory. This correction curve depends upon the receiving angle of the detector; at a particle size of about 6 μm an amount of light flux equal to that incident on the particle is diffracted away from the forward direction and not picked up by the receiver. For particles smaller than 6 μm, the shape of the extinction curve becomes more heavily dependent on the optical properties of the particulate fluid system under investigation. Allen's theoretical approach results in an extinction curve for opaque particles which is applicable to the EEL photosedimentometer (a narrow-angle instrument). On the basis of this work, a wide-angle scanning photo-sedimentometer (WASP) was developed.

Allen proposed an equation, for use with the EEL, whereby volume-specific surface (S_v) could be determined. $S_v' = \rho_s S_W'$ may be determined using equation (9.20) and assuming K_m equal to unity. The true volume-specific surface is then obtained from the equation:

$$S_v = 1.82 S_v'^{0.87} \tag{9.21}$$

With the WASP, equation (9.20) may be used with K_m put equal to unity. The specific surface so obtained will be correct for opaque particles larger than about 6 μm and low for partially transmitting particles. For opaque particles smaller than 6 μm the experimentally determined surface will be high due to loss of light scattered outside the receiving cone of the photocell.

9.4.4 Photosedimentometers

The Wagner photosedimentometer, a single standard turbidimeter for use in the determination of the fineness of cement, has a variable-height platform for the sedimentation tank and a galvanometer with which to record the output of a barrier layer photocell.

The EEL (figure 9.5) uses a white light beam of circular cross-section collimated by a lens and a series of stops. The beam passes through the sedimentation cell and, after another series of stops, to a selenium barrier layer photocell which feeds a galvanometer. Six sedimentation cells are housed in a movable carriage so that the beam may traverse them in turn. Hence, in theory, five samples may be analysed simultaneously, the sixth cell being filled with clear liquid and used as a reference cell. In practice a more realistic number is three.

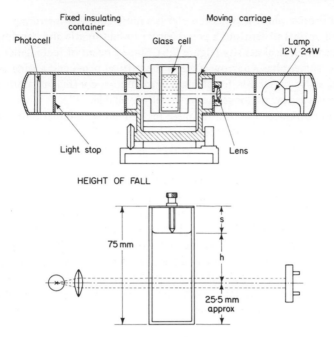

Fig. 9.5 The EEL photosedimentometer, optical system. Height of fall h is selected by setting the height gauge to a known length s.

The small size of the sedimentation cells limits the height of fall to 2 cm, so that only a limited size range may be analysed in one cell. Further, continuous operator attention is needed. Since this is a narrow-angle instrument, correction has to be applied for the breakdown in the laws of geometric optics. The correction curve supplied with the instrument is based on wrong premises and, although it partially corrects for the error it was intended for, the accuracy of the final analysis is very doubtful. Similarly, as a surface area measurer the validity of the correction equations is very doubtful.

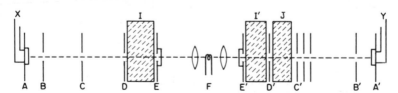

Fig. 9.6 Prototype of the Bound Brooke photosedimentometer as used by Morgan [31]. A 30 watt filament bulb F provides two flat rectangular beams through two plano-convex lenses and two slits E, E'. The intensities of the two beams of light are measured by two photocells A, A' and a suspension of the powder under test is placed in the sedimentation tank J. By interposing heat-absorbing filters at E, E' and dummy tanks I, I' the infra-red component of the light is removed and a variable cut-off arm at D is used to give equal intensities on both arms of the apparatus with the tank J in place and before the sample is introduced.

The Bound Brooke photosedimentometer is a more complex instrument, comprising a sophisticated optical system and a pen recorder for automatic and continuous recording of variations in optical density (figure 9.6). Since a beam of rectangular cross-section is used, the instrument has a narrow angle in the vertical plane and a wide angle in the horizontal plane. No correction for changes in extinction coefficient is provided. These two instruments are now no longer commercially available.

The Seishin Photomicroniser is a Japanese instrument with direct plotting facility.

In the wide-angle scanning photosedimentometer (WASP), the beam of light is split into two components which are passed to separate photocells. One of the beams, after being made parallel, passes through the sedimentation cell and is attenuated by the suspension; the other is a reference beam (figure 9.7).

The two photocells are connected in opposition and thence to a potentiometric recorder. The light flux falling on the reference photocell may be varied (zero set) to balance the electrical output of the two photocells; this zeros the pen recorder. Since the two beams of light originate from the same source, any variation in light intensity affects both photocells equally giving a stable zero. A sensitivity control is then adjusted so that the pen recorder registers full-scale deflection when no light falls on the measurement photocell. The pen recorder chart has an optical density scale to facilitate subsequent evaluation of the size distribution.

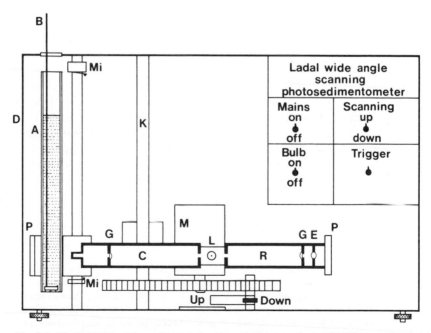

Fig. 9.7 The LADAL wide-angle scanning photosedimentometer (WASP). A, sedimentation tank; B, stirrer; C, collimator; D, light-proof box; E, variable aperture; G, lenses; L, light source; M, motor; K, drive screw; Mi, microswitches; P, photocells.

The sedimentation cell has a capacity of 150 cm^3, allowing a height of fall of up to 20 cm. With the scanning version, the suspension is scanned at a rate of 1 cm min^{-1} after a variable preset time in a static position. This enables accurate determination of the top size to be made together with a rapid complete analysis (usually less than 30 min). Operator time is limited to setting up the instrument and evaluating the results, the time for each of these being of the order of 5 min, although evaluation takes longer if a weight distribution rather than a surface distribution is required. A data-logging device in conjunction with an X-Y plotter gives the surface distribution directly.

Gracias *et al.* [65, 66] describe the application of a laser photosedimentometer to measure the size of water droplets in oil, calculating the extinction coefficient from Mie theory. They also discuss the application of a centrifuge to extend the size range.

9.4.5 Discussion

The particle size limits for photosedimentation techniques are those common to all sedimentation methods. A suitable upper limit for the EEL is a particle that will fall 2 cm in 30 s, whereas for the WASP a settling speed of 40 cm min^{-1} is possible. This is because the extinction readings at times less than 30 s are unreliable due to the effects of stirring. (For particles of s.g. 2.65 in water, these top sizes are 27.3 and 86 μm respectively, and may be increased four- to five-fold by using a more viscous liquid such as ethylene glycol.) The lower size limit is governed by the onset of convection currents at a falling speed of about 14 μm min^{-1}.

Limitation on top size also occurs due to deviations from streamline flow, the breakdown in Stokes' law leading to 5% errors in terminal velocity for particles 61 μm in size for the example above (450 μm in ethylene glycol). Corrections may be applied using Heywood's tables, but these may only partially correct since it is an inherent assumption that the particles fall in random orientation. If flow is not streamlined, the particles may orientate themselves to give maximum drag, thus reducing the cross-sectional area obstructing the light beam. It is, therefore, strongly recommended that this technique should not be used outside the laminar flow range.

It is often suggested that independent settling only occurs for the very low volume concentrations possible with this technique (less than 0.05%). Although not always true, repeat analyses using other sedimentation techniques at volume concentrations of 0.5 and 1.0% give reproducible results with easily dispersed solids. However, it is often necessary and always preferable to use as low a concentration as possible.

9.5 X-ray sedimentation

A natural extension to the use of white light is to use X-rays. In this case the X-ray density is proportional to the weight of powder in the beam

$$I = I_0 \exp(-BC) \qquad (9.22)$$

and D, the X-ray density is defined as:

$$D = \log_{10} (I/I_0) \qquad (9.23)$$

where B is a constant, D the X-ray density and C the concentration of powder in the beam.

Brown and Skrebowski [28] in 1954 first suggested the use of X-rays in an incremental sedimentation technique. This instrument was built by ICI and described by Nonhebel [38] and others [29]. In this ICI instrument a system is used in which the difference between the intensity of an X-ray beam which has passed through the suspension in one half of a twin sedimentation tank, and the intensity of a reference beam which has passed through an equal thickness of clear liquid in the other half, produces an imbalance in the current produced in a differential ionization chamber. This eliminates errors due to the instability of the total output of the source, but assumes a good stability in the beam direction. Since this is not the case the instrument suffers from a zero drift, which affects the results. A water-cooled X-ray tube is used for generation of 18 keV radiation, which, after traversing the tank, is detected in an ionization chamber. The chamber measures the difference in X-ray intensity in the form of an electric current which is amplified and displayed on a pen recorder. The intensity difference is taken as directly proportional to the powder concentration in the beam. The sedimentation curve is manually converted to a cumulative percentage frequency using this proportionality and Stokes' law.

Kalshoven in 1966 described an instrument [41], employing the X-ray absorption technique with a special programme for scanning the sedimentation tank. As the concentration measurement by means of X-rays is very rapid, it is possible to change the height of measurement during the analysis. In this instrument it is done in such a way that the concentration is recorded not as a function of time, but as a function of the Stokes diameter and the cumulative percentage frequency is directly obtained. The analysis is speeded up by scanning and can be as short as a few minutes. An X-ray tube is used as the X-ray source and a scintillation counter as the detector. The difference in intensity between the reference and measuring beams, in which the emitted beam is alternately split, is measured by a rotating wedge that automatically sets the difference to zero. Sub-micrometre particles can be measured if the sedimentation tank is spun in a centrifuge for some time, the time integral of the centrifugal force is measured and the tank is scanned after the centrifugation. The author claims that initial concentrations in the range 0.01 to 1% by volume can be used, depending on the atomic number of the analysed material. The experiments proved that the concentrations can be measured at short distances below the surface of the suspension without seriously affecting the results. When the centrifuge was used the results were independent of the time of centrifugation. No comparison analyses were presented. The principle was patented [42] in 1965.

Oliver, Hickin and Orr [40] in 1969 patented a gravitational X-ray particle size analyser that incorporated the absorption technique and improved the systems used by Kalshoven [41]. The same instrument was also reported by Hendrix and Orr in 1970 [47] and is now available commerically [48]. This instrument automatically presents

results as a cumulative percentage frequency, and the sedimentation tank is driven in such a way that the concentration is recorded directly as a function of the Stokes diameter. An air-cooled, low-power X-ray tube is used for generation of X-rays; these are collimated into a narrow beam and pass through 0.14 in (approx. 3.6 mm) thickness of suspension. The sedimentation tank, only 1.375 in (35 mm) high, is closed at the top, and, in use, completely filled with suspension. Filling and emptying of the tank is accomplished with a built-in circulating pump. The transmitted radiation is detected as pulses by a scintillation detector, amplified, discriminated to eliminate low-energy extraneous noise, clipped to a constant amplitude and fed to a diode pump circuit, which in conjunction with an operational amplifier with a diode feedback, gives a voltage proportional to the logarithm of the X-ray intensity and therefore proportional to the powder concentration. The instrument can analyse powders containing an element with atomic number higher than 13, but rather high initial volume concentrations of powder have to be used (0.5–3.0%). This is due to the need for the initial decrease in X-ray intensity, due to the powder in the suspension, to be greater than 20% of the intensity, with clean liquid. An absolute system is used here in which the initial intensity with clean liquid is first measured and the zero set, and the suspension is then introduced. This system assumes an excellent stability of the source, which is, in the case of X-ray tubes, a rather unreliable assumption. The authors [47] claim a good reproducibility and present several comparison analyses with microscopy, etc. The range of use is said to be 50 to 0.2 μm for most powders. The lower limit is unreal since it is generally accepted that gravitational sedimentation is limited to particle sizes in excess of 1 or 2 μm [62]. The instrument is a good piece of instrumentation engineering, but the results are affected by the necessary damping and the doubtfulness of its lower size limit. The acceptance by the manufacturers of this lower size has affected the design in that only a 35 mm fall-height is possible and this restricts the upper size limit [63]. A review of these and other methods is contained in a thesis by Svarovsky [45] and a paper by Svarovsky and Allen [46].

Allen and Svarovsky developed two versions of the LADAL [43] X-ray sedimentometer (figure 9.8), a gravitational [44] and a centrifugal [49] instrument. In both versions, gravitational and centrifugal, the X-ray beam is generated from a promethium 147/aluminium isotope source with a half-life of 2.6 years, an energy level of 22.6 keV and activity of 3 curies. A xenon-filled proportional counter at 1300 V is used to detect the beam. The gravitational version uses the X-ray absorption technique in the gravity field. The powder is dispersed in a liquid at a volume concentration of less than 1.0%, a necessary condition for free settling. An X-ray beam passes through the suspension at a known depth below the surface and the changing intensity of the transmitted beam is measured as a function of time. To speed up the analysis, after a preset time, the beam is made to scan up to the surface of the suspension. Typically, the analysis will begin with the beam 18 cm below the surface; it then scans to within 1 cm of the surface in 11 min. It is usual to maintain a fixed depth until the largest particles present in the suspension pass through the beam in order to define the top size accurately. For particles of density 3.0 g ml^{-1} in water, using a preset time of 3 min, a full analysis from 30 to 3.3 μm would take 14 min.

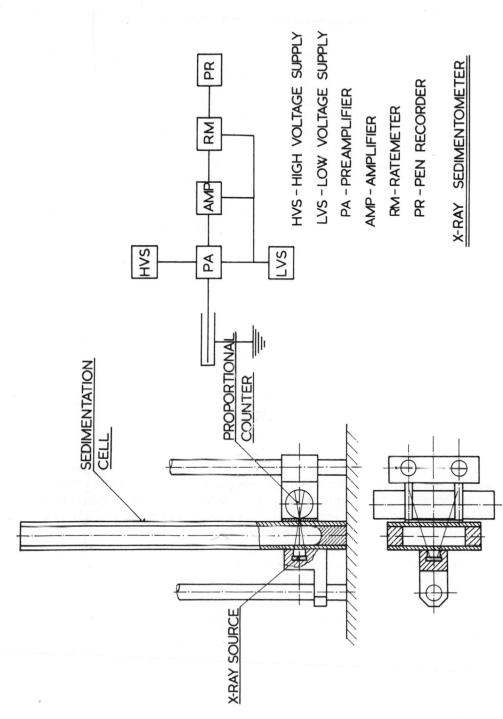

SEDIMENTATION CELL

PROPORTIONAL COUNTER

X-RAY SOURCE

HVS   PA   AMP   RM   PR

LVS

HVS - HIGH VOLTAGE SUPPLY
LVS - LOW VOLTAGE SUPPLY
PA - PREAMPLIFIER
AMP - AMPLIFIER
RM - RATEMETER
PR - PEN RECORDER

X-RAY   SEDIMENTOMETER

Fig. 9.8   Schematic diagram of the sedimentometer and block diagram of the electrical circuit.

Results are presented by a pen recorder as X-ray intensity against time. A typical curve is shown in figure 9.13 (p. 294). Conversion to cumulative percentage undersize by weight against Stokes diameter can be carried out manually in about 10 min or, more conveniently, by digital computer.

The following form of equation (9.22) is used:

$$BC = D = -\log \left(1 - \frac{I_c - I}{I_c}\right) \qquad (9.24)$$

where I_c is proportional to the pen recorder deflection with clear liquid in the sedimentation tank and I is the deflection with suspension in the tank. Adequate sensitivity is obtained by suppressing the zero and expanding the scale by x 1, x 3 or x 10 so that I_c may be made proportional to 20, 60 or 200 cm. The recorder trace suffers a small amount of instability due to statistical variation of the count level. This can be reduced by integrating the count over 3.3 or 10 s.

Improvements in isotope sources have led to the replacement of the 3 curie source by a long-life 100 millicurie source; the radiation emanating from this, after passing through the suspension, is measured by a scintillation counter of high counting efficiency. Two alternatives are available for data presentation; a pen recorder as before or a direct-plotting facility which permits direct evaluation of percentage undersize against Stokes diameter.

9.6 Hydrometers [11, 12]

The variations in density of a settling suspension may be followed with a hydrometer, a method widely used in the ceramic industry. The suspension is made up with a known amount of powder and thoroughly agitated, usually by shaking. The container is then placed in a thermostat and the change in density of the suspension at known depths recorded as the solid phase settles out. Some workers remove the hydrometer after each reading and replace it slowly immediately before the next, others object to the resulting disturbance and reshake the container after each reading. From the series of readings, the size distributions may be determined.

With the hydrometer immersed, its weight W, which is constant, equals the weight of suspension displaced. Let the length of stem immersed in the clear suspending liquid be L, i.e. the same as would be immersed in the suspension at infinite time, the length immersed in the suspension at the commencement of the determination be L_0, and the length immersed at time t be L_t; then:

$$W = V\phi(h_0, 0) + L_0 \alpha\rho_1 \qquad (9.25)$$

$$W = V\rho_f + L\alpha\rho_1 \qquad (9.26)$$

$$W = V\phi(h_t, t) + L_t\alpha\rho_2 \qquad (9.27)$$

where V is the volume of the hydrometer bulb, α the cross-sectional area of the stem and h_t the depth of the hydrometer bulb at time t.

From equation (9.5):

$$\phi = \frac{\phi(h, t) - \rho_f}{\phi(h, 0) - \rho_f}$$

Since the density of the suspension around the stem (ρ_1, ρ_f, ρ_2) varies negligibly compared with the variation in L, this may be written:

$$\phi = \frac{L_t - L}{L_0 - L} = \frac{w_t - w}{w_0 - w} \tag{9.28}$$

where w is the specific gravity marked on the hydrometer.

If the suspension is made up of W g of powder made up to 1 litre of suspension, equation (9.22) may be written:

$$\phi = \frac{1000}{W} \frac{\rho_s(\rho_t - \rho_f)}{\rho_s - \rho_f} \tag{9.29}$$

where ρ_t is the density of the suspension at the given time t.

An equivalent formula for powders which are present as a slurry in water removes the necessity for drying out the slurry. A specific gravity bottle is filled with water and weighed and the water is then replaced by the sample under test and reweighed, the difference in weights being ΔW. The sample is then taken out of the bottle and used for the analysis. The equivalent formula is:

$$\phi = \frac{1000(\rho_t - \rho_f)}{\Delta W} \tag{9.30}$$

With the hydrometer technique, both density and depth of immersion vary with each reading. If the temperature is maintained constant at the calibration temperature for the hydrometer, the determination of the former presents no problems since it may be read directly from the hydrometer stem. If the temperature varies, corrections have to be applied for thermal expansion and misreading of the meniscus [13].

The chief difficulty lies in determining the point of reference below the surface to which this density refers, for when the hydrometer is put into the suspension the level rises in the container, thus giving a false reference point (figure 9.9). If the cross-sectional area of the container is A, the depth to be used in Stokes' equation, from geometrical consideration is [14]:

$$L = L_1 + \tfrac{1}{2}L_2 - \frac{1}{2}\frac{V}{A}$$

This simple formula has been challenged by several workers who claim that corrections have to be applied for the density gradient about the bulb and the displacement of suspension by the stem. Johnson [8], for example, gives:

$$L = L_1 + \tfrac{1}{2}L_2 - \frac{1}{2}\frac{V}{A} - 0.5 \text{ cm}$$

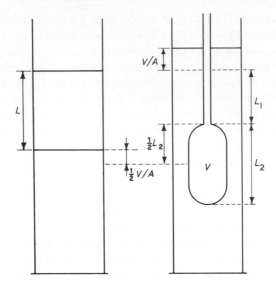

Fig. 9.9 Depth of immersion using a hydrometer.

In order to achieve sufficient accuracy in the specific gravity readings, it is necessary to use a concentration of at least 40 g l^{-1}, which is well into the hindered settling region. The only justification for this that has been advanced is that the method gives reproducible results.

Most hydrometers are calibrated to be read at the bottom of the meniscus and this is usually not possible in a suspension. The readings are therefore taken at the top of the meniscus, and an experimentally determined correction, which is usually of the order of 0.003 g ml^{-1}, applied.

It is usual to disregard calibration errors, although these may be substantial.

Good-quality hydrometers are usually guaranteed to ± 0.0005 g ml^{-1}, which corresponds to a percentage error of around ± 1.5% under normal operating conditions. Johnson [8] recognized this error and suggested that the error should be determined at several points by calibration in a series of dilute suspensions of common salt. A correction to meniscus reading error and density should also be applied if a wetting agent is used.

Although the hydrometer cannot be recommended as an absolute instrument, it is useful for control work. It is of little use with discontinuous size distributions, since these give sharp boundaries in the settling suspension, which lead to peculiar results.

9.7 Divers

An extension of the hydrometer technique which overcomes many of the objections to it is to use miniature hydrometers, which are completely immersed in the suspension. The method was developed by Berg [15] who called the miniature hydrometers 'divers'

(figures 9.10(a) and (b)). The first type was designed for gravitational and the second for centrifugal sedimentation. Once the diver had been constructed, its density was brought as close as possible to the required density by adding mercury. The bulb was then sealed and its density adjusted to the required density by etching with hydrofluoric acid. During the analysis, the diver was inserted into the suspension and sank to the level at which the density of the suspension was equal to its own. A number of divers are required since each one gives only one point on the size distribution curve. Since the divers were not visible in the suspension, Berg located them by drawing them to the side of the sedimentation tube with a magnet. This method of location has the disadvantage that the magnet may displace the diver in a vertical as well as a horizontal direction.

Jarrett and Heywood [16] located their divers (figures 9.10(c) and (d)) with an alternating-current inductance resistance bridge; the balance was destroyed when a thin search coil surrounding the sedimentation tube was moved past the diver. In the roof diver technique which they developed (figure 9.10(d)), the diver is placed at the foot of a vertical column, the locating device being placed just below it. The diver is held on to the column with a magnetic rod which is afterwards withdrawn. Due to the excess press-

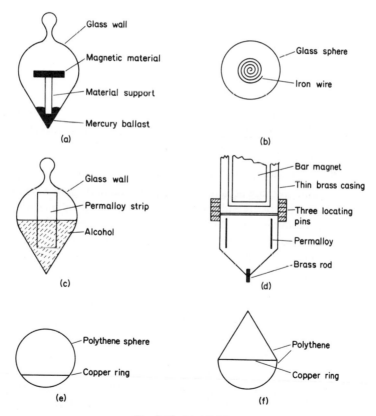

Fig. 9.10 (a)–(f) Divers.

ure, it remains in position until the density of the surrounding suspension equals its own and it then falls. Jarrett and **Heywood** found it necessary to vibrate the vertical column to prevent sticking and this is obviously a disadvantage.

Johnson [17] used six divers, consisting of small sealed bulbs containing iron powder, and located them using a magnet. He found a density range 1.0242 to 1.004 g ml⁻¹ satisfactory for ceramic materials.

The problem with all the above divers is that they are difficult to construct and calibrate. These difficulties were overcome by Kaye and James [18], who used polythene spheres of 0.6 cm diameter, containing copper rings (figures 9.10(e) and (f)). These were located with a thin search coil fed with high-frequency alternating current. When the plane of the search coil coincided with that of the copper ring, there was an increase in the power consumption which was recorded by **pen recorder**.

Kaye and James showed that the diver fell to a greater depth than its density level because of its kinetic energy. It should then have oscillated about its density level with damped simple harmonic motion but did not, due to the picking up and shedding of powder. This fault is confined to divers that are asymmetrical about their centres of gravity, which are therefore unsuitable for density determination. Polythene spheres coated with copper on graphite are symmetrical and develop a continuous spin giving reproducible satisfactory results.

9.8 The specific gravity balance

The changes in density within a settling suspension may be followed by using a specific gravity balance [19]. This instrument comprises a bob on each arm of a beam balance, one bob being immersed in clear suspension fluid and the other in the suspension. The depth of immersion of the bobs may be varied, the change in buoyancy being counterbalanced by means of solenoids which are connected to a **pen recorder**. From the trace on the **pen recorder**, the particle size distribution may be determined.

9.9 Appendix: worked examples

9.9.1 Wide-angle scanning photosedimentometer: analysis of silica

Horizontal scales

Time: recorder chart travels at 1 cm min⁻¹.

Depth: initially 18 cm. After 16 min, delay scanning commences at a rate of 1 cm min⁻¹.

Diameter: Stokes' diameter evaluated by:

$$d_{Stk} = 175 \sqrt{\left(\frac{0.01 \times 18h}{(2.60 - 1.0)\, 18T} \right)}$$

$$= 58.7 \left(\frac{h}{18\,T} \right)^{1/2}$$

Vertical scales
D is optical density, pen recorder scale, cumulative
S is D converted to a percentage, the surface undersize by weight.

Evaluation

Time (min)	1	2	4	8	16	20	24	28	32
Height of fall (cm)	18	18	18	18	18	14	10	6	2
Stokes' diameter (μm)	58.7	41.5	29.4	20.8	14.7	11.6	8.9	6.4	3.45
Optical density D	0.495	0.475	0.407	0.321	0.240	0.194	0.152	0.172	0.060
Cumulative % undersize by surface D (%)	100	96	82.2	64.8	48.5	39.2	30.8	22.6	12.1
Average diameter d	45.7	35.5	25.1	17.7	13.2	10.2	7.65	4.9	1.7
ΔD		0.20	0.68	0.86	0.81	0.46	0.42	0.40	0.52	0.60
ΔDd		9.1	24.2	21.6	14.3	6.1	4.3	3.1	2.5	1.0
Cumulative % undersize by weight ΔDd (%)	100	89.6	61.5	36.4	19.7	12.7	7.7	4.1	1.1

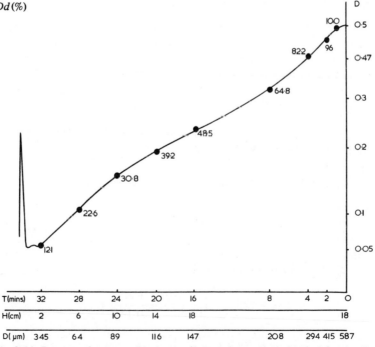

Fig. 9.11 Pen recorder trace of a photosedimentation analysis (optical density D against time T).

9.9.2 *Conversion from surface distribution to weight distribution*

Conversion may be carried out by a tabular method as illustrated above (method 1). An improvement on this is to tabulate with data from the surface distribution curve; this permits smoothing out of experimental variations and extrapolation (method 2). For reliable cumulative undersize weight results, the top size needs to be accurately known since small errors in top size can lead to gross errors when the integration is performed (i.e. when the area ΔDd is evaluated). Such accuracy is not generally required at the lower end of the size spectrum and the analysis may often be terminated when 20% or so by surface remains unmeasured. The most rapid and accurate method is to perform the integration graphically, by counting squares for example (method 3). The results from all three methods are illustrated graphically (figure 9.12).

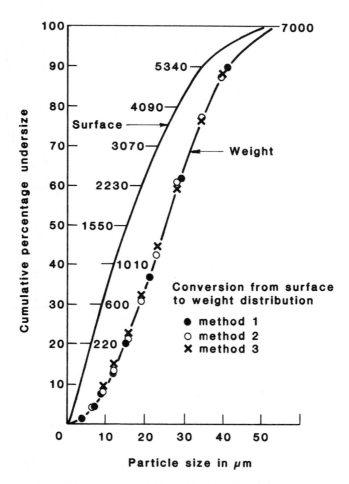

Fig. 9.12 Size distribution using the wide-angle scanning photosedimentometer (WASP). Counting squares, e.g. the area ΔDd between the 20% and 30% surface undersize, is 380 squares.

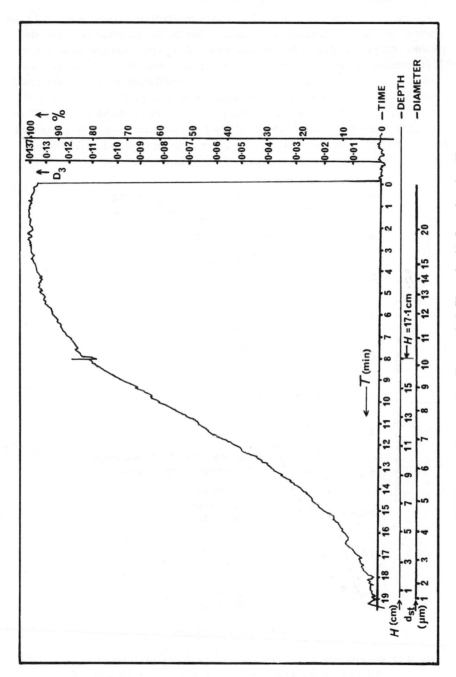

Fig. 9.13 Pen recorder trace of an X-ray analysis (X-ray density D_3 against time T).

9.9.3 The LADAL X-ray sedimentometer: analysis of tungstic oxide

Horizontal scales

Time: recorder chart travels at 1 cm min^{-1}.

Depth: initially 17.1 cm. After 8 min delay, scanning commences which causes a kick on the chart.

With a scanning rate of 1.515 cm min^{-1}, the decrease in depth is evaluated.

Diameter: Stokes' diameter evaluated by

$$d_{Stk} = 175 \sqrt{\left(\frac{0.01 \times 17.1h}{(7.76 - 1.00)\,17.1T} \right)}$$

$$= 27.8 \sqrt{\left(\frac{h}{17.1\,T} \right)}$$

Vertical scales

D_3 is the X-ray density which may be converted to cumulative percentage undersize by weight using equation (9.24): e.g.

Maximum deflection L_{max} = 16.2 cm FSD = 60 cm

$$D_{max} = BC(h, 0) = -\log\left(1 - \frac{16.2}{60}\right) = 0.137$$

Evaluation

Time (min)	2	2.8	4	5.6	8	10	12	14	16	18
Height of fall (cm)	17.1	17.1	17.1	17.1	17.1	14.0	11.0	8.1	5.1	2.0
Stokes' diameter (μm)	2.0	17.5	14.6	12.3	10.0	8.4	6.8	5.3	3.9	2.3
X-ray density D	0.137	0.137	0.135	0.128	0.109	0.084	0.0555	0.0274	0.0092	0.0066
Cumulative % under- size $D\%$ by weight	100	100	98.5	93.5	79.5	61.5	40.5	20.0	8.7	4.8

References

1 States, M.N. (1939), *Proc. Am. Soc. Test. Mat.*, **39**, 795.
2 Heywood, H. (1947), *Symp. Particle Size Analysis, Inst. Chem. Engrs.*, 114.
3 Robinson, G.W. (1922), *J. Agric. Sci.*, **12**, 3, 306–21.
4 Andreasen, A.H.M. (1928), *Kolloid Beith.*, **27**, 405.
5 Andreasen, A.H.M. (1939), *Ingen. Vidensk. Skr.*, 3.
6 BS 3406 (1980), *The determination of particle size of powders*, Part 2.

7 Andreasen, A.H.M. and Lundberg, J.J.V. (1930), *Ber. Dt. Keram. Ges.*, **11**, 5, 312–23.
8 Johnson, R. (1956), *Trans. Ceram. Soc. B*, **55**, 237.
9 Leschonski, K. (1962), Vergleichende Untersuchungen der Sedimentation Analyse, *Staub*, **22**, 11, 475–86.
10 Heywood, H. (1938), *Proc. Inst. Mech. Engrs.*, **140**, 27.
11 BS 1377 (1961), Methods of Testing Soils for Civil Engineering Purposes.
12 ASTM 1272 (1961), Part 4.
13 Orr, C. and Dallavalle, J.M. *Fine Particle Measurement*, Macmillan, NY, p. 64.
14 Bauer, E.E. (1937), *Engng News-Rec.*, No. 118, McGraw-Hill, p. 662.
15 Berg, S. (1940), *Ingeniorvidenskab*, Skrifter, 2, Danish Acad. Tech. Sci., Copenhagen.
16 Jarrett, B.A. and Heywood, H. (1954), *B.J. appl. Phys.*, suppl. No. 3, S21.
17 Johnson, R. (1955), *Trans. Ceram. Soc. B*, **55**, 237.
18 Kaye, B.H. and James, G.W. (1962), *B.J. appl. Phys.*, **13**, 415.
19 Suito, E. *et al.* (1964), *J. Soc. Mat. Sci. Japan*, **13**, 133, 825.
20 Berg, S. (1958), *Symposium on Particle Size Measurement*, ASTM Sp. Publ. No. 234, p. 143.
21 Alex, W. (1970), *G.I.T.*, **11**, 5, 637–40, 643–6.
22 Grinrod, P.S. (1970), ASTM Sp. Tech. Publ. No. 473, 45–70.
23 Grandillo, A.D. (1970), *J. Powder Metall.*, **6**, 1, 3–16.
24 Joos, P., Rvyssen, R. and Haners, Y. (1970), *J. Pharm. Belg.*, **25**, 2, 133–51.
25 Allen, T. (1967), Ph.D. thesis, Univ. of Bradford.
26 Allen, T. (1968), *J. Powder Technol.*, 132–40.
27 Allen, T. (1968), *ibid.*, 141–53.
28 Brown, J.F. and Skrebowski, J.N. (1954), *B.J. appl. Phys.*, suppl. No. 3, S27.
29 Conlin, S.G. *et al.* (1967), *J. Sci. Instrum.*, 44, 606–10.
30 Jarrett, B.A. and Heywood, H. (1954), *B. J. appl. Phys.*, suppl. No. 3, S21.
31 Morgan, V.T. (1954), *Symp. Powder Metallurgy.*, Iron and Steel Institute, preprint Group 1, 38–43.
32 Vouk, V. (1948), Ph.D. thesis, London Univ.
33 Lewis, P.C. and Lothian, G.F. (1954), *B. J. appl. Phys.* suppl. No. 3, S571.
34 Rose, H.E. and Lloyd, H.B. (1946), *J. Soc. Chem. Ind.*, **65**, 52.
35 Rose, H.E. (1946), *ibid.*, **65**, 65.
36 Rose, H.E. (1952), *J. appl. Chem.*, **2**, 80.
37 Allen, T. (1962), M.Sc. thesis, London Univ.
38 Nonhebel, G. (ed.) (1964), *Gas Purification Processes*, Newnes.
39 Kratohvil, J.P. (1964), *Ann. Chem.*, **36**, 35, 485R.
40 Oliver, J.P., Hickin, G.K. and Orr, C. (1969), US Patent 3, 449 567
41 Kalshoven, J. (1967), *Conf. Proc. Particle Size Analysis*, Soc. Analyt. Chem., London.
42 Kalshoven, J. (1965), BP. 1 158 338.
43 Allen, T. (1970), BP Appl., 1764/70; (1971), US Pat. Appl. 106 013.
44 Allen, T. and Svarovsky, L. (1970), *J. Phys. E.*, **3**, 458–60.
45 Svarovsky, L. (1972), Ph.D. thesis, Univ. Bradford.
46 Svarovsky, L. and Allen, T. (1973), Paper presented at Heywood Memorial Symp., Univ. Loughborough.
47 Hendrix, W.P. and Orr, C. (1970), *Conf. Proc. Particle Size Analysis*, Soc. Analyt. Chem., London.
48 Sedigraph. See list of Manufacturers and Suppliers, Appendix 2.
49 Allen, T. and Svarovsky, L. (1974), *J. Powder Technol.* **10** (1/2), 23–28.
50 Jennings, D.S., Thomas, M.D. and Gardner, W. (1922), *Soil Sci.*, **14**, 485–99.
51 Krauss, G. (1923), *Int. Mitt. Bodenk.*, **13**, 1, 2, 147–60.
52 Koettgen, P. (1927), *Z. Pflnern. Düng Bodenk.*, **9**, 35–46.
53 Köhn, M. (1928), *Landw. Jahrb.*, **67**, 1, 485–546.
54 Lehman, H. (1932), *Sprechsaal Keram. Glas. Email*, **65**, 36.
55 Lorenz, R. (1932), *Ber. dt. Keram. Ges.*, **13**, 3, 124–39.

56 Kast, W. (1960), *Staub*, **20**, 8, 253–66.
57 Report from the abrasives laboratory of the MSO-Maschinen und Schleifmittelwerke AG (1954), *Sprechsaal Keram. Glas. Email*, **87**, 19.
58 Kuncewicz, L. and Krzyzewski, Z. (1960), *Staub*, **20**, 2, 47–48.
59 Joos, E. (1954), *Staub*, **35**, 18–34.
60 Oden, S. (1920), *Kolloid Z.*, **26**, 1, 100–21.
61 Gille, F. (1952), *Zement-Kalk-Gips*, **5**, 10, 309–14.
62 Allen, T. and Baudet, M.G. (1977), *Powder Technol.*, **18**, 2, 131–8.
63 Borothy, J. (1975), *Chimia*, **29**, 5, 240–2.
64 Pavlik, R.E. and Sansone, E.B. (1973), *Powder Technol.*, **8**, 159–64.
65 Gracias, A. *et al.* (1975), *C.R. hebd. Séances Acad. Sci.*, Ser. B, **281**, 23, 595–8.
66 Lachaise, J., Martinez, A. and Gracias, A. (1977), *Conf. Proc. Particle Size Analysis*, (ed. M.J. Groves) (1978), Chemical Soc., Analyt. Div., Heyden.

10 Cumulative methods of sedimentation size analysis

10.1 Introduction

There are two approaches to the determination of the size distribution of a sedimenting suspension. The incremental method involves the determination of the rate of change of density or concentration with time or height or both. In the cumulative method the rate at which the powder is settling out of suspension is measured. In either of these techniques, line-start or homogeneous suspensions may be used. Usually it is preferable to use incremental methods since analyses can be carried out more rapidly by these methods. The big advantage of cumulative techniques is that the amount of powder required is small (about 0.5 g), which reduces interaction between particles to a minimum. This is particularly useful when only a small quantity of powder is available or when one is dealing with toxic materials.

10.2 Line-start methods

If the powder is initially concentrated in a thin layer floating on the top of the suspending fluid, the size distribution may be directly determined by plotting the fractional weight settled against the free-falling diameter of the particles. Marshall [1] was the first to use this principle. Eadie and Payne [2] developed the micromerograph, the only method in which the suspending fluid is air.

The Werner [3] and Travis [4] methods also operate on the layer principle but utilize a liquid suspension on top of clear liquid. These methods have found little favour since the basic instability of the system, a dense fluid on a less dense fluid, is responsible for what is commonly known as 'streaming'; some of the suspension settles *en masse* in the form of pockets of particles which fall rapidly through the clear liquid leaving a 'tail' of particles behind (see [5, p. 78]).

Whitby eliminated this fault by using a clear liquid with a density greater than that of the suspension. He also extended the range of the technique by using centrifugal settling for the finer fraction. A description of the apparatus and method is given by Whitby [6, Pt. 1] with procedures [6, Pt. 1] and applications [6, Pt. 2]; Whitby *et al.* [7] describe the equipment in use. The apparatus has been commercialized under the name MSA particle size analyser.

An objection that can be levelled at all these methods, apart from the micromerograph, is that the amount settled is determined by the height of the sediment. As the settled volume is not independent of particle size this introduces errors unless a correction is applied.

The line-start technique has also been used to size and fractionate UO_3 particles [8]. The weight that had sedimented out was determined by a device to measure the radioactivity at the bottom of the tube and the settled powder was washed out at fixed time intervals without disturbing the settling suspension.

10.3 Homogeneous suspensions

The principle of this method is the determination of the rate at which particles settle out of a homogeneous suspension. This may be determined by extracting the sediment and weighing it, allowing the sediment to fall on to a balance pan or determining the weight of powder still in suspension by using a manometer.

The theory outlined below was developed by Oden [9] and later modified by Coutts and Crowthers [10] and Bostock [11].

If one considers a distribution of the form $W = f(D)$ where W is the percentage having a diameter greater than D, the weight per cent P which has settled out at time t is made up of two parts; one consists of all the particles with a free-falling speed greater than that of D_t as given by Stokes' or some related law, where D_t is the size of particle which has a velocity of fall h/t and h is the height of suspension; the other consists of particles smaller than D_t which have settled because they started off at some intermediate position in the fluid column. If the free-falling velocity of one of these smaller particles is v, the fraction of particles of this size that have fallen out at time t is vt/h giving:

$$P = \int_{D_t}^{D_{max}} f(D)dD + \int_{D_{min}}^{D_t} \frac{vt}{h} f(D)dD$$

By differentiating with respect to time and multiplying by t:

$$t \frac{dP}{dt} = \int_{D_{min}}^{D_t} \frac{vt}{h} f(D)dD$$

i.e.

$$P = W + t \frac{dP}{dt} \tag{10.1}$$

where W is the percentage oversize D_t.

Since P and t are known, it is possible to determine W using this equation. It is preferable, however, to use the form of equation (10.1) suggested by Gaudin, Schumann and Schlechter [12].

$$W = P - \frac{dP}{d \ln t} \tag{10.2}$$

Several methods have been suggested for the determination of W using these equations. The most obvious is to tabulate t and P, hence derive dP, dt and finally W.

Alternatively, P may be plotted against t and tangents drawn. A tangent drawn at point (P_1, t_1) will intercept the abscissa at W_1, the weight oversize t_1. This method is particularly useful when pen recorders are used in association with a cumulative method and the data presented as graphs of P against t.

Another tabulation method suggested by Stairmand [13] (see appendix) is dependent on tabulating P against t at times such that the ratio of (t/dt) remains constant, i.e. a time interval in a geometric progression. In order to smooth out experimental and operator errors it is probably better to plot P against t, determining the values of P from the smoothed curve. Stairmand's method is illustrated in Table 10.1. A similar method has been proposed by Kim et al. [71].

Many powders have a wide size distribution and, in such cases, the time axis on a linear plot tends to become cramped at the lower end or unduly extended; the use of

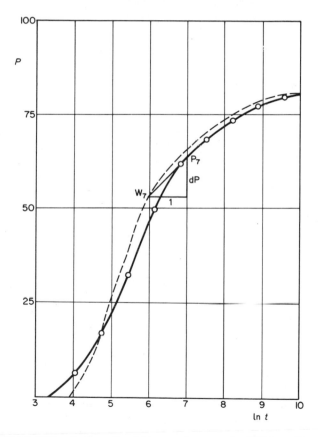

Fig. 10.1(a) Determination of weight percentage oversize (W) from the graph of weight of powder sedimented (P) against ln t by Bostock's method of drawing tangents.

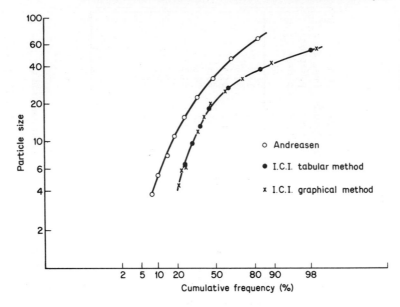

Fig. 10.1(b) Frequency distribution using ICI apparatus using Stairmand's method and also using the theoretical equation. A comparison is made with an Andreasen analysis.

equation (10.2) is then recommended. Evaluation proceeds from a plot of P against ln t; one can then plot $dP/\ln t$ on the same scale, measuring the difference between the abscissae at specific times [6]. The determination is simplified by using a special protractor devised by Edwald [14].

A far less cumbersome technique is to draw tangents. This has been further simplified by Bostock [11] who differentiates graphically every half-unit of ln t; the point where the tangent cuts the ordinate line one ln t unit less than the value at which it is drawn gives the weight per cent oversize W at that value. These methods are illustrated in Table 10.1 and figure 10.1.

The cumulative method of size analysis has been criticized by Nissan [15] and Jarrett and Heywood [16] because of the need for differentiation. Donoghue [17], however, states that these criticisms are baseless, particularly if the weight of sediment is determined at times increasing by a factor of not more than two.

10.4 Sedimentation balances

The great disadvantage of balance techniques is the time required to carry out an analysis since it is necessary to wait until most of the powder in suspension has settled out on to the balance pan. On the credit side, the amount of powder required is small (about 0.5 g usually) and this is important if one is dealing with noxious or toxic material or if the powder is available only in small quantities; the method can be easily automated.

Table 10.1 Illustration of various methods of determining the cumulative percentage oversize by weight from cumulative weight deposited data.

	Experimental data			(a) Stairmand's method (see Table 10.2)			Experimental data		(b) From the graph of P versus ln t (equation (10.2))			(c) Andreasen analysis	
(1) Extraction time (s)	(2) Stokes' diameter (μm)	(3) Incremental weight of sediment (g × 10⁻⁵)	(4) Weight in grade	(5) Percentage in grade	(6) Percentage undersize	(7) ln t	(8) P (%)	(9) ln t	(10) W	(11) Particle size (μm)	(12) Particle size (μm)	(13) Cumulative percentage oversize
57	75	4333			100	4.043	6.5	5.0	1.7	53.7	65.3	15.9
114	53	7064	1243	1.9	98.1	4.736	17.1	5.5	10.0	41.7	45.7	34.9
228	37.5	10154	8754	13.1	85.0	5.429	32.2	6.0	26.0	32.7	31.9	41.1
456	26.5	11554	15107	22.8	62.2	6.123	49.8	6.5	39.5	25.1	22.3	65.6
912	18.8	8001	11528	17.4	44.8	6.816	61.8	7.0	53.2	19.9	15.7	74.8
1824	13.3	4473	5566	8.3	36.5	7.509	68.1	7.5	59.7	15.8	10.9	81.7
3648	9.4	3382	4402	6.6	29.9	8.202	73.7	8.0	65.4	12.1	7.7	85.8
7296	6.6	1625	3097	4.7	25.2	8.895	77.2	8.5	69.4	9.5	5.4	89.6
14592	0	13523*	16773	25.2		9.588	79.6	9.0	73.8	7.4	3.8	91.7
								9.5	76.8	5.8		
Total		66470	66470					10.0	78.7	4.5		

* Extrapolated value.

Columns 6 and 2, 11 and 10 and 13 and 12 are presented graphically in figure 10.1(b).

Theoretically the first value in col. 3 cannot be greater than half the second. However, in the theory a cylindrical tube is assumed and the reduction in cross-sectional area due to the conical portion reduces the weight sedimented at the commencement of the analysis.

The tabular solution (col. 6) agrees well with the graphical solution (col. 10), but these disagree with the Andreasen analysis (col. 13). This is due to the shape of the container which results in an underweighting of the coarse fraction and to loss of powder (ca. 12% in this instance). This loss is added to the fines and biases the analysis in favour of the fines.

Particle size distribution measurements can be made in liquids or gases using either a homogeneous or line-start technique. The use of an analytical beam balance with counterbalancing as sedimentation proceeded was first described by Oden [9, 18]; the first automatic recording sedimentation beam balance was described by Svedberg and Rinde [20]. As sediment collected on the pan, electrical points made contact and current was fed to a motor which acted on the other arm to restore equilibrium. The time–current plot was automatically recorded to be later converted into a time–weight plot. Other instruments have also been described [21-24]. In an ideal system all the powder initially in suspension will eventually settle on to the pan. If the pan is situated in the suspension, allowance has to be made for the particles settling between the rim of the pan and the sides of the sedimentation column, and the particles originally below the pan at the start of the run. The density unbalance as particles settle out from under the pan results in particles streaming over it [10]; this is a size-selective phenomenon, since fine particles are more likely to be lost than coarse ones, hence difficult to correct for [9, 25, 77, 78]. With balance systems using counterpoising, such as the Shimadzu and Sartorius, this effect is aggravated by the pumping action resulting from the periodic movement of the pan.

It is necessary to know the final weight to be expected after complete sedimentation since it is rarely possible to prolong a run until all the suspension has settled out. Particle size measurements above 50 or 60 μm are generally impossible with this type of instrument [64].

Weighing errors also occur as a result of air bubbles trapped on the balance pan, the formation of a meniscus at the pan suspension [30], evaporation and the error due to zeroing with clear liquid [64].

10.4.1 The Gallenkamp balance

In 1942 Gaudin *et al.* [70] proposed placing the balance pan below a sedimentation cylinder with an open bottom, the internal diameter of the pan being larger than the external diameter of the cylinder. The whole arrangement is placed in a second vessel filled with sedimentation liquid. During a run the suspension is located only above the balance pan in the sedimentation cylinder. This arrangement was adopted by Gallenkamp [11, 32].

In the Gallenkamp sedimentation balance (figure 10.2), the weight settled is read at chosen intervals of time directly from the deflection of the torsion wire. The instrument has been automated by the makers by the incorporation of a camera which photographs a scale reading of deflection at fixed intervals of time. A more elegant way of automating this balance was suggested by Ames *et al.* [33], who fastened the core of a linear variable differential transformer to the balance beam. The output of the transformer was amplified and recorded and the weight-sedimented time-curve derived from the data. A similar modification has also been described by Topham [76].

One disadvantage of this design is the method of filling with suspension prior to an analysis: above the sedimentation column there is a premixing tube from where the suspension flows into the cylinder after isolation from the atmosphere and the opening

of two pinch clips. Some powder is retained in the premixing tube; this loss can be reduced by coating with a film of silicone water repellent. It is not possible to clamp the balance pan tightly against the sedimentation column and when the vessel is filled, about 15 cm^3 of the 225 cm^3 charge overflow into the solids-free external vessel.

The balance system tends to stick; tapping with a pencil prior to taking a reading can cause the pointer to move one or two divisions. Some balances are found to have non-linear scales; this may be checked by placing ball-bearings on the pan under operating conditions with water replacing the normal suspension [66]. Although this system incorporates most of the requirements for a good sedimentation balance, its design could be greatly improved.

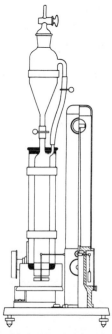

Fig. 10.2 Bostock's torsion balance (Gallenkamp).

To determine the weight of powder still in suspension at the end of the experiment the makers recommend a parallel experiment in which the supernatant is removed at a time and height to give an equal size to that settling at the termination of the balance run. An alternative method suggested by the manufacturers is discussed and rejected by Leschonski and Alex [64].

A breakdown of powder losses, by Allen [66], is given below.

	grams	%
Residue on balance pan	0.4277	85.5
Overflow into clear reservoir	0.0336	6.7
Fines still in suspension	0.0200	4.0
Particles adhering to premixing tube	0.0173	3.5
Weight unaccounted for	0.0014	0.3

These values are in very good agreement with those of other workers [64, 76]. According to BS 3406 these losses should be assumed representative; however agreement with the Andreasen technique is better if the losses are assumed coarse.

10.4.2 The Sartorius balance

The recording Sedibal or Sartorius balance was developed by Bachman and Gerstenberg [28, 77–79]. It is relatively easy to operate, errors in initiating a run are easily detected and rectified and duplicate runs are easily carried out by resuspension of the original sediment. In this instrument, when 2 mg of sediment have deposited, electronic circuitry activates a step-by-step motor which twists a torsion wire to bring the beam back to its original position. A pen records each step on chart paper. The balance pan is situated within the suspension and the manufacturers recommend that an 8% loss be assumed to account for particles settling between the rim of the pan and the sides of the column and for particles below the level of the pan at the commencement of a run. Leschonski [56, 64] however reports losses of 10% to 35% depending on the fineness of the powder and attributes the difference between these values and the expected 8% loss to the pumping action of the pan. For 250 steps and a pan area of about 23 cm^2, about 8.6 cm^3 are pumped to and fro.

Leschonski [56] proposed a modification so that the balance pan was situated at the bottom of the sedimentation column but was otherwise surrounded by clear liquid (figure 10.3). This eliminated powder losses and resulted in much more accurate analyses as confirmed by other workers [57, 60]. This improvement results in a considerable loss in ease and simplicity of operation and makes it impossible to carry out repeat runs on the same sample. Furthermore, the actual loss for coarse powders is close to 8% and

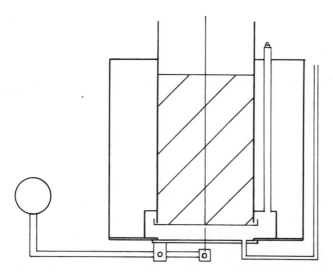

Fig. 10.3 Schematic representation of the Leschonski modification of the Sartorius balance.

the standard method gives results of sufficient accuracy, making it unnecessarily inconvenient to use the Leschonski modification for every sample [59].

Scott and Mumford [59] show that the weight loss is 9.3% at 32 μm rising to 28.8%
at 4.5 μm and suggest that the modified instrument is only necessary for powders finer
than 10 μm. They suggest the following correction procedure, first proposed by Harris
and Jowett [61], for the unmodified balance which they show gives reasonable agreement with results using the modified balance.

Assume that the final approach to the unknown asymptote represented by complete
sedimentation is a simple first-order exponential function:

$$W_t = W_\infty [1 - \exp(-\alpha t)] \tag{10.3}$$

where W_t is the weight collected at time t, and W_∞ is the weight expected at infinite
time; then, as shown by Mengelsdorf [62], a plot of $W_{(t + \Delta t)}$ against W_t with constant
Δt should give a straight line which intersects the 45° line at the point (W_∞, W_∞), the
value of the correct end-weight.

This technique should not be regarded as a panacea for all the ills of current balance
techniques since it is necessary that the losses be representative for the correction to
apply. It is, therefore, recommended that whenever possible accurate experimental
results should be sought.

Improvements to the electronics of the Sartorius balance have also been suggested
[63, 64]. The instrument is described in detail by Friedrich [29] and Rose and
Langmaid [72]; a critical comparison with the Andreasen technique has been carried
out by Ioos [30] and a comparison with the Coulter and Fisher subsieve sizer has been
carried out by Grandillo [65].

10.4.3 The Shimadzu balance

One automatic recording beam balance is sold under the name Shimadzu [25] (figure
10.4). In this balance, as sediment falls on to one pan, electrical points make contact
and operate a solenoid which in turn operates a ratchet. As the ratchet turns, ball-
bearings fall into the other pan, thus causing the electrical circuit to be broken. A subsidiary feed from the ratchet leads to a pen on a recording drum which yields a plot of
sediment against time (P versus t).

The Shimadzu suffers from an inertia in reading due to the excess pressure required
to activate the electrical contact which energizes the ball-bearing release and pen
recorder set-up; this has the effect of moving the resulting distribution curve towards
the fines. Other sources of error are due to the pan being in the suspension and the
pumping action due to rebalancing; the former leads to streaming and powder loss
errors as discussed earlier and the latter to a general instability in the system. Although
reproducible results may be obtained with the Shimadzu instrument, the size distribution is finer than that obtained by other methods due to the reasons mentioned
earlier. The high concentration required (1 to 5% by weight) with the apparatus may

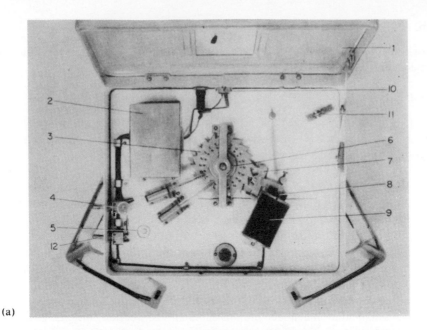

(a)

(b)

Fig. 10.4(a), (b) The Shimadzu sedimentation balance. (1) Case top. (2) Transformer. (3) Rotary disc. (4) Steel ball reservoir (L). (5) Steel ball reservoir (S). (6) Pulley wheel. (7) Ratchet. (8) Iron piece for magnet. (9) Electromagnet. (10) Receptacle for source. (11) Intermediate pulley wheel. (12) Switch. (13) Contact point. (14) Contact point. (15) Stopper. (16) Right pan. (17) Ball receptacle. (18) Balancing weight. (19) Knob for changeover to clockwork. (20) Clockwork drum. (21) Slide pole. (22) Pen holder. (23) Pen. (24) Sedimentation bottle. (25) Sedimentation pan. (26) Balancing weight. (27) Levelling screw. (28) Door.

cause hindered settling which will also act in such a way as to give a finer distribution than actually exists.

10.4.4 Other balances

Palik [31] designed a torsion-type balance, in which alteration in pan height due to particles settling on it is recorded by light and photocell systems. Special precautions are proposed to minimize the error due to flow of liquid of lower density below the pan to its upper region which interferes with the free vertical fall of the particles.

In 1954, Kiffer [26] described a continuous weighing chain-link balance. Rabatin and Gale [27] developed a simple spring balance operated in conjunction with photocells and shutter mechanism to intercept a beam of parallel light. Additional damping circuits were incorporated to prevent excessive oscillations of the system. This instrument needs frequent recalibration to compensate for ageing of the spring.

The manufacturers of the Cahn micro-balance have made available an accessory to convert it into a sedimentation balance. This is described in detail by Kaye and Davies [74] and Leschonski and Alex [64]. The vessel has a weighing balance pan immediately below the sedimentation cylinder and this eliminates convection currents. Leschonski and Alex state that it is impossible to stir the suspension before the start of the experiment. They therefore used a Leschonski-modified Sartorius balance with the Cahn balance system and report that the results are always coarser than those obtained with the Sartorius. This they attribute to the inferior switching speed of the Sartorius balance before the modernization of the electronic gear [63, 64].

The performance of the Mettler H2OE has also been monitored [56, 64] and found satisfactory.

10.5 The granumeter

The granumeter (figure 10.5) is a two-layer balance intended for the particle size range 60 to 1000 μm [67, 68]. Two settling tubes are used for continuous operation. While one run is being made in one tube, a second sample is prepared for the other. The instrument operates above the Stokes range and Brezina [68] presents a comprehensive review of the problems of 'line-start' techniques under these conditions. He concludes that if the mass charge is less than 0.2 kg m^{-2} streaming does not arise. This he attributes, in part, to his method of feeding the particles on to the settling column. The sample release gate consists of a Venetian blind, i.e. a system of parallel, horizontal overlapping, metal shutters. A sample is introduced by 90° rotation of the shutters which are opened electrically and closed manually. The particle size distribution is recorded directly as phi-grain size.

10.6 The micromerograph

The micromerograph is a two-layer sedimentation balance which uses a gas as the sedimentation fluid [2]. The sample is placed in a chamber at the top of the sedimentation column. A blast of nitrogen at high pressure forces the sample through the annulus

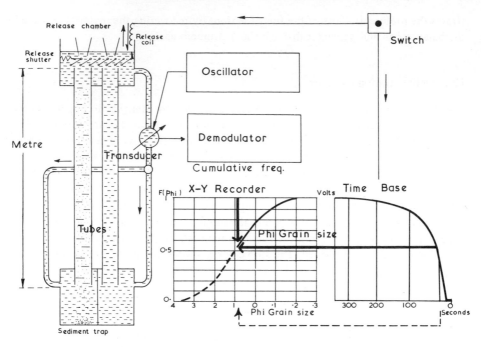

Fig. 10.5 Granumeter, pressure version. Micropressure difference is sensed electrically by a transducer and recorded as cumulative weight frequency (ordinate). Time base controls every *y*-plot to be recorded in phi-grain size scale.

between two cones to break up the aggregates. The particles then settle down the sedimentation tube on to the pan of a servo-electric balance. The accumulating weight is recorded on a strip chart giving a graph of weight against time. By use of a template incorporating Stokes' law a continuous particle size distribution curve is obtained. The makers recommend the instrument for the size range 1 to 250 μm, but the best operating range is 2 to 70 μm. Reproducible results are obtained easily and rapidly without the need for highly trained operators and the instrument is largely automatic. Difficulties arise if a substantial fraction of the powder is smaller than 1 μm since it is not possible to correct for the weight of sample below the lowest measured size. The major difficulty with this instrument is that not all the sample is recorded [5] and losses may be as high as 80%. The assumption that has to be made is that this loss, which is due to powder adhering to the walls of the settling column, is representative. It is more probable, however, that the loss is selective, the fines being more readily attracted to the walls. In such cases the results from this instrument should be accepted with caution. The instrument, is therefore, only suitable for a limited range of material. The micromerograph is the only instrument in which terminal velocities can be measured in a gas, so that it is useful when no dispersing liquids can be found.

A recent study [58] of the micromerograph confirms its reproducibility but demon-

strates the preferential loss of the finer fractions by recovering the material collected on the balance pan and passing it through the instrument again.

10.7 Sedimentation columns

The ICI equipment (figure 10.6) was described in 1947 by Stairmand [13] in its original and in a modified form. The sedimentation tube is connected near its base to a reservoir of clear liquid. The sedimentation tube is one-third filled with clear liquid through which air is gently bubbled. Stopcock B is kept closed and stopcock A is replaced with the stirring funnel. About 0.1 ml of powder is dispersed in about 10 ml of clear liquid with a suitable agent and transferred to the sedimentation tube. The sedimentation tube is filled by opening stopcock B to allow liquid to enter from the reservoir. The stopcock is then closed, and, after 5 min the sampling cock is closed, thus isolating the system from the air supply, and the funnel is replaced by stopcock A. Stopcock A is then closed and cock B opened. The stop clock is started when the air agitation ceases,

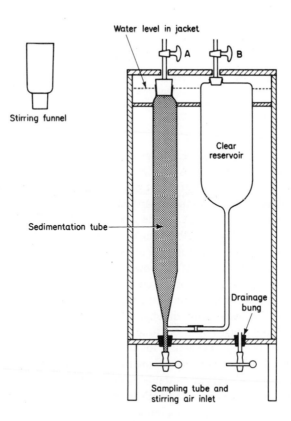

Fig. 10.6 ICI sedimentation column.

this being considered to be the commencement of sedimentation. At recorded times the sedimented powder is flushed out into a centrifuge tube, dried and weighed. If 10 ml of clear fluid is insufficient to remove all the powder, it is recommended that a 25-ml beaker be used instead of a centrifuge tube. If a 2 : 1 time progression is used the size distribution may be calculated (section 10.11), otherwise the size distribution may be derived in the usual way by plotting P against t and differentiating.

The BCURA equipment [34, 35] (figure 10.7) is similar to the ICI sedimentation column, except that a reservoir of clear liquid is not used. Instead the lower end of the tube is constructed of 1.0 mm bore glass tubing. When the tube is isolated from the air supply, surface tension prevents the suspension leaving the tube. A small sampling tube filled with clear liquid is then fitted over the outlet; this breaks the meniscus and allows the sediment to fall into the tube. At convenient time intervals the sampling tubes are replaced. The tubes are then dried and weighed as before. Recommended volume concentration is again 0.1%.

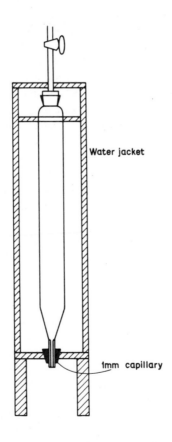

Fig. 10.7 BCURA sedimentation column.

The two types of equipment described above are cheap and may be operated by semi-skilled assistants and, although continuous attention is required, several analyses may be carried out concurrently. The concentration required is far lower than for incremental sedimentation, hence the possibility of hindered settling is greatly reduced. The surrounding water jacket maintains the temperature within the suspension tolerably constant and reduces thermal convection within the suspension. Disturbance of the suspension due to sampling is reduced with the ICI and eliminated with the BCURA equipment.

Since the tubes are not cylindrical, the early samples are deficient in solids and the analysis is finer than the actual distribution. If the analysis is carried out in the tabular way described in section 10.11, an error of opposite sense but similar magnitude is introduced. For this reason the tabular method is recommended but it is suggested that the experimental points should be plotted and a smooth curve drawn through them to reduce operator error. A common way of interpreting the results is to assume that at infinite time all the powder would be collected. This assumption may lead to large errors if much powder is lost in handling or if any powder sticks to the side of the tube. The latter can be considerable if the conical portion of the tube is rippled and any such tube should be rejected. Treatment with an antistatic agent (silicone M.441 as a 10% solution in carbon tetrachloride) can reduce this loss and with a Perspex tube and many powders the loss is negligible. A more accurate way of determining the weight of powder still in suspension is to carry out a subsidiary experiment, decanting at an appropriate time and height and drying and weighing the sediment or the suspension to determine the fractional weight of powder still in suspension at the Stokes diameter for which the final sample was withdrawn in the experiment. This procedure is applicable to all cumulative sedimentation techniques when the loss of powder is small. One problem experienced with the BCURA equipment is some loss of liquid at the beginning of the experiment. The surface-tension force is unable to support the column unless aided by a partial vacuum above the column. The error introduced by this loss is slight and can be easily corrected. Alternatively, agitation may be carried out by using a vacuum pump connected to the top of the column instead of an air supply at the bottom and the residual vacuum when the pump is disconnected helps to support the column of liquid. These methods have some value for comparison work, but no advantage can be gained by using them for routine analyses since the analyses they give, though reproducible, are inaccurate.

Wozniak and Dulcyt [75] used a sedimentation tube 2 cm in diameter and 94 cm long. This was fitted at the bottom with a graduated tube and the height of the sediment was monitored in order to determine the size distribution.

An alternative method is described in [69]. The apparatus consists of a Pyrex glass tube about 120 cm in length and 2.5 cm internal diameter. One end is tapered to an 8 mm internal diameter outlet tube, as shown in figure 10.8. A piece of soft rubber tubing is attached to the outlet tube and this is closed with a spring clip. The tube is calibrated along its length in centimetres, the zero position at the outlet end of the tube being determined as follows. A quantity of water is placed in the tube with the outlet clip closed, sufficient being added to bring the level above the tapered portion of the tube. The level of the water is then marked. A further volume of water is added to the tube and the level again marked. From this difference in volumes and levels, the volume of

water per unit length of tube is established. The tube is then emptied by opening the clip and the quantity of water corresponding to a 5-cm length of tube again added. The water level is permanently marked with the figure 5 and subsequently a linear centimetre scale is marked on the tube from 5 to 100 cm.

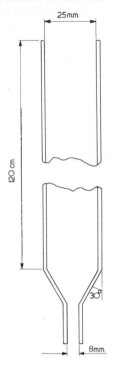

Fig. 10.8 Sedimentation tube.

For an analysis, a sample of powder weighing about 4 g is dispersed in 100 ml of distilled water and added to the tube which is then inverted repeatedly for about a minute to give a uniform dispersion. After some 20–30 inversions a stopwatch is started on the last inversion immediately the air bubble has left the lower end of the tube. The tube is immediately placed in a rack and held in a vertical position. After about 90 s, the spring clip is opened carefully and sufficient of the suspension is allowed to run out under gravity into a 100 ml centrifuge tube (pre-weighed), so that the suspension surface falls from the 100-cm mark to the 86-cm mark. Timing should be such that exactly 100 s should have elapsed when the clip is closed and the suspension level reaches 86 cm. The temperature of this sample is taken and recorded. The process is repeated after 3, 6, 12, 30 and 90 min at levels of 72, 58, 44, 30 and 15 cm. After 180 min, the remaining suspension is withdrawn, down to 5 cm in one centrifuge tube and the remainder in the last tube. The suspension is now distributed among the eight centrifuge tubes used. The amount of sediment is found by centrifuging and removing the supernatant, or filtering and drying. For quantities less than 1 g, the suspended material can be filtered off on a glass-fibre or membrane filter, dried, and the weight again found.

The calculation of results is shown in example 1 (p. 321). Columns 4, 5 and 6 are self- evident. Column 7 is the summation of col. 6, starting at the bottom (i.e. col. 7, line 8, is col. 6, line 8; col. 7, line 7, is col. 6, lines 7 and 8; col. 7, line 6, is col. 6, lines 6 + 7 + 8, etc.). Column 8, the depth factor, is applied to restore the cumulative weight and time to that required for a full 100-cm depth of fall. Column 8 is 100/col. 2. Column 9 is the corrected weight − col. 7 x col. 8. Column 10 is the corrected time − col. 3 x col. 8. The percentage in col. 11 is expressed as a percentage of the total weight. This method is the standard method used at the Hydraulic Research Station [80, 81].

Graphical interpretation of results

The results calculated in columns 10 and 11 of example 1 (p. 321) are plotted on 5-cycle semi-logarithmic graph papers. The points are uniformly spaced, thus allowing the curve to be drawn with maximum accuracy.

A piece of tracing paper, or thin Perspex sheet, is then taken and horizontal lines are drawn to correspond with the upper and lower limits of the graph (0 and 100% cumulative weight). Vertical lines are drawn, their distance apart being equal to the distance between log 1.0 and log e on the graph (i.e. 1.72). If now the tracing paper is placed on the curve and moved horizontally, the lower horizontal line following the axis through 0%, until the line log e is at the time corresponding to the settling time for a selected grain diameter, a tangent to the curve at this point will intersect the line log 1 at a point whose ordinate is the cumulative percentage finer than the chosen size. This process is continued for as many points as desired. The grain size distribution curve is then obtained by plotting these points against particle diameter on another semi-logarithmic sheet.

Particle size is given using Stokes' equation, allowance being made for the fall in height after multiplication by $(100/h)$, where h is the height immediately prior to withdrawal. This factor is necessary in order to restore the cumulative weight deposited to that required for a full 100-cm depth of fall. A plot of corrected weight deposited against corrected time $[(100/h)$ x true time$]$ is essentially P versus t from which the weight undersize may be determined.

10.8 Manometric methods

The manometric method was first described by Weigner [36] and later by Kramer and Stamm [37] who used it with emulsion. In it a vertical capillary side-tube filled with pure dispersion medium is joined to the lower end of a sedimenting column. The pure liquid has a greater height than that of the suspension due to its lower density. Measurement of this excess height at various times gives the variation of mean density of the suspension between the surface of the suspension and the sampling level. From these data the size distribution curve may be determined. This technique was later improved by Dotts [19] by use of a more consistent technique, simplified calculations and simpler apparatus. Dotts claimed an accuracy of 1%. The apparatus is commercially available as the Fisher–Dotts apparatus.

In 1924, Kelly [38] produced a modification of an earlier apparatus by Ostwald and Hahn by constructing the side-tube at a small angle to the horizontal instead of the vertical in order to increase the sensitivity. He claimed an accuracy of 0.5%. A variation was proposed by Goodhue and Smith [39] who increased the sensitivity of a Kelly-type instrument by using two immiscible liquids to magnify the reading zone. Duncombe and Withrow [40] made an extensive study of the above instrument, improved the apparatus further and applied corrections for leakage of clear fluid into the suspension as the manometer level fell. The instrument was automated by Knapp [41] who magnified the pressure change by an optical lever system and recorded it photographically. More recent work has been carried out by Soper [42] who added clear immiscible fluid to the manometer to maintain a constant level. This remedied the fault of the earlier instruments in that manometer fluid entered the sedimentation tube and disturbed the settling powder (Gessner [43]). Other faults of the method are that the change in excess height is small so that accurate measurement is difficult, the manometer tends to be sluggish in action and the fluid in the manometer can be contaminated with powder that has migrated from the sedimentation vessel. The last fault can be minimized by making a constriction in the manometer tube [37] . A review of manometric methods was made by Hawksley in 1951 [44] . The method was not recommended by Jarrett and Heywood who, in 1954, compared various types of size analysis equipment [16] .

10.9 Pressure on the walls of the sedimentation tube

This is a method used by Edwald [45] , who recorded the change in pressure on the wall of the sedimentation vessel as a sedimentation proceeded, and assumed it to be proportional to the amount of sediment.

10.10 Decanting

In this method a homogeneous suspension is allowed to settle for a predetermined time. The supernatant liquid is then decanted off and replaced by fresh dispersing liquid. This process is repeated several times for the same sedimentation time until the supernatant liquid is clear. The decanted liquid will only contain particles finer than size $D = k\sqrt{(h/t)}$, where h is the depth at which the liquid is siphoned off at time t and k is a constant. The process is repeated for shorter times so that the particles removed become progressively coarser. Herdan [46] suggests that six repeats are sufficient to remove substantially all the particles under the selected size and Allen [47] found that fifteen repeats are necessary to produce a clear supernatant liquid.

A theoretical approach is to consider a cylindrical vessel containing a depth h_1 of homogeneous suspension, the siphoning being carried out at a depth h_2. No particles coarser than $D = k\sqrt{(h_2/t)}$ will be removed. Particles of size xD, $x < 1.0$ will have fallen a distance h in the same time, where:

$$xD = k\sqrt{(h/t)}$$

Hence, from these two equations:

$$x^2 = h/h_2$$

The fraction F_1 of particles of size xD removed is

$$(h_2 - h)h_1$$

$$= (h_2/h_1)(1 - x^2)$$

The fraction of particles of this size still in suspension will be:

$$1 - \frac{h_2}{h_1}(1 - x^2)$$

If the suspension is made up to its original volume, redispersed and a second fraction removed after a further time t, the fraction of particles of size xD removed will be:

$$F_2 = \frac{h_2}{h_1}(1 - x^2)\left[1 - \frac{h_2}{h_1}(1 - x^2)\right]$$

The fraction of particles of this size still in suspension will be:

$$\left[1 - \frac{h_2}{h_1}(1 - x^2)\right]^2$$

After n decantings the fraction still in suspension will be:

$$\left[1 - \frac{h_2}{h_1}(1 - x^2)\right]^n \tag{10.4}$$

Therefore, the fraction removed is

$$1 - \left[1 - \frac{h_2}{h_1}(1 - x^2)\right]^n \tag{10.5}$$

The effect of h_2/h_1 is negligible over quite a wide range compared with x, hence there is little to be gained from making $h_2 - h_1$ very small with subsequent risk of disturbing the solids that have settled out. For example, if h_2/h_1 is reduced from 1 to 0.9, the number of decantings to achieve the same separation will be increased from n to m, where:

$$\frac{m}{n} = \frac{\log[1 - (1 - x^2)]}{\log[1 - 0.9(1 - x^2)]} = \frac{\log x^2}{\log(0.1 + 0.9x^2)}$$

for $x = 0.5$, $m/n = 1.23$.

Some values of separation are given in Table 10.2 for $h_2/h_1 = 0.9$.

A large number of decantings are required to remove particles whose size is close to the cut size. Hence the wider the size range of the original suspension, the fewer the decantings required.

Although the method would not normally be used for size analysis, it is a useful means of obtaining closely graded samples. Several descriptions of apparatus are available

[46, 48–51]. Apparatus has also been devised so that the whole operation may be carried out automatically [52].

Table 10.2 Percentage of particles removed after a known number of decantings.

Relative Particle size (x)	Number of decantings (n)					
	1	2	4	8	16	32
0.9	17.1	31.3	52.7	77.6	95.3	97.7
0.8	32.4	54.4	79.3	95.7	98.2	99.7
0.7	45.9	70.6	91.4	99.3	–	–
0.6	57.6	82.0	96.8	99.8	–	–
0.5	68.5	90.1	99.0	–	–	–

10.11 The β-back-scattering method

In this technique the thickness of the deposited solid is determined by recording the intensity of β-radiation scattered from the base of a sedimentation column [53–55]. The apparatus (figure 10.9) consists of a 100-cm^3 Perspex centrifuge tube with a thin flat base (0.015 in), situated above a 3-mCi ^{90}Sr–^{90}Y source, which rests in the wall of a thick lead ring placed on top of a β-scintillation counter, the output of which is fed to a ratemeter and recorder.

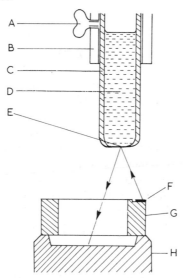

A = Perspex locating screw
B = Perspex tube holder
C = Perspex sedimentation tube
D = Sedimentation liquid

E = Sedimented material
F = 3mC^{90}Sr ^{90}Y source
G = Lead ring
H = β-Scintillation counter or ionisation chamber

Fig. 10.9 Arrangement of source, sedimentation tube and counter for β-back-scattering method.

The intensity of the radiation scattered back from the base of the tube is proportional to both the atomic number and the thickness of the scattered material. The graph of scattered radiation against concentration of deposit exhibits an initial linear portion before saturation is reached. If this portion only is used, the increase in count rate is proportional to the increase in weight of deposit. The maximum thickness of deposit for this condition is of the order of 0.07 g cm^{-2}. The technique is applicable to materials having an effective atomic number $\bar{z}$ greater than 12.5, where:

$$\bar{z} = \frac{n(A_b Z_b) + m(A_c Z_c)}{\text{mol. wt of } B_n C_m}$$

where A_b and A_c are the atomic weights of B and C, and Z_b and Z_c their atomic numbers in a compound $B_n C_m$.

Settling under gravity allows size distributions from approximately 80 to 1 μm and this may be extended to 0.1 μm by centrifuging the tube.

A variation of this technique was developed by Curzio [73] who measured β-ray absorption. The source consisted of a few drops of a ^{90}Sr nitrate solution which were deposited in a slot along the whole circumference near the base of the tube. The activity of the source was about 5 μCi. The detector consisted of a Geiger–Müller counter on which the sedimentation tube stood. The authors reported very good results with this simple apparatus.

10.12 Discussion

The main advantage that most cumulative methods have over incremental methods of sedimentation analysis is that the concentration required is much lower, thus reducing the risk of hindered settling. The technique is most useful for normal distributions, but the need to differentiate leads to errors when the size distribution is irregular in any way (e.g. bimodal).

The BCURA and ICI equipment give similar and reproducible analyses but the latter is simpler in use. This apparatus is of most use when nearly all the powder is finer than about 50 μm since, if there is a substantial weight of powder above this size, repeat analyses produce wide variations in the weight of the initial aliquot sample. The shape of the sedimentation tube limits the method to comparison work and justifies the use of the approximate theory derived in section 10.13. Loss of powder on the conical portion of the tube also leads to errors which are usually negligible but may be significant with some powders.

The cumulative effect of the many disadvantages of the manometric method makes it highly unlikely that it will ever be widely used for routine analysis. Repeated decanting will continue to prove attractive to many industries because of the low cost of equipment and the by-product of closely graded fractions. Manual decanting is too time-consuming to merit general approval but machine decanting is a more attractive proposition.

Balance methods continue to be most attractive; the best of these is probably the Cahn. The Bostock is the cheapest of these systems and a modified version to eliminate

its faults could be most useful. It could also be improved by being made a continuous recording balance. The Sartorius has this advantage, combined with the disadvantage that the pan is immersed in the suspension, which leads to errors in the record. The Shimadzu is completely unsuitable due to its sluggish response.

10.13 Appendix: An approximate method of calculating size distribution from cumulative sedimentation results

All particles larger than 75 μm are removed by sieving. The top sedimentation size of the remainder is taken as 106 μm, due to the overlapping of sieve size and Stokes' diameter. The sieve range is then extended to the following sizes:

Grade size (μm) 106   75   53   37.5   26.5   18.8   13.3   9.4   6.6
                         4.7   3.3
Mean size (μm)   89   63   44.5   31.5   22.3   15.8   11.2   7.9   5.6
                         4.0

It is assumed that the mean size may be attributed to the whole range, e.g. in the 75 to 106 μm range all particles are assumed to have the size 89 μm.

Since the sizes diminish in a $\sqrt{2} : 1$ ratio, the free-falling speed will decrease in a 2 : 1 ratio. Thus when the 89-μm particles have fallen the full length of the sedimentation tube, the 63-μm particles will have fallen half the length, hence half of them will have fallen out of suspension.

The first sample is withdrawn when all the 89-μm particles have fallen out of suspension at time t, the second sample when all the 63-μm particles have fallen out of suspension at $2t$ and so on.

Let:

W_{89} be the weight of solids in the size range 75 to 106 μm,
X_{89} be the total weight of sediment at time t_{89},
γ_{63} be the weight of the increment at time t_{63}.

Figure 10.10 shows the upper levels for the various sizes at the times given below the figures.

The first sample, withdrawn at t_{89}, will be:

$$X_{89} = W_{89} + \tfrac{1}{2}W_{63} + \tfrac{1}{4}W_{44.5} + \tfrac{1}{8}W_{31.5} + \tfrac{1}{16}W_{22.3} + \ldots + \tfrac{1}{512}W_4 \qquad (10.6)$$

The first increment will be:

$$\gamma_{63} = \tfrac{1}{2}W_{63} + \tfrac{1}{4}W_{44.5} + \tfrac{1}{8}W_{31.5} + \tfrac{1}{16}W_{22.3} + \ldots + \tfrac{1}{512}W_4 \qquad (10.7)$$

Therefore:

$$W_{89} = X_{89} - \gamma_{63}$$

The second increment will be:

$$\gamma_{44.5} = \tfrac{1}{2}W_{44.5} + \tfrac{1}{4}W_{31.5} + \tfrac{1}{8}W_{22.3} + \ldots + \tfrac{1}{256}W_4 \qquad (10.8)$$

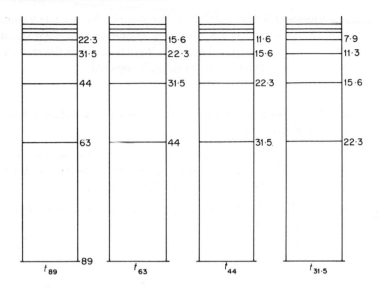

Fig. 10.10 Theory for the ICI technique

From equations (10.5) and (10.6):

$$W_{63} = 2\gamma_{63} - \gamma_{44.5}$$

The third increment will be:

$$\gamma_{31.5} = \tfrac{1}{2}W_{31.5} + \tfrac{1}{4}W_{22.3} + \tfrac{1}{8}W_{15.8} + \ldots + \tfrac{1}{256}W_4 \tag{10.9}$$

From equations (10.8) and (10.9):

$$W_{44.5} = 2\gamma_{44.5} - \gamma_{31.5}$$

This form of solution continues until the final withdrawal, which we shall assume is at t_4. The solids still remaining in suspension after t_4 we shall call γ_0.

Total weight of solids = total weight of increments

$$W_{89} + W_{63} + W_{44.5} + W_{31.5} + \ldots + W_4 + W_{3.3}$$

$$= X_{89} + \gamma_{63} + \gamma_{31.5} + \ldots + \gamma_4 + \gamma_0$$

Substituting in the solutions for W:

$$(X_{89} - \gamma_{63}) + (2\gamma_{63} - \gamma_{44.5}) + (2\gamma_{63} - \gamma_{31.5}) + (2\gamma_{5.6} - \gamma_4) + W_{<3.3}$$

$$= X_{89} + \gamma_{63} + \gamma_{44.5} + \gamma_{31.5} + \ldots + \gamma_4 + \gamma_0 - \gamma_4 + W_{<3.3} = \gamma_4 + \gamma_0$$

$$W_{<3.3} = 2\gamma_4 + \gamma_0$$

These results may be summarized (see also Table 10.3):

Example 1–Sedimentation Tube Result Sheet

1	2	3	4	5	6	7	8	9	10	11
Tube no.	Height (cm) $\dfrac{h_0}{100}$	Time	Weight of tube + sample (g)	Weight of tube (g)	Weight of sample (g)	Cumulative weight (g)(P)	Depth factor $\dfrac{h_0}{h} = \dfrac{100}{h}$	Weight in 100 cm suspension (g)	Corrected time (min) $\dfrac{100}{h}T$	Corrected (%)
1	86	100 s	66.6800	66.5175	0.1625	0.9640	1.000	0.9640	1.94	100
2	72	3 min	66.5791	66.4241	0.1550	0.8015	1.163	0.9322	4.17	96.7
3	58	6 min	67.6784	67.5212	0.1572	0.6465	1.389	0.8980	10.30	93.2
4	44	12 min	67.1532	66.9886	0.1646	0.4893	1.724	0.8436	27.3	87.5
5	30	30 min	73.8796	73.7008	0.1788	0.3247	2.273	0.7380	100	76.6
6	15	90 min	73.1364	73.0001	0.1363	0.1459	3.333	0.4863	600	50.5
7	5	180 min	65.2502	65.2414	0.0088	0.0096	6.667	0.0640	3600	6.6
8		+180 min	73.3064	73.3056	0.0008	0.0008	20	0.0160		1.7

Table 10.3

Size of grade (µm)	Mean size (µm)	Time of sampling	Weight of increment	Weight in grading
106–75	89	t_{89}	X_{89}	$X_{89} - \gamma_{63}$
75–53.3	63	t_{63}	γ_{63}	$2\gamma_{63} - \gamma_{44.5}$
53.5–37.5	44.5	$t_{44.5}$	$\gamma_{44.5}$	$2\gamma_{44.5} - \gamma_{31.5}$
37.5–26.5	31.5	$t_{31.5}$	$\gamma_{31.5}$	$2\gamma_{31.5} - \gamma_{22.4}$
26.5–18.8	22.3	$t_{22.4}$	$\gamma_{22.5}$	$2\gamma_{22.4} - \gamma_{15.6}$
18.8–13.3	15.6	$t_{15.6}$	$\gamma_{15.6}$	$2\gamma_{15.6} - \gamma_{11.2}$
13.3–9.4	11.2	$t_{11.2}$	$\gamma_{11.2}$	$2\gamma_{11.2} - \gamma_{8}$
9.4–6.6	7.9	t_{8}	γ_{8}	$2\gamma_{8} - \gamma_{5.6}$
6.6–4.7	5.6	$t_{5.6}$	$\gamma_{5.6}$	$2\gamma_{5.6} - \gamma_{4.0}$
4.7–3.3	4.0	$t_{4.0}$	$\gamma_{4.0}$	
		t_{0}	γ_{0}	$2\gamma_{4.0} - \gamma_{0}$
			W Total	W Total

γ is obtained by difference. γ_{0} = initial wt of sample − sum of increments up to and including γ_{4}.

t_{89} is determined using Stokes' equation:

$$t_{89} = \frac{18\eta h}{(\rho_{s} - \rho_{f})g(89)^{2}} \times 10^{8}$$

and $t_{63} = 2t_{89}$; $t_{44.5} = 2t_{63}$ and so on.

References

1 Marshall, C.E. (1930), *Proc. R. Soc.*, **A126**, 427.
2 Eadie, F.A. and Payne, R.E. (1954), *Iron Age*, **174**, 99; (1956) *Br. Chem. Eng.*, **1**, 306.
3 Werner, D. (1925), *Trans. Faraday Soc.*, **21**, 381.
4 Travis, P.M. (1940), *ASTM Bull.*, **29**, 102.
5 Irani, R.R. and Callis, C.F. (1963), *Particle Size: Measurement, Interpretation and Application*, Wiley, NY.
6 Whitby, K.T. (1955), *Heat., Pip. Air Condit.*, Jan., Part 1, 231; June, Part 2, 139.
7 Whitby, K.T., Algren, A.B. and Annis, J.C. (1958), ASTM Sp. Publ. No. 234, 117.
8 Imris, P. and Landspersky, H. (1956), *Silikaty*, **9**, 4, 327.
9 Oden, S. (1916), *Kolloid Z.*, **18**, 33–47.
10 Coutts, J. and Crowthers, F.M. (1925), *Trans. Faraday Soc.*, **21**, 374.
11 Bostock, W. (1952), *J. scient. Instrum.*, **29**, 209.
12 Gaudin, A.M., Schumann, R. and Schlechter, A.W. (1942), *J. Phys. Chem.*, **46**, 903.
13 Stairmand, C.J. (1947), *Symp. Particle Size Analysis*, Inst. Chem. Eng., **25**, 110.
14 Edwald, P. (1942), *Ind. Engng Chem., analyt. Edn*, **14**, 66.
15 Nissan, A.H. (1951), *Faraday Soc. (Discussion)*, **11**, 15.
16 Jarrett, B.A. and Heywood, H. (1954), *Brit. J. appl. Phys.*, suppl. 3, 21S.
17 Donoghue, J.K. (1956), *Brit. J. appl. Phys.*, suppl. 5, 7, 333.
18 Oden, S. (1925), *Soil Science*, **19**, 1.
19 Dotts, W.M. (1946), *Ind. Engng Chem., analyt. Edn*, **19**, 326.
20 Svedberg, T. and Rinde, H. (1934), *J. Am. Chem. Soc.*, **45**, 173.
21 Bishop, D.L. (1934), *Bur. Stand. J. Res.*, **12**, 173.
22 Knapp, R.T. (1934), *Ind. Engng Chem., analyt. Edn*, **6**, 66.

23 Muller, R.H. and Garman, R.L. (1936), *Analyt. Chem.,* **10**, 436.
24 Jacobsen, A.E. and Sullivan, W.G. (1947), *ibid.,* **19**, 855.
25 Suito, E. and Arakawa, M. (1950), *Bull. J. Chem. Res., Kyoto University*, **23**, 7.
26 Kiffer, C. (1954), *Bull. Soc. Ceram.,* **17**, 22; *Abs. Trans. Ceram. Soc.,* **53**, 392A.
27 Rabatin, G.J. and Gale, R.H. (1956), *Analyt. Chem.,* **28**, 1314.
28 Bachman, D. and Gerstenberg, H. (1957), *Chem. Ing. Tech.,* **8**, 589.
29 Friedrich, W. (1959), *Staub,* **19**, 281–312.
30 Ioos, E. (1959), *ibid.,* **19**, 392–8.
31 Palik, E.S. (1962), *Ceramic Age,* 78, 8, 49.
32 Cohen, L. (1959), *Instrum Pract.,* **13**, 1036.
33 Ames, D.P., Irani, R.R. and Callis, C.F. (1959), *J. Phys. Chem.,* **63**, 531.
34 Kabak, J. and Loveridge, D.J. (1960), *J. scient. Instrum.,* **37**, 266.
35 Kabak, J. (1955), *J. scient. Instrum.,* **32**, 153.
36 Weigner, G. (1918), *Landw. vers. Sta.,* **91**, 41.
37 Kramer, E.O. and Stamm, A.J. (1924), *J. Am. Chem. Soc.,* **46**, 2709.
38 Kelly, W.S. (1924), *Ind. Engng Chem.,* **16**, 928.
39 Goodhue, L.D. and Smith, C.M. (1936), *Ind. Engng Chem., analyt. Edn,* 8, 469.
40 Duncombe, C.G. and Withrow, J.R. (1932), *J. Phys. Chem.,* **36**, 31.
42 Soper, A.K. (1947), *Symp. Particle Size Analysis,* Inst. Chem. Eng., 110.
43 Gessner, H. (1928), Akademische Verlagsgesellschaft, Leipzig.
44 Hawksley, P.G.W. (1951), *British Coal Utilisation Research Assoc. Bull.,* **15**, 4, 129.
46 Herdan, G. (1960), *Small Particle Statistics,* Butterworths.
47 Allen, T. (1962), M.Sc. thesis, London Univ.
48 Truog, E. *et al.* (1936), *Proc. Soil Sci. Soc. Am.,* **1**, 10.
49 Davies, R.J., Green, R.A. and Donelly, H.F.E. (1937), *Trans. Ceram. Soc.,* **36**, 181.
50 Birchfield, H.P., Gullaston, D.K. and McNew, G.L. (1948), *Analyt. Chem.,* **20**, 1168.
51 Johnson, E.I. and King, J. (1951), *Analyst,* **76**, 661.
52 Horsfall, F. and Jowett, A. (1960), *J. scient. Instrum.,* **37**, 4, 120.
53 Connor, P., Hardwick, W.H. and Laundy, B.J. (1958), *J. appl. Chem.,* **8**, 716.
54 Connor, P., Hardwick, W.H. and Laundy, B.J. (1959), *ibid.,* **9**, 525.
55 Hardwick, W.H. and Laundy, B.J. (1966), *Int. Symp. Particle Size Measurement, Loughborough,* Soc. Analyt. Chem.
56 Leschonski, K. (1962), *Staub,* **22**, 475–86.
57 Pretorius, S.T. and Mandersloot, W.G.B. (1967), *Powder Technol.* **1**, 23–27.
58 Bryant, A.C., Freeman, D.S. and Tye, F.L. (1966), *Int. Conf. Particle Size Analysis,* Soc. Analyt. Chem.
59 Scott, K.J. and Mumford, D. (1970), *CSIR Spec. Rep. Chem.,* 156; Scott, K.J. (1972), *Addendum,* Pretoria, SA.
60 Van Tonder, J.C. *et al.* (1969), *R.A.K. Verslag No. 10.*
61 Harris, C.C. and Jowett, A. (1963), *Nature,* **197**, 4873, 1192.
62 Mengelsdorf, P.C. (1959), *J. appl. Phys.,* **30**, 442.
63 Alex, W. and Putz, R. (1971), *Messtechnik,* **78, 3**, 69–73.
64 Leschonski, K. and Alex, W. (1972), *Proc. Int. Symp. Particle Size Analysis, Bradford 1971,* Soc. Analyt. Chem.
65 Grandillo, A.D. (1970), *J. Powder Metall.,* **6**, 1, 3–16.
66 Allen, T. (1967), Ph.D. thesis, Univ. Bradford.
67 Brezina, J. (1969), *J. Sediment. Petrol.,* 1627–31.
68 Brezina, J. (1972), *Particle Size Analysis 1970,* Soc. Analyt. Chem., 255–66.
69 *Report No. 7* (1943), A study of methods used in measurement and analysis of sediment loads in streams, St. Paul, US Engineer District Sub-office Hydraulic Laboratory, University of Iowa, Iowa City, USA.
70 Gaudin, A.M., Schumann, R. and Schlechten, A.W. (1942), *J. Phys. Chem.,* **46**, 902–4.
71 Kim, S.C., Schlotzer, G. and Palik, E.S. (1967), *Powder Technol.,* **1**, 54–55.

72 Rose, H.E. and Langmaid, R.N. (1957), *Nature,* **179**, 774.
73 Curzio, G. (1974), *Powder Technol.,* **9**, 2/3, 121–4.
74 Kaye, B.H. and Davies, R. (1972), *Proc. Conf. Particle Size Analysis, London* (eds. Groves and Wyatt-Sargent), Soc. Analyt. Chem., pp. 207–22.
75 Wozniak, K. and Dulcyt, B. (1974), *Mater. Ogniotrwale,* **26**, 5, 109–14.
76 Topham, J.D. (1975), *J. Pharm. Pharmacol.,* **27**, 1, 6–12.
77 Bachman, D. (1959), *Dechema Monograph,* **31**, 23–51.
78 Gerstenberg, H. (1959), *ibid.,* 52–60.
79 Weber, A. (1960), *Ziegelind.,* **13**, 12, 427–31.
80 Kiff, P.R. (1972), Paper presented at a Group Meeting, Soc. Analyt. Chem., Salford.
81 Owen, M.W. (1976), Determination of the settling velocities of cohesive muds, Report No. IT 161, Hydraulic Research Station, Wallingford, England.

11 Fluid classification

11.1 Introduction

Fluid classification is a process for separating dispersed materials, based on the movement of suspended particles to different points under the effect of different forces. The fluid is usually water or air and the field force either gravity, as with elutriators, or centrifugal and Coriolis, in rotary classifiers. The other forces of importance are the drag forces due to relative flow between the particles and flow medium, and the inertia forces, due to accelerated particle movement. A wide variety of equipment is available, ranging in capacity from many tons per hour with the larger units to a few grams per hour with laboratory machines.

The results of classification processes may be presented as size distributions, the accuracies of which depend on the sharpness of cut. The ideal is when the cut size is well defined and there are no coarse particles in the fine fraction and vice versa. In practice, however, there is always overlapping of sizes. The cut size may sometimes be predicted from theory, but this usually differs from the actual cut size due to difficulties in accurately predicting the flow patterns in the system. It is, therefore, necessary to be able to predict the performance of classifiers based on their actual performance.

11.2 Assessment of classifier efficiency [18, 20]

Consider a single stage of a classifier, where: W, W_c, W_f are the weights of feed, coarse product and fine product respectively; $F(x)$, $F_c(x)$, $F_f(x)$ are the cumulative fraction oversizes of feed, coarse product and fine product respectively; x is particle size. Then:

$$W = W_c + W_f \tag{11.1}$$

and

$$W \frac{dF(x)}{dx} = W_c \frac{dF_c(x)}{dx} + W_f \frac{dF_f(x)}{dx} \tag{11.2}$$

The total coarse efficiency may be defined as:

$$E_c = \frac{W_c}{W} \tag{11.3}$$

and the total fine efficiency as:

$$E_f = \frac{W_f}{W} \tag{11.4}$$

$$= 1 - E_c \tag{11.5}$$

The value of E_c is meaningless for a general characterization of the classifier since it depends upon the quality of the feed.

The grade efficiency, which is independent of the feed, is defined by:

$$\text{coarse grade efficiency} = \frac{\text{amount of coarse product of size } x}{\text{amount of feed of size } x}$$

$$G_c(x) = W_c \frac{dF_c(x)}{dx} \div W \frac{dF(x)}{dx}$$

$$= \frac{W_c}{W} \frac{dF_c(x)}{dF(x)}$$

$$= E_c \frac{dF_c(x)}{dF(x)} \tag{11.6}$$

Similarly, the fine grade efficiency is defined by:

$$G_f(x) = E_f \frac{dF_f(x)}{dF(x)} \tag{11.7}$$

which, from equations (11.2) and (11.4) may be written:

$$G_f(x) = 1 - G_c(x). \tag{11.8}$$

These equations enable one to determine the grade efficiency of a classification process from the total efficiency and the size distributions of any two streams (Table 11.1). Results are usually plotted as *grade efficiency curves* of $G_c(x)$ or $G_f(x)$ against x [30].

Table 11.1 Example of grade efficiency calculation.

Particle size x (μm)	Feed frequency $F(x)$ (%)	$\frac{dF(x)}{dx}$ (%/μm)	Coarse frequency $F_c(x)$ (%)	$\frac{dF_c(x)}{dx}$ (%/μm)	Grade efficiency $G_c(x)$	$\bar{x}$ (μm)
2	0	1.00	0	0	0	2.8
4	2	2.25	0	0.50	0.133	5.66
8	11	2.75	2	1.25	0.273	11.3
16	33	2.00	12	2.00	0.600	22.6
32	65	0.73	44	1.16	0.945	45.3
64	88.5	0.15	81	0.25	1.00	90.5
128	98	0.02	97	0.02		181
256	100		100			

Mass of feed = 100 g; mass of coarse product = 60 g.

These data are plotted in figure 11.1 The 50% size on the grade efficiency curve is called the *equiprobable* size (e), since particles of this size have an equal chance of being found in the coarse or fine product and are usually taken as the cut size. The grade efficiency curve is best determined by plotting $F(x)$ against $F_c(x)$ and differentiating.

Determination of the cut size (which is independent of the feed material) requires a knowledge of the whole grade efficiency curve. Two other cut sizes are also used [27]. The analytical cut size is that at which an ideal split of the feed would occur with a proportion E_c passing to the coarse and a proportion E_f passing to the fine. Cut size x_p is

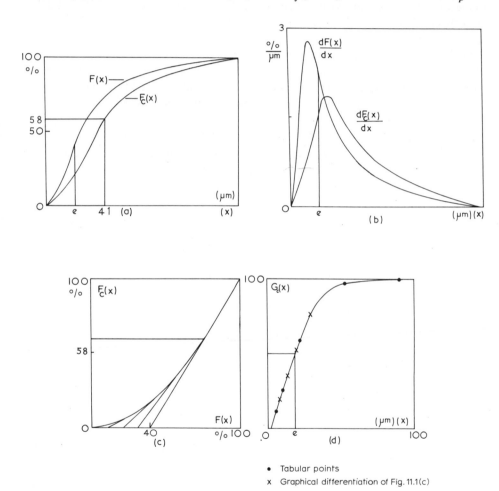

• Tabular points
x Graphical differentiation of Fig. 11.1(c)

Fig. 11.1 Curve (c) is coincident with the line $dF_c(x)/dF(x) = 100/60$ passing through the point (100, 100) for $F_c(x) \geqslant 58\%$ therefore $G_c(x) = 100\%$. From (a) this is for $x = 41\ \mu m$. Differentiating curve (c) and multiplying by 60 gives $G_c(x)$, the relevant diameters being extracted from (a). $e \simeq 20\ \mu m$.

defined as the size at which the cumulative percentage undersize coarse equals the cumulative percentage oversize fines.

The grade efficiency is often expressed as a single number. This number is the sharpness index, ϕ, and is related to the slope of the grade efficiency curve at the equiprobable size:

$$75\phi_{25} = \frac{x_{75}}{x_{25}} \tag{11.9}$$

where x_{75} and x_{25} are the particle sizes at which the grade efficiency is 75% and 25% respectively. For perfect classification $\phi = 1$, while values above 3 are considered poor. Alternatively, $_{90}\phi_{10}$ has been used.

Leschonski [1] is of the opinion that these ratios are not always adequate to define the sharpness of cut. In many cases it is important to keep the proportions finer or coarser than the cut size as small as possible in both, or at least one fraction. For these cases a measure of the effectiveness of the separation process is given by [19]:

the coarse yield: $\psi_c = \dfrac{\text{weight of coarse particles larger than } e \text{ in the coarse fraction}}{\text{weight of coarse particles larger than } e \text{ in the feed}}$

$$\psi_c = E_c \frac{\sum\limits_{e}^{x_{max}} F_c(x)}{\sum\limits_{e}^{x_{max}} F(x)} \tag{11.10}$$

Similarly:

the fine yield: $$\psi_f = E_f \frac{\sum\limits_{x_{min}}^{e} F_f(x)}{\sum\limits_{x_{min}}^{e} F(x)} \tag{11.11}$$

Gibson [28] developed a simple geometric construction to calculate the grade efficiency directly from cumulative particle size distribution data. His method consists of plotting a square diagram of $F_c(x)$ against $F(x)$ directly above a square diagram of $F_f(x)$ against $F(x)$. Consider the coarse versus feed diagram (figure 11.2). Equation (11.6) may be written:

$$G_c(x) = E_c(\tan \alpha)_x \tag{11.12}$$

For large x, i.e. coarse sizes, $G_c(x) = 1$ and:

$$(\tan \alpha)_{x \to \infty} = \frac{1}{E_c} \tag{11.13}$$

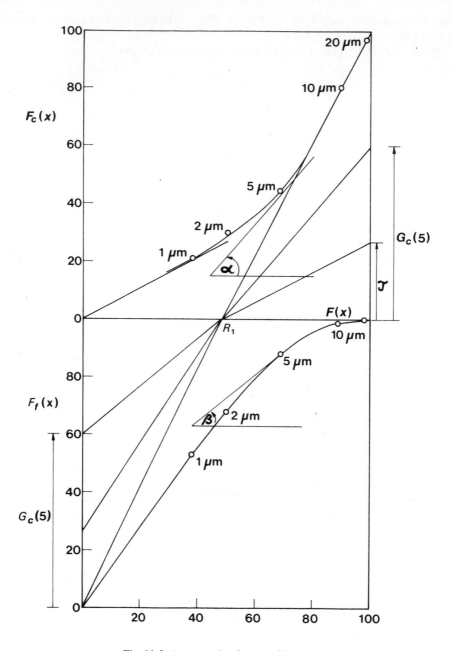

Fig. 11.2 An example of square diagrams.

Hence, a tangent through $F_c(x) = F(x) = 100\%$, intersects the $F(x)$ axis at the point where $F(x) = 1 - E_c$ (= 49%).

Equations (11.5), (11.7) and (11.8) can be combined to give:

$$G_c(x) = 1 - E_f \frac{dF_f(x)}{dF(x)}$$

i.e.

$$G_c(x) = 1 - (1 - E_c)(\tan \beta)_x \qquad (11.14)$$

Consider the fines versus feed diagram. At small sizes $G_c(x) = 0$ and:

$$(\tan \beta)_{x \to \infty} = \frac{1}{1 - E_c} \qquad (11.15)$$

Hence a tangent through $F_f(x) = F(x) = 0$, intersects the $F(x)$ axis at the point where $F(x) = 1 - E_c (= 49\%)$.

Plotting the two square diagrams as shown in figure 11.2, using the data presented in Table 11.2 the total efficiency is represented by a single point R_1 and its straight line connections to $F_c(x) = 100\%, F(x) = 100\%$ for coarse and $F_f(x) = F(x) = 0$ for fines.

Table 11.2 Particle size distribution of feed, coarse and fine together with data derived from square diagrams.

	Particle per cent finer than x				
Size x (μm)	Feed $F(x)$	Coarse $F_c(x)$	Fine $F_f(x)$	Grade efficiency $G_c(x)$	$G_c'(x)$
20	99.0	98	100.0	0.985	0.979
10	89.7	80	99.0	0.905	0.870
5	69.5	44	88.0	0.595	0.445
2	50.8	30	68.0	0.385	0.158
1	38.7	21	53.0	0.345	0.103

Leschonski [29] noted that some classifiers give grade efficiency curves where the grade efficiency does not reach the x axis but runs parallel to it at a constant value τ. He proposed the use of a corrected curve defined by:

$$G_c'(x) = \frac{G_c(x) - \tau}{1 - \tau} \qquad (11.16)$$

In the limit, at small sizes when $x \to 0$, $G_c(x) \to \tau$.

Equation (11.6), for coarse versus feed becomes:

$$\tau = E_c(\tan \alpha)_{x \to 0}$$

therefore

$$(\tan \alpha)_{x \to 0} = \frac{\tau}{E_c} \qquad (11.17)$$

This corresponds to a line passing through the R_1 point and parallel to the tangent to the curve at $F_c(x) = 0, F(x) = 0$. This intersects the $F_c(x) (F(x) = 100\%)$ axis at $\tau(= 27\%)$.

Similarly for the fines versus feed plot for $x \to 0$, $G_c(x) \to \tau$ and equation (11.7) becomes

$$(\tan \beta)_{x \to 0} = \frac{1 - \tau}{1 - E_c} \qquad (11.18)$$

This line passes through $F(x) = 0, F_f(x) = 0$ and is parallel to a line through R_1 which intersects the $F_f(x), F(x) = 0$ axis at a value τ.

The construction for grade efficiency is similar to that for τ and this is illustrated for $x = 5 \ \mu m$. A line drawn parallel to the tangent at $x = 5 \ \mu m$, on the $F_c(x)$ curve, and passing through R_1 intersects the $F(x) = 100\%$ line at $F_c(x) = G_c(5) = 60\%$. A line drawn parallel to the tangent at $x = 5 \ \mu m$, on the $F_f(x)$ curve, and passing through R_1 intersects the $F(x) = 0$ line at $F_f(x) = G_c(5) = 60\%$.

Gibson developed a computer technique to carry out these evaluations. Basically this consisted of fitting a parabola to three consecutive data points, differentiating at the middle point and multiplying by R_1.

11.3 Systems

All classifiers can be divided into two systems, the counterflow equilibrium and the transverse-flow separation.

Counterflow can occur either in a gravitational or centrifugal field. The field force and the drag force act in opposite directions and particles leave the separation zone in one of two directions according to their size. Particles of a certain size are acted upon by two equal and opposite forces, hence stay in equilibrium in the separation zone. In gravitational systems these particles remain in a state of suspension, while in a centrifugal field the equilibrium particles revolve at a fixed radius which is governed by the rate at which material is withdrawn from the system. They would, therefore, accumulate to very high concentrations in a continually operated classifier if they were not distributed to the fine or coarse fraction by a stochastic mixing process.

In a transverse-flow classifier, the feed material enters the flow medium at one point of the classification chamber at a certain angle with a component of velocity transverse to the flow and is fanned out under the action of field, inertia and resistance forces. Particles of the same size describe identical trajectories which differ from the trajectories of particles of a different size, and it is possible to separate them according to size.

11.4 Counterflow equilibrium classifiers in the gravitational field – elutriators

Elutriation is a process of grading particles by means of an upward current of fluid, usually water or air. The process is therefore the reverse of gravity sedimentation and Stokes' law applies. The grading is carried out in a series of vessels, usually of cylindro-

conical form and of successively increasing diameter. Hence the fluid velocity decreases in each stage, the coarsest particles being retained in the smallest vessel and relatively finer particles in the following vessels. The operation is considered completed when, for air elutriation, the rate of change in the weight of the residue is deemed negligible, say 0.2% of the initial weight in half an hour. For water elutriation, the end-point is reached when there are no visible signs of further classification taking place.

In most elutriators Stokes' law does not apply, since the ratio of tube length to tube diameter is too small for laminar flow conditions, i.e. the fluid disturbances at inlet and outlet overlap. Combined with this, the tube shape is not always conducive to laminar flow.

Due to the viscosity of the fluid, a parabolic velocity front exists which is flattened only in the case of large-diameter tubes. The cut is not sharp, therefore, since the upward force on a particle depends on its axial position in the tube. In spite of these defects, elutriation methods are still useful for design or control work. Roller [2] showed that the effect of the uneven cut is the removal of some coarse above the theoretical cut-point, while leaving behind some of the fines. Thus, while the separate fractions are not accurately sized, the final mass fraction is often reasonably close to the correct value. This has been confirmed by Stairmand [3], who points out that the method is not applicable to bimodal distributions.

The parameter by which particles are classified is their falling speed, which is not uniquely related to size, but which is of greater significance than size in many applications, e.g. in determining the respirability of dust or the point of deposition of particulate matter emitted from a chimney stack

A particular advantage of elutriation is the production of closely graded fractions which are often useful for further investigations.

Theory

If it is assumed that the fluid flow is streamline, the velocity profile is parabolic, the velocity at a point at a distance r from the axis of the tube being given by Poiseuille's equation:

$$v = \frac{p}{4\eta L}(a^2 - r^2) \tag{11.19}$$

where p is the pressure drop across the elutriator tube,
η is the fluid viscosity,
L is the tube length,
a is the tube radius.

The flow rate through the tube is:

$$Q = \frac{pa^4}{8\eta L} \tag{11.20}$$

giving an average velocity:

$$v_m = \frac{Q}{\pi a^2} = \frac{pa^2}{8\eta L} \tag{11.21}$$

By putting $r = 0$ in equation (11.19), the maximum velocity may be found:

$$v_{max} = \frac{p}{4\eta L} a^2 = 2v_m \tag{11.22}$$

From equations (11.19) and (11.20):

$$v = v_{max} \left(1 - \frac{r^2}{a^2}\right) = 2v_m \left(1 - \frac{r^2}{a^2}\right) \tag{11.23}$$

If it is assumed that there is no radial flow of particles, the possibility of a particle being elutriated depends on its position in the tube. Since v_m is the velocity usually taken as the elutriation velocity, from equation (11.23):

$$\frac{r^2}{a^2} = 1 - \frac{v}{2v_m} \tag{11.24}$$

Thus a particle of terminal velocity v can ascend if it is in a coaxial circle of radius r. Assuming a homogenous distribution of particles, the fraction of this size ascending is the fraction contained in a cylinder of radius r:

$$F = \pi r^2 / \pi a^2$$

$$= 1 - v/2v_m \tag{11.25}$$

Since, from Stokes' equation, v is proportional to particle size squared:

$$F = 1 - \frac{1}{2} \left(\frac{D}{D_m}\right)^2 \tag{11.26}$$

The collection efficiency may be calculated using equation (11.26).

The cut, however, is better than one would expect from the theoretical grade efficiency curve (Table 11.3) due to a radial flow of particles from the outer to the inner areas. This is due to the pressure difference across the particle in a radial direction.

Table 11.3 Theoretical efficiency of elutriators.

D/D_m	0.1	0.2	0.3	0.4	0.5	0.6	0.7	0.8	0.9	1.0	1.1	1.2	1.3	1.4	$\sqrt{(2)}$
Percentage elutriated (100F)	99.5	98	95.5	92	87.7	82	75.5	68	59.5	50	39.5	28	15.5	2	0

At A (figure 11.3), the particle rotates in the same direction as the gas flow; this increases the gas velocity at A and at B the reverse occurs. The energy to accelerate is drawn from the pressure energy of the gas, hence the pressure is lower at A than B, therefore the particles move from the regions of low velocity to the regions of high velocity.

The cut velocity is therefore v_{max} since, if elutriation is carried out to completion, all particles smaller than $\sqrt{(2)} D_m$ will be removed.

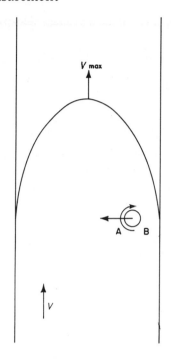

Fig. 11.3 The pressure gradient across a particle.

Experimental results [3] are in general agreement with the above, some typical curves being given in figure 11.4. It can be seen that the top size elutriated is approximately $\sqrt{(2)}D_m$ and the fraction of cut size retained is of the order of 0.5.

11.4.1 Water elutriators

Water elutriators are recommended for the size range 10 to 200 μm for powders with densities greater than 2.0 g ml^{-1}. With the Andrews elutriator [4] (figure 11.5), the sample of known weight is placed in the feed tube in the form of a dispersion. With wet samples, such as pottery slip, the weight of solids is determined by weighing the feed tube full of water W_w, then weighing it full of mixture W_m. The weight of solid is then:

$$W = \frac{W_m - W_w}{\rho_s - 1} \tag{11.27}$$

where ρ_s is the specific gravity of the solids.

A screw pinchcock at the inlet is first adjusted to give the required water velocity, which is evaluated by determining the volume rate of flow. The stopcock N is then closed to stop the upward flow of water; the bung at the top of the instrument is then removed and the feed tube inserted.

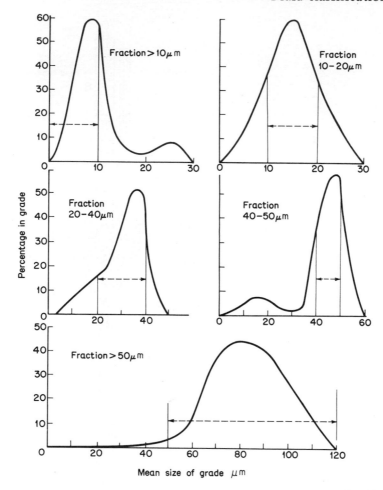

Fig. 11.4 Efficiency curves for the Gonell elutriator (Stairmand [3]). The fractions were analysed using Stairmand's photographic method.

The stopcock N is then reopened. Next the air vent D is opened, thus releasing the air lock between the surface of the water and the bottom of the feed tube, thereby starting the feed. As the particles fall out of the feed tube B, they are replaced by water rising through the inner tube E. The density gradient produces a circulatory motion from the outer vessel H into the central fitting. Aggregates are caused to impinge on the stationary cone F where they are broken up; fines are carried out through the overflow. After the solid material has all passed into H, the tap P is closed and the vent A opened. Closing P allows the feed to fall into L, while opening A allows the water to drain out of B, thus washing out any solid residue.

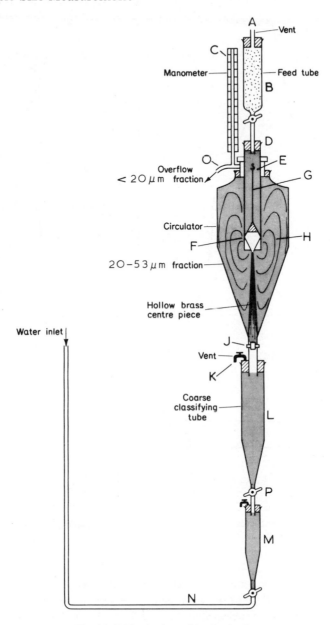

Fig. 11.5 The Andrews kinetic elutriator.

When the solid material has all fallen into L, P is reopened, thus restarting the water flow. When the upper portions of L and H are clear, the clamp at J and the stopcock at N are shut. The coarse fraction then gravitates into M from where it can be removed and weighed. M may then be replaced and the clamp at J opened so that the intermediate

fraction may fall into M and its weight determined. The weight of fines is determined by difference.

Other water elutriators have been described by Schöne [5], Werner [cit. 3], Andreasen [6] and Blythe [7]. The Blythe elutriator (figure 11.6) consists of six beakers almost filled with water and connected in series by siphons; the diameter of the up-going tube of each siphon is $\sqrt{2}$ times that of the preceding one. Each of the beakers is mechanically stirred and water from a constant head tank flows at a controlled rate through the system.

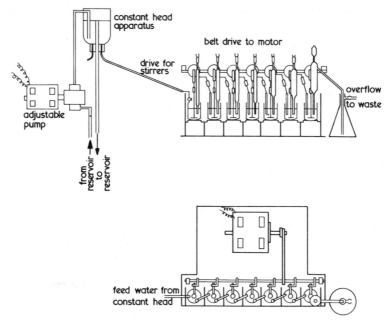

Fig. 11.6 The Blythe elutriator.

A dispersed powder is washed into the first beaker, where the coarsest particles are retained. All particles below a specific size, determined by the water velocity up the first siphon, are carried over to the second beaker. The entraining velocity in the second siphon is only half that of the first, so that again, the coarsest particles remain in the second beaker and the finer ones are carried along. It is thus possible to separate into seven grades. Though the process is lengthy, no attention is needed.

11.4.2 Air elutriators

Air flow is especially useful for powders which are, in practice, subject to grading by air flow, e.g. fine dust, which contains particles of different density with settling velocities which are not uniquely related to particle size. The major difficulties involved in air elutriation are to ensure the breaking up of aggregates of particles and to prevent fine particles from adhering to the sides of the elutriator tube.

The three main types of air elutriator are the up-blast, the down-blast and the circulating type. The disadvantage of the up-blast type is that at low rates of air flow, the
powder tends to choke the air inlet, causing a fluctuation of the velocity.

The Gonell elutriator (figure 11.7) has three cylindrical brass tubes of decreasing
diameter. Ancillary equipment consists of a blower, pressure-stabilizing and air-cleaning
equipment and a rotameter for measuring volume rate of flow. The sample tube has a
down-blast arrangement to prevent choking. The effect of electrostatic charge, which
causes particles to adhere to the walls of the tube, is reduced by continual mechanical
rapping of the tube. This effect may be further reduced by the use of an antistatic agent.

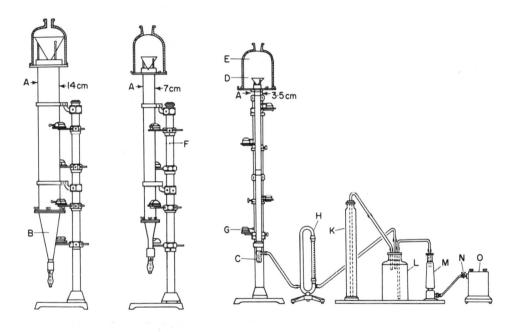

Fig. 11.7 Air-operated elutriator (Gonell [8]). A, air sifter tube (large, diameter
140 mm); B, conical tube, attached to A by a flange; C, glass projection, connected
to B by a rubber sleeve; D, top attachment, for discharging the air stream; E, glass
bell-jar on top place; F, stand with contact-holders for G; G, rappers; H, rotameter;
K, pressure governor with height-adjustable supply tube; L, buffer volume (4–5 l
flask); M, air cleaner and oil separator; N, regulating valve; and O, blower.

Fines are collected in glass containers at the top of the tubes, collection being aided by
the use of deflector cones, though these also create some turbulence in the column. The
analysis times may be greatly reduced if a modification suggested by Hughes [9] is employed. This consists of altering the shape of the dust reservoir and commencing the
analysis with a high-velocity blast to carry all the dust into the tube, but of such duration that the particles do not have time to reach the top of the tube.

The circulating type of elutriator as developed by Roller [10] (figure 11.8) is a modification of the up-blast, but the sample of powder is caused to circulate in a U-tube at the base of the elutriator tube. The elutriator has four chambers, to each of which is attached a paper extraction thimble collector for the fines. Ancillary equipment, including mechanical rappers, is included and the instrument is enclosed in a sound-insulated cabinet. In the miniature elutriator [11] a high-velocity air jet blows downward into a thimble at the base of the elutriating tube containing 0.1 to 0.5 g of the powder under test. The tube is much smaller than that of other elutriators, being 14 in long and of 1 in diameter. The infrasizer described by Haultain [12] consists of six elutriating tubes in series, air-flow entry being to the smallest. Air enters through a conical seating supporting a golf ball, which, by rotation and impact, breaks down agglomerations of particles.

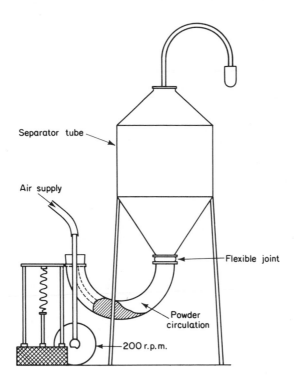

Separator tube

Air supply

Flexible joint

Powder circulation

200 r.p.m.

Fig. 11.8 The Roller particle size analyser.

The velocity distribution in a model Gonell elutriator was investigated by Weilbacher [14]. He found that the flow at the lower end of the tube was characterized by strong turbulence and instability and that separation was governed by this region. Because of this and other disadvantages, he developed a new classifier [1, 13, 14], shown in figure

11.9. The elutriation zone consists of a 1-cm high vertical tube with a height-to-diameter ratio of one-tenth. The material to be classified is dispersed on to an air-permeable medium, for instance, a membrane filter or fine mesh sieve. The emergent air flow is uniform and accelerates immediately after the short separation zone by a conical tube to eliminate redispersion of eluted material. With this classifier, an exact predetermination of cut size is possible. Leschonski [1] showed that to achieve a residue of about 60% for a cut size of 10 μm about 1000 min of separation were required with the Gonell elutriator, and 200 min with the Analysette. This is reduced to 80 min with the addition of a dispersing agent (Aerosil). Because of the long classification times, the method is not recommended below 10 μm. This equipment is now no longer available.

Recently a patent has been issued for an elutriator with perforated baffles to give disaggregation and a flat velocity profile [45].

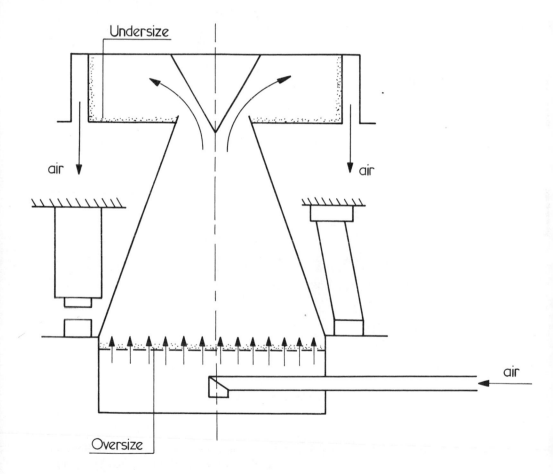

Fig. 11.9 Schematic design of the Fritsch Analysette 8 with a short cylindrical separation zone and accelerated removal of fines.

11.4.3 Zig-zag classifiers

Several versions of the Alpine Multiplex Zig-zag Classifier are available [33, 34] and these may be categorized as gravitational and centrifugal. In the gravitational laboratory classifier 1–40 MZM the material is fed into a zig-zag-shaped sifting tube in which there is an upward air flow (figure 11.10). The fine material is carried upward and the coarse downward and at every change in direction the material is graded to give good grade efficiency. This instrument will cut in the 0.1 to 6 mm range at a feed rate of up to 50 kg h^{-1}.

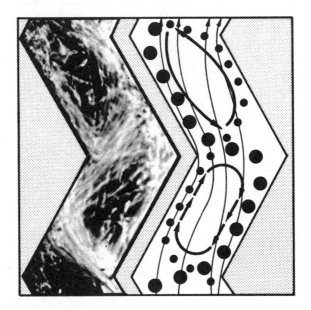

Fig. 11.10 The Zig-zag Gravitational Laboratory Classifier. **Left: Characteristic** material movement in zig-zag tube. There is clearly visible the formation of induced resistance in each limb. Right: Procedure of an individual classification. The small balls are the fine material, the large ones are the coarse material; bolts indicate materials' guide motion.

In the centrifugal version, 100 MZR, gravity is replaced by centrifugal force and the zig-zag chambers are built into a rotor (figure 11.11).

Cut size is in the range 1 to 100 μm at a feed rate of 0.5 to 3 kg h^{-1}.

11.5 Cross-flow gravity classifiers

11.5.1 The Warmain cyclosizer

The design and operating characteristics of the hydraulic cyclone elutriator are described in detail by Kelsall and McAdam [31]. Using inverted cyclones as separators with water as a medium, samples of between 25 and 200 g are reduced to five fractions having cut

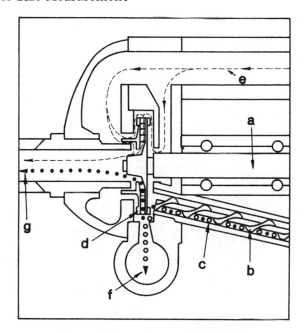

Fig. 11.11 The Zig-zag Centrifugal Laboratory Classifier. A feed-rating worm (b) conveys the unclassified material (c) laterally into the classifying chamber. Radially arranged blades on the outer face of the classifier rotor (d) speed the inflow of material up to the peripheral velocity of the rotor suspending it, at the same time, in the air admitted on either side of the classifying zone (e). The dust–air mixture is now sucked in by the zig-zag-shaped rotor channels. Within the controlled counter-current of fine and coarse particles inside these vortices the separation proceeds continuously, the separated fine material being sucked into the classifier centre (g) where it leaves via a cyclone. The coarse material (f) is expelled from the zig-zag channels by the centrifugal force. At the periphery of the classifier chamber it is flushed again by the entering classifying air before it is finally discharged into the wide-neck bottle.

sizes (for quartz, sp. gr. = 2.7) 44, 33, 23, 15 and 10 μm. The cyclones are arranged in series and during a run only the inlet and undersize outlets are open whilst the oversize for each particular cyclone is trapped and subjected to elutriating action for a fixed time period. At the termination of a run the trapped materials are extracted by opening the valves at the apex of each cyclone in turn and, after decantation, the solids are recovered by filtration or evaporation.

Particles smaller than the finest cut point are contained in 180 litres or more of water and are difficult to recover quantitatively and a procedure has been described by Kelsall [32] to simplify the procedure. This consists of dispersing the material in a 2-l beaker filled to a level at 16 cm above the base. After a settlement period of 1 h the upper 14.5 cm of suspension are removed. The residue is made up to the 16-cm mark and the procedure repeated. The procedure is repeated a further twice and the four decant suspensions are combined for subsequent flocculation, settling, dewatering, drying, weighing and assaying. The coarse solids from the decant are wet-screened on a

37-μm sieve and the undersize is cyclosized. The cyclosizer discharge water is fed directly to a continuous centrifuge and the retained solids (12 to 5.6 μm quartz) are washed from the centrifuge, settled overnight, washed with alcohol, dried and weighed. This procedure is the one adopted at CSIRO, Victoria, Australia.

11.6 Counterflow equilibrium classifiers in the centrifugal field

11.6.1 The Bahco classifier

The Bahco microparticle classifier (figure 11.12) is a combination air centrifuge-elutriator. The sample is introduced into a spiral-shaped air current created by a hollow disc rotating at 3500 rev min^{-1}. Air and dust are drawn through the cavity in a radially inward direction against centrifugal forces. Separation into different size fractions is made by altering the air velocity. Since no two instruments perform identically, instrument calibration is necessary [23, 24]; 5 to 10 g of dust are required for the sample, which can be graded in the size range 5 to 100 μm.

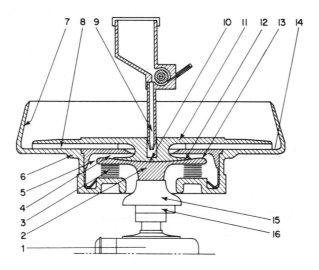

Fig. 11.12 Simplified schematic diagram of a Bahco-type microparticle classifier showing its major components: 1, electric motor; 2, threaded spindle; 3, symmetrical disc; 4, sifting chamber; 5, container; 6, housing; 7, top edge; 8, radial vanes; 9, feed point; 10, feed hole; 11, rotor; 12, rotary duct; 13, feed slot; 14, fan-wheel outlet; 15, grading member; 16, throttle.

As a first step in standardizing the evaluation of particulate removal equipment, the ASME Power Test Code 28 Committee undertook to recommend standard tests for measurement of the significant properties of fly ash. After investigating many devices for particle size evaluation, the committee selected the Bahco microparticle classifier as the standard instrument [35, 36, 38].

11.6.2 The BCURA centrifugal elutriator

The BCURA centrifugal elutriator [25] consists of two parallel concentric discs of equal radius mounted on a motor-driven spindle in a short cylindrical coaxial chamber. The discs are spun at a controlled rate and dusty air is drawn through the system entering the chamber axially and flowing outward over the surface of the first disc, then between the two from the periphery to the centre and leaving axially. Centrifugal forces act outwards and drag forces inwards, hence coarse particles are deposited on the circumference of the chamber and fine particles pass through the exit. There is considerable turbulence in the grading zone so that an empirical calibration has to be made. The range of the instrument is from 5 to 40 μm.

11.6.3 Centrifugal elutriation in a liquid suspension

Particles smaller than about 10 μm tend to agglomerate in air classifiers. This is also the lower limit with gravity water elutriators since at liquid velocities less than 10 μm s^{-1} flow is so irregular that separation is no longer possible. By applying a centrifugal field the lower limit of liquid elutriators can be extended down to 0.10 μm. Several such elutriators have been described by Colon et al. [39] with a maximum capacity of 400 cm^3 min^{-1} of a 0.2% w/v suspension.

11.7 Cross-flow equilibrium classifiers in the centrifugal field

11.7.1 Analysette 9

This system is used preferably in spiral classifiers, investigated for the first time by Rumpf [15] in 1939 and subsequently improved by Rumpf et al. [16, 17].

As a result the Analysette 9 centrifugal transverse-flow classifier was developed (figure 11.13) although, due to constructional difficulties, it has not been commercially available. The feed material is dispersed and accelerated in tube 1 and enters the flat classification chamber almost tangentially at a point on the circumference 2. The air leaves the chamber together with the fine fraction on spiral paths through the orifice 3 at the centre while the coarse fraction gathers at the circumference 4, constituting a ring of coarse material that spins along the peripheral wall of the classification chamber. The rotating ring of material enters the constant-velocity jet of air at 2 transversely at a very acute angle with a definite velocity and is distributed fanwise under the influence of drag and inertia forces.

The procedure repeats itself very frequently, i.e. of the order of 2000 times per minute. This high number of classifications coupled with the dispersion of agglomerates by friction forces in the rotating ring is claimed to lead to a sharpness of cut at very low sizes not previously obtainable in a spiral classifier. The coarse material is removed intermittently through a special outlet 5 and is collected on a small filter 6. The Analysette 9 provides a good sharpness of cut in the size range 2 to 12 μm. In this range it is claimed to be superior to the Bahco [1], described in Section 11.6.1, but the Bahco permits separation at coarser cut sizes with equally good sharpness. The cut size may be

varied by altering the amount of coarse material in the classifier and this can be estimated theoretically [22].

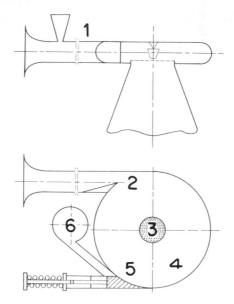

Fig. 11.13 Schematic design of the Analysette 9 centrifugal transverse-flow classifier.

11.7.2 The Donaldson classifier

The Donaldson ACACUT principle was patented in 1970 [46] and its mode of operation was described by Schaller and Lapple [21]. The principle is shown in figure 11.14 [47]. A vaned rotor (a) produces a centrifugal field while, at the same time, air is sucked out of the middle of the rotor. All but about 5% of the air intake, induced by a positive displacement pump downstream of the classifier, enters the classification zone through a very narrow gap formed between the rotor and the stator (b). This leads to an extremely high turbulence in the pre-classification zone (c). The material enters the classifier through a venturi-type nozzle (d) with the remaining 5% of air. The material slowly reaches the tangential velocity of the air. Between planes 1 and 2 the ratio of centrifugal force to drag force is kept very nearly constant by a diverging radial cross-section. This zone is the classifying zone. Equations of motion for such flow [48] and an analysis of particle trajectories [49] have been published together with a detailed description of the system [47]. The smaller particles are carried out through the middle and the larger ones (and agglomerates) move towards the stator, where they undergo disaggregation until they reach the exit (e). This system has several claimed advantages over others [22]. The ACACUT range consists of a laboratory model A12 which can handle 10–30 kg h^{-1}; a pilot plant model B18 to handle up to 400 kg h^{-1} and a production scale model C24 to handle up to 900 kg h^{-1}. A new version of the B18 has recently been described, the HT24 with a 400% increased capacity. The cut size of these machines ranges from 0.5 to 50 μm.

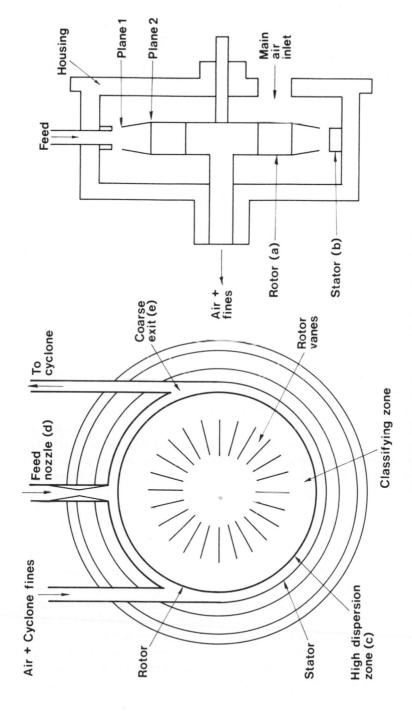

Fig. 11.14 Schematic diagram of the Donaldson classifier.

11.7.3 The Micromeritics classifier

The Micromeritics model 1001 [26] works on the following principle. A deagglomerated stream of particles is sucked from a dispersion device into the centre of a rotor where it divides into two streams. Air of either stream flows radially outward to the rotor wall, while the particles follow curvilinear paths depending on their size and density. At the wall they contact either of two paper or plastic films which can be removed once a sufficient deposit is collected. The film can then be cut into segments, each containing discrete sizes, or the particles can be scraped off into a series of separate containers.

11.8 Other commercially available classifiers

Some other commercially available classifiers, which are intended mainly for separating rather than size analysis, are listed in Table 11.4.

Table 11.4 Commercially available classifiers.

Type	Cut size	Capacity
Walther air vortex classifier	2–20 μm	3–4 tn h^{-1} and 1–60 lb h^{-1}
Multiplex laboratory zig-zag classifier	+ 1 μm	$<$5 kg h^{-1} to tn h^{-1}
Double-cone air separator	$<$30 mesh	tn h^{-1}
Head Wrightson air classifier	75 μm	–
Alpine Microplex spiral-air classifier	3–90 μm	–
Hosokawa micron separator	10–140 μm	50–1500 kg h^{-1}
Self-contained mechanical air separators		up to 100 tn h^{-1}
Revolving-blade air classifier	325–10 mesh	
British Rema microsplit	3–60 μm	ca. 1000 lb h^{-1}

11.9 Hydrodynamic chromatography

Hydrodynamic chromatography (HDC) offers a means of obtaining size information on colloidally suspended particles with the same ease that is characteristic of chromatographic methods for size analysis of molecules in solution. The basic experimental set-up is illustrated in figure 11.15. A fluid, generally an aqueous solution, is pumped through a column packed with impermeable spheres. Means are provided for injecting

about 0.2 cm^3 of colloidal suspension, containing about 0.01% polymer by weight, into the flowing stream at the entrance to the column and monitoring the colloid in the column effluent. Small [40, 41, 43] noted two points regarding the effluent: (1) larger particles elute faster than small ones; (2) the smaller the packing diameter the better the separation. The method is applicable for the size range 0.01 to 1 μm. A marker is added with the latex samples and calibration can be effected by the difference in exit times between the marker spheres and the latex.

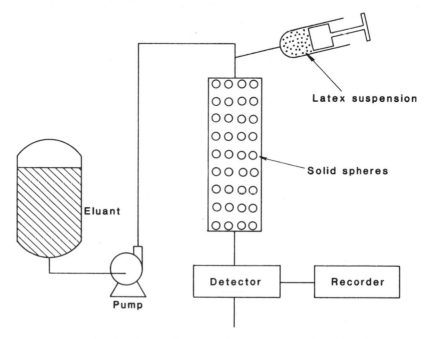

Fig 11.15 Experimental arrangement for HDC experiment.

A theory was later developed to explain this behaviour [42] based on the premise that the large particles could not sample the slower velocities close to the wall and hence move with a greater than average velocity than the fluid or smaller particles. A distribution of elution times due to axial dispersion also occurs leading to a Gaussian distribution of velocities for monosize particles.

An extension of the technique to powders such as cement, flour and chalk has also been described [44]. The packed column consists of 50 to 250 μm diameter particles.

References

1 Leschonski, K. and Rumpf, H. (1968-9), *Powder Technol.*, **2**, 175-85.
2 Roller, P.S. (1931), *Ind. Engng Chem., analyt Edn*, **3**, 212-16.
3 Stairmand, C.J. (1947), *Symp. Particle Size Analysis*, Inst. Chem. Engrs., **25**, 77.
4 Andrews, L. (1927-8), *Proc. Inst. Engng, Inspection*, **25**; also (1929), *Min. Mag., May* 301.

5 Schöne, E. (1867), *Über Schlammanalyse und einen neuen Schlammapparat,* Berlin (*cit.* [2]).
6 Andreasen, A.H.M. (1930), *Ber. dt. Keram. Ges.,* **11**, 675.
7 Blythe, H.N., Pryor, E.J. and Eldridge, A. (1953), *Symp. Recent Developments in Mineral Dressing,* Inst. Min. Metall. London., 23-5 Sept., 1952, p. 11.
8 Gonell, H.W. (1928), *Z. Ver. dt. Ing.,* **72**, 945.
9 Hughes, T.H. (1957), *2nd Conf. Pulverized Fuel,* Inst. Fuel, London, November, Paper 9.
10 Roller, P.S. (1932), *ASTM Proc.,* **32**, 11, 607; (1937), *J. Am. Ceram. Soc.,* **20**, 167.
11 Stairmand, C.J. (1951), *Engineering,* **171**, May, 585-7.
12 Haultain, H.E.T. (1937), *Trans. Canad. Min. Metall.,* **40**, 229.
13 Berns, E.G. (1954), *Br. J. appl. Phys.,* suppl. No. 3, S208.
14 Weilbacher, M. and Rumpf, H. (1968), *Aufbereit. Technik,* **9**, 7, 323-30.
15 Rumpf, H. (1939), thesis, Univ. Karlsruhe, (*cit.* [1]).
16 Rumpf, H. and Kaiser, F. (1952), *Chem. Ing. Tech.* **24**, 129-35.
17 Rumpf, H. and Leschonski, K. (1967), *ibid.,* **39**, 21, 1231.
18 Richards, J.C. (1966), The efficiency of classifiers, *BCURA Mthly Bull.,* **30**, 4, 113.
19 Newton, H.W. and Newton, W.H. (1932), *Rock Prod.,* V35.
20 Wessel, J. (1967), *Aufbereit. Technik,* (2), 53.
21 Schaller, R.E. and Lapple, C.E. (1971), Particle size separation of plastic powders. Paper presented at 162nd Nat. Mtg Am. Chem. Soc., Washington, DC, USA.
22 Ebert, F. (1975), *Habilationsschrift,* T.U., Erlangen (*cit.* [47]).
23 Weilbacher, M. (1968), thesis, Univ. Karlsruhe, (*cit.* [1]).
24 Crandall, W.A. (1964), *Am. Soc. Mech. Engrs,* Winter Ann. Mtg, Publ. 64-WA/PTC-3.
25 Godridge, A., Badzioch, S. and Hawksley, P. (1962), *J. scient. Instrum.,* **13**, 611.
26 Burson, J.H., Keng, E.Y.H. and Orr, C. (1967-8), *Powder Technol.,* 305-15.
27 Svarovsky, L. (Ed.) (1978), *Solid-Liquid Separation,* Butterworths.
28 Gibson, K.R. (1977), *Powder Technol.,* **18**, 2, 165-70.
29 Leschonski, K. (1976), *Proc. Comminution and Air Classification,* Univ. Bradford Short Course.
30 Allen, T. and Baudet, M.G. (1977), *Powder Technol.,* **18**, 2, 131-38.
31 Kelsall, D.F. and McAdam, J.C.H. (1963), *Trans. Inst. Chem. Engrs,* **41**, 84, 94.
32 Kelsall, D.F., Private Communication, CSIRO Div. Chem. Engng, P.O. Box 312, Clayton, Victoria 3168, Australia.
33 Lauer, O. (1969), *Chem. Ing. Tech.,* **41**, 491-6.
34 Special Leaflets available from manufacturer Nos. 44, 58, 604.
35 Crandall, W.A. (1964), Development of standards for determining properties of fine particulate matter. Power Test Codes Committee contribution at the Winter Ann. Mtg, ASME, NY.
36 Todd, W.F., Hagan, J.E. and Spaite, R.A. (1963), Test dust preparation and evaluation, Taft Sanitary Engineering Center, US Public Health Service, Cincinnati.
37 de Silva, S.R. and Gühne, H. (1976), *Aufbereit. Technik,* **17**, 10, 515-19.
38 Stein, F. and Corn, M. (1976), *Powder Technol.,* **13**, 133-41.
39 Colon, F.J., van Heuven, J.W. and van der Laan, H.M. (1970), *Proc. Conf. Particle Size Analysis.* (eds Groves and Wyatt-Sargent), Soc. Analyt. Chem., London (1972), 42-52.
40 Small, H. (1974), *J. Colloid Interfac. Sci.,* **48**, 1, 147-61.
41 Small, H. (1977), *Chemtech.,* 7, 3, 196-200.
42 Stoisists, R.F., Pochlein, G.W. and Vanderhoff, J.W. (1976), *J. Colloid Interfac. Sci.,* **57**, 2, 337-44.
43 Small, H., Saunders, F.L. and Sole, J. (1976), *Adv. Colloid Interfac. Sci.,* **6**, 4, 237-66.
44 Kawahashi, M. *et al.* (1975), *Japan Kokai,* **54**, 391.
45 Neuzil, L., Bafrnec, M. and Bena, J., Czech Pat. 169 912 (cl G01N15100), 15 June 1977.
46 Lapple, C.E., Centrifugal classifier, US Pat. 3 491 879, Jan. 1970.
47 de Silva, S.R. (1979), *Powtech 79,* Birmingham, Org. Inst. Chem. Eng. and Specialist Exhibitions Ltd.
48 Burson, J.H., Cheng, F.Y.H. and Orr, C. (1967/8), *Powder Technol.,* **1**, 305.
49 de Silva, S.R. (1978), *Powder Europa,* Wiesbaden.

12 Centrifugal methods

12.1 Introduction

Gravitational sedimentation techniques have limited worth for particles below about 5 μm in size due to the long settling times involved. In addition, most sedimentation devices suffer from the effects of convection, diffusion and Brownian motion. These difficulties may be reduced by speeding up the settling process by centrifuging the suspension.

As with gravitational methods the data may be cumulative or incremental, homogeneous or two-layer. Calculations of size distributions from centrifugal data are more difficult than calculations from gravitational data since particle velocities increase as they move away from the axis of rotation. That is, the velocity of a particle depends on its position in the sedimenting suspension as well as its size, whereas for gravity sedimentation the settling velocity is dependent solely on size. One method of overcoming this difficulty is to use long-arm centrifuges so that the centrifugal force on all particles is approximately the same. Another solution is to use the line-start technique, in which a thin layer of concentrated suspension is introduced on to the surface of the bulk sedimentation liquid, often referred to as the 'spin fluid'. This technique suffers problems due to 'streaming', which is discussed later, and has been extended to a three-layer or to a 'buffered line-start' in an attempt to overcome these problems. With the line-start technique, all particles of the same size are in the same position in the centrifugal field and, hence, have the same velocity, simplifying the theoretical treatment.

With the cumulative homogeneous technique, no full solution has been available until recently, relating the weight settled out, the weight undersize and time, although partial solutions were published. To overcome this problem investigators tended to keep the time of analysis constant but varied the level of the suspension in the centrifuge.

The earliest applications of the centrifugal principle involved the modifications of existing laboratory centrifuges. There were several problems associated with these. Cylindrical tubes are unsuitable since the direction of motion of the particles is radial and this leads to deposition of particles on the walls of the tubes. This can be overcome by the use of specially shaped tubes and minimized by the use of long-arm centrifuges. Tangential forces set up during starting and stopping cause particles to be deposited on the walls of the centrifuge tubes unless the accelerating and decelerating times are protracted, a procedure which entails the use of correction factors. Convection currents

are set up which cause remixing and there are also problems in measuring the concentration gradient within the suspension. These problems have been overcome with the advent of shallow-bowl or disc centrifuges, which were first suggested in 1934 [1] but have only become popular since 1965.

12.2 Stokes' diameter determination

A particle settling in a centrifugal field is acted upon by two forces in opposition, a centrifugal force and a drag force. In the laminar-flow region this leads to the following equation:

$$\frac{\pi}{6}(\rho_s - \rho_f)D^3 \frac{d^2x}{dt^2} = \frac{\pi}{6}(\rho_s - \rho_f)D^3 \omega^2 x - 3\pi D\eta \frac{dx}{dt}$$

where x = distance from the axis to the particle.
dx/dt = outward velocity of particle,
$\rho_s;\rho_f$ = density of particle and suspension medium,
η = coefficient of viscosity of medium,
D = equivalent spherical diameter of particle,
ω = speed of rotation of centrifuge in radians per second.
At the terminal velocity this equation becomes:

$$3\pi D\eta \frac{dx}{dt} = \frac{\pi}{6}(\rho_s - \rho_f)D^3 \omega^2 x$$

$$\frac{dx}{dt} = u_c = \frac{(\rho_s - \rho_f)}{18\eta} D^2 \omega^2 x \qquad (12.1)$$

Comparing with Stokes' equation for gravitational settling:

$$u_{Stk} = \frac{(\rho_s - \rho_f)}{18\eta} gD^2$$

therefore

$$u_c = \frac{\omega^2 x}{g} u_{Stk} = \frac{V_t}{xg} = Gu_{Stk}$$

where V_t is the tangential velocity at distance x from the axis and G, the separation factor, is a measure of the increased rate of settling in a centrifugal field.

12.3 Line-start technique

12.3.1 Theory

Rewriting equation (12.1) in integral form:

$$\int_{S}^{r} \frac{dx}{x} = \int_{0}^{D_m} \frac{\rho_s - \rho_f}{18\eta} D^2 \omega^2 \, dt$$

$$\ln \frac{r}{S} = \frac{\rho_s - \rho_f}{18\eta} D_m^2 \omega^2 t \qquad (12.2)$$

$$D_m = \sqrt{\left[\frac{18\eta \ln(r/S)}{(\rho_s - \rho_f)\omega^2 t} \right]}$$

where t is the time for a particle of size D_m to settle from the surface of the fill liquid at distance S from the axis of the centrifuge to r, the measurement zone, which, for cumulative techniques, is equal to R, the distance from the axis to the bottom of the centrifuge. Hence at time t, all the particles at r will be of size D_m or for cumulative techniques, all particles greater than D_m will have settled out at time t.

12.3.2 Line-start technique using a photometric method of analysis

Treasure [34] found that the simple theory for the line-start technique was unsuitable when the concentration of the suspension was estimated by a photoextinction method, and developed the theory below.

For an injected layer of infinitesimal thickness with a light beam of thickness $2Z$ at distance r from the axis of the centrifuge, let $D_1 > D_m > D_2$ be the diameters of the particles reaching $(r + Z), r, (r - Z)$ from the surface of radius S in time t. Then, from equation (12.2):

$$\ln \frac{r}{S} = kD^2 t$$

where

$$k = \frac{(\rho_s - \rho_f)}{18\eta} \omega^2$$

$$\frac{1}{D_1^2} \ln \frac{r + Z}{S} = \frac{1}{D_m^2} \ln \frac{r}{S} = \frac{1}{D_2^2} \ln \frac{r - Z}{S}$$

Writing $D_1 = (1 + \theta) D_m$ and $D_2 = (1 - \beta) D_m$ gives

$$(1 + \theta)^2 = 1 + \frac{\ln [1 + (Z/r)]}{\ln (r/S)}$$

$$(1 - \beta)^2 = 1 + \frac{\ln [1 - (Z/r)]}{\ln (r/S)}$$

For both θ and β both small, it may be assumed that over the range $(\theta + \beta)D_m, n_D,$ the relative number of particles per unit micrometre range centred on D, may be expressed by the linear function $n_D = f + (g/D_m)D$ (see figure 12.1).

Thus the total particle cross-section in the beam is proportional to:

$$\int_{D_m (1 - \beta)}^{D_m (1 + \theta)} n_D D^2 \, dD$$

is proportional to $(\theta + \beta)(f + g)D_m^3$ if high-order terms are neglected (i.e. θ and β are small and approximately equal).

Now the optical density of suspension is proportional to the total particle scattering cross-section in the beam, i.e.

$$I \propto (f + g)(\theta + \beta)D_m D_m^2 K_m$$

where K_m is the extinction coefficient (see Chapter 11), i.e:

$$I/K_m \propto (f + g)D_m^2 \text{ since } \theta, \beta \text{ are constant}$$

$$\propto n_{D_m} \times \text{mass of one particle}$$

$$\propto m_{D_m}, \text{the relative mass of particles in unit range centred on } D_m.$$

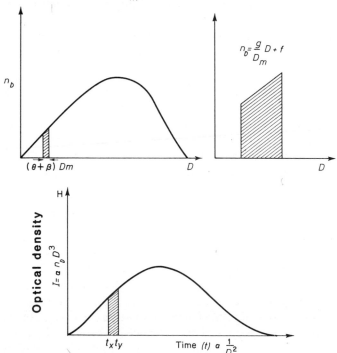

Fig. 12.1 Theory for the Kaye disc centrifuge.

If K_m is considered constant, the plot of optical density against time represents the weight–size frequency distribution of the suspension plotted with an inverse D^2 scale.

From the experimental curve,

$$\text{the area } (t_x - t_y) = \int_{t_x}^{t_y} I \, dt$$

$$\propto \int_{D_x}^{D_y} (n_D D^3) \left(-\frac{dD}{D^3} \right), \quad \text{since } t \propto \frac{1}{D^2}$$

$$\propto \int_{D_y}^{D_x} n_D \, dD$$

the relative number of particles in the range D_x to D_y. If each ordinate is multiplied by D^3 then the area:

$$(t_x - t_y) \propto \int_{D_y}^{D_x} n_D D^3 \, dD$$

the relative mass of particles in the range D_x to D_y.

Similarly multiplication by D^2 gives the relative surface of particles in the range. The same relationship holds also for an injected layer of finite thickness. For a layer thickness of about 3% of the radius of the curved liquid surface, the resultant error is about 1%.

Treasure also shows that the relationship for the gravitational case is similar:

$$I/K_m \propto m_{D_m}$$

Brugger [81] gives a detailed analysis of photometric measurements for both the line-start and homogeneous techniques. An empirical relation is pointed out which can simplify the evaluation for the former. It is also shown that this method can be used for the determination of complex refractive index. It is further established that a density gradient is necessary to guarantee turbulence-free sedimentation. Finally an automatic data acquisition system is described.

12.3.3 Early instruments: the Marshall centrifuge and the MSA particle size analyser

One method of simplifying the calculation is to float the suspension to be centrifuged on a column of denser liquid so that all particles settle essentially the same distance to reach the bottom of the centrifuge tube. This method was first used by Marshall [2]. Since all the particles commence sedimentation from the same point, the sediment will only contain particles greater than D_m (equation (12.2)). Also, the solids concentration at any point is proportional to the fraction of particles of a calculable size in the original suspension. Marshall simplified his technique further by making the settling distance $R - S$ small in comparison with the distance from the axis of rotation, thus permitting the approximation of a constant centrifugal force. It also allows the use of a

cylindrical settling vessel, since the walls subtend a smaller angle at the axis of rotation the larger their distance from this axis. Interest in the two-layer technique was re-awakened by Whitby [3], who used the special centrifuge tube shown in figure 12.2; the weight oversize is determined by the height of the sediment in the capillary.

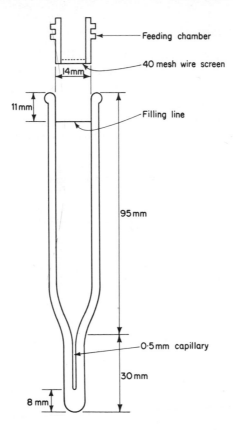

Fig. 12.2 Special centrifuge tube and feeding chamber. Sediment height is read with a projector-type viewer (Whitby [5]).

Several papers have been published on applications of this equipment, which is available commercially as the MSA particle size analyser [4–7]. Although sediment height is proportional to sediment weight for monosize particles and only one size settles at a time, criticism has been levelled at the instrument on the ground that many particles consist of tight aggregates. In these cases, compression of the sediment column with increasing speed takes place, thus necessitating the use of correction factors [8].

Other disadvantages are: the tube is the wrong shape to eliminate wall effects; hindered settling in the neck of the capillary prevents the analysis of materials with a narrow size range, which would all settle at about the same time; the loss of sedimentation height as sediment builds up in the capillary [9].

Provided it is followed precisely the method is very reproducible. Zwicker [88] found the method highly unsatisfactory and recommended that it should no longer be used. Its main advantage is that it is suited to both the gravitational and centrifugal range, hence a size range from about 0.2 to 80 μm may be analysed.

12.3.4 The photocentrifuge

The first photocentrifuge was developed by Kaye [10, 11] (figure 12.3). In this instrument, concentration changes within a suspension are followed using a light beam (see also [48]). The instrument is usually used in the two-layer mode and evaluated as a weight distribution using the theory developed by Treasure, but the technique has also been used with a homogeneous suspension [20]. The original instrument has been widely and successfully used by Burt and modifications have been used by others [57].

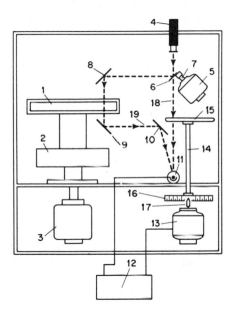

Fig. 12.3 The Kaye disc centrifuge. A hollow centrifuge tank (1) is mounted on the flywheel (2) of a centrifuge. A source of light (4) is mounted at the top of the container. A synchronous electric motor (5) has a mirror (6) eccentrically mounted to its spindle (7). Mirrors (8), (9) and (10) are arranged so as to reflect light on to a photocell (11) which has its output connected to an ac amplifier (12). This amplifier can energize an ac servomotor (13) whose spindle (14) carries an annular optical wedge (15), lying between the light source (4) and the photocell (11) and a scale (16) calibrated relative to a fixed pointer (17). Broken lines indicate the alternative paths of the light beam. When the motor (5) operates it swings the mirror (6) round causing it to reflect light periodically along path (19). When the intensities of the light reaching the photocell via the two paths are equal, there is no ac output from the photocell. If the intensity along path (19) is reduced to the presence of particles in the light beam, then the intensities are unequal, causing an energizing current to flow from the photocell to (13) causing (14) to rotate. As the optical wedge (15) rotates it balances out the intensities.

In the instrument described by Bayness *et al.* [52], the light source comprises a 12 V tungsten lamp and the receiver an ORP 12 photocell which forms part of a balanced electrical bridge. The authors had limited success with the two- or three-layer start method and used the homogeneous suspension in their reported results, making no correction for the breakdown in the laws of geometric optics. This modification is now incorporated in the Joyce–Loebl disc centrifuge. Statham [53] used a prototype instrument developed by Coulter Electronics, the Photofuge. He found streaming and distortion with the two-layer technique at concentrations above 0.01% w/w of PVC in the injection fluid, using 2 ml of fluid. At higher concentrations the three-layer technique [54] was used, but this resulted in a coarsening of the analysis. The use of 55 to 70% v/v of methanol in water as injection fluid with water with 0.00075% Nonidet P42 wetting agent as spin fluid improved the results. The results were corrected for the breakdown in the laws of geometric optics using Allen's method [55], which considerably altered the relative weight percentage distributions to give analyses which were in reasonable agreement with known distributions.

Provder and Holsworth [95] compared the Joyce–Loebl disc centrifuge used as a photocentrifuge with other techniques and found good agreement. However they used the two-layer technique and assumed that the attenuation was proportional to surface area. They therefore multiplied by particle size to obtain a weight distribution. An involuntary correction was applied for extinction coefficient whereby it was assumed that K was proportional to particle size. This assumption was investigated by Khalili [96], using both the homogeneous mode and the two-layer technique, and he found it to be valid.

Wood *et al.* [82–84] describe a photocentrifuge which they used to analyse herbicide in the size range 0.2 to 7 μm; they compared their results with those from electron microscopy and the Coulter counter.

A laser centrifugal photocentrifuge has also been described [85–87] (Technord). The value of such a system is, in my experience, questionable. Experiments carried out in my laboratory using combinations of a white-light source, a laser, a photodiode and a barrier layer photocell gave similar results; hence the cheapest system, a white-light source and a barrier layer photocell, is preferred. For very fine powders, where the height of fall $(r - S)$ is small, a photodiode may be preferred since this can be used with a narrower beam and thus improve the resolution. The size distributions using line-start techniques and a homogeneous suspension were very different. Although in theory, corrections can be made for extinction coefficient, these were found to be of such a magnitude that slight differences in the uncorrected experimental data led to gross differences in the derived curves. The method is therefore only recommended for comparison purposes.

The Photo Micronsizer (Seishin) operates with a white-light source and photodiode. The suspension is placed in a cell which may be monitored under gravity or in a centrifugal field.

12.3.5 The Joyce–Loebl disc centrifuge

This instrument (figure 12.4) was developed for the determination of the size distribution of organic dyestuffs [58, 59]. It is used with a buffered line-start, and the under-

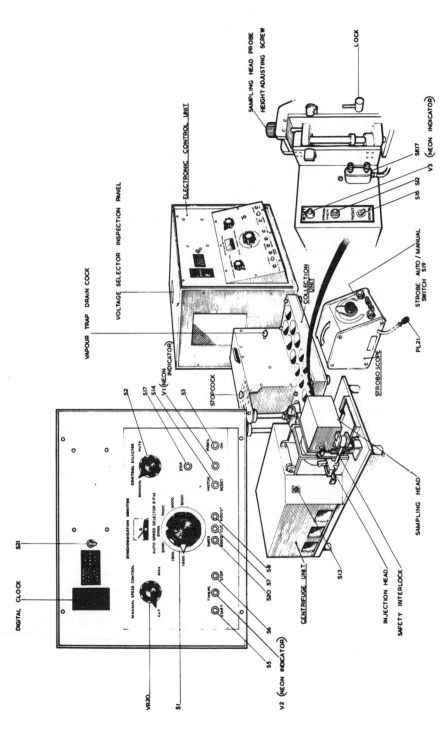

DISC CENTRIFUGE CONTROL POSITIONS

Fig. 12.4 The Joyce–Loebl disc centrifuge.

size fraction determined by removing the supernatant with a vacuum probe after fixed times of rotation and determining the concentration calorimetrically. Although gravimetric analyses can be made, this procedure is not recommended because of the small amount of sample used [12, 13]. Since each withdrawal gives only one point on the size analysis graph, only one or two analyses can be carried out per day. In later instruments the manufacturers have incorporated a photocell and detector for quick routine comparisons and to analyse powders which the unmodified instrument could not handle. It must be stressed, however, that the curve so produced is not a size distribution as is demonstrated by Statham [53]. The instrument has been widely described [60–67]. Lombard and Carr [95] determined a calibration curve for the photocell and detector for a pigment in order to convert the photocentrifuge results to absolute values.

It is usual with the line-start method to disperse the solids in a liquid which is not as dense as the fill liquid in order to get stable conditions. However, it has been pointed out [15] that as soon as the particles break through the interface, the uppermost region is denser than the rest, and consequently breaks up and streamers occur. This effect can be demonstrated with the use of a stroboscope. Statham [53] and Burt [68] report a lower concentration below which streaming does not occur, but Beresford [66] found streaming at concentrations of 0.004% w/v.

Scarlett et al. [54] found the following method satisfactory: 10% aqueous sucrose spin fluid, add 1 cm³ of distilled water to the inner surface while the disc is in uniform motion and then inject the aqueous dispersion. Jones [69, 70], in the 'buffered layer-start' technique, introduces a transient acceleration after the introduction of water to give a gradual density gradient instead of an interface. Beresford [66] tested the various techniques and reported that the buffered layer-start was the only technique to eliminate streaming. As stated earlier, this is at variance with the results of other researchers.

12.4 Homogeneous suspension

12.4.1 Sedimentation height small compared with distance from centrifuge axis

The simplest procedure with a homogeneous suspension is to make $r - S$ small compared with S and assume that the particles fall with constant velocity:

$$u_c = \frac{(\rho_s - \rho_f)}{18\eta} D^2 . \omega^2 \left(\frac{r + S}{2} \right)$$

from equation (12.1).

This method was used by Norton and Spiel [16], who used hydrometers to follow density changes within the settling suspension; by Jacobsen and Sullivan [17] and Menis et al. [18], who used a cumulative method and Oden's method of tangential intercepts; by Gupta [24] who withdrew samples perpendicular to the plane of rotation and determined the concentration changes gravimetrically; by Conner et al. [19], who determined the sediment weight by β-back-scattering and by Hildreth and Patterson [20], who used the attenuation of a light beam in a disc centrifuge similar to Kaye's

and dealt with their experimental data according to a treatment by Musgrove and Harner [21].

12.5 Cumulative sedimentation theory for a homogeneous suspension

Equation (12.2) may be written (for $r = R$):

$$S = R \, \exp(-kD_m^2 \, t) \qquad (12.3)$$

where

$$k = \frac{\rho_s - \rho_f}{18\eta} \, \omega^2$$

At the end of t seconds, all particles greater than D_m will have reached the bottom of the tube. In addition, partial sedimentation will have taken place for all particles smaller than D_m. For each of these smaller sizes, a starting point x_0 exists, beyond which all the smaller particles will have reached R where, from equation (12.1):

$$x_0 = R \, \exp(-kD^2 t) \qquad (12.4)$$

The volume fraction of the suspension lying between R and x_0 for a shallow bowl or flat sector-shaped tubes is equal to:

$$\frac{R^2 - x_0^2}{R^2 - S^2} = \frac{R^2}{R^2 - S^2} \, [1 - \exp(-2kD^2 t)] \qquad (12.5)$$

If the particle size distribution is defined such that the weight fraction in the size range D to $D + dD$ is $f(D)dD$ then the weight fraction of particles with diameters greater than D_m that have completely settled is:

$$W = \int_{D_m}^{\infty} f(D)dD$$

The weight fraction of particles smaller than D_m that have completely settled is:

$$I = \frac{R^2}{R^2 - S^2} \int_0^{D_m} [1 - \exp(-2kD^2t)] \, f(D)\,dD \qquad (12.6)$$

The total weight fraction deposited is:

$$P = W + \frac{R^2}{R^2 - S^2} \int_0^{D_m} [1 - \exp(-2kD^2t)] f(D)\,dD \qquad (12.7)$$

The weight fraction oversize can be evaluated if the weight fraction deposited is measured for different values of the variables S, R and t.

Similarly, the fraction of particles still in suspension at time t will consist of particles smaller than D_m that have originated in the volume between radius S and x_0. By comparison with equations (12.5) and (12.6) this fraction is:

$$P(D_m) = (1 - I) = \int_0^{D_m} \left(\frac{x_0^2 - S^2}{R^2 - S^2} \right) f(D) \, dD$$

$$P(D_m) = \frac{R^2}{R^2 - S^2} \int_0^{D_m} [\exp(-2kD^2t) - \exp(-2kD_m^2 t)] f(D) \, dD$$

$$= \frac{1}{1 - \exp(-a)} \int_0^{D_m} \left[\exp\left(\frac{aD^2}{D_m^2} \right) - \exp(-a) \right] f(D) \, dD \qquad (12.8)$$

where

$$a = 2 \ln \left(\frac{R}{S} \right) \qquad (12.9)$$

12.6 Variable-time method (variation of P with t)

Romwalter and Vendl [25] derived a solution to equation (12.7) by differentiating with respect to time and substituting back in the original equation. Brown [26] drew attention to an error in their derivation and stated that an exact solution for the distribution function appeared to be difficult, if not impossible, to obtain by the above method.

The following approximate solutions were derived by Robison and Martin [27, 28], who used sector-shaped tubes. Their analyses agreed closely with those obtained by the variable-height method:

$$\int_0^{D_m} f(D) \, dD = 1 - \left[\frac{M(6 - M)}{8} P + \frac{MD_m}{8} \frac{dP}{dD} + \frac{M(M - 2)(M - 4)}{8} I(D_m) \right] \qquad (12.10)$$

where

$$I(D_m) = \frac{1}{D^M} \int_0^{D_m} PD^{M-1} \, dD$$

and

$$M = \frac{4(R^2 - S^2)}{S^2 \ln(R/S)}$$

An exact solution to equation (12.10) as given by Kamack [71], is as follows:

$$F(D_m) = \left(\frac{\exp(a) - 1}{a}\right) \left(q(D_m) - \int_0^\infty h_1(x)q\,(D_m \exp(-x))\right)$$

where

$$q(D) = p(D) + \tfrac{1}{2}D\frac{dP}{dD} \tag{12.11}$$

and

$$h_1(x) = h(x)\exp(-2x)$$

$$x = \ln\left(\frac{D_m}{D}\right)$$

and the 'resolvent kernel' $h(x)$ is a function that depends on the apparatus constant (a).

Muschelknautz [90-92] designed a centrifuge in which the displacement of two diametrically opposed bodies floating in a dispersion was measured. The bodies are fixed on a common rod and are immersed at different depths in two chambers. The differential force yields the size distribution directly.

Sokolov [93] described a centrifugal sedimentometer with a float measurement system.

12.7 Variable inner radius (variation of P with S)

Brown [26] avoided the complications arising from the differentiation of equation (12.7) with respect to time by considering the fraction sedimented in a given time interval with the centrifuge tubes filled with suspension to a series of levels. On increasing S an increasingly large fraction of suspended particles will be deposited in a given time.

Differentiating equation (12.7) with respect to S and substituting back with $\delta P/\delta S$ gives:

$$\int_{D_m}^\infty f(D)\,dD = P - \frac{R^2 - S^2}{2S}\frac{\delta P}{\delta S} \tag{12.12}$$

Thus a rigorous solution is obtained by keeping t and R constant and determining the weight fraction deposited for varying heights of suspension.

Just as in gravity sedimentation, second derivatives of the fraction sedimented are required to obtain the distribution function itself. In order to obtain $F(D_m)$ in terms of the second derivatives of P, it is necessary to differentiate equations (12.3) and (12.7) with respect to S, eliminating $\delta D_m/\delta S$ from the two resulting equations:

$$\frac{\delta D_m}{\delta S} = \frac{-D_m}{2S\ln(R/S)} \quad \text{from equation (12.3)}$$

giving, in combination with the differentials of equation (12.7):

$$\frac{\ln(R/S)}{D_m} \left[\frac{R^2 + 3S^2}{S} \frac{\delta P}{\delta S} - (R^2 - S^2) \frac{\delta^2 P}{\delta S^2} \right] = F(D_m) \qquad (12.13)$$

Similarly the distribution function may be derived in terms of $\delta^2 P/\delta S \delta t$ and $\delta P/\delta t$ by differentiating equations (12.3) and (12.7) with respect to time:

$$\frac{2t}{D_m} \frac{\delta P}{\delta t} - \frac{R^2 - S^2}{2S} \frac{\delta^2 P}{\delta S dt} = F(D_m) \qquad (12.14)$$

Three distinct methods are therefore available for calculating the distribution of particle sizes in a suspension, if the weight fraction sedimented is determined with sector-shaped centrifuge tubes filled to a series of levels. First, the weight fraction of particles larger than a known diameter may be calculated from equation (12.13) and the distribution function determined from the slope of the cumulative weight per cent deposited curve. Secondly, the distribution function may be calculated directly in terms of the first and second derivatives of the fraction sedimented with respect to the length of the column of suspension centrifuged by use of equation (12.13). Thirdly, from the sedimentation-time curve at a series of levels, the distribution functions may be calculated by use of equation (12.14). In all cases the range of particle size covered is that of D_m as calculated from equation (12.3).

12.8 Shape of centrifuge tubes

The use of cylindrical tubes instead of sector- or conoidal-shaped tubes (figure 12.5) has advantages in that they are easier to construct and may be used in ordinary laboratory centrifuges. The disadvantages of a cylindrical tube are that particles will strike the walls of the tube, agglomerate with other particles on the wall and reach the bottom more quickly than freely sedimenting particles, and convection currents will be set up due to the oblique force of the suspension on the walls of the tube.

Brown [26] states that for control tests sufficient accuracy is obtained if $S/R < 0.50$ A figure of $S/R < 0.88$ is given by Bradley [9]. However, as $R - S$ becomes smaller, it becomes more difficult to measure accurately and may produce large errors in the analysis.

The theory is modified if cylindrical tubes are used, equation (12.5) becoming:

$$\frac{R - x_0}{R - S} = \frac{R}{R - S} [1 - \exp(- kD^2 t)]$$

This modifies equation (12.13) to:

$$\int_{D_m}^{\infty} f(D) \, dD = P - (R - S) \frac{\delta P}{\delta S} \qquad (12.15)$$

Instead of varying the quantity of suspension in the tubes S, the above equation may be used with a fixed quantity of suspension and time but with changes of R with $R - S$

constant. This is readily accomplished with cylindrical tubes, by placing blocks of known thickness under the tubes [26].

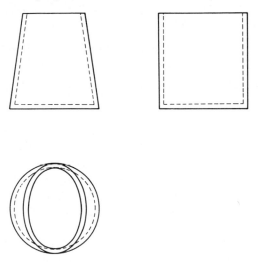

Fig. 12.5 Conoidal centrifuge tube.

12.9 Alternative theory (variation of P with S)

An alternative approach is given by Murley [29] as follows:

If the inner radius of the centrifuging suspension is decreased by a small amount dS, then the extra weight of sample introduced into the centrifuge is $2\pi STC \, dS$, where T is the thickness of the suspension in an axial direction and C is the weight of solid per unit volume of suspension. For this extra amount of material added, all that with a particle size less than D_m will reach the collecting plane at R and all that greater than D_m will be at a smaller radius than R at the end of the running time t. The diameter D_m is given by equation (12.3).

The extra weight of sample deposited at a place of radius R due to this change in radius dS of the top surface of the sample is therefore:

$$\Delta P = - \, 2\pi STC \, dS \int_{D_m}^{\infty} f(D) \, dD \qquad (12.16)$$

The negative sign occurs because the added layer causes a decrease in S. This equation applies to an apparatus where the liquid is run off and the deposited layer is retained for analysis, whereas in the type of apparatus where the overlying liquid layer is removed for estimation of the weight of solids, the following equation is applicable:

$$\Delta P = - \, 2\pi STC \, dS \int_{D_m}^{0} f(D) \, dD \qquad (12.17)$$

By plotting P against S and finding the slope of the curve, $f(D)\,dD$ may be evaluated.

Equations (12.16) and (12.17) can also be derived by differentiating equation (12.7) which may be written:

$$P = \left\{ 1 - \int_0^{D_m} f(D)\,dD \right\} + \int_0^{D_m} \frac{R^2}{R^2 - S^2} \left\{ 1 - \exp(-2kD^2 t) \right\} f(D)\,dD$$

$$1 - P = \int_0^{D_m} \frac{1}{R^2 - S^2} \left\{ R^2 \exp(-2kD^2) - S^2 \right\} f(D)\,dD$$

where $1 - P$ is the weight fraction still in suspension. The weight fraction of powder that has sedimented is P, where $P = \pi(R^2 - S^2)\,TC(1 - P)$, i.e.

$$P = \pi TC \int_0^{D_m} [R^2 \exp(-2kD^2) - S^2]\,f(D)\,dD \qquad (12.18)$$

Differentiating this with respect to S leads directly to equation (12.17).

12.10 Variable outer radius (variation of P with R)

Donoghue and Bostock [30] differentiated equation (12.7) with respect to R, giving:

$$\int_{D_m}^{\infty} f(D)\,dD = P + R \left(\frac{R^2 - S^2}{2S^2} \right) \frac{dP}{dR} \qquad (12.19)$$

Slope measurements of the P-R curve permit calculation of the weight fraction oversize.

The apparatus developed for this determination consists of a stepped centrifuge so that the suspension is contained in a space consisting of a number of rings, each of which has the same inner radius S, but progressively larger outer radii R from top to bottom. Particles are deposited on detachable surfaces which are removed and dried before weighing, the supernatant being removed before the centrifuge is stopped.

The advantages of this instrument are that six points on the distribution curve are determined simultaneously and the quick acceleration removes the need for correction terms for accelerating time. The main disadvantage which has prevented this centrifuge from becoming accepted is loss of sediment due to movement of the supernatant liquid during its withdrawal.

12.11 Incremental analysis with a homogeneous suspension

12.11.1 The Simcar centrifuge

The disc centrifuge has replaced earlier types and is now available in a variety of forms. The pipette method has long been used as a reliable method of gravity sedimentation and one of the earliest commercial disc centrifuges was a centrifugal pipette withdrawal

technique [33, 56] (figure 12.6). With this instrument, concentration changes are monitored at a fixed depth below the surface. This concentration is related to the weight
undersize by an integral equation for which no usable mathematical solution was known
(equation (12.25)). This difficulty is avoided if the variation in centrifugal force over the
settling distance is made small, i.e. $r - S \ll S$, where r is the distance from the axis of
rotation to the sampling zone, but this creates design problems in the centrifuge construction. The method becomes feasible, however, with the introduction of the approximate method described below.

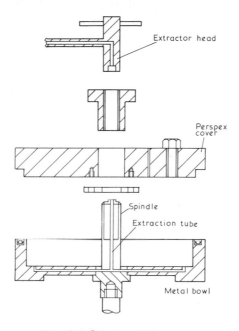

Fig. 12.6 Simcar centrifuge.

12.11.2 General theory

The largest particle, of size D_m, present in the measurement zone, at radius r and time
t, will have started from the surface at radius S. From equation (12.2) the following
relationship will hold:

$$\left(\frac{r}{S} \right) = \exp\left[\left(\frac{\rho_s - \rho_f}{18\eta} \right) D_m^2 \, \omega^2 t \right] \tag{12.20}$$

$$= \exp\left(kD_m^2 \, t \right) \tag{12.21}$$

where

$$k = \left(\frac{\rho_s - \rho_f}{18\eta} \right) \omega^2 \tag{12.22}$$

Particles in the measurement zone of size D_i will have originated from radius r_i where $r > r_i \geqslant S$ and:

$$\frac{r}{r_i} = \exp\left(kD_i^2 t\right) \tag{12.23}$$

The particles originally at radius r_i, in an annular element of thickness Δr_i, move in diverging (radial) paths and at radius r occupy an annular element of thickness Δr. There will be a fall in the concentration of particles of this size in the measurement zone therefore, since the same number of particles will occupy a greater volume.

The fractional increase in volume is given by:

$$\frac{r \Delta r}{r_i \Delta r_i} = \left(\frac{r}{r_i}\right)^2 \tag{12.24}$$

since $\Delta r / \Delta r_i = r/r_i$ from equation (12.23).

For a polydisperse system with a weight fraction in the size range D to $D + dD$ of $f(D)\,dD$, the concentration dQ of this weight fraction at r is given by:

$$dQ = \left(\frac{r_i}{r}\right)^2 f(D)\,dD$$

hence:

$$Q(D_m) = \int_0^{D_m} \left(\frac{r_i}{r}\right)^2 f(D)\,dD \tag{12.25}$$

Combining with equation (12.23) gives:

$$Q(D_m) = \int_0^{D_m} \exp\left(-2kD_i^2 t\right) f(D)\,dD \tag{12.26}$$

Substituting for k from equation (12.21):

$$Q(D_m) = \int_0^{D_m} \exp\left[-2\left(\frac{D_i}{D_m}\right)^2 \ln \frac{r}{S}\right] f(D)\,dD \tag{12.27}$$

Two methods of analysis are available.

(1) Variable height method. The concentration is measured as a function of (r/S) and all other variables are kept constant.
(2) Variable time method. The concentration is measured as a function of $(\omega^2 t)$ and all other variables are kept constant.

The use of the variable time method requires a solution to equation (12.27). A practical approximate solution may be derived which depends on the variable height method formulated in equation (12.26) which has a simple exact solution as follows:

$$\frac{dQ}{dD} = \exp(-2kD^2t)f(D) \tag{12.28}$$

where

$$f(D) = \frac{dF}{dD}$$

The boundary conditions are that $Q = 1$ when $t = 0$ for all r; $Q = 0$ when $r = S$ for $t > 0$; and the additional condition that $F = 0$ when $D = 0$. Thus:

$$F(D_m) = \int_0^{D_m} \exp(2kD^2t)\,dQ$$

$$= \int_0^{Q_m} \left(\frac{r_i}{S}\right)^2 dQ \tag{12.29}$$

D_m is the diameter of the particle that settles from the surface, radius S, to radius r_i in time t. This expression was first developed by Berg [31] and later by Kamack [32].

If Q is plotted as a function of $y = (r_i/S)^2$ with $t' = \omega^2 t$ as parameter, a family of curves is obtained whose shape depends on the particle size distribution function. The boundary conditions are that $Q = 1$ when $t' = 0$ for all r_i and $Q = 0$ for $r_i = S$ when $t > 0$, hence all the curves except that for $t' = 0$ will pass through the point $Q = 0$, $y = 1$, and they will all be asymptotic to the line $t' = 0$, which has the equation $Q = 1.0$. Furthermore, from equation (12.23), the area under the curve is equal to $F(D_m) = F_m$.

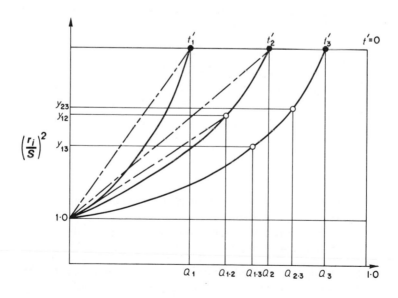

Fig. 12.7 Theoretical diagram for the homogeneous centrifuge technique.

Let Q_1 be the lowest experimentally determined concentration so that $t_1 > t_2 > \ldots$, and let Q be determined at a fixed sampling distance r for various values of t'. Then one point is known on each curve in addition to the common point $y = 1$, $Q = 0$. Such a set of points is illustrated by the black circles in figure 12.7. To each such point corresponds a known value D_m obtained from equation (12.2). Further, the area included between each curve and the concentration axis and the ordinates $Q = 0$ and Q_m is equal to F_m. Thus F_m may be approximated by the trapezoidal rule, for, first of all, approximately $F_1 = \frac{1}{2}(1 + y)Q_1$. Now considering the curve for t_2', a point can be found on it corresponding to D_1, i.e. a point such that the area under the curve up to this point is F_1, which is now known. If the ordinate at this point is called y_{12} and the abscissa Q_{12}, then by equation (12.2):

$$D_1 = \sqrt{\left[\frac{9\eta \ln y_{12}}{(\rho_s - \rho_f)\, t_2'}\right]} \quad \text{and} \quad D_2 = \sqrt{\left[\frac{9\eta \ln y}{(\rho_s - \rho_f)t_2'}\right]}$$

$$\text{so } y_{12} = y^{(D_1/D_2)^2} \tag{12.30}$$

Also, equating areas:

$$Q_1(1 + y) = Q_{12}(1 + y_{12}) = 2F_1 \tag{12.31}$$

so both y_{12} and Q_{12} are known. Hence, by the trapezoidal rule:

$$F_2 - F_1 = \frac{1}{2}(y + y_{12})(Q_2 - Q_{12})$$

Substituting for Q_{12} from equation (12.31):

$$F_2 = \frac{1}{2}(y + y_{12})Q_2 + \left[1 - \frac{y + y_{12}}{1 + y_{12}}\right] F_1$$

Proceeding in a like manner gives the general formulae:

$$F_n - F_{n-1} = \frac{1}{2}(y + y_{n-1,\, n})(Q_n - Q_{n-1,\, n}) \tag{12.32}$$

$$F_{n-1} - F_{n-2} = \frac{1}{2}(y_{n-1,\, n} + y_{n-2,\, n})(Q_{n-1,\, n} - Q_{n-2,\, n}) \tag{12.33}$$

and so on. By considering this series of equations with successive elimination of the Q functions, there obtains a general solution in recursive form:

$$F_i = \frac{1}{2}(y + y_{i-1,\, i})Q_i + \sum_{j=1}^{i-1} \left[\frac{y + y_{i-1,\, i}}{y_{j+1,\, i} + y_{ji}} - \frac{y + y_{i-1,\, i}}{y_{ji} + y_{j-1,\, i}}\right] F_j \tag{12.34}$$

where

$$y_{ij} = y^{(D_i/D_j)} \tag{12.35}$$

$$D_n = \sqrt{\left(\frac{9\eta \ln y}{(\rho_s - \rho_f)t_n'}\right)} \tag{12.36}$$

$$i = 1, 2, 3, \ldots m$$

$$y_{0,i} = 1$$

Equations (12.34) are a set of linear equations which express the desired values of F_t explicitly in terms of the measured values of Q_i. The coefficients of the equations depend on the value of D_i (corresponding to the values of t_i') at which the concentrations are measured; more exactly, the coefficients depend on the ratios of the values of D_i as shown by equation (12.35). Consequently, if the values of D_i are chosen in a geometric sequence when making particle size analysis, the coefficients of equation (12.34) are considerably easier to calculate and the equations themselves are also simplified. A ratio of $\sqrt{2}$ is recommended. The coefficients in equation (12.32) depend also on the value of y, that is, on the dimension of the centrifuge bowl employed.

The modified form of equations (12.34) for experimental points in a $2:1$ progression in time giving a $\sqrt{2}:1$ progression in diameter is:

$$i > 1, F_i = \tfrac{1}{2} y_B Q_i + \sum_{j=1}^{i-1} \left[\frac{1}{y^{\frac{1}{2}(i-j-1)} + y^{\frac{1}{2}(i-j)}} - \frac{1}{y^{\frac{1}{2}(i-j)} + A} \right] F_j \qquad (12.37)$$

where

$$A = 1 \text{ for } j = 1,\ A = y^{\frac{1}{2}(i-j-1)} \text{ for } j \neq 1$$

$$y_B = y + y^{\frac{1}{2}}$$

When $i = 1$,

$$F_1 = \tfrac{1}{2}(1 + y^{\frac{1}{2}})Q_1$$

The method of using these equations to compute size distributions is as follows. The experiment data consist of a series of fractional concentrations, Q, measured at known values of ω and t, from which a value of D for each value of Q is calculated using equation (12.2). The values of Q versus D are plotted (usually on logarithmic-normal paper) and a smooth curve is drawn through the points. Values of Q are read from this curve at any convenient set of values of D in a $\sqrt{2}$ progression: $D_1, D_2 = \sqrt{(2)}D_1, D_3 = 2D_1$ etc. The corresponding values of Q are called Q_1, Q_2, Q_3, etc. These values are substituted in equation (12.36) to give the values of equation (12.36) or by extrapolating the curve of Q versus D to get additional points to use in equation (12.36).

Berg solved equation (12.25) graphically by plotting r_i^2/S^2 against Q and determining the area under the curve, deriving the following formulae:

$$\int_0^{D_m} F(D)\,dD = C_y + \frac{y}{4a}\,5C_y - 4F\left(\frac{D}{2}\right) \quad \text{for } y < \tfrac{1}{5}\,a \qquad (12.38)$$

$$\int_0^{D_m} F(D)\,dD = C_y \left(1 + \frac{2y}{3S}\right) \quad \text{for } C_y < 0.15 \qquad (12.39)$$

where $R = S + y$ and C_y is the concentration at depth y at the time required for a particle of size D_m to fall from the surface to depth y.

Ordinarily the approximate formulae are used, the integration formula being required for nearly monodisperse powders when C_y varies greatly with y.

Equation (12.39) is used for the calculation of the smallest value of $F(D)$, that is the smallest C_y, and equation (12.38) for other values of $F(D)$. $F(D) = 0$ when $D = 0$, hence $F(D/2)$ may be estimated by joining $F(D)$ to 0 for a small value of D as most functions are linear towards the origin. The $F(D)$ curve is then built up step by step. Concentration is determined by pipette withdrawal or, alternatively, by the use of divers. An alternative approximation of equation (12.26) has been given by Prochazka [75].

In a later paper Kamack [71] replaced his approximate formulae by an exact solution which is equally practical to use and which has more accuracy and generality. Starting with equation (12.28) he derived the exact solution:

$$F(D_m) = \exp\left[2\ln\left(\frac{r}{S}\right)\left\{Q(D_m) - \int_0^\infty h(x)Q(D_m \exp(-x))\,dx\right\}\right] \quad (12.40)$$

in which $h(x)$ is a function which depends on the ratio (r/S) and $x = \ln(D/D_m)$. For full details readers should refer to the original paper.

Svarovsky and Svarovska [72] derived an alternative version of equation (12.40) and later [73, 74] developed a new data analyser for data evaluation. Their equation was:

$$F(T) = e^a\left(C(T) - \int_T^\infty H(Z)C(Te^Z)\,dZ\right) \quad (12.41)$$

where $Z = \ln(t/T)$, (t is time as a variable, T is the time at which evaluation is made.)

$$a = 2\ln\left(\frac{r}{S}\right)$$

$$H(Z) = K(Z) - \int_0^Z K(Z - Z')H(Z')\,dZ'$$

$$K(Z) = a\exp(a)\exp[-Z - a\exp(-Z)] \quad \text{for } Z \geq 0$$

and

$$K(Z) = 0 \text{ for } Z < 0$$

Function $H(Z)$ is a convolution integral which is readily computed digitally for different values of a and then used in equation (12.41) for evaluation.

Alternative solutions to equation (12.26) are available if the shape of the distribution curve is assumed. Svarovsky and Friedova [76] assumed a fit by the three-parameter equation of Harris [50]; the parameters being found by means of a direct curve-fitting technique applied to the measured concentrations. This method is applicable to variable height, variable time and variable height and time. The main disadvantage of this method, apart from the need for a computer, is that it is unsuitable for multimodal distributions.

Other methods have also been proposed. Alex [78] suggested iteration by Neumann's series, Langarm approximation or substitution of polynomial functions for the distri-

bution curves. Lloyd *et al.* [79] suggested an extremely complicated solution involving high-order differentials; they also speculated on the 'on-line' evaluation of experimental data. Truly, on-line evaluation is, however, impossible and only a delayed quasi-on-line evaluation can be performed [74].

It is preferable when using the Kamack equation to smooth out the experimental Q values. A recommended procedure is to plot Q against D on log-probability paper.

The approximation due to Kamack can be modified for the scanning mode of operation by replacing the constant (r/S) by the variable (r_i/S), where r_i is the position of the source and detector at time t_i, i.e. equation (12.2) becomes:

$$D_i = \sqrt{\left(\frac{9\eta \ln y_i}{(\rho_s - \rho_f)\omega^2 t}\right)} \tag{12.42}$$

where

$$y_i = \left(\frac{r_i}{S}\right)^2 \tag{12.43}$$

Equation (12.29) becomes:

$$F_i = \int_0^{Q_i} \left(\frac{r_i}{S}\right)^2 dQ \tag{12.44}$$

i.e.

$$F_i = \int_0^{D_i} \exp(2kD^2 t) dQ \tag{12.45}$$

where D_i is the largest particle present at radius r_i at time t and can be calculated using equation (12.42). Kamack's equation can also be modified for variable inner radius; the relevant equations are developed in section 12.13.1.

12.12 The LADAL X-ray centrifuge

The LADAL X-ray centrifuge [49] is an extension of the X-ray gravitational technique. The X-rays are generated by an isotope source, and, after passing through the suspension, are detected by a scintillation counter. The signal from the counter passes to a pre-amplifier and thence to a ratemeter and a trace is recorded by a pen recorder.

The attenuation of the X-ray beam is proportional to the concentration of the suspension at the measurement radius. The largest size present in the beam can be calculated using Stokes' equation and the concentration undersize can be determined using Kamack's equation.

The LADAL instrument consists of the centrifuge body and the associated electronics (figure 12.8). The centrifuge itself comprises the bowl (3) which is mounted directly onto the shaft of an electric motor (2), the base plate (1), the source and detector holders (4) and (5). The base plate is mounted on a housing which absorbs the radiation that escapes from the measurement zone and on which the controls for the motor are

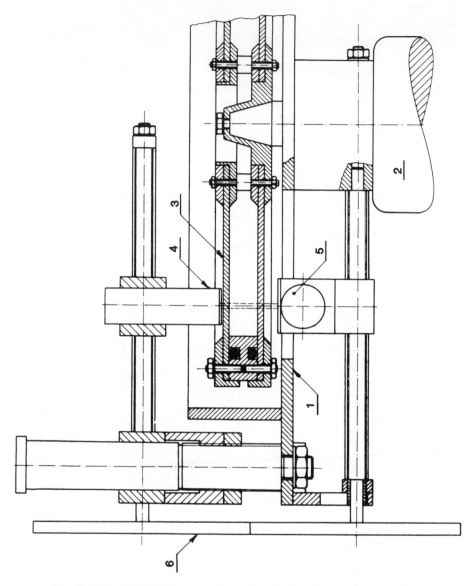

Fig. 12.8 The LADAL X-ray centrifuge: 1, casing; 2, motor; 3, disc centrifuge; 4, isotope source; 5, detector; 6, driving cogs for scanning mechanism.

fastened. A small axial fan is mounted on the housing so that the temperature of the suspension in the bowl does not rise due to heat convection from the motor, which is directly underneath. The assembly of the centrifuge can be seen in figure 12.9, which also shows the counting electronics consisting of a pre-amplifier, ratemeter and corresponding power supplies.

The centrifuge bowl is mounted horizontally on the steel body, surrounded with a metal safety ring and driven from underneath by an induction electric motor with speeds of 750 or 1500 rev min^{-1}. The radioactive source is placed in a brass holder and this is mounted on an arm above the bowl. The bowl is shown in cross-section in figure 12.8. It consists of a stainless steel ring with two rubber 'O' rings for sealing the fluid. Two resin transparent windows are clamped to it from each side by means of two steel rings and a number of screws equally spaced on the circumference. The windows are attached to the centrifuge shaft by another set of clamping rings. A constant axial distance between both windows is maintained by means of four spacers. The clamping rings and the windows are centred on the outer surface of the spacing ring which is used as a register.

The X-ray beam is collimated by lead slits in front of the source and the detector into a section of the annulus of about 20 mm^2. The scintillation counter is the same

Fig. 12.9 The LADAL X-ray centrifuge.

type as is used with the gravitational unit. The counting electronics are common to both versions and have been described in Chapter 9.

The ratemeter features a zero suppression control, which together with a range switch makes it possible to enlarge the output by a factor of three or ten. The range of available time constants facilitates smoothing of statistical fluctuations in detector output.

The operational procedure is very similar to the one used with the gravitational instrument, except that the powder cannot be simply stirred into the clean dispersing liquid, but the suspension must be prepared beforehand and poured into the bowl through the central opening. The bowl is first filled with clean liquid and the electronics set up with the source in position and the bowl spinning.

The emergent intensity, I_c, with clean liquid in the bowl is recorded. A suitable range is selected (a magnification of 1, 3 or 10) depending on the mass absorption coefficient of the material under examination, and the zero is suppressed to get the deflection on to the pen recorder paper. The arm with the source is then moved to the rest position while the bowl is emptied. It is then spun to the required speed. The arm is then swung back to the original position and the suspension is poured into the bowl through the central opening using a funnel attached above it. The filling usually takes about 5 s. The end of filling is taken as the starting point at which the recorder drive is switched on. The reading starts dropping immediately due to the continuous dilution occurring at every point in the suspension due to radial settling. The instrument may be left until most of the powder has settled out. For a fully automated evaluation the signal from the instrument can be recorded on a magnetic tape which is then fed to a computer.

Owing to the high initial powder concentration required (0.2 to 1.0% v/v), the homogeneous technique must be used. The emergent intensity of the X-ray beam with clear liquid in the disc is first recorded to provide a zero base line. The clear liquid is then removed and replaced with the suspension while the disc is spinning. The reading starts dropping immediately due to the continuous dilution occurring at every point due to radial settling.

The emergent intensity of the radiation is given by:

$$I = I_c \exp(-BC) \tag{12.46}$$

where B is a constant, C is the solids concentration and I_c is the emergent intensity with clear liquid in the bowl.

$$I_{max} = I_c \exp(-BC_{max}) \tag{12.47}$$

hence

$$BC_{max} = -\ln \frac{I_{max}}{I_c} \tag{12.48}$$

$$BC = -\ln \frac{I}{I_c} \tag{12.49}$$

Referring to figure 12.10:

$$DN \ = L_0 \ = K(I_c - I_0)$$

$$LM \ = L \ \ = K(I_c - I)$$

$$CX \ \ = 3AV = L_c = KI_c$$

where K is the constant of proportionality.

The above equations may be written in terms of the pen recorder deflection.

$$BC = \ln \frac{L_c}{L_c - L} \tag{12.50}$$

$$BC_{max} = \ln \frac{L_c}{L_c - L_0} \tag{12.51}$$

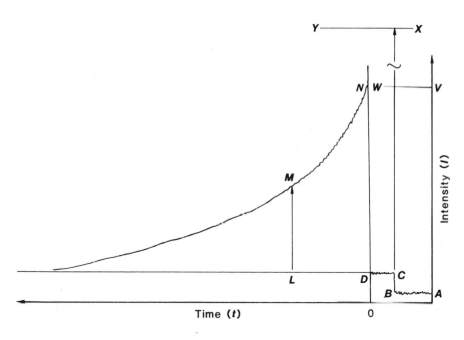

Fig. 12.10 Pen recorder trace from X-ray centrifuge.

AB – Deflection with clear liquid in the bowl.
VW – Zero intensity reading.
CD – After changing to (say) × 3 magnification AB goes off the scale. The zero suppression control brings AB back on to the scale to give a new maximum intensity reading CD. This effectively moves VW to position XY.
AV is proportional to I_c at magnification 1.
$L_c = CX = 3AV$ is proportional to I_c at magnification 3.
$L_0 = DN$ – Deflection at zero time with suspension in the bowl.
$L \ = LM$.

As an alternative to using a pen recorder the signal can be processed by a data-recording unit and presented as concentration against particle size. This can be further processed either manually or with an optional analogue evaluation unit to give cumulative percentage undersize. A note on data reduction using this technique has been published [89].

12.13 The LADAL pipette withdrawal centrifuge

The Simcar centrifuge was developed to conform with Kamack's theory. A major problem with this instrument is the amount of suspension required (approximately 2500 cm^3 of liquid and 5 cm^3 of powder), so that the liquid level does not alter appreciably during an analysis. The amount removed at each withdrawal is about 40 cm^3, hence the error due to assuming a constant liquid level increases as more samples are withdrawn. It is, therefore, not advisable to withdraw more than four samples. An eight-point analysis, in the size range 5 to 0.2 μm approximately, will take up to a full day. The equation can, however, be modified to take account of this fall, so that more points can be obtained with a single run. The author was faced with the problem of analysing material of which there were only about 2 g available and, with a colleague, designed and had constructed a modified pipette withdrawal centrifuge and developed a modified theory [80].

12.13.1 Theory for the LADAL pipette withdrawal technique

(a) Calculation of particle size
Let the time of the first withdrawal be t_1; the largest particle present in the withdrawn sample at this time will have fallen from the surface at radius S to the measurement zone at radius r.

Equation (12.2) will apply and may be written:

$$D_1^2 t_1 = k \ln \frac{r}{S} \tag{12.52}$$

The liquid level will then fall to S_1, where:

$$S_1^2 - S^2 = \frac{v}{\pi h} \tag{12.53}$$

where v is the volume extracted (10 cm^3) and h the thickness of the centrifuge disc (1.02 cm).

The fall in the inner radius can therefore be determined:

$$\Delta S_1 = S_1 - S \tag{12.54}$$

Let the time for the second withdrawal be t_2; then the largest particle present in the withdrawn sample will have fallen from S to x_{12} in time t_1, then a distance Δx_{12} due to the withdrawal of the first sample. Hence:

$$D_2^2 t_1 = k \ln \frac{x_{12}}{S} \tag{12.55}$$

$$D_2^2 (t_2 - t_1) = k \ln \frac{S}{x_{12} + \Delta x_{12}} \tag{12.56}$$

Adding equations (12.55) and (12.56) gives:

$$D_2^2 t_2 = k \ln \frac{r}{S} \left(1 + \frac{\Delta x_1}{x_{12}} \right)^{-1} \tag{12.57}$$

For the third withdrawal:
In time t_1 particles of size D_3 will fall from the surface at radius S to x_{13}, hence:

$$D_3^2 t_1 = k \ln \frac{x_{13}}{S} \tag{12.58}$$

These particles will then fall a distance Δx_{12} due to the withdrawal of the first sample, where from equation (12.53):

$$(x_{13} + \Delta x_{13})^2 - x_{13}^2 = 3.1207 \tag{12.59}$$

In the next time increment particles of size D_3 will fall from radius $x_{13} + \Delta x_{13}$ to radius x_{23}, hence:

$$D_3^2 (t_2 - t_1) = k \ln \frac{x_{23}}{x_{13} + \Delta x_{13}} \tag{12.60}$$

These particles will then fall a distance Δx_{23}, due to the withdrawal of the second sample, as before:

$$(x_{23} + \Delta x_{23})^2 - x_{23}^2 = 3.1207 \tag{12.61}$$

and

$$D_3^2 (t_3 - t_2) = k \ln \frac{r}{x_{23} + \Delta x_{23}} \tag{12.62}$$

Adding equations (12.58), (12.60) and (12.62) gives:

$$D_3^2 t_3 = k \ln \frac{r}{S} \left(1 + \frac{\Delta x_{13}}{x_{13}} \right)^{-1} \left(1 + \frac{\Delta x_{23}}{x_{23}} \right)^{-1} \tag{12.63}$$

The bracketed terms are the correction terms for the fall in level due to each extraction.

These equations are best solved by computer using iteration techniques.

Numerical solution. Using a feed volume of 150 cm^3 gives $S = 4.146$ cm; hence, for withdrawals in a 2 : 1 progression in time, equation (12.52) becomes :

$$D_1^2 t_1 = k \ln (7/4.146)$$

Letting (k/t_1) be equal to 1.91 makes D_1 equal to unity.

Second extraction: $(t_2 = 2t_1)$
Equation (12.55) gives $D_2^2 = 1.91 \ln (x_{12}/4.146)$.
Equation (12.56) gives $D_2^2 = 1.91 \ln [7/(x_{12} + \Delta x_{12})]$.

Hence

$$x_{12}^2 + x_{12}\Delta x_{12} = 29.022$$

and, from equation (12.53),

$$(x_{12} + \Delta x_{12})^2 - x_{12}^2 = 3.1207$$

Solving simultaneously gives $x_{12} = 5.244$ and $\Delta x_{12} = 0.2896$; hence

$$D_2 = 0.670$$

Assuming that this progression of sizes continues, $D_3 = 0.449$.
Substituting in equation (12.58) gives:

$$x_{13} = 4.146 \exp(0.449^2/1.91)$$

$$x_{13} = 4.608$$

From equation (12.59) : $\Delta x_{13} = 0.327$.
From equation (12.60) :

$$x_{23} = 4.935 \exp(0.449^2/1.91)$$

$$x_{23} = 5.484$$

From equation (12.61) : $\Delta x_{23} = 0.278$
Substituting these values into equation (12.63) gives a more accurate value for D_3.
$(t_3 = 4t_1)$.

$$D_3^2 = (1.91/4) \ln \left[\frac{7}{4.146} \left(1 + \frac{0.327}{4.608}\right)^{-1} \left(1 + \frac{0.278}{5.484}\right)^{-1} \right]$$

$$D_3 = 0.44$$

This process is best carried out by computer which gives the diameter ratios tabulated below (Table 12.1).

Table 12.1 The LADAL pipette centrifuge : Ratio of particle sizes for extractions in a 2 : 1 progression in time.

Ratio of times:	1	2	4	8	16	32	64
Ratio of sizes:	1	0.67	0.438	0.282	0.179	0.112	0.070

Other time differences are best handled with a computer programme.

(b) Calculation of frequency undersize
Let the concentration of the final sample withdrawn be Q_1 and let the surface be at radius S_1 immediately prior to this withdrawal (figure 12.11): then

$$F_1 = \tfrac{1}{2}(1 + y_1)Q_1 \tag{12.64}$$

where

$$y_1 = \left(\frac{r}{S_1}\right) \tag{12.65}$$

$$F_1 = \tfrac{1}{2}(1 + y_{12})Q_{12} \tag{12.66}$$

where

$$y_{12} = y_2 (D_1/D_2)^2 \tag{12.67}$$

Hence, by the trapezoidal rule:

$$(F_2 - F_1) = \tfrac{1}{2}(y_2 + y_{12})(Q_2 - Q_{12}); \quad y_2 = \frac{r}{S_2}$$

Substituting for Q_{12} gives:

$$F_2 = \tfrac{1}{2}(y_2 + y_{12})Q_2 + \left[1 - \frac{y_2 + y_{12}}{1 + y_{12}}\right] F_1 \tag{12.68}$$

Proceeding in a like manner gives the general formula:

$$F_n - F_{n-1} = \tfrac{1}{2}(y_n + y_{n-1,\, n})(Q_n - Q_{n-1,\, n})$$

$$F_{n-1} - F_{n-2} = \tfrac{1}{2}(y_{n-1,\, n} + y_{n-2,\, n})(Q_{n-1,\, n} - Q_{n-2,\, n})$$

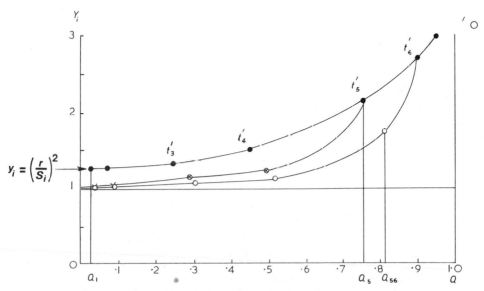

Fig. 12.11 Theoretical figure for variable inner radius centrifuge. Black circles are experimental points; open circles are derived points.

By successively eliminating the Q-functions, this gives a general equation in recursive form as before:

$$F_i = \tfrac{1}{2}(y_i + y_{i-1,\,i})Q + \sum_{j=1}^{i-1}\left[\frac{y_i + y_{i-1,\,i}}{y_{j+1,\,i} + y_{ji}} - \frac{y_i + y_{i-1,\,i}}{y_{ji} + y_{j-1,\,i}}\right]F_j \quad (12.69)$$

where

$$y_i = (r/S_i) \text{ and } y_{i-1,\,i} = y_i{}^{(D_{i-1}/D_i)^2}.$$

A numerical solution to equation (12.69) is given below for a feed volume of 150 cm^3 and a 2 : 1 progression in time. y values are given in Table 12.2 and these are inserted into the general equation to give the F values presented in Table 12.3. The dimensions of the centrifuge bowl are such that $y = 1.364$.

Table 12.2 Tabulated y values for pipette centrifuge.

$y_{12} = 1.1654$ $y_{23} = 1.2216$ $y_{34} = 1.2863$ $y_{45} = 1.3616$ $y_{56} = 1.4596$ $y_{67} = 1.6005$
$y_{13} = 1.0793$ $y_{24} = 1.1057$ $y_{35} = 1.1352$ $y_{46} = 1.1715$ $y_{57} = 1.2237$
$y_{14} = 1.0391$ $y_{25} = 1.0519$ $y_{36} = 1.0632$ $y_{47} = 1.0882$
$y_{15} = 1.0195$ $y_{26} = 1.0263$ $y_{37} = 1.0353$
$y_{16} = 1.0099$ $y_{27} = 1.0140$
$y_{17} = 1.0053$

Numerical solution. Using a feed volume of 150 cm^3 as before.
Calculation of F values.
Equation (12.64) gives $F_1 = \tfrac{1}{2}(1 + 1.364)Q_1$ $F_1 = 1.182Q_1$

Equation (12.68) gives $F_2 = \tfrac{1}{2}(1.494 + 1.1654)Q_2 + \left(1 + \dfrac{2.6596}{2.1654}\right)F_1$

$$F_2 = 1.330\,Q_2 - 0.228F_1$$

Equation (12.69) gives:

$$F_3 = \tfrac{1}{2}(1.651 + 1.222)Q_3 + \left(1 - \frac{2.873}{2.301}\right)F_2 + \left(\frac{2.873}{2.301} - \frac{2.873}{2.079}\right)F_1$$

$$F_3 = 1.4366\,Q_3 - 0.249F_2 - 0.133F_1$$

and so on, giving the general equations for the conditions $i = 7$, $V = 150$ cm^2 presented in Table 12.3.

Table 12.3 Table of F values for the LADAL pipette centrifuge operating in the normal mode.

$F_1 = 1.1822Q_1$
$F_2 = 1.330Q_2 - 0.228F_1$
$F_3 = 1.4366\,Q_3 - 0.2486F_2 - 0.133F_1$
$F_4 = 1.5658\,Q_4 - 0.3093\,F_3 - 0.1509F_2 - 0.0757F_1$
$F_5 = 1.7255\,Q_5 - 0.3836\,F_4 - 0.1950F_3 - 0.0877F_2 - 0.0427F_1$
$F_6 = 1.9373\,Q_6 - 0.4713F_5 - 0.2595\,F_4 - 0.1200F_3 - 0.0521F_2 - 0.0249F_1$
$F_7 = 2.2255\,Q_7 - 0.5755\,F_6 - 0.3489\,F_5 - 0.1717F_4 - 0.0760F_3 - 0.0323F_2 - 0.0153F_1$

12.14 The supercentrifuge

The supercentrifuge rotates at several thousand revolutions per minute and may be used to determine the size distribution of particles too small to be analysed with conventional centrifuges.

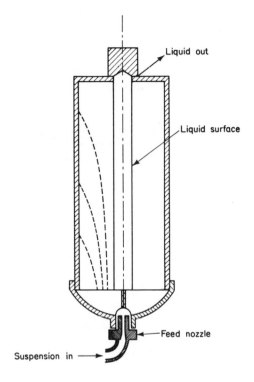

Fig. 12.12 Diagrammatic section of the Hauser–Lynn centrifuge [37].

Several ways of using Hauser's [35, 36] supercentrifuge (figure 12.12) for particle size analysis have been described in the literature. The usual procedure involves successive fractionation of the suspension and weighing of the fractions collected on a removable liner in the bowl [38–40]. The suspension is fed into the bottom of the bowl and the particles move in a spiral path until they reach the wall. The liquid is then discharged in an annular layer over the overflow dam.

If q is the rate of flow of suspension that causes a particle of size D to be deposited at a height h above the feed inlet, it can be shown [36] that:

$$q = khD^2 \tag{12.70}$$

If the removable liner is divided into identical strips, dried and weighed, the weight deposited on each strip may be used to find the size distribution [22, p. 86]. The constant k may be evaluated from curves given by Hauser and Lynn [37], or a nomograph developed by Saunders [41] may be used.

This method of determining size distributions cannot be recommended for routine analysis and a method developed by Bradley is to be preferred [9]. This method is applicable to the Sharples supercentrifuge (figure 12.13).

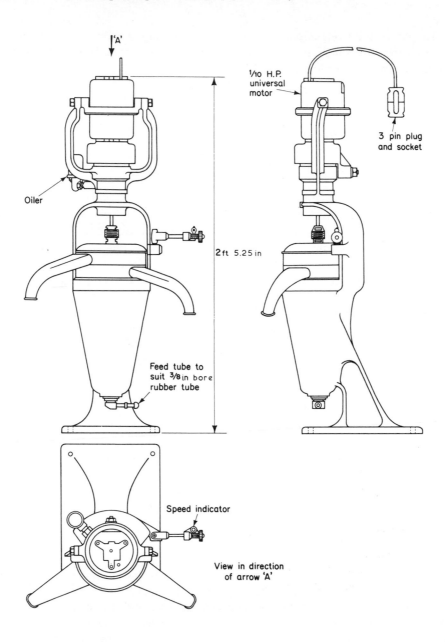

Fig. 12.13 The Sharples no. 1A open-type laboratory supercentrifuge. Motor drive.

If D_m is the smallest particle retained in the centrifuge, it can be shown that:

$$P = W + \int_{0}^{D_m} \frac{R^2 - x_0^2}{R^2 - S^2} \, f(D) \, dD \tag{12.71}$$

This is identical to equation (12.7), but the relationship between x_0 and D is more complex than that for a centrifuge without flow. Bradley derives the empirical solution for the Sharples supercentrifuge as:

$$\frac{D^2 (\rho_s - \rho_f) \omega^2}{18q} = 4.1 \times 10^{-3} \, x_0^{-1.2} \tag{12.72}$$

This gives, for the supercentrifuge as well as for batch centrifuges:

$$W = P - \frac{R^2 - S^2}{2S} \frac{dP}{dS} \tag{12.73}$$

which is the same as equation (12.12).

Hence, the weight fraction oversize is calculable by measurement of P for different values of S at constant W and q. The quickest analytical procedure is to calculate P from gravimetric or chemical analysis of feed and overflow suspensions. Choice of flow rate and speed can be made in accordance with prior knowledge of approximate size and use of derived theoretical expressions, or by trial and error to establish the rate at which P approaches unity with maximum S.

The main disadvantages of the technique are the need for large samples and the uncertainty of end-effects in the bowl. A big advantage is the ability to use an item of standard equipment without modification for a size below the range of most of the specially designed centrifuges, since the speeds vary between 8000 and 50 000 rev min^{-1}.

12.15 The ultracentrifuge [42–44]

The rotor of the ultracentrifuge is spun at speeds of up to 60 000 rev min^{-1} in a vacuum to minimize air drag. It may be used, therefore, to measure the size distribution of very fine particles. McCormick [45], for example, describes its use for determining the size distribution of polystyrene ($0.088 < d < 0.511$ μm) and Brodnyan [46] uses it for determining emulsion particle size. It has also been used in combination with light scattering for polymer size distribution determination [94].

12.16 Conclusion

Historically, centrifugal analyses were first carried out using ordinary centrifuge tubes. The errors involved in using these techniques are considerable: settling particles move in a radial path, impinge on the tube walls, agglomerate, and slide down the walls – hardly ideal settling conditions; problems arise due to disturbance in the suspension during acceleration and deceleration; difficulties arise in removing supernatant suspension or determining the amount of deposit without disturbing the deposited powder. This

latter problem is overcome with the Whitby technique, in which the height of the deposit in a capillary at the bottom of the tube is measured. However, problems arise due to non-uniform packing, and acceleration and deceleration have to be slow and controlled in order to reduce the deposit of particles on the walls of the tube.

The introduction of sector-shaped tubes [28] and conoidal-shaped tubes [31] removed many of the objections listed above and the advent of disc centrifuges [11, 30] rendered the earlier designs obsolete.

Centrifugal sedimentation may be carried out by line-start or homogeneous techniques, cumulatively or incrementally. The two-layer techniques give rise to streaming, a phenomenon most likely to affect cumulative analyses, in which the fraction sedimented against time is determined.

The Kaye disc centrifuge is the only instrument in which the two-layer technique is used incrementally, so that the presence of streaming is unimportant, provided the streams are representative of the bulk. The concentration is determined by the attenuation of a light beam passing through the suspension, a perfect sampling device, since it does not disturb the suspension. However, the laws of geometric optics break down as the size of the particles approaches the wavelength of light, thus making it very difficult to interpret correctly the analysis results. An improvement is to use X-rays, a technique that has been used to analyse lead glass, using a Mo $K\alpha$ source and a proportional counter for the size range 2 to 0.1 μm [47]. More solids may be used with homogeneous suspensions and the problem of streaming is removed. This technique may be applied to both the Kaye and Joyce–Loebl centrifuges and may probably be the best way of operating the latter.

The Joyce–Loebl disc centrifuge operates using the cumulative line-start technique and the fraction deposited is determined by removing the supernatant liquid, while the centrifuge is in motion, and examining it. However, it is probable that removal of the sediment occurs, due to disturbance propagated ahead of the pick-up [14] probe and gravimetric analysis of the supernatant liquid is difficult because of the small weights involved. The method is time-consuming since each experiment yields only a single point on the size distribution curve and is not really suitable for gravimetric array due to the small amounts of powder used.

The Simcar disc centrifuge is an incremental pipette withdrawal method using a homogeneous suspension.

The above instrument has been superseded by the LADAL pipette centrifuge with which a full analysis can be performed in less than an hour and a half as opposed to a full day. The amount of powder required is about one-tenth of the requirement for the Simcar, and a side-effect of the increasing inner radius when samples are extracted is a speeding up of the analysis. The equipment is relatively cheap and easy to operate. The main advantage of this particular instrument is that, if a powder can be dispersed in a liquid, it can be analysed. The results are in line with the Andreasen technique and indicate that gravity sedimentation overweighs the percentage of sub-micron particles.

The LADAL X-ray centrifuge is more expensive than the modified pipette centrifuge, but it can be easily automated to give size distributions with the minimum of operator involvement. It is limited in its need for powders that are opaque to X-rays. Determinat-

ion of the amount sedimented against time has found favour only with Martin and Robison [28], who derived a partial solution to the equation for this case, earlier work by Martin being based on an incorrect solution of Romwalter and Vendl [25]. Since Robison and Martin found good agreement between their analyses and analyses carried out by other techniques, it is possible that this method may become more popular.

Because of the difficulty of finding a solution with respect to time of the general equation for centrifugal settling, experimenters have found solutions with respect to the outer and inner radius and based techniques on them. The apparatus for the former, the Gallenkamp stepped-disc centrifuge, was not a success since there was a tendency for the sediment to be removed with the supernatant liquid, whereas the latter technique is so laborious, one run providing only one point on the graph, that it has not found commercial favour.

Some of these techniques have been used in conjunction with a low height of settling compared to the distance of the suspension from the axis of the centrifuge so that the centrifugal force could be considered constant. This has usually been done with cylindrical or sector-shaped centrifuge tubes, although Hildreth and Patterson [20] used a disc and determined concentration changes photometrically.

The ranges of sizes measurable using the above techniques vary from 80 to 0.1 μm, using the MSA particle size analyser, gravitational sedimentation being used for the coarse particles; 30 to 0.3 μm with the Kaye disc centrifuge; 5 to 0.1 μm with the Simcar centrifuge; to 10 to 0.01 μm with the Joyce–Loebl centrifuge. The modified pipette centrifuge operates in the range 5 to 0.05 μm; the full range covered by this technique and the gravitational pipette method is of the order of four decades using essentially the same technique. The LADAL X-ray gravitational and centrifugal techniques cover a similar range. The Simcar and LADAL X-ray centrifuges require about 10 to 20 g of powder, the LADAL pipette about 2 g and the other techniques a few milligrams.

12.17 Appendix: Worked examples

12.17.1 Simcar centrifuge

(a) Determination of F factors

These factors are constant, provided a constant volume of suspension is used, and samples are extracted in a 2 : 1 progression in time.

Let volume of suspension $V = 2410 \text{ cm}^3$, so that: $\quad y = \left(\dfrac{r}{S}\right)^2 = 2$

Applying equation (12.31):

$$F_1 = \tfrac{1}{2}(1 + 2)Q_1$$

$$F_1 = 1.5Q_1.$$

Applying equation (12.37):

$$F_2 = \tfrac{1}{2}(2 + \sqrt{2})Q_2 + (2 + \sqrt{2})\left[\left(\frac{1}{2 + \sqrt{2}}\right) - \left(\frac{1}{\sqrt{2} + 1}\right)\right]F_1$$

$$F_2 = 1.71Q_2 - 0.62Q_1$$

Similarly:

$$F_3 = \tfrac{1}{2}(2+\sqrt{2})Q_3 + (2+\sqrt{2})\left[\left(\frac{1}{2+2^{\frac{1}{2}}}\right) - \left(\frac{1}{2^{\frac{1}{2}}+2^{\frac{1}{4}}}\right)\right]F_2$$

$$+ \left[\left(\frac{1}{2^{\frac{1}{4}}+2^{\frac{1}{4}}}\right) - \left(\frac{1}{2^{\frac{1}{4}}+1}\right)\right]F_1$$

$$F_3 = 1.71Q_3 - 0.31F_2 - 0.25F_1$$

$$F_3 = 1.71Q_3 - 0.53Q_2 - 0.18Q_3$$

Proceeding in a like manner gives:

$$F_4 = 1.71Q_4 - 0.53Q_3 - 0.15Q_2 - 0.03Q_1$$

$$F_i = 1.71Q_i - 0.53Q_{i-1} - 0.15Q_{i-2} - 0.03Q_{i-3}$$

(b) Experimental results

Material: zinc oxide
Dispersant: 0.1% Calgon in distilled water
Weight of powder: 10 g
Powder density: 5610 kg m^{-3}
Liquid viscosity: 0.001 N s m^{-2}
Liquid density: 1000 kg m^{-3}
Centrifuge speed: $N = 480$ rev min^{-1}, $\omega = 16\pi$ rad s^{-1}.

Applying equation (12.2):

$$D^2 t = 5.34 \times 10^{-4} \text{ m}^2 \text{ s}^{-1}$$

$$D = \frac{2.98}{\sqrt{T}} \ \mu m \ (T \text{ in min})$$

Table 12.4 Simcar centrifuge analysis.

Time (T) (min)	Particle size (D) (μm)	Sample weight in 25 cm^3 (g)	Concentration (Q) (%)	Percentage undersize (F)
0		0.0830*	100	–
2	2.11	0.0822	99.1	–
4	1.49	0.0800	96.5	–
8	1.05	0.0713	86.0	99.9
16	0.75	0.0610	73.5	95.4
32	0.53	0.0407	49.0	67.8
64	0.37	0.0227	27.4	42.2
128	0.26	0.0078	9.4	14.8
256	0.18	0.0027	3.2	4.8

* Determined from original concentration.
Column 3 is the sample weight after drying and cooling in a desiccator, due allowance being made for the weight of dispersant.

The experiment is carried out twice, with four extractions per run. Alternatively, an allowance can be made for the fall in height of the interface, as in section 12.13.1.

12.17.2 X-ray centrifuge

(a) Determination of F factors

Centrifuge dimensions: R = 8 cm, h = 0.70 cm, r = 7 cm. Hence, for a fill volume of 100 cm^3, S = 4.304 cm. Applying equation (12.31) gives, for a 2 : 1 progression in time:

$$
\begin{aligned}
F_1 &= 1.823 Q_1 \\
F_2 &= 2.136 Q_2 - 1.142 Q_1 \\
F_3 &= 2.136 Q_3 - 1.008 Q_2 - 0.201 Q_1 \\
F_4 &= 2.136 Q_4 - 1.008 Q_3 - 0.174 Q_2 \\
F_5 &= 2.136 Q_5 - 1.008 Q_4 - 0.174 Q_3 - 0.020 Q_2 \\
F_i &= 2.136 Q_i - 1.008 Q_{i-1} - 0.174 Q_{i-2} - 0.020 Q_{i-3}
\end{aligned}
$$

(b) Experimental results

Material: iron oxide

Initial concentration: 3.48 g in 100 cm^3

Full scale deflection: 96 divisions

Magnification (x 3), hence L_c = 288 divisions

$$
\begin{aligned}
\rho_s &= 5130 \text{ kg m}^{-3} \\
\rho_f &= 1000 \text{ kg m}^{-3} \\
\eta &= 0.001 \text{ N s m}^{-2} \\
N &= 750 \text{ rev min}^{-1} \\
\omega &= 78.5 \text{ rad s}^{-1}
\end{aligned}
$$

$$
D_1 = \sqrt{\left(\frac{18 \times 0.001 \times \ln (7.00/4.304)}{4130 \times (78.5)^2 \times 60} \right)}
$$

Table 12.5 X-ray centrifuge analysis.

i	Time (T) (min)	Deflection (L) divisions	$(L_0 - L)$ (divisions)	100 BC*	Concentration (Q) (%)	Particle size (D)† (μm)	Percentage undersize (F)
9	1	L_0 = 57.5	230.5	967	100.0		
8	2	55.5	232.3	933	96.5	2.39	99.7
7	4	54.0	234.0	902	93.3	1.69	99.4
6	8	51.2	236.8	850	87.9	1.20	97.2
5	16	46.0	242.0	756	78.2	0.85	88.4
4	32	37.5	250.5	606	62.7	0.60	65.5
3	64	25.5	262.5	403	41.7	0.42	34.5
2	128	13.5	274.5	209	21.6	0.30	19.1
1	extrapolated	6.8	281.2	104	10.8	0.21	6.4
					3.8	0.15	

* Equation (12.50).

† Equation (12.2).

12.17.3 LADAL pipette centrifuge

Experimental results

Material : zinc oxide
Dispersant : 0.1% Calgon in distilled water
Volume of suspension : 170 cm^3
Weight of powder : 1.70 g
Powder density : 5600 kg m^{-3}
Temperature : 20°C
Liquid viscosity : 0.001 N s m^{-2}
Liquid density : 1000 kg m^{-3}
Centrifuge speed : 700 rev min^{-1} ($\omega = 70\pi/3$ rad s^{-1})
Measurement radius : 7 cm
Bowl radius : 8 cm; hence
Initial surface radius : 4.146 cm
Bowl thickness : 1.02 cm
Diameter at first withdrawal:

$$D_1 = 1.660 \, \mu m$$

Provided the extraction times are in a 2 : 1 progression it is only necessary to calculate D_1 since the ratio of sizes remains constant.

Table 12.6 LADAL centrifuge analysis.

i	Extraction time (T) (min)	Weight extracted (g)*	Concentration (Q) (%)	Particle size (D) (μm)	Percentage undersize (F)†
7	2	0.0931	93.1	1.66	100
6	4	0.0872	87.2	1.11	100
5	8	0.0770	77.0	0.732	100
4	16	0.0502	50.2	0.473	67.8
3	32	0.0230	23.0	0.303	30.9
2	64	0.0062	6.2	0.191	7.9
1	128	0.0012	1.2	0.118	1.4

* Weight of powder minus weight of dispersing agent.
† Using F values in Table 12.3.

References

1 Steel, J.G. and Bradfield, R. (1934), *Am. Soil Survey Assoc. Rep., 14th Ann. Mtg Bull.*, No. 15, 88.
2 Marshall, C.E. (1930), *Proc. R. Soc.*, **A 126**, 427.
3 Whitby, K.T. (1955), *Heat., Pip. Air Condit.*, **61**, 449.

4 Whitby, K.T. (1955), *J. Air Poll. Control Ass.*, **5**, 120.
5 Whitby, K.T. Algren, A.B. and Annis, J.C. (1958), ASTM Sp. Publ. No. 234, 117.
6 Cartwright, L.M. and Gregg, R.Q. (1958), *ibid.*, 127.
7 Dewell, P. (Sept., 1966), *Particle Size Analysis Conf.*, Loughborough, 1966. Soc. Analyt. Chem.
8 Irani, R.R. and Fong, W.S. (1961), *Cereal Chem.*, **38**, 67.
9 Bradley, D. (1962), *Chem. Proc. Engng*, **43**, 591, *et seq.*, 634, *et seq.*
10 Groves, M.J., Kaye, B.H. and Scarlet, B. (1964), *Br. Chem. Eng.*, **9**, 11, 742.
11 Kaye, B.H. (1962), BP 895 222.
12 Atherton, E. and Tough, D. (1965), *J. Soc. Dyers Colour.*, 624.
13 Tough, D. (1965), *Am. Dyestuffs Rep.*, **54**, 17, 34.
14 Treasure, C.R.G., private communication.
15 Marshall, C.E., Keen, B.A. and Schofield, R.K. (1930), *Nature*, **126**, 94.
16 Norton, F.H. and Spiel, S.J. (1938), *J. Am. Ceram. Soc.*, **21**, 89.
17 Jacobson, A.E. and Sullivan, W.F. (1946), *Ind. Engng. Chem., analyt. Edn*, **18**, 360.
18 Menis, O., House, H.P. and Boyd, C.M. (1957), *Oak Ridge National Laboratory Rep. 2345*, **22**, 86; (1958), **23**, 87.
19 Conner, P., Hardwick, W.M. and Laundy, B.J. (1958), *UKAEA Rep.* AERE, CE/R2465.
20 Hildreth, J.D. and Patterson, D. (1964), *J. Soc. Dyers Colour.*, **80**, 474.
21 Musgrove, J.R. and Harner, H.R. (1947), *Turbimetric Particle Size Analysis*, Eagle Pilcher Co., Cincinnati, Ohio, USA.
22 Irani, R.R. and Callis, C.E. (1963), *Particle Size Measurement*, Wiley, NY.
23 Orr, C. and Dallavalle, J.M. (1960), *Fine Particle Measurement*, Macmillan, NY.
24 Gupta, A.K. (1959), *J. appl. Chem.*, **9**, 487.
25 Romwalter, A. and Vendl, M. (1935), *Kolloid Z.*, **72**, 1.
26 Brown, C. (1944), *J. Phys. Chem.*, **48**, 246.
27 Martin, S.W. (1939), *Ind. Engng Chem., analyt. Edn*, **11**, 471–5.
28 Martin, S.W. and Robison, H.E. (1948), *J. Phys. Colloid Chem.*, **52**, 854–81; (1949) *ibid.*, **53**, 860–86.
29 Murley, R.D. (1965), *Nature*, **207**, 1089.
30 Donoghue, J.K. and Bostock, W. (1955), *Trans, Inst. Chem. Engrs*, **33**, 72.
31 Berg, S. (1940), *Ingen, Vidensk. Skr. B.*, no. 2.
32 Kamack, H.J. (1951), *Analyt. Chem.*, **23**, 6, 844–50.
33 Slater, C. and Cohen, L. (1962), *J. scient. Instrum.*, **39**, 614.
34 Treasure, C.R.G. (1964), Whiting and Industrial Powders Research Council, Tech. Paper No. 50, 11 White Lion House, Town Centre, Hatfield, Herts.
35 Hauser, E.A. and Read, C.E. (1936), *J. Phys. Chem.*, **40**, 1169.
36 Hauser, E.A. and Schachman, H.K. (1940), *ibid.*, **44**, 584.
37 Hauser, E.A. and Lynn, J.F. (1940), *Ind. Engng Chem.*, **32**, 660.
38 Fancher, G., Oliphant, S.C. and Houssiere, C.R. (1942), *Ind. Engng Chem., analyt. Edn*, **14**, 552.
39 McIntosh, J. and Seibie, F.E. (1940), *Br. J. exp. Path.*, **21**, 143.
40 Schachman, H.K. (1948), *J. Phys. Colloid Chem.*, **52**, 1034–45.
41 Saunders, E. (1948), *Analyt. Chem.*, **20**, 379.
42 Svedberg, T. (1938), *Ind. Engng Chem., analyt. Edn*, **10**, 113.
43 Svedberg, T. and Peterson, K.O. (1940), *The Ultracentrifuge*, Oxford Univ. Press.
44 Alexander, J.(ed.), (1926), *Colloid Chemistry*, Chemical Catalogue Co., NY, Chapter 6.
45 McCormick, H.W. (1964), *J. Colloid Sci.*, **19**, 173.
46 Brodnyan, J.G. (1960), *ibid.*, **15**, 563.
47 Martin, J.J., Brown, J.H. and de Bruyn, P.L. (1963), In: *Ultrafine Particles* (ed. L Kuhn), Wiley, NY.
48 Moser, H. and Schmidt, W. (1957), *Das Papier*, II, **189**; (1963), *ibid.*, 377.
49 Allen, T. and Svarovsky, L. (1972), *Proc. Soc. Analyt. Chem.*, **9**, 2, 38–40.
50 Harris, C.C. (1969), *AMIE Trans.*, **244**, 187.
51 Svarovsky, L. and Friedova, J. (1972), *Powder Technol.*, **5**, 5, 273–7.
52 Bayness, J.E., Attaway, A.V. and Young, B.W. (1972), *Proc. Soc. Analyt. Chem.*, **9**, 4, 83–6.

53 Statham, B.R. (1972), *Proc. Soc. Analyt. Chem.,* 9, 2, 40–3.
54 Scarlett, B., Rippon, M. and Lloyd, P.J. (1967), *Particle Size Analysis,* Soc. Analyt. Chem. 242.
55 Allen, T. (1968), *Powder Technol.,* 2, 133.
56 Vaughan, G.N., Ford, R.W. and West, H.W.H. (1969), *Proc. Br. Ceram. Soc.,* 13, 47–56.
57 Naumann, D. and Seydel, K.J. (1969), *Plaste Kaut,* 15, 2, 136–8.
58 Atherton, E. and Cooper, A.C. (1962), BP 983 760.
59 Atherton, E., Cooper, A.C. and Fox, M.R. (1964), *J. Soc. Dyers Colour.,* 26, 62.
60 McDonald, D.P. (1969), *Chem. Proc.,* 15, 3, 22–23.
61 Carr, W. (1970), *Paint Oil Colour J.,* 157, 37/34, 8, 82–83.
62 Carr, W. (1970), *J. Oil Colour Chem. Assoc.,* 53, 1, 81.
63 Carr, W. (1971), *ibid.,* 54, 155–73.
64 Carr, W. (1971), *Paint Tech.,* 35, 1, 16–23.
65 Carr, W. (1972), *Proc. Symp. Particle Size Analysis,* Bradford (1970), Soc. Analyt. Chem.
66 Beresford, J. (1967), *J. Oil Colour Chem. Assoc.,* 50, 7, 594–614.
67 Toyoshima, Y. (1970), *J. Jap. Soc. Col. Mat.,* 43, 7, 325–32, 364–9.
68 Burt, M.W.F. (1964), *AWRE Rep. 0-76/64.*
69 Jones, M.H. (1966), *Proc. Soc. Analyt. Chem.,* 3, 116.
70 Jones, M.H. (1969), US Pat. 3 475 968.
71 Kamack, H.J. (1972), *Br. J. appl. Phys.,* 5, 1962–8.
72 Svarovsky, L. and Svarovska, J. (1975), *J. Phys. D,* 5, 1962–8.
73 Svarovsky, L. and Svarovska, J. (1976), *J. Phys. E,* 9, 959–62.
74 Svarovsky, L. and Svarovska, J. (1976), *Dechema Monogram,* Nuremberg 1975, Numbers 1589–1615, 293–308.
75 Prochazka, D. (1970), *Proc. Conf. Dispersoidal Analysis,* Pardubice, Dum Techniky CVTS Pardubice 60/577/70, 61–65.
77 Alex, W. (1972), Dissertation, Univ. Karlsruhe, W. Germany.
78 Lloyd, P.J., Scarlett, B. and Sinclair, I. (1972), *Proc. Symp. Particle Size Analysis,* Bradford (1970), Soc. Analyt. Chem., London, 267–75.
79 Allen, T. and Svarovsky, L. (1976), *Dechema Monogram,* Nuremberg 1975, Numbers 1589–1615, 279–92.
80 Brugger, K. (1976), *Powder Technol.,* 13, 215–21.
81 Kanellopoulos, A.G. and Wood, R.J. (1976), *Pestic. Sci.,* 7, 75–85.
82 Wood, R.J. *et al.* (1976), *Powder Technol.,* 13, 143–9.
83 Kanellopoulos, A.G. and Wood, R.J. (1978), *ibid.,* 19, 283–5.
84 Groves, M.J. and Yalabik, H.S. (1974), *Pharm. Pharmacol. Suppl.* 26, 77–78.
85 Groves, M.J. and Yalabik, H.S. (1975), *Powder Technol.,* 11, 3, 245–56.
86 Groves, M.J. and Yalabik, H.S. (1975), *ibid.,* 12, 233–8.
87 Zwicker, J.D. (1972), *ibid.,* 6, 133–8.
88 Wnek, W.J. (1978), *ibid.,* 19, 1, 129–32.
89 Muschelknautz, E. (1974), Ger. Offen., 2 324 421.
90 Muschelknautz, E. (1967), *Verh. dt. Ing. Z.,* 109, 17, 757–61.
91 Muschelknautz, E. (1975), *Dechema Monogram,* Nuremberg, 1979, Numbers 1589–1615, Part B, 267–77.
92 Sokolov, V.I. *et al.* (1975), *Zh. Prike. Khim.* (Leningrad), 48, 7, 1651.
93 Wallace, T.P. *et al.* (1975), *J. Colloid Interfac. Sci.,* 51, 2, 283–91.
94 Lombard, G.A. and Carr, W. (1975), *J. Oil Colour Chem. Assoc.,* 58, 7, 246–51.
95 Provder, T. and Holsworth, R.M. (1976), *Am. Chem. Soc. Div. Org. Coat. Plast. Chem. Prep.,* 36, 150–6.
96 Khalili, M. (1979), M.Sc. Thesis, Univ. Bradford, England.

13 The electrical sensing zone method of particle size distribution determination (the Coulter principle)

13.1 Introduction

The Coulter technique is a method of determining the number and size of particles suspended in an electrolyte by causing them to pass through a small orifice on either side of which is immersed an electrode. The changes in resistance as particles pass through the orifice generate voltage pulses whose amplitudes are proportional to the volumes of the particles. The pulses are amplified, sized and counted and from the derived data the size distribution of the suspended phase may be determined. The technique was originally applied to blood cell counting [1, 2]. Kubitschek [3, 4] introduced modifications which permitted counting of bacterial cells, and pointed out that this principle could be applied to the measurement of cell-volume distributions as well as number counting. Modified instruments were soon developed with which particles could be sized as well as counted.

Since analyses may be carried out rapidly with good reproducibility using semi-skilled operators, the method has become popular in a very wide range of industries, [6]. This type of counting device is designated in ASTM 3365–74T as a 'Tentative method of test for concentration and particle size distribution of airborne particulates collected in liquid media'. It is specified that the method is suitable for particulate matter greater than 0.6 μm in diameter collected in a Greenburg-Smith or a midget impinger. The original development of the method was carried out by Anderson *et al.* [90] and was extended by others [91, 92]. In ASTM method C–21 it is stated that the experience of several laboratories indicates that the instrument is capable of a repeatability of 1% and a reproducibility of 3% at the 95% confidence level. The method is also standard for dry toners (ASTM F577–78) and an acceptable method for aluminium oxide powder (F7–70).

13.2 Operation

The operating principle of the instrument may be followed by referring to figure 13.1. A controlled vacuum initiates flow through a sapphire orifice let into a glass tube and unbalances a mercury siphon. The system is then isolated from the vacuum source by closing tap A and flow continues due to the balancing action of the mercury siphon. The advancing column of mercury activates the counter by means of start and stop

probes, so placed that a count is carried out while a known volume of electrolyte passes through the orifice (0.05 ml, 0.5 ml or 2.0 ml). The resistance across the orifice is monitored by means of immersed electrodes on either side. As each particle passes through the orifice it changes this resistance, thus generating a voltage pulse which is amplified, sized and counted, and from the derived data the size distribution of the suspended phase is determined.

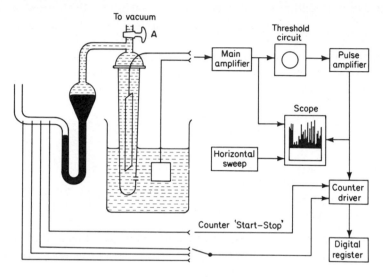

Fig. 13.1 Diagram of the Coulter counter.

The amplified voltage pulses are fed to a threshold circuit having an adjustable threshold level. The threshold level is indicated on an oscilloscope screen by a brightening of the pulse segment above the threshold setting and the pulse pattern also serves as a monitor. All pulses above the threshold level are counted and this count represents the number of particles larger than some determinable volume proportional to the appropriate threshold setting. Some instruments have upper and lower threshold circuits which permit sizing between two determinable volumes, i.e. a relative-frequency distribution. By taking a series of counts at various amplifications and threshold settings, data are directly obtained for determining number frequency against volume.

13.3 Calibration

Calibration is effected using particles having a narrow size range. The particles are placed in suspension in a concentration so low that the primary coincidence effects are less than 2%. Since the particles are nearly monosize the pulses on the oscilloscope will be nearly uniform in height. The instrument is adjusted so that this height is about 40% of maximum. This is effected by varying the gain and current selector settings, calibration preferably being carried out at a gain setting of 3. The threshold dial is then set at half the average pulse height and several full (n_f) counts taken and averaged. Next the

dial is set at one and a half times the average pulse height and several oversize (n_o) counts taken and averaged. The dial setting (t_c) to give the half-count $n_h = \frac{1}{2}(n_o + n_f)$ is found; this is proportional to the median size for the distribution (d_c). Then:

$$d_c = k \sqrt[3]{t_c} \qquad (13.1)$$

where k is the calibration factor.

An alternative procedure is to plot the number count against the instrument response (say $F_n t' \sqrt{2^{G-3}}$ for model A) and differentiate to find the mode which is assumed to occur at t_c [67].

The counter may also be calibrated using the powder under analysis if the whole size range of the sample is covered [11, 15, 30, 31]. The volume of particles in a metered volume of suspension will be:-

$$v_\rho = \frac{v}{V_s} \cdot \frac{w}{\rho_s} \qquad (13.2)$$

where:

v = volume of suspension metered for each count,
w = total weight of powder used,
V_s = total volume of suspension,
ρ_s = density of particles.

A size analysis consists of decreasing number of pulses against increasing threshold settings. If $\bar{t}$ is the average threshold setting as the pulse count changes by Δn, then:

$$v_p = \frac{\pi}{6} k^3 \sum \Delta n \bar{t} \qquad (13.3)$$

$$v_p = \frac{\pi}{6} \sum \Delta n d^3 \qquad (13.4)$$

Hence:

$$k^3 = \frac{6}{\pi} \frac{w}{\rho_s} \frac{v}{V_s} \frac{1}{\sum \Delta n \bar{t}} \qquad (13.5)$$

If the calibration constant as determined by equation (13.5) differs from that determined by equation (13.1), it is likely that the whole range of powder has not been examined. That is, there are some particles present in suspension which are too small to be detected by the system. In this case, equation (13.5) may be used to determine the fraction undersize by comparing the experimental value of $\sum \Delta n \bar{t}$ with the expected value.

For the example in Table 13.1, applying equation (13.5):

$$\sum_{t=0}^{t=max^m} \Delta n \bar{t} = \frac{6}{\pi} \frac{0.0168}{200} \frac{2}{2.62} \frac{1}{(20.6)^3}$$

$$= 14\ 100\ (\mu m)^3$$

which agrees with the experimental value (column 15).

Table 13.1 Coulter counter data and weight conversion.

Sample silica
Aperture diameter = 280 μm
Aperture resistance = 6350 Ω

Manometer volume = 2.0 ml
Calibration and zero data 26.4 pollen
$t' = 64$ at $I6$, 127 at $I7$

Source –
Coincidence factor $(F) = 13.72$
Notes conc. = 0.0168 g in 200 ml
$\rho_s = 2.62$ g/ml

Calibration factor $(k) = 20.6$

Electrolyte 0.9%
Dispersant 1% Calgon

w/v NaCl
Operator –
Date –

Gain index G	t'	I	F	n' (raw counts)	$\bar{n}'$	$n'' = P\left(\dfrac{\bar{n}'}{1000}\right)^2$	$-\sqrt{}$	$\bar{n} = \overline{n'+n''}$	$t = t'(F)$	$d = k\sqrt[3]{t}$	Δn	$\bar{t}$	$(\Delta n)\bar{t}$	$\sum(\Delta n)\bar{t}$ Progress	Cum. weight (%)
3	132	1	1.00	1.1.1.1 0.0.0.0	0.5	—		0.5	132	105	0.5	105	53	53	0.4
	48.5	1	1.00	2.2.1.2 0.1.0.0	1.0	—		1.0	48.5	75	0.5	90	45	98	0.7
	136.5	4	0.126	35 38	36.5	—		36.5	17.2	53	35.5	32.8	1 165	1 263	9.0
	46	4	0.126	255 213	234	8		242	5.8	37	205.5	11.5	2 370	3 633	25.8
	135	7	0.0167	1 004 1 047	1 026	14		1 038	2.25	27	796	4.02	3 200	6 833	48.5
	47.6	7	0.0167	2 629 2 690	2 660	97		2 755	0.796	19	1 717	1.52	2 610	9 443	66.8
	28.5	8	0.00891	6 086 6 135	6 111	512	78	6 545	0.252	13	3 790	0.524	1 988	11 431	81.2
	30.5	10	0.00311	11 591 11 569	11 580	1 840	204	13 216	0.095	9.4	6 671	0.173	1 150	12 581	89.2
													14 100		100

Alternatively it is possible that the instrument response is not proportional to particle volume or that the assumed particle density is incorrect (see section 13.6).

The larger apertures may be calibrated by a 'tie-in' procedure using data from smaller apertures, e.g. having calibrated the 100-μm tube one can determine the k value which must be assigned to the 200-μm tube in order for size distribution curves of the same material to fit exactly on top of one another.

An expression has been quoted for the calculation of k for any size of aperture D (μm) in the same electrolyte [31]:

$$\log k = 3.46 \log D - 5.66 \tag{13.6}$$

It is claimed to be linear for $30 < D < 560$ although errors in particle diameter are said to vary between 0 and 8% when the method is used.

Two-point calibrations have been used but the value of this additional exercise is limited since both calibration materials should have a diameter between 5% and 20% of aperture diameter. Multipoint calibration is useful for checking linearity of scale [16, 36].

In a series of papers, Brotherton examined calibration procedures with Coulter counter models A, B, F and C [7-9], and found wide variations in k. It is interesting to note that while Brotherton found an increase of calibration constant with particle size, the converse was found by Matthews [10].

Calibration materials in general use consist of pollens, latex spheres and glass spheres. Several authors [37, 38, 48] have discouraged the use of pollens due to non-sphericity, surface irregularities and changing size in dispersing media. Various pollens, latices and glass beads are available from Coulters, Dow, Duke Standards and CTI-TNO [68]. A recently introduced red bead latex with a mean size of 18.99 μm and a standard deviation of 0.18 μm (Coulter, England) has been described as a particularly promising latex [39]. It is generally agreed that the standard deviations of the size distributions of the latex samples measured by the Coulter counter are greater than those measured by microscopy and those quoted by Dow [10, 39, 40]. This may be due partly to the quality control methods used by Dow [41] and partly to the effects discussed in the section on pulse shape. Spherical hollow carbon particles (1-300 μm) have also been used for calibration purposes [71].

13.4 Evaluation of results

Table 13.1 shows a Coulter analysis using a model A Coulter counter. The amplification of the pulses is controlled by a gain selector switch numbered from 1 to 6 (column 1), each step being an amplification of $\sqrt{2}$. The pulse height is further controlled by a current selector switch I (column 3) numbered from 1 to 10 which puts known resistances in series with a 300 V dc supply and the electrodes. This causes the current to approximately double with each step so that the voltage pulse is a maximum at $I10$, $G6$; these settings are used for counting the smallest particles present in the suspension. The usual procedure is to calibrate with the gain setting at 3 and carry out the first count, on the powder suspension, in duplicate, at settings $G = 3, I = 10, t' = 30$ to count all particles oversize (D) where:

$$D = k\sqrt[3]{(t' F_n \sqrt{2}^{(G - 3)})}$$

For the tabulated example, $D = 20.6\sqrt[3]{(30.5 \times 0.00311 \times 1)} = 9.4\ \mu m$. The minimum settings before spurious counting occurs due to electrical interference are $t' = 15$, $G = 1, I = 10$ to give $D_{min} = 5.89\ \mu m$ (i.e. approximately 2% of tube aperture size). It is then recommended that t' be increased to 60 to increase the threshold size to $11.8\ \mu m$. Counts are then carried out, in duplicate, at $I = 9$ to $I = 1$ keeping all other settings the same; t' is then increased to 120 and 240 to give a maximum size $D = 128\ \mu m$. (The maximum size that can be counted at, for example, $t' = 240$ and $G = 1$ is greater than 40% of aperture size and will therefore be in error. Even at the suggested settings $D_{max} \cong 45\%$ of aperture size.) Different instrument settings can be selected for the same threshold size. Optimum values are rapidly selected by experienced operators but a computer programme is available to aid selection [62]. In the tabulated example the F and t' values are chosen to give a root-two progression of size based on $75\ \mu m$.

The use of these three controls permits the sizing of a wide range of volumes $(8000 : 1)$ equivalent to a size range of $20 : 1$, roughly 2% to 40% of aperture diameter. The lower limit arises due to electrical (background) interference, the latter due to non-linearity of response and aperture blocking.

A range of orifice diameters is available (10 to 1000 μm). The upper particle volume diameter, which may be measured with any orifice, is limited to about 40% of the orifice diameter, since frequent blocking of the orifice occurs if larger particles are present in the suspension. Hence, with a 100 μm orifice, a size range $40 > D > 2\ \mu m$ may be measured since $40^3 : 2^3 = 8000 : 1$.

The raw count is corrected for background, that is, particles which are present in the electrolyte before the addition of the powder sample, and coincidence. The latter factor has been derived empirically by the manufacturers to compensate for loss of count when two particles go through the orifice together and are counted as one, and for the gain in count due to two particles below threshold size being in such close proximity in the orifice that the pulse generated is above threshold.

Since the corrected count $\bar{n}$ of particles greater than threshold t, where particle size $d = k\sqrt[3]{t}$, is known, the size distribution may be determined on a number or weight basis.

13.5 Theory

The basic assumption underlying the operation of the Coulter counter is that the response, i.e. the voltage pulse generated when a particle passes through the orifice, is directly proportional to particle volume. The reliability of the instrument depends upon the accuracy of this assumption.

The relationship between response and particle size may be determined in the following manner:

Figure 13.2(a) shows a particle passing through the orifice.
Figure 13.2(b) shows an element of the particle and orifice.
Resistance of element without a particle, $\delta R_0 = (\rho_f \delta l)A$.

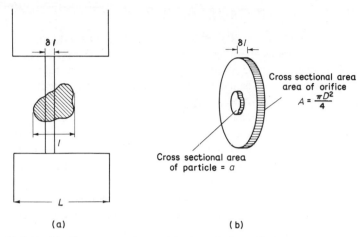

(a)   (b)

Fig. 13.2 (a), (b) The passage of a particle through the orifice of a Coulter counter.

Resistance of element with a particle included is that of two resistors in parallel:

$$\delta R = 1 \left/ \left[\frac{A-a}{\rho_f \delta l} + \frac{a}{\rho_s \delta l} \right] \right.$$

where ρ_f, ρ_s are the resistivities of the particle and fluid respectively. Thus the change in the resistance of the element due to the presence of the particle $\delta(\Delta R)$ is given by:

$$\delta(\Delta R) = \delta R_0 - \delta R$$

$$= \frac{-\rho_f a \delta l}{A^2} \left(1 - \frac{\rho_f}{\rho_s} \right) \frac{1}{\left\{ 1 - \left(1 - \frac{\rho_f}{\rho_s} \right) \frac{a}{A} \right\}} \quad (13.7)$$

The external resistance in the circuit is sufficiently high to ensure that the small change ΔR in the resistance of the orifice due to the presence of a particle will not affect the current I; the voltage pulse generated is therefore $I\Delta R$.

In practice it is found that the response is independent of the resistivity of the particle. In fact, if this were not so the whole technique would break down, since a different calibration factor would be required for each electrolyte–solid suspension. Berg [5] suggested that this may be due to oxide surface films and ionic inertia of the Helmholtz electrical double layer and associated solvent molecules at the surface of the particles, their electrical resistivity becoming infinite. The terms involving ρ_f/ρ_s may therefore be neglected.

Thus, equation (13.7) becomes:

$$\delta(\Delta R) = -\frac{\rho_f a \delta l}{A^2} \left/ \left(1 - \frac{a}{A} \right) \right. \quad (13.8)$$

The response, therefore, is not proportional to the volume of the particle, but is modified due to the term a/A.

With rod-shaped particles, this leads to an oversizing of about 6% in terms of diameter at the top size with distortion of the distribution [11, 16]. This error decreases as a/A decreases.

For a spherical particle of radius b, the change in resistance due to an element of thickness δl at a distance l from the centre of the sphere may be determined and this can be integrated to give the resistance change due to the particle [16, 42, 45].

$$\Delta R = -\frac{2\rho_f \pi^2}{A} \int_0^b \frac{(b^2 - l^2)}{\left[1 - \frac{\pi(b^2 - l^2)}{A}\right]} \, dl$$

$$= \frac{8\rho_f d^3}{3\pi D^4} \left[1 + \frac{4}{5}\left(\frac{d}{D}\right)^2 + \frac{24}{35}\left(\frac{d}{D}\right)^4 + \frac{169}{280}\left(\frac{d}{D}\right)^6 + \ldots\right] \quad (13.9)$$

where d is the diameter of the sphere. This equation gives a limiting value of two-thirds the Maxwellian value. Recognizing this fact Gregg and Steidley [45] multiplied their solution by three-halves. This procedure has been questioned [42].

The complete solution is:

$$\Delta R = -\frac{4\rho_f}{\pi D} \left[\frac{\sin^{-1}(d/D)}{\sqrt{[1 - (d/D)^2]}} - \frac{d}{D}\right] \quad (13.10)$$

This equation may be written:

$$\Delta R = \rho_f \frac{v}{A^2} F_1 \quad (13.11)$$

Hence the instrument response is proportional to the volume of the sphere modified by the function F_1. This equation results from a simple integration of the cross-section available for conduction. De Blois and Bean [42] derived an approximation which employs the solution to Laplace's equation for a sphere in an infinite medium, using only those streamlines which do not cross the sphere wall:

$$F_2 = 1 + 1.26_8 \left(\frac{d}{D}\right)^3 + 1.1_7 \left(\frac{d}{D}\right)^6 + \ldots \quad (13.12)$$

De Blois and Bean derived the following equation as the best fit to their experimental results.

$$F_3 = 1 + 0.73 \left(\frac{d}{D}\right)^3 \quad (13.13)$$

Smythe [cit. 43, 46, 47] solved the problem numerically by using an integral formula deriving:

$$F_4 = \frac{2}{3} C_0 \left(\frac{d}{D}\right) \quad (13.14)$$

with C_0 determined to an accuracy of 1 part in 10^7 for diameter ratios from 0.1 to 0.95.

Grover *et al.* [44] reported measurements on polystyrene spheres and pollen using the equation:

$$F_5 = 1 + \frac{2}{3} \left(\frac{d^3}{D^2 L_e} \right) + \frac{2}{3} \left(\frac{d^3}{D^2 L_e} \right)^2 + \ldots \qquad (13.15)$$

where L_e is the effective length of the aperture. In the earlier derivations the length is assumed infinite and hence is not a parameter.

Anderson and Quinn [43] compared the above equations and concluded that F_5 agrees well with experimental results at small $(d/D)^3$, F_2 does not coincide with F_5 as $(d/D)^3$ approaches unity and that F_1 converges with the numerical results as $(d/D)^3$ approaches unity.

The initial experimental data obtained by De Blois and Bean were with PVC spheres down to 0.09 μm in diameter using a pore in a plastic sheet. They further conclude that the technique is applicable down to 0.015 μm. In a later paper [59] they analysed down to 0.06 μm using a pore in a Nuclepore filter and also measured the electro-osmotic velocity of the fluid in the pore.

The error, assuming a linear relationship between resistivity change and particle volume for spherical particles, is about 5.5% at the top size for the instrument. The technique may be applied to higher values of (d/D) than 0.4 provided corrections are applied and aperture blocking does not become too troublesome. For non-spherical particles F_1 is modified by the inclusion of a shape factor [43]. The general conclusion is that as (d/D) increases the resistance pulse generated is greater than predicted by assuming proportionality, and thus oversizing of the larger particles occurs.

13.6 Effect of particle shape and orientation

The instrument response essentially is to particle volume. It has been claimed that particle shape, roughness and the nature of the material have little effect on the analysis [32, 33] but there is considerable evidence that the size parameter measured is in fact the envelope of the particle. Comparison with other techniques has been found to be good for spherical particles; for non-spherical particles results may differ [93, 94]. With porous materials such as nylon the measured volume may be several times the skeletal volume [11]. Thus porous materials are unsuitable for this method since the effective densities are not known. Anomalous results have also been reported with fly ash [11, 16]. This particular effect has been used to measure the particulate matter within a floc [34] and the porosity of porous samples [35]. Ratios of 1.31 : 1 have been quoted for non-extreme shapes with higher values for flaky particles [41].

Model experiments have been carried out by Marshall [13], Lloyd [14] and Eckhoff [15], but no firm conclusions may be drawn from them since the models used differed widely from the commercial instruments.

The instrument has been used for fibre length analysis [49, 50] in which pulse duration is used in a sensing zone that was longer than the fibres. An alternative approach to fibre measurement involves a flow collar upstream of the sensing zone to provide select-

ive screening and alignment of the fibres with the aperture axis [41, 51, 52]. Fibre dimensions are obtained via fibre volumes for a series of maximum diameters as determined by sequentially used screens. The Elzone solution to this problem [53] involves the use of a long flow tube upstream of the sensing zone to provide laminar flow and fibre pre-alignment. This flow is caused to join a clear liquid sheath and the flow is removed from the exit by a sheathing stream.

13.7 Coincidence correction

If it is assumed that perfect data result if particles traverse the orifice singly, two types of error result due to deviations from this ideal situation.

(a) 'Primary coincidence' or horizontal interaction'. Two particles in the sensitive zone about the orifice at the same instant in time will give rise to two overlapping pulses. There must be, therefore, some limit of separation at which the pulses cannot be resolved giving rise to a loss in count. This coincidence loss is minimized by using extremely dilute suspension and by using correction formulae.

(b) 'Secondary coincidence' or 'vertical interaction'. Two particles which individually give rise to pulses below threshold level collectively give rise to a single pulse above threshold level. For this to arise the particles must be of similar size (which must be close to the threshold limit) and also in close proximity.

The correction for primary coincidence may be theoretically determined by assuming a Poisson probability of finding a number of particles concurrently in the sensing zone of volume s. This yields the relationship between the true count N and the observed count n:

$$n = \left(\frac{v}{s}\right)\left\{1 - \exp\left(-\frac{s}{v}N\right)\right\} \tag{13.16}$$

giving:

$$N = \left(n + \frac{s}{2v}N^2\right) \tag{13.17}$$

if higher-order terms are neglected.

The equation developed on the basis of transit time distribution takes the form:

$$n = N \exp\left(-\frac{s}{v}N\right) \tag{13.18}$$

giving:

$$N = \left(n + \frac{s}{v}N^2\right) \tag{13.19}$$

The equation presented by the manufacturers of the Coulter Counter is [2]:

$$N = n + pn^2 \tag{13.20}$$

where:

$$p = 2.5\left\{\frac{D}{100}\right\}^3\left\{\frac{500}{v}\right\} \times 10^{-6} \tag{13.21}$$

D is the aperture volume in micrometres and v the volume of the suspension in micro-litres monitored for each count. The factor 2.5 was determined experimentally by the manufacturers using a 100-μm aperture and v = 500 μl. The equations are stated to be accurate to within 1%.

For practical considerations it is immaterial whether p is equal to s over v or s over $2v$. It is also of little practical importance whether the correction term is a function of N^2 or n^2, except that the second form is easier to manipulate.

Mattern [78] and Grant et al. [79] used equation (13.17) and determined p by serial dilution; i.e. denoting a dilute concentration by 1, doubling this concentration as 2 etc., a plot of n/C against C will be a straight line of negative slope intercepting the n/C axis; at concentration C equals 0, n equals N.

Ruhenstroth [87] used this method and found agreement with equations (13.20) and (13.21). Wales and Wilson [17, 96] using a similar approach obtained results which were not in agreement. Edmundson [20] found that the slopes of the lines were positive for high values of n and negative for low values. Since none of the equations admit of a positive slope, the coincidence effect must be swamped by other effects. A disadvantage of this method is that its accuracy depends on the count accuracy of the most dilute suspensions [86].

Harvey [21], in an article comparing the Coulter counter model B, the Celloscope 101, the Nuclear Chicago particle measurement system (Nuclear Chicago Corp., Des Plaines, IL) and an instrument described by Harvey and Marr [22], examined coincidence errors in detail comparing their analysis of Dow latices with electron microscopy and X-ray analysis. With a 30 μm aperture the mean transit time of particles through the sensing zone is of the order of 20 μs. The overall pulse duration will be of this order of magnitude. The Coulter counter model B employs simple amplification of pulses by a vacuum tube amplifier having a time constant of about 30 μs, and this increases the duration of the pulses about tenfold. Under conditions where coincidence in the sensing zone is eliminated, coincidence in the amplifier can still occur. Its effect on pulse amplitude differs from that of physical coincidence. The greater part of the 200 to 300 μs pulse duration is due to a long decay from peak to zero voltage. If a second pulse occurs within this period its apparent peak amplitude will increase by an amount equal to the residual amplitude of the first pulse. As counting rate increases the apparent frequency of large particles will also increase. Pisani and Thomson found that size distributions with a total pulse duration of about 30 μs were still skewed and this they attributed to a distribution of transit times through the aperture due to the velocity profile across it. Harvey stated that in order to prevent distortion of the size distribution the transit time should be four or five times the amplifier rise time. In order to examine this effect Priem [23] constructed 50 μm apertures of length 50–250 μm. These were tested by Glover and the results supported Harvey's statement.

An alternative explanation [25] is that particles passing near the walls of the tube generate a trimodal pulse because of the disposition of the electric field. This abnormality is reduced with apertures of increasing length-to-diameter ratio.

The problem with these long apertures is that they degrade the signal-to-noise ratio and increase coincidence problems [24].

Wales and Wilson [17] assumed that primary coincidence predominated and re-placed the N^2 in equation (13.19) by nN. Helleman [86] stated that this equation was valid only for low concentrations and a long resolving power with regard to pulse durat-ion (i.e. small aperture).

Pfeiffer [65] obtained results in agreement with Coulter's (equations (13.20) and (13.21)). Herben [80], using a Celloscope with a tube length twice that of the Coulter, was also in agreement.

Princen and Kwolek [18], Gutmann [81] and Mercer [83] found that these equat-ions over-corrected, finding better agreement with equation (13.17). Strackee [82] found that these equations under-corrected and equation (13.19) best fitted his results. Adams [89] and Pisani and Thomson [19] also found best agreement with equation (13.19). Edmundson [20], however, believed that p depended on the discriminator setting. Helleman [98] found that the coincidence was the same for the Toa as for the Coulter. Kalling et al. [84] obtained results in agreement with those of Wales and Wilson. Walstra et al. [85, 95], using serial dilution, found s values in agreement with those obtained by Princen and Kwolek. They found that this equation was only suitable at low concentrations when secondary coincidence was negligible.

Helleman [86] obtained p from equation (13.18) which, by re-arranging, takes the form:

$$p \frac{\ln N - \ln n}{N} \tag{13.22}$$

Helleman plotted log (n/C) against C and found that p decreased with increasing con-centration due to a reduction in pulse amplitude as the count rate increased. This reduction ceased at concentrations when the pulse duration became less than the resolving power of the counting device and this is dependent on the discriminator level and aperture size.

Equation (13.16), putting $\alpha = s/v$ may be written [89, 54]:

$$N = -\frac{1}{\alpha} \ln (1 - \alpha n)$$

Diluting to half concentration:

$$\tfrac{1}{2} N = \frac{1}{\alpha} \ln (1 - \alpha n_2)$$

therefore

$$\alpha = \frac{2n_2 - n}{n_2^2} \tag{13.23}$$

α can therefore be calculated, and knowing α the true count N can be obtained. The probable error in α is given by [54]:

$$\frac{\sigma_\alpha}{\alpha} = \frac{1}{\sqrt{n}} \frac{(4k - 1/k - 3)^{1/2}}{1/k - 1} \tag{13.24}$$

where

$$k = n/2n_2 .$$

The error in determining α is as likely to be greater than E as it is to be smaller, where:

$$E = \frac{2\sigma_\alpha}{3\alpha} \qquad (13.25)$$

Non-linear effects have been attributed to 'shadowing' [16, 30, 29]. This effect can be demonstrated if a full count is available since this falls as the threshold is reduced, e.g. the counter gives n = 19 000 at t equivalent to 8 μm but only 18 000 at t equivalent to 6 μm. According to Hellerman this is due to the time constant. A small pulse in the wake of a large one will be reduced in size. This error can be reduced by a reduction in transit time and a decrease of concentration, particularly of fines.

According to Auerbach [55] the electrical resistance of a tube of length L and diameter D, where L and D are comparable, is:

$$R = \frac{4\rho_f}{\pi D^2}\left(L + kD\right) \qquad (13.26)$$

where k = 0.80, and ρ_f is the electrical resistivity.

The equivalent length L_e = $L + kD$, hence the volume of the sensing zone is given by:

$$s = A\,(L + kD) \qquad (13.27)$$

The sensing zone volume may be determined from the mean transit time or by determining p experimentally. Adams [89] found s = 2.7 v_a from mean transit time for a 30-μm aperture and s = 3.0 v_a from a determination of p where v_a is aperture volume.

For a 30 μm aperture with v = 50 μl, equation (13.21) yields p = 0.675 $\times$ 10^{-6} and assuming s = pv this gives s = 3.375 $\times$ 10^{-8} cm^3. For an aperture tube with a length-to-diameter ratio of 0.75, the aperture volume $v_a = (3\pi/16)D^3 = 1.592 \times 10^{-8}$ cm^3; hence $s/v_a = 2.12$. Allen's data [16] give $s/v_a = 2.33$. Other published data give values ranging from 1.55 to 4 [86].

Published data on coincidence correction are confused and contradictory. One reasonable assumption that might be made is that the correction varies from instrument to instrument and from aperture tube to aperture tube. It is therefore recommended that the coincidence correction be kept as low as possible and certainly lower than the 10% level suggested by Coulter's.

Muntwyler [57] corrects for coincidence by using a device to generate additional pulses according to an empirical correction formula; these pulses are then counted in addition to the particle-generated pulses.

13.8 Pulse shape

Kubitscheck [56] found that the output pulse due to the passage of a particle through the apertures was round-topped, not rectangular as expected theoretically (the pulse amplitude should remain constant as long as the particle is contained within the orifice).

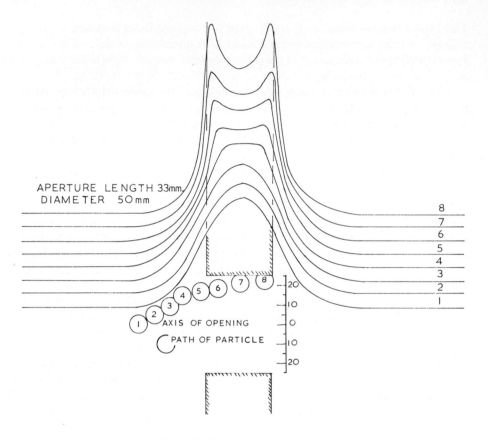

APERTURE LENGTH 33mm,
DIAMETER 50mm

AXIS OF OPENING

PATH OF PARTICLE

Fig. 13.3 Shape of pulses generated.

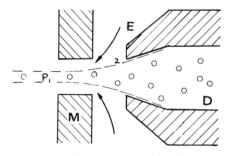

Fig. 13.4 Schematic diagram. Focusing of suspension on the axis of the measurement opening M with a probe D. The suspension streams through the axis after dilution with particle-free electrolyte E.

This implies that the amplifier does not reach full response to the presence of each particle before it leaves the orifice and will result in an undersizing of the coarse fraction. Kubitscheck found it necessary to construct tubes five times as thick as they were wide in order to correct this defect.

Eckhoff [15] stated that pulses were round-topped due to the shape of the electric field and rejected earlier assertions that round-topped pulses were produced due to the instrument not reaching full response. Distorted pulses were reported by Allen [16] who suggested that they arose due to coincidence effects. He also reported variable residence time within the orifice. Spielman and Goren [40] suggested that these effects were due to particles travelling in the vicinity of the orifice walls. They introduced hydrodynamic focusing, a flow directional device which directed the particles along the central axis.

Grover et al. [44] conducted theoretical and experimental studies of the pulses generated in the orifice. They theoretically determined the potential field in and around the orifice and showed that it was dense at the inlet and outlet edges. Particles travelling near to the walls therefore pass through two regions of high potential gradient and generate M-shaped pulses.

Thom et al. [25, 26, 69] used a magnified version of a Coulter counter and by drawing spheres through the orifice on nylon threads, mapped out the generated pulses on different streamlines (figure 13.3). When they rounded the edges of the orifice to form a conical entrance and exit pulse distortion was eliminated. They then made a central filament capillary for the injection of particles into the centre of the orifice (figure 13.4). A secondary benefit of this solution is that the transit time is approximately constant for all particles. This work resulted in the development of an instrument available commercially as the Telefunken particle detector MS PD1 1105/1 and this has been examined by Polke [35].

In 1969 Coulter's patented a conical entrance and exit but in 1970 they filed another patent which indicated a preference for cylindrical orifices over the contoured variety possibly because of orifice plugging difficulties. They also considered channelling the particles down the centre of the aperture but rejected this approach as impracticable. In 1973 they were issued a patent covering a rounded orifice. In 1973 IITRI patented a trumpet-shaped orifice system which included a flow straightener and contoured orifice. Orifice plugging was eliminated by the use of a screen in the flow straightener.

The use of a flow directional collar protected by an accurately produced micromesh sieve whose aperture is equal to 40% of the orifice has two advantages. By protecting the orifice it permits the use of a multi-tube system to operate in the same suspension thus eliminating the need for the conventional two-tube analysis method for wide-ranging powders. It permits the detection and measurement of particles having different aspect ratios; this is particularly useful for the detection of fibres in the presence of particles of other shapes.

A solution adopted by Coulter's was an electronic edit system which they fitted to their channelysers. This system rejects distorted pulses which comprise about half the data. The P series channelysers can be fitted to the Z_{BI} or ZF Coulter counters to improve their sizing accuracy. The edit system is built into the channelyser and can be

switched out. The P models contain the same facility at a lower cost since data storage is in the memory of the plotter and is not displayable on a scope.

Two types of pulse discrimination system have been described; one type operates by rejecting all pulses having a rise time greater than a preset minimum; another type accepts only those pulses generated by particles passing through the aperture on nearly axial paths.

The Coulter H4 system, available in America, examines each pulse, integrates it and normalizes it so that no misshapen pulse data are rejected. The resolution of the H4 system has been discussed by Alliet and Behringer [38]. This 'high resolution' circuitry is also incorporated in model ZH which is basically a model Z_{BI} with the high-resolution circuitry added. A 'hydrodynamic focusing' system has been adopted by Berg [53] in the Elzone system. In 1973 Coulter's purchased the rights to manufacture the Telefunken instrument which was made available as the model TF Coulter counter.

The effect of eliminating distorted pulses is the production of a narrower size distribution. The rejected pulses are oversize; hence a tendency for the distribution to skew to the right, which is apparent in instruments without this facility, is removed. A full description of the IITRI instrument and a discussion on the above lines have been written by Davies *et al.* [41, 51, 52].

A comparison between the TF system, the edit system and the standard Coulter has been given by Lines [70] who also discusses the effect of using a long aperture.

13.9 End-point determination

At the coarse end of the size spectrum it is necessary to continue counting to zero count. At the fine end it is often found that the final value $\Delta n\bar{\imath}$ does not represent the 100% volume level, i.e. the counts are still increasing at the lowest threshold level. Several proposals have been made for assessing total volume in these cases [11, 27, 28]. All are essentially extrapolation techniques.

(a) The simplest method is the purely arbitrary addition of half the last value of $\Delta n\bar{\imath}$, i.e. 100% volume = $\Delta n\bar{\imath}$ + $\Delta n\bar{\imath}$ (last)/2.

(b) Plotting of cumulative corrected number n against t or d on log–log axes with extrapolation of the curve until it becomes asymptotic with the size axis; the t or d value corresponding to this is incorporated into the calculation.

(c) For expected normal or log-normal distributions, plotting the weight percentage against size should give a straight line; in these cases the 100% value may be arrived at by trial and error by the addition of an estimated percentage smaller than the smallest measured size.

(d) Plotting $\Sigma \Delta n\bar{\imath}$ against d on log–linear paper [88]; a straight line is drawn through the last two points on the graph to intercept the $\Sigma \Delta n\bar{\imath}$ axis; a horizontal line is drawn through the final experimental point to intercept the axis; the 100% value is then taken as one-third the distance between the lower and upper intercepts.

(e) More sophisticated extrapolation techniques have been proposed by Harris and Jowett [27] based on the Gates–Gaudin–Schumann or the Rosin–Rammler distribution; the uncertain validity in this assumption was emphasized by Eckhoff [28], who suggested the following expression:

$$100\% \text{ volume} = \Delta n\bar{t} + \Delta n\bar{t} \text{ (last)} \Big/ \left[\left(\frac{\text{penult. } d}{\text{final } d} \right)^2 - 1 \right]$$

(f) Where there is a considerable amount of material unmeasured, equations (13.2) and (13.3) may be equated to give an estimate of 100% volume.

13.10 Upper size limit

The lower size limit with the Coulter principle is generally considered to be restricted only by thermal and electronic noise. There is no theoretical upper limit and for particles having a density similar to that of the electrolyte no difficulty is experienced using the largest aperture (2000 μm) normally available.

When the density difference is considerable the upper size limit is reached when the particles can no longer be kept in suspension. One solution is to increase greatly the viscosity of the suspension by the addition of a thickening agent such as glycerol or sugar. To suspend sand particles of density 2650 kg m^{-3}, of up to 1 mm in size, a 90% solution of glycerol in 20% saline has been used [72].

Other proposed solutions have been round-bottomed beakers [73], baffles [74] and double-bladed stirrers [75]. In other solutions the whole of the sample is caused to pass through the aperture [76].

A novel device proposed by Coulter's [77] analyses a complete sub-sample of the material, removes the particles from the sensing zone so that they cannot recirculate, allows reprocessing of the sample material and in addition uses hydrodynamic focusing. Using it, particles of up to 1.5 mm diameter and 7.8 sg can be sized.

13.11 Commercial equipment

Coulter counter

Model A (medical)	
Model B (industrial)	Obsolete
Model D	A simplified version of model A (medical) with facility for red and white cell counting and rudimentary sizing
Model D (industrial)	A simple model for sizing in the range 1 to 75 μm
Model D$_1$	Transistorized version of model D
Model D$_3$	Model D$_1$ extended to include platelet counting

Model C	Twelve-channel model designed for fast analysis and computer print-out. Obsolete
Model B	Has facility for both cumulative and differential counts
Model M_2 converter	Facilitates easier weight conversion direct from model Z_B
Model M_3 data converter	A keyboard programmable calculator to interface with model TA II to produce mean, median, mode and standard deviation with facility for handling two-tube data, performing extrapolation and comparisons
Model J plotter	Plots directly the frequency histogram
Model H plotter	A more sophisticated version of model J
Model F_N	A fully transistorized version of model A (medical)
Model ZF	A solid-state version of model F_N
Model Z_B	Solid-state version of model B
Model Z_{BI}	Biological version of model Z_B
Model S	A fully automatic medical model, giving red cell (RBC) and white cell (WBC), count, haemoglobin (Hb), haemotecrit (Hct), mean cell haemotecrit and mean cell haemoglobin in 20 s
Model S Sr	Updated version of model S
Model S Jr	Three-parameter (RBC, WBC, Hb) simplified version of model S
Model S5	Five-parameter (RBC, WBC, Hb, MCV, Hct) version of model S
Model S Plus	Up-market version of model S series of haematological analysers which also gives information on red cell distribution width as well as platelet parameters
Model T	A solid-state version of model C with 15 sizing channels with automatic number-to-weight conversion
Model TA	Updated version of model T giving volume results in up to 16 channels (obsolete)
Model TA II	Improved version of model TA which, with Population Count Accessory (PCA), also gives number distribution as well as volume
On-line monitor	Automatic sampling system for prediluted suspensions; fits onto TA II or Z_B electronics consoles
Milk cell	Fully automatic analyser designed specifically for counting somatic cells (white cells) in milk. Samples, dilutes, incubates to remove fat globules, counts, computes and presents corrected results
Channelysers – C.1000	Accumulate data from the counter and separate it into 100 size channels; contain 'edit' circuits
Channelysers P64 and P128	Accumulate and separate in 64 size channels; can be fitted to models Z_{BI} and ZF; contain 'edit' circuits

Channelyser H-4	256-channel solid-state instrument which accepts pulse data from Z_B, Z_{BI} or ZH counters

Celloscope

Model 30	Basic industrial model with twin simultaneously operated discriminators
Model 401	Blood-cell counter

Elzone

Model 110	Fixed count-level settings for red and white cells and platelets
Model 111	Single-threshold instrument
Model 112	Dual fixed ratio (2 : 1) threshold with 15 channels
Model 112C	A second register includes recording count above upper level
Model 112 LTS/ADC	Updated version; solid-state 128-channel counter
Models 111L, 112L	Logarithmic conversion of particle amplitudes with mode selector for linear or log output
Granulometer	(References 63–66)
TUR ZG1	Eastern European equivalent of model A
TUR ZG2	Eastern European equivalent of model B
TOA microcell counter	Japanese instrument with a polypropylene aperture tube at the bottom of a cylindrical tube, i.e. facing downwards. In contrast to the resistance-sensitive instruments this is sensitive to capacitance changes [86].

Telefunken particle detector

MS PD1 1105/1	An improved version of the Coulter principle with hydrodynamic focusing. Now available as the Coulter counter model TF
Microscale	Blood cell counter (see [86])
Picoscale	Blood cell counter
Biotronics	Blood cell counter
Digicell 100	Blood cell counter

13.12 Conclusions

The foregoing remarks comprise a basic statement of the Coulter principle. This technique is eminently suitable for powders in the sub-sieve range down to about 1 μm in size and a size range in the ratio of 20 : 1. The size limits for the technique may be ex-

tended with care, and an element of luck, from 400 to 0.6 μm. The size range may also be extended, using a multiple-tube technique or the Coulter technique in conjunction with some other sizing method. In this case, however, the analysis may become tiresome due to blocking of the aperture which occurs frequently if large particles are present in the electrolyte.

The technique has the unique capability of rapidly differentiating between two distributions with very similar peaks. With the basic counters, a single analysis can be completed in an hour or so; multiple analyses may be carried out more rapidly.

Problems can arise due to the 'black box' nature of the system, which is no more intelligent than the operator. For example, if the sample is made up of sub-micrometre powder, an analysis in the micrometre range will result due to the instrument counting multiple pulses as single particles.

The associated equipment, available with the basic glassware, is many and varied and enables the time for a single count to be reduced to 20 s.

The most widely used models in general use are the Coulter model Z_B and the Elzone model 111. More rapid sizing and evaluation is possible with the model **TA II** Coulter counter and Elzone model 112. The model T is faster and easier to calibrate than the model A [58]. Groves compared eleven Coulter counters using electronically generated pulses and found a standard error of ± 0.3%. Using PVC latex and 12 instruments he obtained a standard error of ± 2.0% [60]. Davidson [61] using eleven operators, four counters and four laboratories found a mean size for a phosphate sample of 12.95 μm ± 0.08 sd.

The popularity of the technique is evidenced by the Coulter Industrial Bibliography which contains 1100 'selected' references and the medical bibliography which contains over 1700 references (available from Coulter's UK).

References

1 Coulter, W.H. (1956), *Proc. Nat. Electronic Conf.,* **12**, 1034.
2 Morgan, B.B. (1957), *Research,* **10**, 271.
3 Kubitschek, H.E. (1958), *Nature,* **182**, 234–5.
4 Kubitschek, H.E. (1960), *Research,* **13**, 128.
5 Berg, R.H. (1958), **ASTM** Publ. No. 234.
6 *Coulter Industrial Bibliography,* Coulter Electronics Ltd.
7 Brotherton, J. (1969), *Cytobios,* **1B**, 95.
8 Brotherton, J. (1969), *ibid.,* **3**, 307.
9 Brotherton, J. (1971), *Proc. Soc. Analyt. Chem.,* 264–71.
10 Matthews, B.A. and Rhodes, C.T. (1969), Paper presented at the 7th Users Conference, London, Coulter Electronics Ltd.
11 Batch, B.A. (1964), *J. Inst. Fuel,* 455.
12 Allen, T. and Marshall, K. (1972), The electrical sensing zone method of p.s.a., available from the author.
13 Marshall, K. (1969), M.Sc. thesis, Bradford Univ.
14 Lloyd, P.J., Scarlett, B. and Sinclair, I. (1972), *Proc. Conf. Particle Size Analysis,* Bradford (1970), Soc. Analyt. Chem., London.

15 Eckhoff, R.K. (1969), *J. Phys. E*, **2**, 973–7.
16 Allen, T. (1967), Loughborough, *Proc. Conf. Particle Size Analysis*, Loughborough (1966), Soc. Analyt. Chem., London.
17 Wales, M. and Wilson, J.N. (1961), *Rev. scient. Instrum.*, **32**, 1132.
18 Princen, L.H. and Kwolek, W.F. (1965), *ibid.*, **36**, 646.
19 Pisani, J.F. and Thomson, G.M. (1971), *J. Phys. E*, **4**, 359–61.
20 Edmundson, I.C. (1966), *Nature*, **212**, 1450.
21 Harvey, R.J. (1972), Methods in colloidal physics, (*cit.* [12]).
22 Harvey, R.J. and Marr, A.G. (1966), *J. Bact.*, **92**, 805.
23 Priem, M. (1970), *J. Phys. E*, **3**, 402–3.
24 Lines, R. (1967), *Proc. Conf. Particle Size Analysis*, Loughborough (1966), Soc. Analyt. Chem., London, 135.
25 von Thom, R. (1971), Available from A.E.G. Telefunken 79 Ulm (Donau), Elisabethstrasse 3, Postfach 830, W. Germany.
26 von Thom, R. (1971), Ger. Pat. 1 955 094.
27 Harris, C.C. and Jowett, A. (1965), *Nature*, **208**, 175.
28 Eckhoff, R.K. (1966), *Proc. Coulter Counter Conf.* Cardiff, Coulter Electronics Ltd., pp. 80–94.
29 Samyn, J.C. and McGee, J.P. (1965), *J. Pharm. Sci.*, **54**, 12, 1794–9.
30 Rigby, O. and Thornton, M.J. (1965), *2nd Br. Coulter Counter Users Mtg*, Nottingham.
31 Palik, S.E. (1967), *Anachem Conf.*, Detroit.
32 Sub-Committee on Sedimentation, December 1964, A Study of Methods Used in Measurement and Analysis of Sediment Loads in Streams; Electronic Sensing of Sediment, Prog. Rep., Interagency Committee on Water Resources, St. Anthony's Fall Hydraulic Lab., Minneapolis, Minnesota.
33 Eckhoff, R.K. and Soelberg, P. (1967), *Betontek. Publik.*, **7**, 1.
34 Treweek, G.P. and Morgan, J.J. (1977), *Environ. Sci. Technol.*, **11**, 7, 707–14.
35 Polke, R. (1975), *Dechema monogr.* 79 (1589–1615), Part B, 361–76.
36 Edmundson, I.C. (1967), In *Advances in Pharmaceutical Sciences 2*, Academic Press, pp. 95–179.
37 Harfield, J.G. and Wood, W.M. (1972), *Proc. Conf. Particle Size Analysis*, Bradford, (ed. M.J. Groves and Wyatt-Sargent) (1970), Soc. Analyt. Chem., 293.
38 Alliet, D.F. and Behringer, A.J., *ibid.*, 353.
39 Alliet, D.F. (1976), *Powder Technol.*, **13**, 3–7.
40 Spielman, L. and Goren, S.l. (1968), *J. Colloid Interfac. Sci.*, **26**, 2.
41 Karuhn, R. *et al.*, *Powder Technol.*, **11**, 157–71.
42 De Blois, R.W. and Bean, C.P. (1970), *Rev. scient. Instrum.*, **41**, 7, 909–16.
43 Anderson, J.L. and Quinn, J.A. (1971), *ibid.*, **42**, 8, 1257–8.
44 Grover, N.B. *et al.* (1969), *Biophys. J.*, **9**, 1398, 1415.
45 Gregg, E.L. and Steidley, K.D. (1965), *ibid.*, **5**, 393.
46 Smythe, W.R. (1961), *Phys. Fluids*, **4**, 756.
47 Smythe, W.R. (1964), *ibid.*, **7**, 633.
48 Fils, F. (1972), *J. Pharm. Belg.*, **27**, 2, 227–32.
49 Valley, R.B. and Morse, T.H. (1965), *TAPPI*, **48**, 6, 372–6.
50 Kominz, D.R. (1971), *Biophys. J.*, **2**, 47–65.
51 Davies, R., Karuhn, R. and Graf. J. (1975), *Powder Technol.*, **12**, 157–66.
52 Davies, R. *et al.* (1976), *Powder Technol.*, **13**, 193–202.
53 Karuhn, R.F. and Berg, R.H. (1978), *Powder and Bulk Solids Conf.*, Fine Particle Soc., Chicago, IL.
54 Bader, H., Gordon, H.R. and Otis, B. (1972), *Rev. scient. Instrum.*, **43**, 10, 1407–12.
55 Auerbach, F. (1921), *Handbook of Electricity and Magnetism* (ed. Graetz), Johann Ambrosius Barth, Leipzig, p. 110 (*cit.* [42]).
56 Kubitschek, H.E. (1960), *J. Res. natn. Bur. Stand.*, *A*, **13**, 128.
57 Muntwyler, F. (1974), Ger. Offen. 2 343 363, 28 March.
58 Bonekowski, N.R. (1974), Mount Lab. Miamisburg, Ohio, Rep. 1974–MLM-2098, NTIS.

59 De Blois, R.W., Bean, C.P. and Wesley, R.K.A. (1977), *J. Colloid Interfac. Sci.*, **61**, 2, 323–35.
60 Groves, M.J. (1972), *J. Pharm. Pharmac.*, Suppl. 23, 238S–9S.
61 Davidson, H.I. (1968), *Proc. Coulter Counter Conf.*, Chicago, Illinois, January 22–23, Coulter UK.
62 Maguire, B.A., Seaney, R.J. and Halpin, R.K. (1972), *Powder Technol.*, **6**, 61–63.
63 Heidenreich, E. (1966), *Die Technik*, **21**, 1, 5, 28.
64 Heidenreich, E. and Schuldes, (1968), *Medizintechnik*, **8**, 2.
65 Pfeiffer, G. (1962), *Z. Med. Labortechnik*, **3**, 2, 5, 57.
66 Wussow, S. and Schuster, R. (1972), *ibid.*, **13**, 5, 290–1.
67 King, A.E. and Vali, G. (1975), *J. Colloid Interfac. Sci.*, **53**, 2, 337–9.
68 Colon, F.J. *et al.* (1973), *Powder Technol.*, **8**, 5/6, 307–10.
69 von Thom, R., Hampe, A. and Sauerbrey, G. (1969), *Z. ges. exp. Med.*, **151**, 331.
70 Lines, R. (1975), *Proc. Coulter Counter Conf.*, London, Coulter UK, 372–88.
71 Suganuma, G. and Tasaka, A. (1975), *Jap. Kokai*, **37**, 495.
72 McCave, I.N. and Jarvis, J. (1973), *Sedimentology*, 305–15.
73 Kinsman, S. (1961), Coulter Counter Users Conf., Chicago, IL., 30–36.
74 Eckhoff, R.K. (1967), *J. scient. Instrum.*, **44**, 648–9.
75 Yarde, H.R. (1968), *J. Phys. E.*, **1**, 711.
76 Coulter, W.H. and Morgan, C.T. (1968), BP 1 125 289.
77 Harfield, J.G. *et al.* (1977), *Int. Conf. Particle Size Measurement* (ed. M.J. Groves), Chem. Soc. Anal. Div., Heyden (1978).
78 Mattern, C.F.T., Bracket, F.S. and Olson, B.J. (1957), *J. appl. Physiol.*, **10**, 56.
79 Grant, J.L., Britton, M.C. and Kurz, T.E. (1960), *Am. J. Clin. Path.*, **33**, 138.
80 Herben, J.G.H.M. (1964), *The Electronic Blood Cell Counter*, Dissertation, Amsterdam.
81 Gutmann, J. (1966), *Electromedizin*, **11**, 80.
82 Strackee, J. (1966), *Med. Biol. Engng.*, **4**, 97.
83 Mercer, W.B. (1966), *Rev. scient. Instrum.*, **37**, 1515.
84 Kalling, L.O., Lantorpe, K. and Gunne, I. (1969), *Acta path. microbiol. scand.*, **76**, 447.
85 Walstra, P. and Oortwijn, H. (1969), *J. Colloid Interfac. Sci.*, **29**, 426.
86 Helleman, P.W. (1972), *The Coulter Electronic Particle Counter*. Druk: Koninklijke Drukkerij C.C. Callenbach N.V., Nijkerk.
87 Ruhenstroth Bauer, G. and Zang, D. (1960), *Blut*, Suppl. 6, 3.
88 Emonet, A.L.D. (1966), *Proc. Coulter Counter Conf.*, Cardiff, Coulter UK, 80–94.
89 Adams, R.B., Voelker, W.H. and Gregg, E.C. (1967), *Phys. Med. Biol.*, **12**, 79.
90 Anderson, F.G., Tomb, T.F. and Jacobson, M. (1968), US Bur. Mines R.I., 7105.
91 Dunson, J.B. Jr. (1970), Ann. Mtg Am. Inst. Chem. Engrs., Paper No. 5e.
92 Tomb, T.F. and Raymond, L.D. (1970), US Bur. Mines R.I., 7367.
93 Schrag, K.R. and Corn, M. (1970), *Am. ind. Hyg. Assoc. J.*, July/Aug., 446–53.
94 Simecek, J. (1967), *Staub Reinhalt. Luft*, **27**, 6, 33–37.
95 Walstra, P., Oortwijn, H. and Graaf, J.J. de (1969), *Neth. Milk Dairy J.*, **23**, 12.
96 Wales, M. and Wilson, J.N. (1962), *Rev. scient. Instrum.*, **33**, 575.
97 Eckhoff, R.K. (1965), *Nature*, **208**, 175.
98 Helleman, P.W. and Benjamin, C.J. (1969), *Scand. J. Haemat.*, **6**, 128.

14 Radiation scattering methods of particle size determination

14.1 Introduction

When a beam of radiation is interrupted by the presence of a particle, the interaction yields particle size information. Many 'stream-scanning' (section 20.2) and light-scattering (section 3.8) devices are based on this principle. Theoretical solutions are possible for particular systems but it is commercial practice to calibrate instruments using standard powders. In the photosedimentation technique (section 9.4) the interaction of a beam of radiation with a sedimenting suspension is used for particle size distribution determination.

The far-field diffraction pattern of an assembly of particles yields information concerning their size distribution. This dependence is utilized in some 'field-scanning' devices (section 20.3). The mean size of a closely graded powder made up of very small particles may also be determined.

When a beam of radiation strikes an assembly of particles, some of it is transmitted, some absorbed and some scattered. The scattered radiation includes the diffracted, refracted and reflected parts of the original beam and the absorbed radiation is retransmitted at a longer wavelength and this is usually not picked up by the detecting device.

The mean particle size of the assembly may be found by measuring the variation in the intensity of the scattered radiation with scattering angle or, more frequently, the ratio of the intensities of the scattered radiation at fixed angles to the direction of the incident beam:

$$D_i = \frac{I(90 + \beta)^\circ}{I(90 - \beta)^\circ} \qquad (14.1)$$

with β usually equal to 30, 45 or 60 (dissymmetry methods) [2–4].

Alternatively, the flux per unit angle at 90° to the incident beam may be measured [5]. Since the vertically polarized component of the scattered light is more sensitive to size changes, it is usual to accept only this component in the detector.

The right-angle scattering lobe method is more sensitive to small differences in particle size than the forward scattering lobe method and may be applied to smaller particles but the latter has advantages over the former in that [40, 41]:

(1) An average size may be determined without recourse to theoretical scattering diagrams.
(2) The method is insensitive to refractive index.

(3) The method is independent of concentration over a wide range.
(4) The method is not too sensitive to particle shape.
(5) Rayleigh–Gans and Fraunhofer theory can be used.
(6) A measure of spread of distribution may be determined.

The angular positions of maxima and minima in the scattered light have also been used to determine particle size, as well as the state of polarization of the scattered light, usually defined by the ratio of the horizontal and vertical components. Theoretical expressions for the attenuation of a beam of light when it traverses an assembly of particles are also available when the light is monochromatic [1]. A practical method of applying these equations is to determine the attenuation for two wavelengths. It is then possible to read off the size from theoretically based diagrams [5, 6].

This has been extended to three wavelengths and the Mie scattering function determined by computer. A relation was developed between ratio of turbidities at two wavelengths and the diameter of spherical particles. The aggregate size of various carbon blacks was then determined by turbidity and the average number of primary particles correlated with the average aggregate cross-section [58].

The intensity of the scattered radiation is a function of $x = \pi D / \lambda_m$, where D is the diameter of the particles, λ_m the wavelength of the radiation in the medium and m the refractive index ratio of scattering particles to medium. Rigorous solutions to the various phenomena are found with the electromagnetic theory of light scattering (the Mie theory), but since these are in a form difficult to interpret, partial solutions are more favoured. These are found by using boundary conditions and the theory of physical optics.

The boundary conditions limit the use of scattering phenomena to molecular weight determinations and spherical monosize colloids and aerosols in the range $0.1 < D < 4.00 \ \mu m$, the refractive index m of the particles relative to that of the surrounding medium lying between 1 and 2, although size information about non-spherical particles and heterogeneous distribution may also be evaluated. Outside the boundary regions, some solutions to the exact theory are available in the form of tables of scattering functions prepared by the National Bureau of Standards [7].

Van de Hulst [1] discusses these boundary conditions, deriving or reproducing the appropriate equations in terms of the m, x domain as m varies from 1 to ∞ and x varies from 0 to ∞. Applications of these equations are reviewed in [1, 8, 9].

When white light is incident on a dilute suspension of sufficiently large, monodisperse, spherical particles, vivid colours appear at various angles to the primary beam and the angular positions of the spectra may be used to determine particle size.

Light-scattering methods are most effective for particles of the same order of size as the incident radiation. Very small particles (e.g. air molecules) exhibit Rayleigh scattering, and the scattered field is proportional to the polarizability of the particles, i.e. to the number of electrons or essentially to the volume of the particles. Therefore, the scattered-light intensity is proportional to the square of the particle volume. The scattering pattern for such a particle is easily recognizable since it is symmetrical about a scattering angle of $90°$. In figure 14.1, the Rayleigh approximation is valid for particles of diameter less than about a twentieth of the wavelength of light.

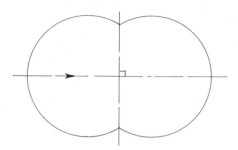

Fig. 14.1 Polar light-scattering diagram for the Rayleigh region.

For somewhat larger particles (e.g. macromolecules) the **Rayleigh–Gans** approximation applies. The boundary conditions for this approximation to hold require the radiation to undergo only a small phase shift in passing through a particle; hence it applies to small particles or particles with a similar refractive index to their surroundings. The scattered field is, as in the Rayleigh approximation, the sum of contributions from each point within the particle; however, in this case the phase shift from point to point must be taken into account. The Rayleigh–Gans equation reduces to the Rayleigh equation when this is negligible. The addition of the phase shift term modifies the scattering pattern, making it asymmetric about 90°. Since the phase shift between contributing points is a minimum in the forward direction, the scattered intensity is greater towards this direction and increases with increasing particle size (figure 14.2).

Particle size may be determined by finding the positions of maxima or minima in the polar-scattering pattern, dissymmetry methods (figure 14.3) or forward-angle scattering.

For Rayleigh scattering $(m - 1 \rightarrow 0, x < 0.3)$ the vertical component of scattered intensity (I_θ) is constant. As D increases, the Rayleigh–Gans region is entered in which I_θ is inversely proportional to θ and passes through a minimum; this condition also holds for $m > 1.0$, provided x is small. However, for large x, several maxima and minima occur. Particle size may be determined using these properties for $0.18 < D < 4.0$ μm using visible light.

For particles of size comparable to the wavelength of the incident radiation, the complex Mie scattering theory is required to describe the scattering pattern. Interaction between the particle and the radiation is very strong (resonance), leading to pronounced maxima and minima (figure 14.2). The angular positions of these maxima and minima can be used for size determination.

For particles larger than the wavelength of the incident radiation, the contribution of the radiation refracted within the particle diminishes in comparison to the radiation diffracted external to the particle. For particle size to wavelength ratios of the order of four or five, the latter becomes dominant and chiefly in the forward direction. For this regime, the Mie theory reduces to the Fraunhofer theory for geometric optics and the expression for the scattered intensity is the one for diffraction by a circular disc:

$$\left[\frac{J_1 (x \sin \theta)}{x \sin \theta} \right]^2 \tag{14.2}$$

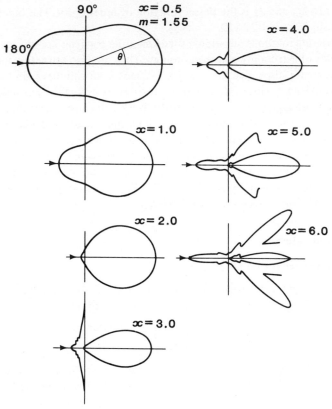

Fig. 14.2 Polar light-scattering diagrams (Vouk [39]). The outer curve magnifies the inner by a factor of 10 in order to show fine detail.

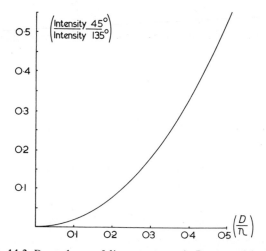

Fig. 14.3 Dependence of dissymmetry ratio D_i on particle size.

where $x = \pi D/\lambda_m$ and J_1 is the Bessel function of order unity. The function has its first zero at $\sin \theta_{min} = 1.22\lambda/D$.

Light-scattering methods are widely used when the particles are not readily accessible, as in the fields of astronomy and meteorology. They are also attractive for determining the size of particles suspended in a gas, e.g. effluent from chimneys, atmospheric pollution and aerosol sprays, since other methods involve collecting the particles and this may alter their state of aggregation. Optical methods have the advantages that the measurements can be made accurately, suspensions are not disturbed by the insertion of an external sampling device, the sample required is small, tests can be carried out rapidly and experimental measurements are obtained in a form which lends itself to automatic recording and remote-control techniques. A comprehensive review of light scattering, with 519 references is given by Kratohvil [37]. Light scattering is also reviewed by Kerker [30], who gives over a thousand references.

14.2 Scattered radiation

14.2.1 The Rayleigh region ($D \ll \lambda$)

In the Rayleigh region, the intensity of the scattered radiation in a direction making an angle θ with an incident beam of unit intensity is given by [1]:

$$I_\theta = |\alpha|^2 \left(\frac{2\pi}{\lambda_m} \right)^4 \frac{(1 + \cos^2 \theta)}{2r^2} \tag{14.3}$$

where α is the particle polarizability, r is the distance from the particle to the point of observation and λ_m the wavelength of the incident radiation in the medium surrounding the particle.

The intensity is the sum of two terms:

$$|\alpha|^2 \left(\frac{2\pi}{\lambda_m} \right)^4 \frac{1}{2r^2} \quad \text{and} \quad |\alpha|^2 \left(\frac{2\pi}{\lambda_m} \right)^4 \frac{\cos^2 \theta}{2r^2}$$

which refer respectively to the intensities of the vertically and horizontally polarized components.

For spheres:

$$\alpha = \frac{3(m^2 - 1)}{(m^2 + 2)} \frac{V}{4} \tag{14.4}$$

Equation (14.4) has been applied to very small particles, but is more relevant to the size determination of molecules. For:

$$m - 1 \rightarrow 0, \quad \alpha = (m^2 - 1) \frac{V}{4}$$

$$\simeq 2(m - 1) \frac{V}{4} \tag{14.5}$$

where m is the index of refraction of the particle relative to that of the surrounding medium and V is the particle volume.

The size of colloidal particles [55, 87] and the size distribution of pores [59, 70] have been determined using low-angle X-ray scattering. Low-angle X-ray scattering has also been used for surface area determination and shows good agreement with nitrogen and water adsorption [57].

14.2.2 The Rayleigh–Gans region ($D < \lambda$)

The Rayleigh–Gans equation for the angular dependence of the intensity of the scattered light I_θ is given for spherical particles by the equation [1, p. 89]:

$$I_\theta = I_0 \left[\frac{k^4 V^2 (m-1)^2}{8\pi^2 r^2} \right] \left[\frac{3}{u^3} (\sin u - u \cos u) \right]^2 (1 + \cos^2 \theta) \qquad (14.6)$$

Again, 1 and $\cos^2 \theta$ in the final term refer respectively to the intensities of the vertically and horizontally polarized components, and the incident beam is of intensity I_0:

$$u = \frac{2\pi D}{\lambda_m} \sin \frac{\theta}{2} \qquad (14.7)$$

$$k = 2\pi/\lambda_m \qquad (14.8)$$

and D is the particle diameter.

Equation (14.6) reduces to equation (14.3) when the middle term is equal to one. The scattering pattern is however modified by this second term, thus enabling size determination to be carried out in the Rayleigh–Gans region.

Differentiating equation (14.6) with respect to u and putting $dI_\theta/du = 0$, for minimum intensity:

$$\sin u - u \cos u = 0 \qquad (14.9)$$

and for maximum intensity:

$$3u \cos u - u^2 \sin u - 3 \sin u = 0 \qquad (14.10)$$

The first minimum is at $u = 4.4934$ rad, corresponding to:

$$\frac{D}{\lambda_m} \sin \left(\frac{\theta_1}{2} \right) = \frac{4.4934}{2\pi}$$

$$= 0.715 \qquad (14.11)$$

Similarly, the first maximum occurs at:

$$\frac{D}{\lambda_m} \sin \left(\frac{\phi_1}{2} \right) = 0.916 \qquad (14.12)$$

A graphical solution for all maxima and minima has been determined by Pierce and Maron [10], who, together with Elder [11, 12], extended equations (14.11) and (14.12) beyond the Rayleigh–Gans region ($m - 1 \to 0$) to $1.0 < m < 1.55$, deriving the following formulae:

$$\frac{D}{\lambda_m} \sin \left(\frac{\theta_1}{2} \right) = 1.062 - 0.347 m \tag{14.13}$$

$$\frac{D}{\lambda_m} \sin \left(\frac{\phi_1}{2} \right) = 1.379 - 0.463 m \tag{14.14}$$

These equations give the positions of the first intensity minimum and maximum respectively. Tabulated data allow the calculation of various maxima and minima, provided m is known. These tables are particularly useful for determining the order of the maxima and minima, since the ratios of the sines of the half-angles are 1.72, 1.41, 1.29, etc., for the angles θ_1/θ_2, θ_2/θ_3, θ_3/θ_4, etc. and 1.58, 1.35, 1.26 for the angles ϕ_1/ϕ_2, ϕ_2/ϕ_3, ϕ_3/ϕ_4 for $m = 1$; ratios for other m are derivable from equations (14.12) and (14.13). The range of validity of this method has been investigated further by Kerker et al. [13], who present their results in the form of 2% and 10% error contour charts. The required apparatus consists of a lamp-house, a lens compartment and a phototube compartment, with the optical parts mounted on an optical bench. The light source may be a 100-W mercury vapour lamp with associated filters. The suspension is placed in a cylindrical cell at the centre of the phototube compartment. The phototube is mounted on a turntable and its output is fed to a micrometer. The observed intensities require correction because of the variation with θ of the volume of scattering suspension scanned by the phototube. This correction is made by multiplying the observed readings by sin $\theta/2$. A plot of I/I_0 against sin $\theta/2$ converts the minima into maxima, θ being found from the position of the first peak. The θ's so obtained vary with concentration and, since the value required is for infinite dilution, when multiple scattering is absent, θ's have to be found for several concentrations and extrapolated to zero. This concentration dependence on scattering has also been noticed with depolarization and dissymmetry methods and becomes negligible only when particle separation exceeds 200 radii [14].

Polymer latex particles, in the size range 0.3 to 11 μm, have been characterized by measurement of the 360° light-scattering pattern from individual particles. The pattern is recorded in 20 ms and the data are stored on computer tape and compared with Mie theory for best fit. The instrument used was a Gucker photometer with a 632.8 nm laser beam [62].

A light-scattering apparatus has been described in which a laser beam provides the incident light and a photomultiplier mounted on a rotating arm is used to measure the scattered light. The variation of intensity against scattering angle is plotted automatically and examples are given of how to treat the recorded data to investigate crystal growth [69].

14.3 State of polarization of the scattered radiation

When a system containing isotropic and monosize spherical particles is irradiated with unpolarized incident radiation, the horizontal and vertical components of the scattered light are in general functions of the three parameters, relative refractive index m, angle of observation θ and $x = \pi D/\lambda_m$. If H is the intensity of the horizontal component of

the scattered light, V the vertical component and $R = H/V$, then R is also a function of m, x and θ. For given m and θ, R depends only on x; hence polarization measurements can be used to determine particle size. Again, if m and x are fixed, then R depends only on θ, hence particle size can be determined from the position of maxima and minima in the angular dependence of R on θ.

For Rayleigh scattering $R = 0$ at $\theta = 90°$. As x increases, this theory shows that R is a periodic function of diameter for monosize particles. Sinclair and La Mer [15] used this theory for plots prior to the first maximum to measure the particle sizes of aerosols ranging in size from 0.1 to 0.4 μm. La Mer, Inn and Wilson [16] used the same procedure at angles of 50° to 120° to obtain particle sizes of sulphur solutions. In this work, transmission and polarization methods yield results in accord with higher-order Tyndall spectra for sizes in the range 0.35 to 0.62 μm. Graessley and Zufall [17] have shown that in a limited region ($0.45 < D \ll 2.8$ μm, $1.06 < m < 1.12$) the fluctuations in R at $90°$ are smoothed out and the identity:

$$\bar{R} = 1.89(m - 1)^2 \, \pi\bar{D}/\lambda_m \qquad (14.15)$$

results where:

$$\bar{D} = \frac{\int_0^\infty D^{7/2} N(D)\mathrm{d}D}{\int_0^\infty D^{5/2} N(D)\mathrm{d}D} \qquad (14.16)$$

and there are $N(D)$ particle in the light beam in the size range D to $D + \mathrm{d}D$. Maron Elder and Pierce [18] review and extend earlier work on R-measurements at $90°$ on monodisperse polystyrene latices and find appreciable differences between theoretical and experimental values. They show that the discrepancy is due to inherent anisotropy of the latex particles believed to be due to non-random orientation of the polymer chains in the colloidal latex particle. The size range of applicability they give to this technique is 0.135 to 1.117 μm.

Takahashi and Iwai [34] also use the monotonic variation in R at 90°.

14.4 Turbidity measurement

If a light beam falls upon an assembly of macroscopic particles the attenuation is given by:

$$I = I_0 \exp[- aNL] \qquad (14.17)$$

where I is the transmitted intensity when a light beam of intensity I_0 falls on a suspension of particles of projected area a and number concentration N and traverses it by a path of length L. In the general equation relating attenuation and concentration aN is replaced by Tc where T is the turbidity and c the weight concentration per unit volume. For dilute suspensions of particles smaller than 0.04 μm in diameter, the turbidity can be calculated from the equation:

$$T = \frac{32\pi^3 \, (m - 1)^2}{3N\lambda_m c^2} f \qquad (14.18)$$

where $m = \lambda_m/\lambda_f$ is the ratio of the refractive indices of the particles and fluid; $\lambda_m = \lambda_0/\lambda_f$ is the wavelength of the light in the fluid and λ_0 is the wavelength in vacuum; N is Avogadro's number and f is very nearly equal to one [1, p. 396].

For particles larger than this:

$$T = Ka \qquad (14.19)$$

where K is the extinction coefficient defined as the effective particle cross-section divided by its geometric cross-section. This may be evaluated theoretically, thus permitting the determination of particle size. If c is the volume concentration:

$$c = \frac{\pi}{6} Nd_v^3 \qquad (14.20)$$

and the projected area for particles in random orientation is:

$$a = \frac{\pi}{4} d_s^2 \qquad (14.21)$$

Hence:

$$I = I_0 \exp\left[-\frac{3Kc}{2d_{sv}}\right] \qquad (14.22)$$

$$I = I_0 \exp\left[-\tfrac{1}{4} KcS_v\right] \qquad (14.23)$$

K has been given by van de Hulst [1, p. 132] in terms of particle size refractive index domain (figure 14.4).

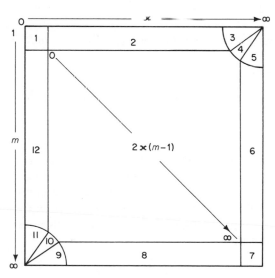

Fig. 14.4 Particle size refractive index domain [1].

Mie theory applies to the whole domain. In the boundary region, simpler equations have been derived. These equations are even simpler in the corner regions (odd numbers), which are all for a limited range of m and x. Thus, for small spheres (regions 9, 10, 11, 12, 1) K is proportional to x^4, hence turbidity is proportional to x^4 (c/d_{sv}), i.e. T is proportional to $cD^3/4$. Similarly for very large spheres T is proportional to $4c/3D$. In the intermediate region, the approximate size must be known and the extinction coefficient determined experimentally over a sufficiently wide range of wavelengths to give a recognizable portion of the extinction curve. The difficulties are overcome with non-adsorbing (dielectric) spheres with m approximately equal to one, since the equation:

$$K = 2 - \frac{4}{\rho} \sin \rho + \frac{4}{\rho^2} (1 - \cos \rho) \qquad (14.24)$$

is found to hold where $\rho = 2x(m - 1)$ and the first maximum occurs at $d = (4.09/2\pi\delta n)$, λ_{max} is the wavelength *in vacuo* of maximum turbidity and δn the difference in refractive indices of the particle and medium (figure 14.5) [1, p. 176]. Variations of this technique are also possible (Hawksley [8, p. 200]).

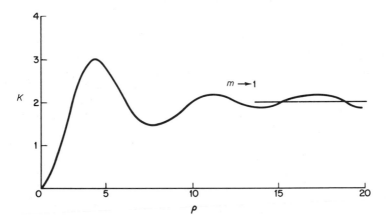

Fig. 14.5 extinction curve for m tending to unity [1].

An essential requirement for the detector is that it should subtend an angle smaller than $D/20$ rad at the suspension to avoid the collection of scattered light. This requirement becomes of increasing importance with increasing values of D [19], but a simple formula may be applied to correct for acceptance of scattered light.

Turbidity measurements have been carried out with non-uniform latices and it is suggested that these are the most useful of the light-scattering techniques for average size determination [68].

14.5 High-order Tyndall spectra (HOTS)

When a dilute suspension of sufficiently large, monodisperse, spherical particles is irradiated with white light, vivid colours appear at various angles to the primary beam. The

angular positions of the spectra depend on m and D, hence they may be used to determine particle size in colloidal suspensions.

Since the red and green bands predominate it is usual to observe, as a function of angle θ, the ratios of the intensities of the vertical components of the red and green light in the scattered radiation. When these ratios, $R = I_{red}/I_{green}$, are plotted against θ, curves showing maxima and minima occur, the maxima being the red order, the minima green. The smaller diameters yield only one order but the number of orders increases with increasing particle size.

High-order Tyndall spectra have been studied extensively in monodisperse sulphur solutions by La Mer et al. [21–25], by Kenyon [26], in aerosols by Sinclair and La Mer [15], in polystyrene latices by La Mer and Plessner [27] and in butadiene latices by Maron and Elder [28]. Equations derived by Maron and Elder for the angular positions of the first red and green order r_1 and g_1,

$$D \sin(r_1/2) = 2300 \tag{14.25}$$

and

$$D \sin (g_1/2) = 3120 \tag{14.26}$$

are particular examples of equations (14.11) to (14.13).

Pierce and Maron [10] show that the angular positions of the red orders are identical with the angles at which minima occur in the intensity of the scattered light when the incident light [20] has a wavelength λ'_g. Similarly, the angular positions of the green orders coincide with the angles at which minima occur with incident light of wavelength λ'_r. Consequently for the same effective incident wavelengths, minimum intensity and high-order Tyndall spectra represent equivalent measurements.

For equations (14.25) and (14.26) respectively, solving with equation (14.13) gives:

$$1.062 - 0.347\,(1.17) = 2300\lambda'_g \quad \therefore \lambda'_g = 3506 \text{ Å } (4673 \text{ Å } in\ vacuo)$$

$$1.062 - 0.347\,(1.17) = 3120\lambda'_r \quad \therefore \lambda'_r = 4756 \text{ Å } (6340 \text{ Å } in\ vacuo)$$

It is pointed out that this method is qualitative unless λ'_g and λ'_r are known, e.g. La Mer's results yield a value $\lambda'_g = 3930$ Å compared with 3506 Å above.

The above equations yield weight average diameters. With increasing m the evaluation of D becomes more difficult due to reduction in the intensity of the maximum and broadening of the peak. The method has been used for the size range 0.26 to 1.01 μm with apparatus similar to that described in section 14.2, the mercury vapour lamp being replaced with a 6 V tungsten lamp and Wratten filters to isolate the red and green radiation and a polarizer in front of the phototube [29].

14.6 Particle size analysis by light diffraction

The far-field diffraction pattern of an assembly of particles yields information concerning their size distribution. When the particles are dispersed on a transparent slide, the geometrically scattered part of the incident beam may be eliminated by coating the slide with aluminium and then removing the particles. The relationship between the

far-field diffraction pattern formed by an aluminium slide prepared in this way and the size distribution of the particles has been derived [42–45]. In the Talbot DISA [46] the slide is illuminated by a sodium discharge lamp and light transmitted by the apertures falls on a wave vector or 'spatial' filter in the far field. A photomultiplier receives the light transmitted by the filter. By using a sequence of interchangeable filters of suitable design the size distribution of the apertures may be determined. The wave vector filter is constructed so that its transparency is a function of the modulus of the diffraction vector. The light received by the photomultiplier is proportional to the relative number frequency, each filter giving one point on the size distribution curve. Talbot later described spatial filters which transmit an amount of light proportional to particle volume [47]. He also incorporated this principle in a slurry sizer [48. 49], the Talbot Spatial Period Spectrometer which houses a sensing unit containing a diffractometer with a rectangular flow cell; 5 dm^3 s^{-1} of slurry can flow through the unit and 100 g of solids can be processed in 30 s. This principle has been utilized in several 'on-line size analysers' (see Chapter 20).

Gucker et al. [63, 64] developed equations, from the Lorentz–Mie theory, relating the size of the Airy points to particle size. Davidson and Haller [65] applied these equations to 0.07 to 0.50 μm latices deposited on microscope slides and obtained poor agreement; this they attributed to strong particle–slide interaction. With a similar experimental set-up Robillard et al. [66, 67] used Mie scattering at two different wavelengths to determine the mean diameter, the polydispersity and the refractive index of a Dow latex. The experimental intensity curves were compared with computer-generated curves and gave good agreement with expected values.

14.7 Light-scattering equipment

Most of the equipment described in the preceding sections has been individually designed, but some commercial equipment is available. The Sinclair-Phoenix aerosol photometer brings the light to a focus at the sample cell, with a diaphragm stop placed in the optical path so that a diverging cone of darkness encompasses the light-collecting lens of the photomultiplier housing. Hence the only light reaching the detector is that which is scattered in the forward direction by particles in the aerosol under test. Solids concentration as low as 10^{-3} μg^{-1} l^{-1} may be detected, the mass concentration at any instant being displayed on a meter or plotted on a recorder.

The Brice-Phoenix light-scattering photometer and the Absolute light-scattering photometer (figure 14.6) are designed for the determination of attenuation, dissymmetry and depolarization. The former has also been adapted to determine the particle size distribution of an aerosol at various levels in a system flowing through a column [33].

The model 7 photonephelometer has two photocells which receive light at 90° to the incident beam, the combined output being fed to a galvanometer (figure 14.7).

The model 9 nephocolorimeter has a third photocell in line with the transmitted beam so that the transmitted light may also be measured. Other instruments which are used for aerosol sizing are discussed in Chapter 3.

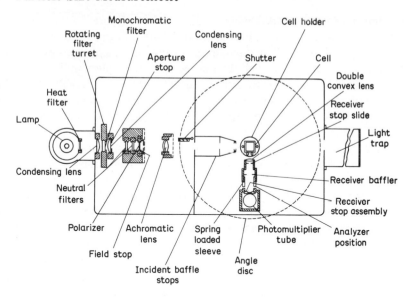

Fig. 14.6 The Absolute light-scattering photometer.

The Differential I [31] light-scattering photometer is designed for the study of mono-size particles in suspension. A cuvette containing the suspension is illuminated by an argon-ion laser beam and a scanning detector system records the intensity of the scattered light as a function of the scattering angle θ. The light-scattering patterns are matched against computer-generated curves which are available in an *Atlas of Light Scattering Curves* [38]. The analytical approach is described in [35, 36].

Where the particle size is large or the size distribution is broad, very low-angle measurement of scattering provides useful information. When the size distribution is so broad that no identifiable maxima remain in the scattering curve, particles may still be measured, one at a time, in the Differential II single-particle light-scattering photometer. This instrument is designed to measure the light scattered from individual particles suspended in air to determine particle size. Particles are injected into the instrument in an airstream and introduced pneumatically into the scattering cell. Selected particles are then brought to the centre of the laser beam by means of plates and point electrodes by manual control and held in position by an automatic servomechanism which permits time-variation monitoring. The compilation of sizes determined for a sample will yield the size distribution [32, 54, 60, 61]. Chang and Davies [85] examined the evaporation kinetics of a single droplet suspended in an electric field.

14.8 Holography

Conventional methods of sampling aerosols are frequently unsatisfactory because they are too slow to monitor dynamic aerosols, non-representative samples are often collected and the aerosol may be modified during collection. Hologram systems, which overcome

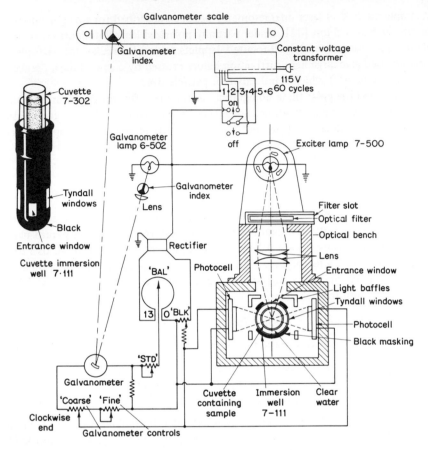

Fig. 14.7 Schematic diagram of model 7 photonephelometer. Optical and electrical systems.

these objections, record and reconstruct relatively large volumes of particles in the size range 3–1000 μm. These holograms are called Fraunhofer or far-field holograms [71–73] because they are recorded at a distance from the object, effectively in its far field. The effective sampling depth is $49(d^2/\pi)$ which, for 50-μm particles and a ruby laser, gives a depth of 18 cm which is over three orders of magnitude greater than the microscope. Prototype instruments based on the use of Fraunhofer holograms have been described [74–76].

The visual comparison of the holographically recorded radiation pattern of a particle with Mie scattering theory has also been used for particle sizing [79]. Holography has also been used to locate sub-micron particles in a three-dimensional volume [78] and, in conjunction with an **image-analysing computer**, to size the droplets in sprays [77].

In-line holography has been used to characterize the spray produced by a commercial rotary device: a description of the optical systems used to record and reconstruct the images has been given [76].

A simple method of laser diffractometry has been described for sizing droplets with radii greater than 1.5 μm [82]. Under partially polarized laser illumination, at a 90° angle to a receiver (camera), well-focused droplets are small circular dots. Droplets that are out-focused give large light diffraction haloes crossed by a row of dark parallel stripes, the number of which is indicative of particle size.

Yule *et al.* [81] suggest these holographic techniques offer no significant advantage over the relatively simple two-dimensional spark photography technique for measuring particles in sprays. Recently Tyler and Thompson [86] reassessed the application of Fraunhofer holography to particle size determination.

14.9 Miscellaneous

A relationship has been derived between the average particle size and concentration and the back-scatter intensity of a light beam entering a slurry of non-transparent particles [50]. Concentrations used were 0.3 to 2.5% and the average sizes were 40 to 1200 μm. The light source and detector were built into one tube which was mounted on a wall of a 1-l vessel which contained an agitated suspension of the particles under test. The derived relationship took the form:

$$(I_{bt} - I_{bw}) = (I_{bm} - I_{bw}) \exp [kc^p d^q]$$

where, for aluminium silicate, experiment yielded $p = 0.77, q = 1.67, k = 0.20$, where

I_{bt} = total back-scatter intensity
I_{bw} = back-scatter intensity of container walls
I_{bm} = maximum possible back-scatter intensity

In a second paper [51] the technique was extended to on-line. A similar technique to the above, but using ultrasonics, has also been described [52, 53]. A description of a crossed laser beam technique for particle sizing and its application to shock tube experiments was presented by Waterston and Chou [56].

The surface of a specimen may become heated through irradiation, the degree of heating depending on the material's adsorption coefficient at the particular wavelength of radiation. A wavelength scan across a suitable part of the electromagnetic spectrum thus causes temperature changes that reflect the absorption spectra at the point of illumination.

In photo-acoustic spectroscopy (PAS) the heating serves to increase the pressure inside a small chamber in which the sample is situated, being irradiated through a window. A recording of the pressure variations versus wavelength of illumination reflects the absorption spectrum of the material. In photothermal spectroscopy (PTS) one records the variations in thermal emission from the sample induced by radiation. PAS techniques require samples to be placed in a spectrophone for analysis. PTS methods allow contact-free on-stream inspection of a powder at a distance. Particle size may be inferred from the signal level. Since specific surface increases with decreasing size the PAS and PTS signals decrease with increasing size. A discussion of these techniques has been presented by Kanstad and Nordal [83].

Transient electric birefringence has been used for evaluating the size distribution of sols in solution. In this technique a colloidal system, in random orientation, is illuminated with polarized light. The system is subjected to an electric field which aligns the particles due to the interaction between the field and any permanent dipole or electrical polarizability of the particles. The birefringence grows as the particles align; when the field is removed the birefringence decays as the particles revert to random orientation. Under two specific experimental conditions, these decay rates can be used to characterize the particle size distribution in terms of standard two-parameter distributions [84].

Small particles suspended in a fluid move in a random manner due to Brownian motion. The frequency of this movement is inversely related to particle size. The Coulter Nano-Sizer detects the change in the scattered light pattern as the particles vibrate and by analysis of the frequency beat (auto-correlation spectroscopy) the particle size is calculated. A dilute suspension of particles is placed in a low-power laser beam and the light scattered at 90° is detected by a photomultiplier. The average size between 40 and 3000 nm is presented typically in two to four minutes. An indication of the width of the distribution is also presented in the form of a Polydispersity Index.

Stallhofen et al. [88] carried out an investigation of sizing a single particle in a sedimentation cell. The settling chamber consisted of a 3-mm cylindrical channel in the centre of a copper block. The particle was illuminated with a laser beam and examined using an ultramicroscope technique.

A method of size determination has been described in which a sedimenting particle passes through the probe volume of a velocimeter to generate a burst signal which contains the Doppler beat frequency corresponding to its velocity [89].

References

1 van de Hulst, H.C. (1957), *Light Scattering by Small Particles*, Wiley, NY.
2 Jennings, B.R. and Jerrard H.G. (1965), *J. Colloid Sci.*, **20**, 448.
3 Jerrard, H.G. and Sellen, D.B. (1962), *Appl. Opt.*, **1**, 243.
4 Nakagaki, M. and Shimoyama, T. (1964), *Bull. Chem. Soc., Japan*, **37**, 11, 1634.
5 Sakurada, I., Hosono, M. and Tamamura, S. (1964), *Bull. Inst. Chem. Res., Kyoto Univ.*, **42**, 23, 145.
6 Goulden, D. (1960), *Spectroturbidity of Emulsions*, National Institute for Research in Dairying, Reading University.
7 Lowan, A.N. (1949), National Bureau of Standards USA, Applied Maths., Series No. 4.
8 Hawksley, P.G.W. (1952), *British Coal Utilisation Research Assoc. Bull.*, **16**, 5, 6.
9 Orr, C. and Dallavalle, J.M. (1960), *Fine Particle Measurement*, Macmillan, NY.
10 Pierce, P.E. and Maron, S.H. (1964), *J. Colloid Sci.*, **19**, 658.
11 Maron, S.H. and Elder, M.E. (1963), *ibid.*, **18**, 107.
12 Maron, S.H., Pierce, P.E. and Elder, M.E. (1963), *ibid.*, 391.
13 Kerker, M. *et al.* (1964), *ibid.*, **19**, 193.
14 Napper, D.H. and Ottewell, R.H. (1964), *ibid.*, 72.
15 Sinclair, D. and La Mer, V.K. (1949), *Chem. Rev.*, **44**, 245.
16 La Mer, V.K., Inn, E.C.Y. and Wilson, J.B. (1950), *J. Colloid Sci.*, **5**, 471.
17 Graessley, W.W. and Zufall, J.H. (1964), *ibid.*, **19**, 516.

18 Maron, S.H., Elder, M.E. and Pierce, P.E. (1964), *ibid.*, 213.
19 Heirwegh, K.P.M. (1966), *ibid.*, **21**, 1–8.
20 Walstra, P. (1965), *Br. J. appl. Phys.*, **16**.
21 Kerker, M. and La Mer, V.K. (1950), *J. Am. Chem. Soc.*, **72**, 3516.
22 Johnson, I. and La Mer, V.K. (1947), *ibid.*, **69**, 1184.
23 Barnes, M.D. *et al.* (1947), *J. Colloid Sci.*, **2**, 349.
24 La Mer, V.K. (1948), *J. Phys. Colloid. Chem.*, **52**, 65.
25 Kenyon, A.S. and La Mer, V.K. (1949), *J. Colloid Sci.*, **4**, 163.
26 Kenyon, A.S. (1947), *Trans. NY Acad. Sci.*, **9**, 234.
27 Plessner, I.V. and La Mer, V.K. (1957), *J. Polymer Sci.*, **24**, 147.
28 Maron, S.H. and Elder, M.E. (1963), *J. Colloid Sci.*, **18**, 199.
29 Bricaud, J. *et al.* (1964), *Staub*, **24**, 8, 287.
30 Kerker, M. (1969), *The Scattering of Light and other Electromagnetic Radiation,* Academic Press, NY.
31 Farone, W.A. and Kerker, M. (1966), *J. opt. Soc. Am.*, **56**, 481.
32 Cooke, D.D. and Kerker, M. (1973), *J. Colloid Interfac. Sci.*, **42**, 1, 150–5.
33 Cooke, D.D., Nicolson, G. and Kerker, M. (1973), *ibid.*, **42**, 3, 535–8.
34 Takahashi, K. and Iwai, S. (1967), *J. Polymer Sci.*, **23**, 113.
35 Wallace, T.P. and Kratohvil, J.P. (1970), *ibid.*, **A2**, 8, 1425.
36 Wallace, T.P. and Kratohvil, J.P. (1968), *ibid.*, **C25**, 89.
37 Kratohvil, J.P. (1964), *Ann. Chem.*, **36**, 5, 485R.
38 *Atlas of Light Scattering Curves* : see List of Manufacturers and Suppliers.
39 Vouk, V. (1948), Ph.D. thesis, London Univ.
40 Wims, A.M. (1973), *J. Colloid Interfac. Sci.*, **44**, 361.
41 Wims, A.M. (1974), *ibid.*, **49**, 2, 259–67.
42 Talbot, J.H. (1967), *J. Min. Vent. Soc. S.A.*, **20**, 21.
43 Talbot, J.H. (1966), *J. scient. Instrum.*, **43**, 744.
44 Talbot, J.H. (1966), *Proc. Phys. Soc.*, **89**, 1043.
45 Talbot, J.H. (1965), *J. Min. Vent. Soc. S.A.*, **18**, 8.
46 Talbot, J.H. (1970), *Proc. Conf. Particle Size Analysis.*, London (eds.M.J. Groves and J.L. Wyatt-Sargent), Soc. Analyt. Chem., 96–100.
47 Talbot, J.H. (1974), *J. Min. Vent. Soc. S.A.*, **27**, 11, 161–7.
48 Talbot, J.H. (1970), S.A. Pat N. 703652.
49 Talbot, J.H. and Jacobs, D.J. (1970), *Proc. Conf. Particle Size Analysis,* London (eds. M.J.Groves and J.L. Wyatt-Sargent), Soc. Analyt. Chem., 72–79.
50 Bemer, G.G. (1978), *Powder Technol.*, **20**, 133–6.
51 Bemer, G.G. (1979), *ibid.*, **22**, 143–4.
52 Pfane, B. (1974), *Verfahrenstechnik*, **8**, 54.
53 Pfane, B. (1974), *ibid.*, **8**, 258.
54 Wyatt, P.J. and Phillips, D.T. (1972), *J. Colloid Interfac. Sci.*, **39**, 1, 125–35.
55 Sosfenor, N.I. and Feigin, L.A. (1974), *Apparatura i metody rentgen analiza*, 13, 69–73.
56 Waterston, R.M. and Chou, H.P. (1975), Modern developments in shock tube research, In *Proc. Int. Shock Tube Symp.,* (ed. G. Kaminto), Shock Tube Res. Soc. Japan, Kyoto, pp. 788–95.
57 Winslow, D.N. and Diamond, S. (1973), *J. Colloid Interfac. Sci.*, **45**, 2, 425–6.
58 Soci. N.T. (1975), *Angew. Makromol, Chem.*, **44**, 1, 165–80.
59 Shchurov, A.F., Ershova, T.A. and Kalinin, V.R. (1976), *Kristallografiya*, **21**, 4, 688–95.
60 Roth, C., Gebhart, J. and Hergwer, G. (1976), *J. Colloid Interfac. Sci.*, **54**, 2, 265–77.
61 Stull, V.R. (1973), *Conf. Particle Technology,* IIT Research Inst., Chicago, Illinois 60616, pp. 52–58.
62 Marshall, T.R., Parmenter, C.S. and Seaver, M. (1976), *J. Colloid Interfac. Sci.*, **55**, 3, 624–36.
63 Gucker, F.T. and Tuma, J. (1968), *ibid.*, **27**, 402.
64 Gucker, F.T. and Rose, D.G. (1955), *Proc. 3rd Nat. Air Pollution Symp.,* Pasadena, Calif., **155**, 120–30.

65 Davidson, J.A. and Haller, H.S. (1976), *J. Colloid Interfac. Sci.*, 55, 1, 170-80.
66 Robillard, F. and Patitsas, A.J. (1974), *Powder Technol.*, 9, 247-56.
67 Robillard, F. Patitsas, A.J. and Kaye, B.H. (1974), *ibid.*, 10, 307-15.
68 Cheeseman, G.C.N. (1977), *Proc. Conf. Particle Size Analysis* (ed. M.J. Groves), Chem. Soc., Analyt. Div. (1978), Heyden.
69 Mosley, L.R., Nobbs, J.H. and Patterson, D. (1977), *Proc. Conf. Particle Size Analysis,* (ed. M.J. Groves), Chem. Soc., Analyt. Div. (1978), Heyden.
70 Krautwasser, P. (1975), Report 1202, 155 pp., NTIS (US sales only).
71 Thompson, B.J. (1965), *Jap. J. appl. Phys.* Suppl., 14, 302.
72 Thompson, B.J. (1964), *Soc. Photo-Opt. Instrum. Eng.*, 43.
73 Parrent, G.B. Jr. and Thompson, B.J. (1964), *Opt. Acta*, II, 183.
74 Silverman, B.A., Thompson, B.J. and Ward, J. (1964), *J. appl. Meteor.*, 3, 792.
75 Thompson, B.J. *et al.* (1965), 237th Nat. Mtg Ann. Met. Soc., Washington, DC.
76 Dunn, P. and Walls, J.M. (1977), *Proc. Conf. Particle Size Analysis* (ed. M.J. Groves), Chem. Soc., Analyt. Div. (1978), Heyden.
77 Bexon, R., Bishop, G.D. and Gibbs, J., Aerosol Sizing by Holography Using the Quantimet, Cambridge Instruments.
78 Silverman, B.A., Thompson, B.J. and Parrent, G.B. Jr. (1969), US Pat. 3 451 755, 24th June, C1. 356-102.
79 Tschudi, T., Herziger, G. and Engel, A. (1974), *Appl. Opt.*, 13, 2, 245-8.
80 Thompson, B.J. and Zinky, W.R. (1968), *ibid.*, 7, 12, 2426-8.
81 Yule, A.J., Chigier, N.A. and Cox, N.W. (1977), *Proc. Conf. Particle Size Analysis* (ed. M.J. Groves), Chem. Soc., Analyt. Div. (1978), Heyden.
82 Kozhenkov, V.I. and Fuchs, N.A. (1975), *J. Colloid Interfac. Sci.*, 52, 1, 120-1.
83 Kanstad, S.O. and Nordal, P.E. (1978), *Int. Symp. In-Stream Measurement of Particulate Solid Properties,* Bergen, Norway (to be published in Powder Technol.).
84 Foweraker, A.R., Morris, V.J. and Jennings, B.R. (1977), *Proc. Conf. Particle Size Analysis* (ed. M.J. Groves), Chem. Soc., Analyt. Div. (1978), Heyden.
85 Chang, R. (1976), *J. Colloid Interfac. Sci.*, 54, 3, 352-63.
86 Tyler, G.A. (1976), *Opt. Acta,* 23, 9, 685-700.
87 Plestil, J. and Baldrian, J. (1976), *Czech. J. Phys. B.,* B26, 5, 514-27.
88 Stallhofen, W., Armbruster, L. and Gebhart, J. (1975), *Atmos. Environ.*, 9, 9, 851-7.
89 Kamatsu, S., Watabe, A. and Saito, H. (1977), *Oyo Butsuri,* 46, 9, 879-85.

15 Permeametry and gas diffusion

15.1 Flow of a viscous fluid through a packed bed of powder

The original work on the flow of fluids through packed beds of powders was carried out by Darcy [1], who examined the rate of flow of water from the local fountains through beds of sand of various thickness. He showed that the average velocity, as measured over the whole area of the bed, was directly proportional to the driving pressure and inversely proportional to the thickness of the bed, i.e.

$$u = K \frac{\Delta p}{L} \qquad (15.1)$$

Equation (15.1) may be compared with the expression for the mean velocity u_m of a fluid of viscosity η, flowing in a pipe of circular cross-section and of diameter d:

$$u_m = \frac{d^2}{32\eta} \frac{\Delta p}{L} \qquad (15.2)$$

This expression was derived by Hagen [2] and independently by Poiseuille [3].

Blake [4], and later Kozeny [5], found it necessary to use an equivalent diameter to relate flow rate with particle surface area for flow through a packed bed of powder, where:

$$\text{equivalent diameter } (d_E) = 4 \times \frac{\text{cross-sectional area normal to flow}}{\text{wetted perimeter}} \qquad (15.3)$$

For a circular pipe,

$$d_E = \frac{4\pi d^2/4}{\pi d} = d$$

Hence d_E may be regarded as a mean pipe diameter. Kozeny assumed that the pore space of a packed bed of powder could be regarded as equivalent to a bundle of parallel capillaries with a common equivalent radius, and with a cross-sectional shape representative of the average shape of the pore cross-section. For a packed bed, equation (15.3) may be written:

$$\text{mean equivalent diameter} = 4 \times \frac{\text{volume of voids}}{\text{surface area of voids}}$$

$$d_E = \frac{4V_v}{S} \qquad (15.4)$$

The surface of the capillary walls is assumed to be the same as the surface of the powder S. By definition:

$$\text{porosity} = \frac{\text{volume of voids}}{\text{volume of bed}}$$

$$\epsilon = \frac{V_v}{V_v + V_s}$$

giving

$$V_v = \left(\frac{\epsilon}{1 - \epsilon}\right) V_s \qquad (15.5)$$

where V_s is the volume of solids.

From equations (15.3), (15.4) and (15.5):

$$d_E = 4 \left(\frac{\epsilon}{1 - \epsilon}\right) \frac{V_s}{S} = d$$

Substituting in equation (15.2):

$$u_m = \frac{\epsilon^2}{(1 - \epsilon)^2} \frac{V_s^2}{S^2} \frac{\Delta p}{2\eta L} \qquad (15.6)$$

The measured velocity is the approach velocity, that is, the volume flow rate divided by the whole cross-sectional area of the bed: $u = Q/A$. Knowing this, it is necessary to estimate the fluid velocity in the pore spaces. It can be shown [6, p. 392] that, in a bed of randomly distributed particles of voidage ϵ, the average free cross-sectional area in any plane is the total cross-sectional area multiplied by ϵ. The velocity in the pore spaces is greater than the approach velocity since the area available for flow is smaller, $u_1 = Q/\epsilon A$, therefore:

$$u = \epsilon u_1 \qquad (15.7)$$

Further, the path of a capillary through the bed is tortuous with an average length L_e, which is greater than the bed thickness L, but it is to be expected that L_e is proportional to L. Thus the velocity of the fluid in the capillary u_m will be greater than u_1 due to the increase in path length:

$$u_m = \left(\frac{L_e}{L}\right) u_1 \qquad (15.8)$$

From equations (15.7) and (15.8):

$$u = \epsilon \left(\frac{L}{L_e}\right) u_m \qquad (15.9)$$

Noting also that the pressure drop occurs in a length L_e we have, from equations (15.6) and (15.9):

$$u = \epsilon \left(\frac{L}{L_e}\right) \frac{\epsilon^2}{(1-\epsilon)^2} \frac{V_s^2}{2\eta S^2} \frac{\Delta p}{L_e}$$

$$S_w^2 = \frac{1}{k\eta\rho_s^2 u} \frac{\epsilon^3}{(1-\epsilon)^2} \frac{\Delta p}{L} \tag{15.10}$$

where $S = \rho_s V_s S_w$; S_w is the weight specific surface of the powder and ρ_s is the powder density. In general $k = k_0 k_1$ where $k_1 = (L_e/L)^2$; for circular capillaries, $k_0 = 2$.

k is called the aspect factor
k_1 is called the tortuosity factor
k_0 is a factor which depends on the shape and size distribution of the cross-sectional areas of the capillaries, hence of the particles which make up the bed.

15.2 Alternative derivation of Kozeny's equation using equivalent capillaries

Replacing the packed bed of thickness L and cross-section A with N uniform capillaries of radius r gives:

$$\text{volume specific surface } S_v = \frac{2N\pi r}{(A - N\pi r^2)}$$

$$\text{porosity } \epsilon = \frac{N\pi r^2}{A}$$

Hence

$$r = \frac{2}{S_v} \frac{\epsilon}{(1-\epsilon)}$$

Substituting in equation (15.2):

$$u_m = \frac{1}{2\eta S_v^2} \frac{\epsilon^2}{(1-\epsilon)^2} \frac{\Delta p}{L} \tag{15.11}$$

As before, the measured velocity is the approach velocity which is less than the velocity in the capillaries, since the cross-sectional area available for flow is greater. The actual velocity in the capillaries is greater than the apparent velocity due to the tortuous path of the capillaries. Combining equation (15.11) with equation (15.9) to include these factors gives:

$$u = \epsilon \left(\frac{L}{L_e}\right) \frac{1}{2\eta S_v^2} \frac{\epsilon^2}{(1-\epsilon)^2} \cdot \frac{\Delta p}{L} \left(\frac{L}{L_e}\right)$$

which reduces to equation (15.10).

For compressible fluids, equation (15.10) is modified to:

$$u = \frac{\bar{p}}{p_1} \cdot \frac{1}{k} \cdot \frac{\epsilon^3}{(1-\epsilon)^2} \cdot \frac{\Delta p}{L} \cdot \frac{1}{\eta \rho_s^2 S_w^2} \tag{15.12}$$

where $\bar{p}$ is the mean pressure of the gas in the porous bed and p_1 is the inlet pressure. This correction becomes negligible if Δp is small and p/p_1 is near to unity [8, p. 2].

15.3 The aspect factor k

Carman [13] carried out numerous experiments and found that k was equal to 5 for a wide range of particles. In the derivation above, k_0 for uniform circular capillaries is found to equal 2. Carman suggested [8, p. 13] that capillaries in random orientation arranged themselves at a mean angle of 45° to the direction of flow, thus making L_e/L equal to $\sqrt{2}$ and k_1 equal to 2.

Fowler and Hertel [32] found theoretically and Sullivan and Hertel [31] found experimentally, that for spheres $k = 4.5$, for cylinders arranged parallel to flow $k = 3.0$, and for cylinders arranged perpendicular to flow, $k = 6.0$. Muskat and Botsel [51] obtained values of 4.5 to 5.1 for spherical particles and Schriever [52] obtained a value of 5.06.

Essenhigh [7] replaced the set of uniform capillaries by a set with a distribution of radii given by:

$$dN = k \exp(-r/r_0)\, dr$$

(i.e. a Rosin–Rammler distribution) where dN is the number of capillaries of radii between r and $(r + dr)$, k and r_0 are constants. The value of S_w derived was greater than that obtained from equation (15.9) by a factor of $\sqrt{3}$, i.e. $k_0 = \frac{2}{3}$.

Carman [8, p. 35] points out that the permeability equation is not valid if the pore space is made up of widely varying radii, since the mean equivalent radius is not the correct mean value for permeability calculation. Large capillaries give disproportionately high rates of flow which swamp the effect of the small capillaries. However, if the pore size range is not too great, say less than 2 : 1, the results should be acceptable. It is, nevertheless, advisable to grade samples by sieving as a preliminary to surface area determinations by permeability, and to determine the surface of each of the gradings independently in order to find the specific surface of the sample. Even if the range of void size is wide, the same type of distribution will always lead to the same ratio between two average values and results are acceptable for comparative purposes. An extreme case is found with bimodal distributions. For spheres with a size ratio of 4 : 1 or more, small spheres may be added to large ones by occupying voids. The porosity of the bed decreases and the pore texture is non-uniform since filled voids give smaller channels than unfilled voids. Initially the rate of fall of the porosity function is greater than the decrease in flow rate, which leads to an apparent fall in specific surface, as the powder becomes finer. When all the voids are filled, the value of k falls to its correct value. Fine dust clinging to larger particles takes no part in the flow and may lead to enormous errors. When aggregation of particles occurs, the surface measured is the envelope of the

aggregates leading to a low value of surface area. As the porosity decreases, the bed becomes more uniform, leading to the correct value. It is usually recommended, therefore, to reduce high porosities by compression to values between 0.4 and 0.5 to reduce this error. In practice this may cause particle fracture which results in high experimental values for surface area.

The value of k_0 also depends on the shape of the pores [30, p. 138] lying between 2.0 and 2.5 for most annular and elliptical shapes.

It can also be argued that k depends on the porosity since the specific surface determined at different porosities tends not to be constant. On the other hand, these discrepancies may be attributed directly to inadequacies in the porosity function. Empirical corrections to the Carman–Kozeny equation have been proposed based on both these arguments.

Wasan *et al.* [57, 58] discuss the tortuosity effect and define a constriction factor, which they include to account for the varying cross-sectional area of the voids through the bed. They develop several models and derive an empirical equation for regularly shaped particles. The equation is equivalent to replacing the porosity function in the Carman–Kozeny equation as follows:

$$\phi(\epsilon) = 0.2 \exp(2.5\epsilon - 1.6) \quad 0.3 \leqslant \epsilon \leqslant 0.6 \tag{15.13}$$

15.4 Other flow equations

At low fluid velocities through packed beds of powders the viscosity term predominates, whereas at higher velocities both viscous and kinetic effects are important.

Ergun and Orning [28] found that, in the transitional region, the equation relating pressure gradient and superficial fluid velocity u_f was:

$$\frac{\Delta p}{L} = 150 \frac{(1-\epsilon)^2}{\epsilon^3} \frac{\eta u_f}{d_{sv}^2} + 1.75 \left(\frac{1-\epsilon}{\epsilon^3} \right) \frac{\rho_s u_f^2}{d_{sv}} \tag{15.14}$$

For Reynolds number less than 2, the second term becomes negligible compared with the first. This is the Carman–Kozeny equation with an aspect factor of $4\frac{1}{6}$. Above a Reynolds number of 2000, the second term predominates and the ratio between pressure gradient and superficial fluid velocity is a linear function of fluid mass flow rate G, where $G = u_f \rho_f$. The constant, 1.75, was determined experimentally by plotting $\Delta p / L u_f$ against G since, at high Reynolds number:

$$\frac{\Delta p}{L u_f} = \frac{1.75}{d_{sv}} \left(\frac{1-\epsilon}{\epsilon^3} \right) G \tag{15.15}$$

It has been found that some variation between specific surface and porosity occurs. Carman [15] suggested a correction to the porosity function, to eliminate this variation. The correction may be written:

$$\frac{(\epsilon - \epsilon'')^3}{(1-\epsilon)^2} \quad \text{for} \quad \frac{\epsilon^3}{(1-\epsilon)^2}$$

where ϵ'' represents the absorbed fluid that does not take part in the flow. Later, Keyes [29] suggested the replacement of ϵ'' by $a(1 - \epsilon)$. The constant a may be easily determined by substituting the above expression in equation (15.27) and plotting $(h_1/h_2)^{1/3} (1 - \epsilon)^{2/3}$ against ϵ. Neither of these corrections, however, is fully satisfactory.

Harris [59] discussed the role of adsorbed fluid in permeametry but prefers the term 'immobile fluid'. He stated that discrepancies usually attributed to errors in the porosity function or non-uniform packing are in truth due to the assumption of incorrect values for ϵ and S. Associated with the particles is an immobile layer of fluid which does not take part in the flow process. The particles therefore have a true volume (V_s) and an effective volume (V_s'); a true surface, (S) and an effective surface (S'); a true porosity (ϵ) and an effective porosity (ϵ'); a true density (ρ_s) and an effective density (ρ_s'). The true values may be determined experimentally and, applying equation (15.10), values of S are derived which vary with porosity; usually S_v increases with decreasing porosity.

Equation (15.10) is assumed correct but the true values are replaced by effective values yielding:

$$\left(\frac{S'}{V_s'}\right)^2 = \frac{\Delta p}{k\eta Lu} \frac{(\epsilon')^3}{(1 - \epsilon')^2} \tag{15.16}$$

The effective porosity is defined as follows:

$$\epsilon' = \left(1 - \frac{w}{\rho_s' \, AL}\right)$$

$$= \left(1 - \frac{\rho_B}{\rho_s'}\right) \tag{15.17}$$

where ρ_B is the bed density.

Combining equations (15.16) and (15.17) the **Carman–Kozeny equation** takes the form:

$$\left(\frac{S'}{\rho_s' V_s'}\right)^2 = \frac{\Delta p}{k\eta Lu} \left\{ \rho_B \left[\frac{1}{\rho_B} - \frac{1}{\rho_s'}\right]^3 \right\}$$

This equation can be arranged in the form suggested by Carman [8, p. 20]:

$$\left(\frac{B}{\rho_B}\right)^{1/3} = \left(\frac{1}{S_w'}\right)^{2/3} \left[\frac{1}{\rho_B} - \frac{1}{\rho_s'}\right] \tag{15.18}$$

where

$$B = \frac{k\eta Lu}{\Delta p}$$

Equation (15.18) expresses a linear relationship between the two experimentally measurable quantities $(B/\rho_B)^{1/3}$ and $(1/\rho_B)$. A linear relationship between these quantities means that the effective surface area and mean particle density are constant and

can be determined from the slope and the intercept. The fraction of the effective particle volume not occupied by solid material, ϵ_p, apparent particle porosity, is related to densities.

$$\epsilon_p = 1 - \frac{\rho_s'}{\rho_s} \qquad (15.19)$$

Schultz [65] examined these equations and found that the effective system surface area was a constant whereas the surface area determined from equation (15.10) varied with porosity. The basic difference between these areas is illustrated by figure 15.1. The standard surface area (Blaine number) at a porosity of 0.50 agrees well with the Bureau of Standards value of 3380 cm^3 g^{-1} for SRM 114L, Portland cement. Further measurements at other bed porosities yield areas of 3200 and 4000 cm^2 g^{-1} at bed porosities of 0.6 and 0.4 respectively.

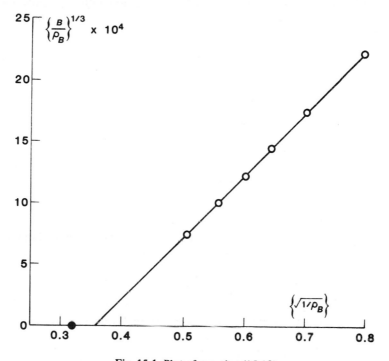

Fig. 15.1 Plot of equation (15.18).

The effective surface area, which is independent of porosity, is 2780 cm^2 g^{-1} at an effective density of 2.78 g cm^{-3} as opposed to a true density of 3.16 g cm^{-3}. The immobile layer of fluid associated with the bed comprises about 12% of the bed volume ($\epsilon_p = 0.12$).

The basic difference between true and effective surface area is illustrated in figure 15.1. The data points lie on a straight line whose intercept on the abscissa determines the effective particle density. The true surface areas are in effect determined for each

successive bed porosity by drawing straight lines from each point to the assumed intercept on the abscissa as indicated by the black circle where the bulk density equals the
true solid density. Since the slope decreases with decreasing bed porosity the true surface area increases with decreasing bed porosity.

Replacing ϵ' by $(V_B - V_s')/V_B$ yields an alternative form of equation   (15.16)

$$\left[\frac{\eta L u V_B}{\Delta p}\right]^{1/3} = \frac{1}{k^{1/3}} \left(\frac{V_B - V_s'}{(S')^{2/3}}\right)$$ (15.20)

Using the identity $V_B = [V_S/(1 - \epsilon)]$, equation (15.20) becomes:

$$\left[\frac{L u \eta (1 - \epsilon)^2}{\Delta p}\right]^{1/3} = \left[\frac{\epsilon - (1 - V_s/V_s')}{k^{1/3}(S_v')^{2/3}(V_s/V_s')}\right]$$ (15.21)

Harris [59] examined published data which were available in various forms.

(1) If $(L u \eta/\Delta p)$ is known for a range of ϵ values then, assuming $k = 5$, equation
(15.21) can be used to determine (V_s/V_s') and S_v'.
(2) If it is assumed that the aspect factor varies with porosity, it can be determined as
a function of porosity using a previously measured value of S. Equation (15.10)
may be written:

$$\frac{L u \eta (1 - \epsilon)^2}{\Delta p} = \frac{\epsilon^3}{k' S_v^2}$$

Inserting in the left-hand side of equation (15.21) and re-arranging gives:

$$\left(\frac{k'}{k}\right) = \left(\frac{S_v'}{S_v}\right)^2 \left[\frac{\epsilon}{1 - (1 - \epsilon) V_s'/V_s}\right]^3$$ (15.22)

The ratio (k'/k) can then be calculated for several values of ϵ and of the parameters
(S_v'/S_v) and (V_s'/V_s).
(3) S_v can be determined directly from equation (15.10) assuming $k = 5$. Equation
(15.21) can be adapted for handling such data by replacing the left-hand side by
$(\epsilon/kS_v^2)^{1/3}$ where k is the assumed value mentioned above.

Harris found that most published data give $(k'/k) = 1 \pm 0.10$ and this could be accounted for by modest changes in (S'/S) and V_s'/V_s). The tabulated data show that as ϵ
increases (k'/k) decreases at a rate increasing with (V_s'/V_s). Increasing k' with decreasing porosity is a common experimental finding. Harris found that the effective volume
specific surface, calculated assuming a constant aspect factor, remained sensibly constant over a range of porosity values.

Equation (15.21) is analogous to equation (15.20) derived from Fowler and Hertel's
model [32] expressed in substantially the same form as Keyes [29] and Powers (see
Chapter 7).

The measured specific surface is found to decrease with increasing porosity. One way
of eliminating this effect, using a constant-volume permeameter, is to write equation
(15.33) as

$$S_w = \sqrt{\left(\frac{k}{\rho^2 L}\right)} \cdot \frac{\epsilon^{3/2}}{(1-\epsilon)} t^{1/2}$$

$$S_w = Z \frac{\epsilon^{3/2}}{(1-\epsilon)} t^{1/2}$$

$$\log t = 2 \log S_w - 2 \log Z - \log \frac{\epsilon^{3/2}}{(1-\epsilon)} \qquad (15.23)$$

Usui [20] replaced the last term with $C + D\epsilon$ and showed that the relationship between $\log t$ and ϵ was linear, proving that C and D were constants independent of ϵ. A plot of $\log t$ against ϵ therefore yields a value for surface area, the calculation being simplified if comparison is made with a standard powder.

Rose [54] proposed that an empirical factor be introduced into the porosity function to eliminate the variation of specific surface with porosity.

$$(S')^2 = S^2 \left\{ \frac{1.1}{X} + 140 \frac{\epsilon^3}{(1-\epsilon)^2} S' \frac{0.2\epsilon^2}{(1-\epsilon)^2} \cdot \epsilon^{10} \right\} \qquad (15.24)$$

where S' and S are respectively the effective area and the surface as determined with equation (15.10) and X the ratio of porosity function proposed by Carman [13].

15.5 Experimental applications

The determination of specific surface by permeability methods was suggested independently by Carman [13] and Dallavalle [14] and elaborated experimentally by Carman [15]. The cement industry published the first test method for the determination of surface area by permeametry, the Lea and Nurse method and this method is still described in the British Standard on permeametry.

The simple Carman–Kozeny equation gives a linear relationship between pressure drop and flow rate. Commercial permeameters can be divided into constant pressure, such as the Fisher sub-sieve sizer and the Permaran or constant-volume such as the Blaine and the Rigden.

The Blaine method is the standard for the cement industry in the United States (ASTM C204–68) and, although based on the Carman–Kozeny equation, it is normally used as a comparison method using a powder of known surface area as standard reference (see figure 15.1).

The assumptions made in deriving the Carman–Kozeny equation are so sweeping that it cannot be argued that the determined parameter is a surface area. First, in many cases the determined parameter varies with porosity. The general tendency is for the specific surface to increase with decreasing porosity; low values at high porosity may be due to channelling, i.e. excessive flow through wide pores; high values at low porosities could be due to particle fracture. It thus appears that the Carman–Kozeny equation is only valid over a limited range of porosities. Attempts have been made to modify the equation, usually on the premise that some fluid does not take part in the flow process. The deter-

mined surface areas are usually lower than those obtained by other surface area measuring techniques and it is suggested that this is because the measured surface area is the envelope surface of the particles. Assuming a stagnant layer of fluid around the particles decreases the measured surface even further.

The equation applies only to monosize capillaries. Underestimation of surface area occurs if the distribution is other than monosize. Thus the method is only suitable for comparison purposes using similar material. Because of its simplicity the method is ideally suitable for control purposes on a single product and that is probably the limit of its usefulness. Finally, the method is not suitable for fine powders since, in these cases, the flow is predominantly molecular.

15.6 Preparation of powder bed

Various methods of packing the powder bed have been recommended. Constant-volume cells and the cell of the Fisher sub-sieve sizer should be filled in one increment only. It is often advantageous to tap or vibrate such cells before compaction, but if this is overdone size segregation may occur.

With other cells, the powder should be added in four or five increments, each increment being compacted with a plunger before adding the next, so that the bed is made up in steps. This procedure largely avoids non-uniformity of compaction down the bed which is liable to occur if the whole of the powder required is compacted in one operation.

In some laboratories a standard pressure is applied ($1 \, \text{MN m}^{-2}$) to eliminate operator bias. In order to test bed uniformity, the specific surface should be determined with two different amounts of powder packed to the same porosity.

15.7 Constant-pressure permeameters

The earliest equipment designed for routine service is that of Lea and Nurse [10, 16] (figure 15.2). The powder is first compressed to a known porosity in a special permeability cell of cross-sectional area A. Air flows through the bed, for which the pressure drop is measured on a manometer as h_1 and then through a capillary flowmeter, for which another pressure drop, given by h_2, is recorded.

The liquid in both manometers is the same and has a density ρ'. The capillary is designed to ensure that both pressure drops are small compared with atmospheric pressure, so that compressibility effects are negligible.

The volume rate of flow of air through the flowmeter is given by:

$$Q = \frac{ch_2 \rho'}{\eta}$$

(15.25)

where c is a constant for a given capillary.

The pressure drop across the bed as measured on the manometer is:

$$p = h\rho'g$$

(15.26)

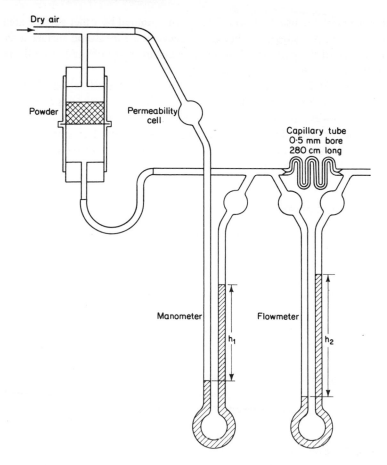

Fig. 15.2 The Lea and Nurse permeability apparatus with manometer and flowmeter [10].

Substituting equations (15.25) and (15.26) in equation (15.10) gives:

$$S_w = \frac{\sqrt{(g/k)}}{\rho_s(1 - \epsilon)} \sqrt{\left/\left(\frac{\epsilon^3 A h_1}{cLh_2}\right)\right.}$$

Taking Carman's value of 5.0 for k, this becomes:

$$S_w = \frac{14}{\rho_s(1 - \epsilon)} \sqrt{\left/\left(\frac{\epsilon^3 A h_1}{cLh_2}\right)\right.} \qquad (15.27)$$

Since the terms on the right-hand side of the equation are known, S_w may be determined.

Gooden and Smith [17] modified the Lea and Nurse apparatus by incorporating a self-calculating chart which enabled specific surface to be read off directly. This is in-

corporated into the Fisher sub-sieve sizer (figure 15.3). The Gooden and Smith equation is a simple transform of the permeametry equation and is developed as follows. The porosity of the powder bed may be written:

$$\epsilon = \frac{V_B - M/\rho_s}{V_B}$$

The volume specific surface may be replaced by the surface-volume mean diameter:

$$\frac{S}{V_s} = \frac{d_s^2}{(\pi/6)d_v^3}$$

$$= \frac{6}{d_{sv}}$$

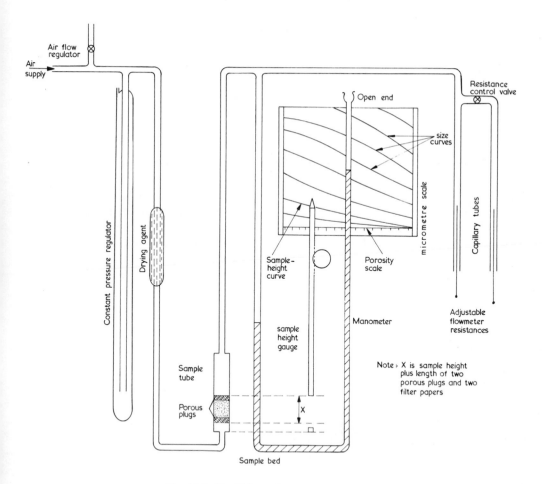

Fig. 15.3 The Fisher sub-sieve sizer.

Also:

$$\Delta p = (P - F)g$$

$$u = Fc/A$$

and $\quad AL = V_B$

Applying these transformations to equation (15.10) gives:

$$d_{sv} = \frac{60\,000}{14} \sqrt{\left[\frac{\eta c F \rho_s L^2 M^2}{(V_B \rho_s - M)^3 (P - F)} \right]} \qquad (15.28)$$

where d_{sv} = surface-weight mean diameter in micrometres,
   c = flowmeter conductance in ml s^{-1} per unit pressure (g force cm^{-2}),
   F = pressure difference across flowmeter resistance (g force cm^{-2}),
   M = mass of sample in grams,
   ρ_s = density of sample in g cm^{-3},
   V_B = apparent volume of compacted sample in ml,
   P = overall pressure head in g force cm^{-2}.

The instrument chart is calibrated to be used with a standard sample volume of 1 cm^3. It is therefore calibrated according to the equation:

$$d_{sv} = \frac{CL}{(AL - 1)^{3/2}} \sqrt{\left(\frac{F}{P - F} \right)} \qquad (15.29)$$

where C is a constant.

The chart also indicates the bed porosity ϵ in accordance with the equation:

$$\epsilon = 1 - 1/AL \qquad (15.30)$$

If a different sample is used so that $M/\rho_s = X$ is not unity, the average particle diameter can nevertheless be calculated from the diameter indicated on the chart [39]. Since the chart only extends to a porosity of 0.40, this is necessary for powders that pack to a lower porosity.

Many authors [18, 19, 22, 29, 39] have observed that the average particle diameter varies with porosity and usually passes through a minimum. Since this minimum value is more reproducible than the value at a fixed porosity, some authors prefer this value. For this purpose the Fisher sub-sieve sizer is more convenient than other types of apparatus since it incorporates a device for compressing the powders to successively lower porosities.

If X cm^3 of powder is used instead of 1 cm^3, comparison of equations (15.28) and (15.29) shows that the average particle diameter d_{sv} will be related to the indicated particle diameter d'_{sv} by:

$$d_{sv} = X \left[\frac{1 - X/AL}{1 - 1/AL} \right]^{3/2} d'_{sv} \qquad (15.31)$$

Similarly the bed porosity ϵ can be calculated from the indicated porosity ϵ':

$$\epsilon = 1 - X(1 - \epsilon') \qquad (15.32)$$

A recommended volume to use in order to extend the range to the minimal attainable porosity is $X = 1.25 \text{ cm}^3$ [39].

The ASTM method for cement [40] standardizes operating conditions by stipulating a porosity of 0.5. This is satisfactory since cement is flowing and non-cohesive; the range of porosities achievable with cement is limited.

Some values obtained with griseofulvin are shown in Table 15.1 The low initial values are probably due to the tendency of fine powders to form lumpy aggregates. Until the bed is packed uniformly air will pass more readily round them than through them. Thus the experimental values will be too low. Most workers accept the maximum value due to its higher reproducibility [39].

Table 15.1

Porosity	0.6	0.55	0.5	0.45	0.4	0.35	0.30	0.25
Specific surface cm^2 g^{-1}	4500	4820	5150	5460	7300	9080	9080	7760

15.8 Constant-volume permeameters

In the apparatus devised by Blaine [53; 30, p. 142] the inlet end of the bed is open to atmosphere (figure 15.4).

Since, in this type of apparatus, the pressure drop varies as the experiment proceeds, equation (15.12) is modified in the following manner. Let the time for the oil level to fall from start to A be t, and let the time for the oil level to fall a distance δh when the imbalance is h be δt. Then: $\Delta p = h\rho'g$ where ρ' is the density of the oil and:

$$u = \frac{1}{A} \frac{dV}{dt} = \frac{1}{A} \frac{adh}{dt}$$

where dV is the volume of air displaced by the oil as it falls.

Substituting in equation (15.12) and putting $k = 5$:

$$S_w^2 \frac{a}{A} \cdot \frac{dh}{dt} = \frac{1}{5\eta\rho_s^2} \frac{\epsilon^3}{(1-\epsilon)^2} \cdot \frac{h\rho'g}{L}$$

$$S_w^2 \int_{h_1}^{h_2} \frac{dh}{h} = \frac{A}{5a\eta\rho_s^2} \frac{\epsilon^3}{(1-\epsilon)^2} \cdot \frac{\rho'g}{L} \int_0^t dt$$

$$S_w = \sqrt{\left[\frac{kt\epsilon^3}{\rho_s^2 L(1-\epsilon)^2} \right]} \qquad (15.33)$$

where k, an instrument constant, is equal to

$$\frac{A\rho'g}{5a\eta\ln(h_2/h_1)}$$

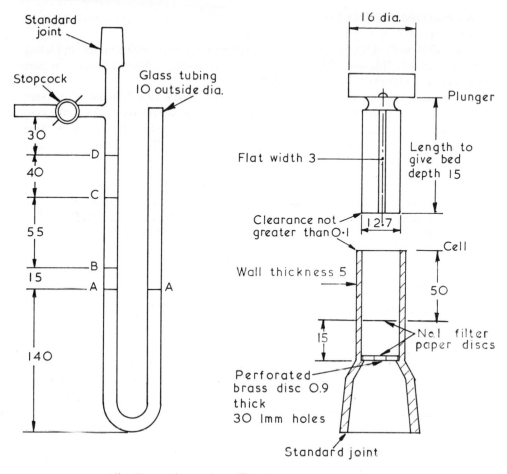

Fig. 15.4 (a) Blaine apparatus. (b) Cell and plunger for Blaine apparatus.

A simplified form of the air permeameter was developed by Rigden [19]. In his apparatus, air was caused to flow through a bed of powder by the pressure of oil displaced from equilibrium in two chambers which were connected to the permeability cell and to each other in U-tube fashion. An instrument working on this principle is available in England as the Griffin surface area of powder apparatus (figure 15.5). The oil is brought to the start position using bulb E with two-way tap C open to the atmosphere. The taps C and D are then rotated so that the oil manometer rebalances by forcing air through the powder bed F. Timing is from start to A for fine powders and start to B for coarse powders.

Another variation of the variable flow technique is the Reynolds and Branson auto-permeameter, in which air is pumped into the inlet side to unbalance a mercury mano-meter. The tap is then closed and air flows through the packed bed to atmosphere. On rebalancing, the mercury contacts start–stop probes attached to a timing device. The pressure difference (Δp) between these probes and the mean pressure $\bar{p}$ are instru-ment constants. The flow rate is given by:

$$\frac{dv}{dt} = \frac{1}{\bar{p}} \frac{\Delta p}{t}$$

Substituting this in the Carman–Kozeny equation yields a similar equation to the Rigden equation. The automatic timing device on this instrument makes it preferable to the Rigden.

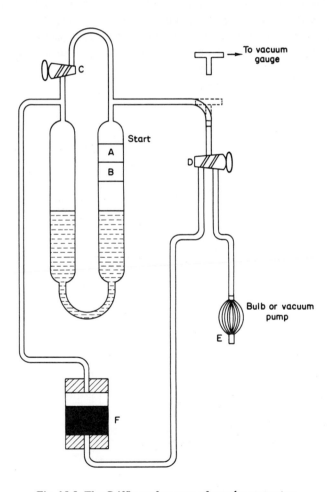

Fig. 15.5 The Griffin surface area of powder apparatus.

15.9 Fine particles

Pechukas and Gage [21] designed a permeameter for surface area measurement in the
0.10 to 1.0 μm size range. In deriving their data they made two fundamental errors in
that (a) they did not correct for slip, and (b) although the inlet pressure was near atmospheric and the outlet pressure was low no correction was made for compressibility.
Their permeameter was modified and automated by Carman and Malherbe [22]. The
plug of material is formed in the brass sample tube A (figure 15.6). Clamp E controls
the mercury flow into the graduated cylinder C, the pressure being controlled at atmospheric by the manometer F. The side arm T_1 is used for gases other than air. Calculations are carried out according to section 15.14. The plug is formed in a special press by
compression between hardened steel plungers. By taking known weights of powder the
measurements may be carried out at a known and predetermined porosity, e.g. 0.45.
Carman and Malherbe recommended that the final stages of compression be carried out
in small increments and that the plungers be removed frequently to prevent jamming.

Carman and Malherbe used equation (15.34) as outlined in section 15.11, keeping the
inlet and outlet pressure constant whilst a known volume of gas passed through the bed.

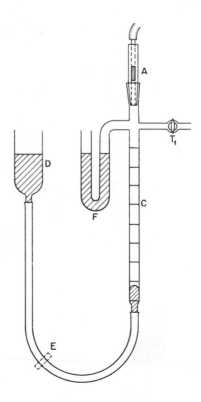

Fig. 15.6 Modified Pechukas and Gage apparatus for fine powders [22].

15.10 Types of flow

With compacted beds of very fine powders and gases near atmospheric pressure or with coarse powders and gases at reduced pressure, the mean free path of the gas molecules is of the same order as the capillary diameter. This results in slippage at the capillary walls so that the rate of flow is higher than that calculated from Poiseuille's law. If the pressure is further reduced until the mean free path is much greater than capillary diameter, viscosity takes no part in flow, since molecules collide only with capillary walls and not with each other. Such free molecular flow is really a process of diffusion and takes place for each constituent of a mixture against its own partial pressure gradient even if the total pressure at each end of the capillary is the same.

There are, therefore, three types of flow to consider. In the first, the flow is viscous and equation (15.10) may be applied; in the transitional region, in which the mean free path λ of the gas molecules is of the same order as the capillary diameter, the slip term is of the same order as the viscous term and both have to be evaluated; in the molecular region, the slip term predominates.

15.11 Transitional region between viscous and molecular flow

Poiseuille's equation was developed by assuming that the velocity at the capillary walls was zero. If it is assumed that the velocity does not reach zero until a distance $x\lambda$ beyond the capillary walls, the modified equation for the average velocity becomes:

$$u_m = \frac{\Delta p}{k\eta LS_v^2} \frac{\epsilon^3}{(1-\epsilon)^2} + \frac{\Delta p}{k\eta LS_v} \frac{\epsilon^2}{(1-\epsilon)} Z\lambda \qquad (15.34)$$

where $Z = 2x$.

Rigden [9] assumed a value for x of:

$$x = \frac{2-f}{f} \qquad (15.35)$$

where f is the fraction of molecules undergoing diffuse reflection from the capillary walls. If the capillary walls are smooth, molecules striking them at any angle rebound at the same angle with the same average velocity and the component of velocity perpendicular to the wall reversed. This is termed *specular reflection*. The surfaces of packed beds of powder are not smooth and molecules striking them rebound in any direction. This is termed *diffuse reflection* or *inelastic collision*. The maximum value of f is unity which makes $x = 1$ for molecular flow conditions.

Lea and Nurse [10] modified the Poiseuille equation (15.2) by assuming a slip velocity at the capillary walls so that:

$$u_m = \frac{d^2}{32\eta} \frac{\Delta p}{L} \left[1 + \left(\frac{8M}{d}\right)\lambda\right] \qquad (15.36)$$

where M is a constant, i.e.

$$x = M \qquad (15.37)$$

Rigden [11] accepted Millikan's value of 0.874 for M, making f greater than unity. The required compensating factor to bring equation (15.36) into line with other equations is $(16M/d)$, making:

$$x = 1.748 \quad \text{and} \quad f = 0.73$$

Carman [8] added an extra term to Poiseuille's equation to take account of slip:

$$u_m = \frac{d^2}{32\eta} \cdot \frac{\Delta p}{L} + \frac{d}{4\xi} \frac{\Delta p}{L} \tag{15.38}$$

where the coefficient of external friction is defined by Millikan as:

$$\xi = \tfrac{1}{2}\rho_g \bar{v} \left\{ \frac{f}{2-f} \right\} \tag{15.39}$$

where $\bar{v}$ is the mean thermal velocity and ρ_g is the density of the gas.

Substituting for $\rho_g \bar{v}$, from equations (15.40) into equation (15.39) and inserting in equation (15.38) reduces the equation to (15.34), but quoted values of the constant in equations (15.40) vary $(0.5, 0.31, 0.35, \pi/10, 1/3)$ (see [8, 34–37]). Alternative forms of equation (15.34) have been used and may be derived by substituting from the equations below:

$$\bar{v} = \left(\frac{8RT}{\pi M} \right)^{1/2}; \quad \rho_g = \frac{M}{RT}p; \quad \eta = \tfrac{1}{2}\rho_g \bar{v} \lambda \tag{15.40}$$

15.12 Experimental techniques for determining Z

Carman and Arnell [12] used the following form of equation (15.34):

$$\frac{p_1}{\bar{p}} u = \frac{1}{k\eta S_v^2} \frac{\epsilon^3}{(1-\epsilon)^2} \frac{\Delta p}{L} + \frac{1}{k\bar{p}S_v} \frac{\epsilon^2}{1-\epsilon} \frac{\Delta p}{L} \sqrt{\left(\frac{2RT}{\pi M} \right)} . \delta k_0 . \frac{8}{3} \tag{15.41}$$

They found $\delta k_0/k \to 0.45$ by plotting $(\bar{p}/\Delta p)(V/At)$ against $\bar{p}$, where $V/At = u$. At the intercept $\bar{p} = 0$, $\delta k_0/k$ can be found. Using the identities (15.40), this yields a value $Z = 3.82$.

Rigden [9] measured the flow rate and the pressure drop across a packed bed of powder using oil manometers.

The volume rate of flow $Q = ACh_2\rho'$ is given by the oil manometer flowmeter with a flowmeter constant C and oil of density ρ'. Also the pressure drop across the bed is given by $\Delta p = h_1\rho'$.

Substituting these values into equation (15.34) gives:

$$\frac{h_2}{h_1} = \frac{A}{k\eta LCS_v^2} \frac{\epsilon^3}{(1-\epsilon)^2} + \frac{A}{k\eta LCS_v} \frac{\epsilon^2}{(1-\epsilon)^2} \left(\frac{2-f}{f_1} \right) \lambda \tag{15.42}$$

Rigden plotted h_2/h_1 against $100/\bar{p}$. At the intercept $100/\bar{p} = 0$.

$$\lambda = 0 \quad \text{since } \lambda \propto (1/\bar{p})$$

Denoting the intercept value as $(h_2/h_1)_1$ gives:

$$S_K = \frac{A\epsilon^3}{k\eta LC(1-\epsilon)^2} \left(\frac{h_1}{h_2}\right)_1 \tag{15.43}$$

where S_K is the surface area independent of 'slip'. The slope of the graph is:

$$\frac{A}{k\eta LCS_v} \frac{\epsilon^2}{(1-\epsilon)} \cdot 2\left(\frac{2-f}{f}\right)\lambda \tag{15.44}$$

Hence:

$$2\left(\frac{2-f_1}{f_1}\right)\lambda = \frac{\epsilon}{S_K(1-\epsilon)} \cdot \frac{\text{slope}}{\text{intercept}} \tag{15.45}$$

Rigden found that S_K was approximately equal to S_v, so it was in order to interchange them. However, the uncorrected value of S as derived directly from the viscous term was appreciably smaller than the intercept value from the graph. The average experimental value of Z was found to be 3.80, making $f = 0.69$, but a great deal of scatter was found, i.e. $3.0 < Z < 4.2$.

15.13 Calculation of permeability surface

Although the above graphical method may also be used to determine specific surface, a rather elaborate apparatus is needed and several experiments are required for one value of specific surface. For practical purposes it is preferable to make a single measurement with the simplest form of apparatus.

If the viscous term predominates, the specific surface is determined using equation (15.12) taking the aspect factor k to be equal to 5. If the compressibility factor is negligible, this equation takes the form of equation (15.10).

When the molecular term predominates the specific surface is obtained from the second term in equation (15.34) with $Z = 3.4$.

When the two terms are comparable the specific surface is obtained as follows.
The specific surface using the viscous flow term (equation (15.10)) is:

$$S_K^2 = \frac{\Delta p}{5\eta Lu} \frac{\epsilon^3}{(1-\epsilon)^2} \tag{15.46}$$

The specific surface using the molecular flow term is:

$$S_M = \frac{\Delta p}{5\eta Lu} \frac{\epsilon^2}{(1-\epsilon)} 3.4\lambda \tag{15.47}$$

Substituting these equations in equation (15.34) gives:

$$\frac{S_K^2}{S_v^2} + \frac{S_M}{S_v} = 1 \tag{15.48}$$

This is a quadratic in S_v having the following solution:

$$S_v = \frac{S_M}{2} + \sqrt{\left(\frac{S_M^2}{4} + S_K^2\right)} \tag{15.49}$$

Crowl [38] carried out a series of experiments using pigments comparing Carman and Malherbe's equations ((15.34) with $Z = 3.4$), Rose's equation (15.24) and nitrogen adsorption. He found good agreement between the Carman and Malherbe figures and nitrogen adsorption, a ratio of 0.6 to 0.8 being obtained with a range of surface areas from 1 to 100 m² g⁻¹. The Rose figures were considerably lower, ranging from 0.2 to 0.5, and being particularly poor with pigments of high surface area. With fine pigments, surface area above about 10 to 12 m² g⁻¹ by nitrogen adsorption, the agreement was less satisfactory but the order of fineness agreed with nitrogen adsorption results.

From equations (15.46) and (15.47)

$$S_M = 3.4 \left(\frac{1 - \epsilon}{\epsilon}\right) \lambda S_K^2 \tag{15.50}$$

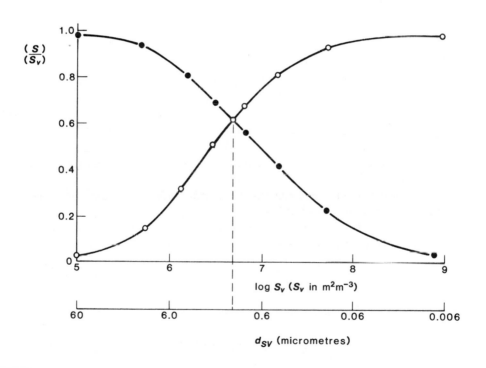

Fig. 15.7 Comparison between the surface area evaluated using one term of the flow equation (S) and the 'true' surface obtained by using both terms (S_v). The curves represent the fraction of true surface obtained using the Carman–Kozeny term only (black circles) and using the slip term only (open circles). Atmospheric conditions assumed.

Using typical values for the variables as an illustration: ϵ = 0.50 and, for air at atmospheric pressure, λ = 96.6 nm,

$$S_M = 3.28 \times 10^{-7} \, S_K^2 \qquad (15.51)$$

with the specific surface in m^2/m^3.

Substituting in equation (15.47) gives the solutions:

$$S_K^2 = \frac{S_v}{1 + 3.28 \times 10^{-7} \, S_v} \qquad (15.52)$$

$$S_M = \frac{3.28 \times 10^{-7} \, S_v}{1 + 3.28 \times 10^{-7} \, S_v} \qquad (15.53)$$

Solutions to these equations are presented in figure 15.7 [60]. The two terms are equal (i.e. $S_M = S_K$) at an apparent d_{sv} of 1.22 μm and yield the value d_{sv} = 1.97 μm. At 16 μm, and a porosity of 0.5, the error in assuming that the contribution due to slip is negligible is 5%.

15.14 Diffusional flow for surface area measurement

The rate of transfer of a diffusing substance through unit cross-sectional area is proportional to the concentration gardient and is given by Fick's laws of diffusion [41]:

$$\frac{1}{A} \frac{dm}{dt} = D \frac{dC}{dx} \qquad (15.54)$$

where (dm/dt) is the mass flow rate (kg mole s^{-1}) across area A where the concentration gradient is (dC/dx). D, the diffusion constant, has dimensions of $m^2 \, s^{-1}$ in SI units.

For uni-directional flow into a fixed volume, the increase in concentration with time is given by:

$$\frac{dC}{dt} = D \frac{d^2 C}{dx^2} \qquad (15.55)$$

If one face of the powder bed is kept at constant concentration, i.e. infinite volume source ($C = C_2$ at x = 0), while at the other face the initial concentration ($C_1(0)$ at $x = L, t$ = 0) changes, i.e. fixed volume (V) sink, a finite time will pass before steady-state conditions are set up and:

$$C_1 = \frac{\epsilon A D_S}{LV} \left\{ [C_2 - C_1 \, (0)] \; t - \frac{L^2}{6 D_t} \right\} \qquad (15.56)$$

Rewriting in terms of pressure, after Babbit [42]:

$$p_1 = \frac{\epsilon A D_S}{LV} \left\{ [p_2 - p_1 \, (0)] \; t - \frac{L_e^2}{6 D_t} \right\} \qquad (15.57)$$

where p_1 is the (variable) outlet pressure
           p_2 is the (constant) inlet pressure
           p_1 (0) is the initial outlet pressure
           V is the outlet volume
           L_e is the equivalent pore length through the bed
           D_t is the unsteady-state diffusion constant and
           D_S is the steady-state diffusion constant and these are not necessarily the
same. This can be explained by adsorption into pores during the unsteady-state period
so that the pore volume in the two flow regimes may be different.

Graphs of outlet pressure p_1 against time t can be obtained at various fixed inlet
pressures p_2. These will be asymptotic to a line of slope:

$$\left(\frac{dp_1}{dt}\right)_{p_2} = \frac{\epsilon AD}{LV} \, p_2 \tag{15.58}$$

$$(p_1 \, (0) \ll p_2)$$

These lines will intersect the line through p_1 (0) and parallel to the abscissa at times:

$$t_L = \frac{L_e^2}{6D_t} = \frac{k_1^2 \, L^2}{6D_t} \tag{15.59}$$

15.15 The relationship between diffusion constant and specific surface

Knudsen [43, 44] deduced that the energy flow rate G through a capillary with a
pressure drop across its ends of Δp is:

$$G = \frac{4}{3} r \sqrt{\left(\frac{2RT}{\pi M}\right)} \cdot \frac{A\Delta p}{L} \left(\frac{2-f}{f}\right) \tag{15.60}$$

where R, T and M are the molar gas constant, the absolute temperature and the gas
molecular weight.

Knudsen's equation can be expressed in terms of the diffusion constant since:

$$G = \frac{eA\,\Delta p}{L} D = V \frac{dp}{dt} \tag{15.61}$$

since $p_1 \ll p_2$, $\Delta p = p_2$ and

$$r = \frac{2}{S_v} \left(\frac{\epsilon}{1-\epsilon}\right)$$

Hence, inserting in equation (15.58) gives for steady-state molecular flow:

$$V\left(\frac{dp}{dt}\right)_V = \frac{8}{3} \frac{\epsilon A \, \Delta p}{LS_v} \left(\frac{\epsilon}{1-\epsilon}\right) \sqrt{\left(\frac{2RT}{\pi M}\right)} \left(\frac{2-f}{f}\right) \tag{15.62}$$

Equation (15.62) is equivalent to equation (15.47) with the constant 3.4 replaced by
8/3 (it being assumed that $f = 1$ for diffusional flow).

Inserting in equation (15.59) gives for non-steady-state molecular flow:

$$S_v = 16 \left(\frac{\epsilon}{1-\epsilon}\right) \sqrt{\left(\frac{2RT}{M}\right)\frac{t_L}{k_1^2 L^2}} \left(\frac{2-f}{f}\right) \qquad (15.63)$$

Derjaguin [33] showed that the numerical constant 4/3 in equation (15.60) should be replaced by 12/13 for inelastic collisions, and Pollard and Present [45] use a constant of $\pi/2$.

A similar equation was derived by Kraus et al. [24] who neglected the tortuosity factor on the grounds that it was already accounted for in the derivation of the diffusion equation.

The general form of equation (15.62) is:

$$\frac{1}{A}\left(\frac{dp}{dt}\right)_V = \beta \left(\frac{\epsilon^2}{1-\epsilon}\right)\frac{\Delta p}{LS_v V}\sqrt{\left(\frac{2RT}{\pi M}\right)} \qquad (15.64)$$

The values for β derived by the various authors are:

Barrer and Grove	8/3 = 2.66
Pollard and Present	π = 3.14
Kraus and Ross	48/13 = 3.70
Derjaguin	8/3 = 2.66

In a recent analysis of Derjaguin's treatment of gas diffusion Henrion [63] suggests that molecular diffusion is best interpreted in terms of elastic collisions against the pore walls.

15.16 Non-steady-state diffusional flow

Barrer and Grove [23] applied equation (15.36) to obtain:

$$S_v = 8 \left(\frac{\epsilon}{1-\epsilon}\right) \sqrt{\left(\frac{2RT}{\pi M}\right)\frac{t_L}{L}} \qquad (15.65)$$

assuming $k_1 = \sqrt{2}$ after Carman.

Equation (15.65) has been applied experimentally by Kraus et al. [24] and Krishnamoorthy [46].

The apparatus of Kraus and Ross (figure 15.8) consists essentially of two 4-l reservoirs connected through the cell holding the powder. A mercury manometer was used to measure the pressure on the high-pressure side and a calibrated thermocouple vacuum gauge was used to measure the pressure on the discharge side. Before an experiment, the whole system was evacuated and flushed with the gas being used. The system was then pumped down to 1 or 2 μm of mercury and shut off from the pumps with stopcock G. Stopcocks E and F were closed and the desired inlet pressure established by bleeding gas into reservoir A through H. At zero time, stopcock F was opened and the gas was allowed to diffuse through the cell C into reservoir B. Figure 15.9 shows a typical flow-rate curve. The time lag t_L is determined by extrapolation of the straight-line, steady-state portion of the curve to the initial pressure in the cell and discharge reservoir.

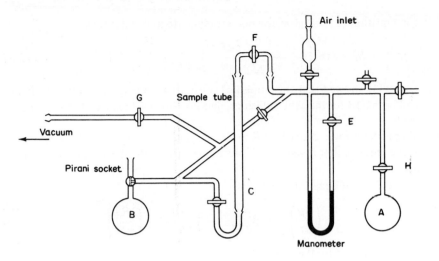

Fig. 15.8 Transient flow apparatus (Kraus and Ross).

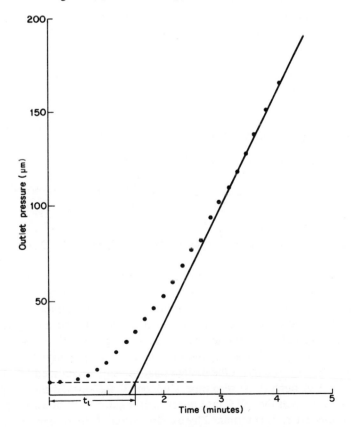

Fig. 15.9 Flow-rate curve for the transient flow apparatus.

Krishnamoorthy [46, 47] found that the time lag remained constant at 1.48 min for rutile titanium dioxide over a range of inlet pressures (54 to 103 mm Hg) with $L =$ 15.3 cm, $\epsilon = 0.726$, $\rho_s = 4.26$ g cm^{-3}, $T = 20°$ C, $R = 8.314 \times 10^7$ erg deg^{-1} mol^{-1} and $M = 29.37$ (air); equation (15.63), using Krauss and Ross's constant (144/13) gave $S_w = 6.05$ m^2 g^{-1} compared with the BET nitrogen gas-adsorption value of 14.5 m^2 g^{-1}. Kraus et al. [24] reported that in all five samples they used, comparable results with the BET method were obtained. Krishnamoorthy attributed his low values to specular reflection.

With zinc oxide, Krishnamoorthy found that the time lag increased with decreasing inlet pressures. Extrapolation to zero inlet pressure after Barrer [48] and Barrer and Grove [23] gave a value which produced a surface area in agreement with BET. Since there is no theoretical justification for such an extrapolation and since there is some difficulty in determining t_L accurately, this method is not recommended for routine analyses.

15.17 Steady-state diffusional flow

Early work [49] was carried out using the apparatus shown in figure 15.8. The slope of figure 15.9 is given by equation (15.58) which, in conjunction with equations (15.60) and (15.61), gives:

$$G = V \left(\frac{dp_1}{dt}\right)_{p_2} = \frac{8}{3} \frac{Ap_2}{LS_v} \left(\frac{\epsilon^2}{1-\epsilon}\right) \sqrt{\left(\frac{2RT}{\pi M}\right)} \tag{15.66}$$

A graph of energy flow rate $V(dp_1)/dt$ against p_2 gives a straight line from which S_v can be determined (figure 15.10). These results were found to be in good agreement with BET gas-adsorption values. On the basis of this work the simplified apparatus shown in figure 15.11 was constructed [50, 61]. The system is first evacuated with tap 4 closed. Taps 1, 2 and 3 are then closed and gas is allowed in on the inlet side by opening tap 4, thus unbalancing the mercury manometer. Opening tap 1 allows the gas to flow through the powder plug, the flow rate being monitored by the changing inlet pressure which is recorded as a deflection θ on a pen recorder graph.

Since, in this system, $p_2 \gg p_1$ then $\Delta p = p = p_2$ and equation (15.62) may be written:

$$\frac{1}{p} \left(\frac{dp}{dt}\right)_V = \frac{8}{3} \cdot \frac{A}{LS_v V} \frac{\epsilon^2}{1-\epsilon} \sqrt{\left(\frac{2RT}{\pi M}\right)}$$

i.e.

$$\frac{1}{\theta} \frac{d\theta}{dt} = \frac{8}{3} \frac{A}{LS_v V} \frac{\epsilon^2}{1-\epsilon} \sqrt{\left(\frac{2RT}{\pi M}\right)} \tag{15.67}$$

where θ is the deflection of the pen on the recorder.

For coarse powders it is necessary to correct for the effect of the support plug and filter paper.

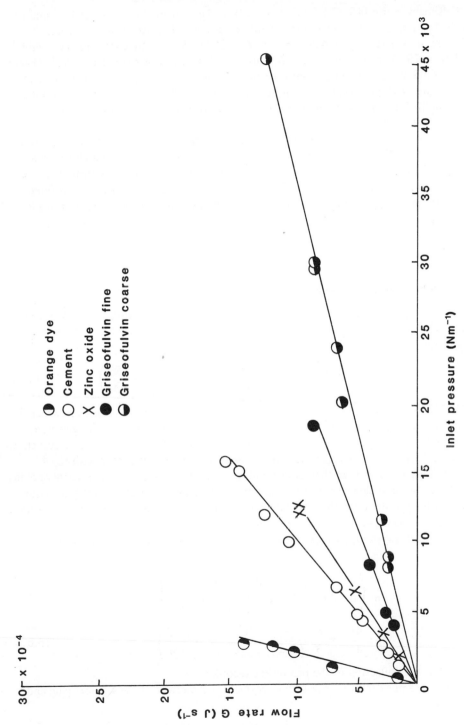

Fig. 15.10 Graph of gas flow rate against inlet pressure.

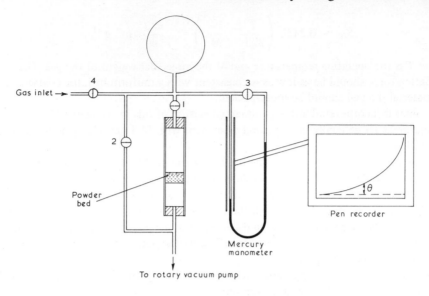

Fig. 15.11 Simple gas diffusion apparatus.

If ψ is the slope of the graph $(1/\theta)(d\theta/dt)$ with the powder present, and ψ_0 the slope with an empty cell, then from equation (15.67)

$$S_v = \frac{8}{3} \frac{A}{LS_vV} \frac{\epsilon^2}{(1-\epsilon)} \sqrt{\left(\frac{2RT}{\pi M}\right)} \left\{ \frac{1}{\psi} - \frac{1}{\psi_0} \right\} \qquad (15.68)$$

Equation (15.67) may be further simplified if the time is recorded for the pressure to fall from a preset high pressure (p_H) to a preset low pressure (p_L).

$$S_v = \frac{8}{3} \frac{A^2}{V} \sqrt{\left(\frac{2RT}{\pi M}\right)} \frac{1}{\ln(p_H/p_L)} \frac{\epsilon^2}{V_p} t$$

$$= K \frac{\epsilon^2}{V_p} t \qquad (15.69)$$

Here the volume of powder in the bed $V_p = AL(1-\epsilon)$ and K is the product of an instrument constant and the average velocity of the gas molecules.
For dry air at 273 K

$$\bar{u} = \left(\frac{8 \times 8.314 \times 273 \times 10^3}{29\pi} \right)^{1/2}$$

$$= 446 \text{ m s}^{-1}$$

Under standard operating conditions $A = 5.005 \text{ cm}^2$, $V = 1000 \text{ cm}^3$, $p_H = 40$ torr, $p_L = 20$ torr. Replacing V_p in equation (15.69) by w, the weight of powder in grams gives:

$$S_W = 0.215 \left(\frac{T}{273} \cdot \frac{29}{M} \right)^{1/2} \frac{\epsilon^2 t}{w} \ \mathrm{m^2 \ g^{-1}}$$

where T is the operating temperature and M the molecular weight of the gas. The consolidating force should be as low as is consistent with a uniform bed; for coarse granular material the bed should be loose-packed.

A timer is incorporated in the commercial version of this instrument (Alstan). The surface areas are generally in very good agreement with BET gas adsorption values [60–62].

An instrument based on this design has been examined by Henrion [63, 64] who states that the above equation breaks down for non-random voidage. A classic case is with porous material where diffusion through the wide voids between particles completely swamps the diffusion through the narrow pores within particles. In cases such as this there is good agreement between diffusion and mercury porosimetry [60, 64].

A similar instrument has also been described by Derjaguin et al. [66] who use a vibrator to compact the powder bed.

Orr [55] developed an apparatus, the Knudsen-Flow permeameter, which is based on the following form of Derjaguin's equation:

$$S_v = \frac{24}{13} \left(\frac{2}{\pi} \right)^{1/2} \frac{A \epsilon^2 \Delta p}{Q (MRT)^{1/2} L} \tag{15.70}$$

where Q is in mol cm^{-2} s^{-1}.

The flow rate of helium passing through a packed bed of powder is measured together with the upstream pressure p and pressure drop across the bed which gives, on rearrangement:

$$S_v = 0.481 \frac{A \epsilon}{q [M(273 + \theta)]^{1/2}} \frac{\Delta p}{L} \left[\frac{760}{p} \left(\frac{273 + \theta}{273} \right) \right] \tag{15.71}$$

where q is in cm^2 s^{-1}.

The relationship found by Orr between specific surface and pressure drop was semi-logarithmic.

15.18 The liquid phase permeameter

In the early stages of development, liquid permeameters were favoured. As long as there is no appreciable size fraction below 5 μm, this is still the easiest technique. Below 5 μm, the use of liquids becomes unsatisfactory, due to settling and segregation, the difficulty of removing air bubbles, aggregation and wetting problems. Gas permeametry was also more attractive due to the higher permeabilities of air and other gases. However, the surface areas determined by air permeametry were less than those determined by liquid permeametry and the difference increased with decreasing size. Though gas permeameters were introduced, they were not placed on a satisfactory footing until the difference

between liquid and gas permeabilities was shown to be due to slip in gases and corrections to the Kozeny equation were derived.

The apparatus used by Carman [13], Walther [25] and Wiggins *et al.* [26] is shown in figure 15.12. The powder is formed into a bed in a uniform tube A, and rests on the metal gauze B. The gauze is supported horizontally by a loosely wound spiral. The flow of liquid is adjusted to a steady rate with stopcock G, and the difference in level between D and the constant level in A gives the pressure drop causing flow. Air bubbles enter the tube H, causing a constant level to be maintained in A and the volume of liquid supplied in known time is given by the graduated reservoir J. The bed is formed by washing a known weight of powder into A, using small quantities at a time and allowing each to settle into place with the assistance of gentle suction. The thickness is measured with a cathetometer.

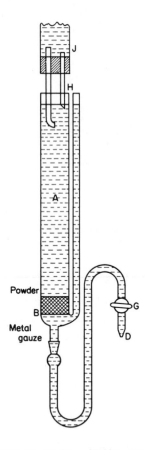

Fig. 15.12 Liquid phase permeameter [13].

Dodd, Davis and Pidgeon [27] used the apparatus shown in figure 15.13, in which the head decreased during an experiment.

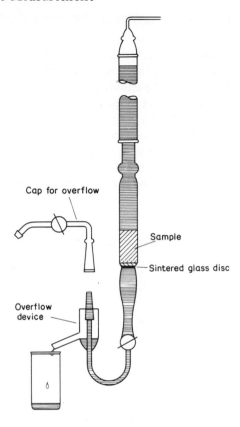

Fig. 15.13 The variable-head permeameter [27].

15.19 Application to hindered settling

The settling of particles, constrained to fixed positions, in a stagnant liquid is analogous to the permeametry situation where the liquid is moving and the bed is fixed. For a sedimenting suspension, the pressure head may be replaced by the gravitational minus the buoyant force on the particles:

$$\frac{\Delta p}{L} = (\rho_s - \rho_f)(1 - \epsilon)g \tag{15.72}$$

Replacing also the volume specific surface by $6/d_{sv}$ modifies equation (15.10) to:

$$u = 0.10\, u_{Stk}\ \frac{\epsilon^3}{(1 - \epsilon)} \tag{15.73}$$

This equation is very similar to the ones derived for particles settling in a concentrated suspension (equations (7.49) and (7.52)) with the replacement of d_{sv} with d_{Stk}.

References

1 Darcy, H.P.G. (1856), *Les Fontaines Publiques de la Ville de Dijon*, Victor Dalamont.
2 Hagen, G. (1839), *Ann. Phys. (Pogg. Ann.)*, **46**, 423.
3 Poiseuille, J. (1846), *Inst. de France Acad. Des Sci.*, **9**, 433.
4 Blake, F.C. (1922), *Trans. Am. Inst. Chem. Engrs*, **14**, 415.
5 Kozeny, J. (1927), *Ber. Wien. Akad.*, **136**A, 271.
6 Coulson, J.M. and Richardson, J.F. (1955), *Chemical Engineering*, Pergamon Press.
7 Essenhigh, R.H. (Nov. 1955), *Safety in Mines Res. Est.*, Report No. 120.
8 Carman, P.C. (1956), *Flow of Gases through Porous Media*, Butterworths.
9 Rigden, P.J. (1954), *Road Res. Tech.*, *Paper No. 28* (HMSO).
10 Lea, F.M. and Nurse, R.W. (1947), Symp. Particle Size Analysis, *Trans. Inst. Chem. Engrs*, **25**, 47.
11 Rigden, P.J. (1956), *Nature*, **157**, 268, 694.
12 Carman, P.C. and Arnell, J.C. (1948), *Can. J. Res.*, **26**, 128.
13 Carman, P.C. (1938), *J. Soc. Chem. Ind. (Trans.)*, **57**, 225.
14 Dallavalle, J.M. (1938), *Chem. Met. Engng*, **45**, 688.
15 Carman, P.C. (1941), *ASTM, Symp. New Methods for Particle Size Determination in the Sub-sieve Range*, 24.
16 Lea, F.M. and Nurse, R.W. (1939), *J. Soc. Chem. Ind.*, **58**, 277.
17 Gooden, E.L. and Smith, C.M. (1940), *Ind. Engng Chem., analyt. Edn*, **12**, 479.
18 Pendleton, A.G. (1960), *Chem. Proc. Engng*, **41**, 147-8.
19 Rigden, R.J. (1947), *J. Soc. Chem. Ind. (Trans.)*, **66**, 191.
20 Usui, K. (1964), *J. Soc. Mat. Sci., Japan*, **13**, 828.
21 Pechukas, A. and Gage, F.W. (1946), *Ind. Engng Chem. analyt. Edn*, **18**, 370-3.
22 Carman, P.C. and Malherbe, P. le R. (1950), *J. Soc. Chem. Ind.*, **69**, 134.
23 Barrer, R.M. and Grove, D.M. (1951), *Trans. Faraday Soc.*, **47**, 826, 837.
24 Kraus, G., Ross, R.W. and Girifalco, L.A. (1953), *J. Phys. Chem.*, **57**, 330.
25 Walther, H. (1943), *Kolloid-Z.*, **103**, 233.
26 Wiggins, E.J., Campbell, W.B. and Maas, O. (1948), *Can. J. Res.*, **26**A, 128.
27 Dodd, C.G., Davis, J.W. and Pidgeon, F.D. (1951), *Ind. Engng Chem.*, **55**, 684.
28 Ergun, S. and Orning, A.A. (1949), *ibid.*, **41**, 1179.
29 Keyes, W.F. (1946), *ibid.*, **18**, 33.
30 Orr, C. and Dallavalle, J.M. (1959), *Fine Particle Measurement*, Macmillan, NY.
31 Sullivan, R.R. and Hertel, K.L. (1940), *J. appl. Phys.*, **11**, 761.
32 Fowler, J.L. and Hertel, K.L. (1940), *ibid.*, **11**, 496.
33 Derjaguin, B. (1946), *C. r. Acad. Sci., USSR*, **53**, 623; (1956) *J. appl. Chem., USSR*, **29**, 49.
34 Kaye, G.W.C. and Laby, T.H. (1966), *Physical and Chemical Constants*, 13th ed., Longmans, London.
35 Millikan, R.A. (cited in [16]).
36 Partington (1958), *General and Organic Chemistry*.
37 *Encyclopaedia of Science and Technology* (1960), McGraw-Hill.
38 Crowl, V.T. (1959), Paint Research Station, Teddington, Middlesex, *Research Mem., No. 274*, **12**, 7.
39 Edmondson, I.C. and Toothill, J.P.R. (1963), *Analyst*, October, 805-8.
40 ASTM Standard (1961): Part 4, p. 149.
41 Crank, J. (1946), *Mathematics of Diffusion*, Clarendon Press, Oxford.
42 Babbit, J.D. (1951), *Can. J. Phys.*, **29**, 427, 437.
43 Knudsen, M. (1909), *Ann. Physik*, **4**, 28, 75, 999.
44 Knudsen, M. (1911), *ibid.*, **4**, 34, 593-656.
45 Pollard, W.G. and Present, R.D. (1948), *Phys. Rev.*, **73**, 762.
46 Krishnamoorthy, T.S. (1966), M.Sc. thesis, Univ. Bradford.
47 Allen, T., Stanley-Wood, N.G. and Krishnamoorthy, T.S. (1966), *Particle Size Analysis Conference*, Loughborough, Soc. Analyt. Chem., London.

48 Barrer, R.M. (1954), *Br. J. appl. Phys.*, suppl. 3.
49 Allen, T. (1971), *Silic. Ind.*, **36**, 718, 173–85.
50 Stanley-Wood, N.G. (1969), Ph.D. thesis, Univ. Bradford.
51 Muskat and Botsel (*cit.* [8]).
52 Schriever (*cit.* [8]).
53 Blaine, R.L. (1943), *ASTM Bull No.* 12B.
54 Rose, H.E. (1952), *J. appl. Chem.,* **2**, 511.
55 Orr, C. (1967), *Analyt. Chem.,* **39**, 834.
56 BS 4359 (1971): Part 2, *Determination of Specific Surface of Powders.*
57 Wasan, D.T. *et al.* (1976), *Powder Technol.* **14**, 209–28.
58 Wasan, D.T. *et al.* (1976), *ibid.*, **14**, 229–44.
59 Harris, C.C. (1977), *ibid.,* **17**, 235–52.
60 Allen, T. (1978), *Proc. Conf. Particle Size Analysis,* Salford (1977) (ed. M.J. Groves), Chem. Soc. Analyt. Div. Heyden.
61 Stanley-Wood, N.G. (1972), *Proc. Conf. Particle Size Analysis,* Bradford (1970) (ed. M. J. Groves and J.L. Wyatt-Sargent), Soc. Analyt. Chem., London, 390–400.
62 Stanley-Wood, N.G. and Chatterjee, A. (1974), *Powder Technol.,* **9**, 1, 7–14.
63 Henrion, P.N. (1977), *ibid.,* **16**, 2, 159–66.
64 Henrion, P.N., Greenwen, F. and Leurs, A. (1977), *ibid.,* **16**, 2, 167–78.
65 Schultz, N.F. (1974), *Int. J. Min. Proc.,* **1**, 1, 65–80.
66 Derjaguin, B.V., Fedoseev, D.V. and Vnukov, S.P. (1976), *Powder Technol.,* **14**, 1, 169–76.

16 Gas adsorption

16.1 Introduction

When a solid is exposed to a gas, the gas molecules impinge upon the solid and may reside upon the surface for a finite time. This phenomenon is called adsorption as opposed to absorption which refers to penetration into the solid body.

The amount adsorbed depends upon the nature of the solid (adsorbent) and the gas (adsorbate) and the pressure at which adsorption takes place. The amount of gas adsorbed can be calculated by determining the increase in weight of the solid (gravimetric method) or by determining, using the gas laws, the amount of gas removed from the system due to adsorption (volumetric method).

The graph of the amount adsorbed (V), at constant temperature, against the adsorption pressure (P) is called the adsorption isotherm. If the gas is at a pressure lower than the critical pressure, i.e. if it is a vapour, the relative pressure $x = P/P_0$, where P_0 is the saturated vapour pressure, is preferred.

A commonly used method of determining the specific surface of a solid is to deduce the monolayer capacity (V_m) from the isotherm. This is defined as the quantity of adsorbate required to cover the adsorbent with a monolayer. Usually a second layer may be forming before the monolayer is complete, but V_m is determined from the isotherm equations irrespective of this. There are also other gas-adsorption methods in which the surface area is determined without determining the monolayer capacity. The fact that gases are adsorbed on solid surfaces was known as early as the late eighteenth century, but systematic studies of this phenomenon have been carried out only in the past sixty years.

Adsorption processes may be classified as physical or chemical, depending on the nature of the forces involved. Physical adsorption, also termed van der Waals adsorption, is caused by molecular interaction forces; the formation of a physically adsorbed layer may be likened to the condensation of a vapour to form a liquid. This type of adsorption is therefore of importance only at temperatures below the critical temperature for the gas. Not only is the heat of physical adsorption of the same order of magnitude as that of liquefaction, but physically adsorbed layers behave in many respects like two-dimensional liquids. On the other hand, chemical adsorption (chemisorption) involves some degree of specific chemical interaction between the adsorbate and the adsorbent and, correspondingly, the energies of adsorption may be quite large and comparable to those of chemical bond formation.

Since physical adsorption is the result of relatively weak interaction between solids and gases, almost all the gas adsorbed can be removed by evacuation at the same temperature at which it was adsorbed. The quantity of physically adsorbed gas at a given pressure increases with decreasing temperature. Consequently, most adsorption measurements for the purpose of determining surface areas are made at low temperatures. Gas that is chemisorbed may be difficult to remove merely by reducing the pressure, and when it does occur, it may be accompanied by chemical changes.

Mathematical theories to describe the adsorption process must, of necessity, be based on over-simplified models since the shapes of the isotherms depend not only on the specific surface of the powder but also upon the pore volume.

Various boundary conditions limit each of the theories, hence a range of equations have been developed to cover the various phenomena.

16.1 Theories of adsorption

16.2.1 Langmuir's isotherm for ideal localized monolayers

The first theoretical equation relating the quantity of adsorbed gas to the equilibrium pressure of the gas was proposed by Langmuir [4]. In his model, adsorption is limited to a monolayer and his equation has limited applicability to physical adsorption with wider application to chemical adsorption and the adsorption of solute, including dye molecules, from solution.

His method was to equate the number of molecules evaporating from the surface with the number condensing on the surface. Since surface forces are short-range, only molecules striking a bare surface are adsorbed; molecules striking a previously adsorbed molecule are elastically reflected back into the gas phase.

From kinetic theory, the number of molecules striking unit area in unit time is given by:

$$Z = \frac{P}{\sqrt{(2\pi mkT)}}$$

where k is Boltzmann's constant, m is the mass of a molecule, P is the pressure and T is the absolute temperature.

The number evaporating n depends upon the energy binding the molecules to the surface. If Q is the energy evolved when a molecule is adsorbed and τ_0 the molecular vibration time, residence time is given by:

$$\tau = \tau_0 \exp\left(+\frac{Q}{RT}\right)$$

where τ_0 is of the order of 10^{-13} s and, for physical adsorption, Q has a value between about 6 and 40 J mol^{-1} [3, p. 463].

The number of molecules evaporating from unit area per second is therefore given by $(1/\tau)$.

If the fraction of surface covered with adsorbed molecules at pressure P is θ, then the rate of adsorption on an area $(1 - \theta)$ equals the rate of desorption from an area $(1 - \theta)$

$$\frac{P}{\sqrt{(2\pi mkT)}} \cdot \alpha_0 (1 - \theta) = \left[\frac{1}{\tau_0} \exp \left(-\frac{Q}{RT} \right) \right] \theta$$

where α_0, the condensation coefficient, is the ratio of elastic to total collisions with the bare surface (α_0 tends to unity under conditions of dynamic equilibrium).

If the gas adsorbed at pressure P is V, and the volume required to form a monolayer is V_m, then:

$$\theta = \frac{V}{V_m} = \frac{bP}{1 + bP}$$

where

$$b = \frac{\alpha_0 \tau_0}{\sqrt{(2\pi mkT)}} \exp \left[\frac{Q}{RT} \right] \tag{16.1}$$

The equation is usually written in the form:

$$\frac{P}{V} = \frac{1}{bV_m} + \frac{P}{V_m} \tag{16.2}$$

A plot of P/V against P yields the monolayer capacity V_m. To relate this to surface area it is necessary to know the area occupied by one molecule, σ. Surface area from the monolayer capacity can be calculated by:

$$S_w = \frac{N\sigma V_m}{M_v} \tag{16.3}$$

where S_w = specific surface area in $m^2 \, g^{-1}$
N = Avagadro number, 6.023×10^{23} molecules/g mol
σ = area occupied by one adsorbate molecule, usually taken as $16.2 \times 10^{-20} \, m^2$, for nitrogen
V_m = monolayer capacity cm^3 per gram of solid
M_v = gram molecular volume = $22\ 410 \, cm^3$/g mol

$$S_w = \frac{(6.023 \times 10^{23})(16.2 \times 10^{-20})}{(22\ 410)} V_m$$

$$= 4.35 \, V_m \text{ for nitrogen at liquid nitrogen temperature} \tag{16.4}$$

One basic assumption in deriving the Langmuir equation is that the energy of adsorption Q is constant thus making b constant. This, in turn, implies that the surface is entirely uniform although this is not supported by experimental evidence.

It is usually assumed that molecules are adsorbed as wholes (discrete entities) on to definite points of attachment on the surface and each point can accommodate only one adsorbed molecule. If adsorption takes place first on high energy-level sites, this must be compensated for by lateral interaction increasing the energy of adsorption of the molecules

adsorbed later. Alternatively, if there are no high energy-level sites, the energies of the adsorbed molecules are independent of the presence or absence of other adsorbed molecules on neighbouring points of attachment.

From equation (16.2), at low pressures $(1 + bP)$ tends to unity, bP may be neglected and Henry's law [13, p. 104] is obeyed:

$$V = bPV_m \tag{16.5}$$

If, in deriving the Langmuir equation, it is assumed that adsorption is not localized, the rate of condensation is proportional to the total surface and not the bare surface, thus:

$$\frac{P}{\sqrt{(2\pi mkT)}} \; \alpha_0 = \left[\frac{1}{\tau_0} \exp\left(-\frac{Q}{RT}\right) \right] \theta$$

i.e. Henry's Law is obeyed at all pressures.

At high pressures bP is large compared with unity and $V = V_m$, therefore the isotherm approaches saturation. If these requirements break down and Q is a linear function of θ, the following equation develops [102].

$$mRT \ln P/P_0 = 1 - V/V_m \tag{16.6}$$

V is plotted against $\log P$, intersecting the ordinate at $V = V_m$. This equation has been applied to the adsorption of carbon dioxide on to alumina at $22°C$, $(80 < P < 400$ mmHg) and $P_0 = 4500$ mmHg. V was found to equal V_m when $x = 0.10$, the same relative pressure as found for adsorption of carbon dioxide at $-78°C$ and the same value for V_m [103].

If Q is a logarithmic function of θ, the Freundlich equation develops:

$$mRT \ln P/P_0 = \ln \theta \tag{16.7}$$

This has been applied to the adsorption of hydrogen on metallic tungsten [102]. Sips [104] considered a combination of the Langmuir and Freundlich isotherms:

$$\theta = \frac{AP^{1/n}}{1 + AP^{1/n}} \tag{16.8}$$

which has the proper limits for monolayer adsorption but reduces to equation (16.7) at low pressures.

In a later paper, Sips [105] revised his theory and arrived at:

$$\theta = \left(\frac{P}{a+P}\right)^c \qquad \text{where } a \text{ and } c \text{ are constants.}$$

The Langmuir equation has also been derived from a thermodynamic [5] and statistical [6] basis.

16.2.2 BET isotherm for multilayer adsorption

The most important step in the study of adsorption came with a derivation by Brunauer, Emmett and Teller [9], for the multilayer adsorption of gases on solid surfaces. This

multilayer adsorption theory known generally as the BET theory has occupied a central position in gas-adsorption studies and surface-area measurements ever since.

On the assumption that the forces that produce condensation are chiefly responsible for the binding energy of multimolecular adsorption, they proceeded to derive the isotherm equation for multimolecular adsorption by a method that was a generalization of Langmuir's treatment of the unimolecular layer. The generalization of the ideal localized monolayer treatment is effected by assuming that each first layer adsorbed molecule serves as a site for the adsorption of a molecule into the second layer and so on. Hence, the concept of localization prevails at all layers and the forces of mutual interaction are neglected.

$S_0, S_1, S_2, \ldots S_i$ represent the areas covered by $0, 1, 2, \ldots i$ layers of adsorbate molecules. At equilibrium, the rate of condensation on S_0 is equal to the rate of evaporation from S_1 giving:

$$a_1 PS_0 = b_1 S_1 \exp(-Q_1/RT). \tag{16.9}$$

where P = pressure,
Q_1 = heat of adsorption of the first layer,
a_1, b_1 = constants.

$$a_1 = \frac{\alpha_1}{\sqrt{(2\pi mkT)}}$$

$$b_1 = \frac{1}{\tau}$$

it being assumed that the molecular vibration time differs from layer to layer.

This is essentially Langmuir's equation, involving the assumption that a_1, b_1, Q_1 are independent of the number of adsorbed molecules already present in the first layer.

Similarly, at the first layer in equilibrium:

$$a_2 PS_1 = b_2 S_2 \exp(-Q_2/RT) \tag{16.10}$$

and so on.

In general for equilibrium between the $(i-1)$th and ith layers

$$a_i PS_{i-1} = b_i S_i \exp(-Q_i/RT) \tag{16.11}$$

The total surface area of the solid is given by:

$$A = \sum_{i=0}^{i=\infty} S_i \tag{16.12}$$

and the total volume of the adsorbate by:

$$V = V_0 \sum_{i=0}^{i=\infty} iS_i \tag{16.13}$$

where V_0 = the volume of gas adsorbed on unit surface to form a complete monolayer.

Dividing equation (16.13) by equation (16.12) gives:

$$\frac{V}{A V_0} = \frac{V}{V_m} = \frac{\sum\limits_{i=0}^{i=\infty} i S_i}{\sum\limits_{i=0}^{i=\infty} S_i} \tag{16.14}$$

An essentially similar equation had been arrived at earlier by Baly [10], who could proceed further only by empirical means. Brunauer *et al.* [9] proceeded to solve this summation by two simplifying assumptions, that:

$$Q_2 = Q_3 = \ldots = Q_i = Q_L \tag{16.15}$$

where Q_L is the heat of liquefaction of the bulk liquid, and:

$$\frac{b_2}{a_2} = \frac{b_3}{a_3} = \ldots = \frac{b_i}{a_i} = g, \quad \text{a constant} \tag{16.16}$$

In other words, the evaporation and condensation properties of the molecules in the second and higher adsorbed layers are the same as those of the liquid state. Rewriting:

$$S_1 = y S_0, \quad \text{where, from equation (16.9):}$$

$$y = \frac{a_1}{b_1} P \exp\left(\frac{Q_1}{RT}\right) \tag{16.17}$$

$$S_2 = X S_1 \text{ where, from equations (16.10) and (16.16):}$$

$$X = \frac{P}{g} \exp\left(\frac{Q_L}{RT}\right) \tag{16.18}$$

$$S_3 = X S_2 = X^2 S_1 \tag{16.19}$$

and in the general case for $i > 0$:

$$S_i = X S_{i-1} = X^{i-1} S_1 = y X^{i-1} S_0 = c X^i S_0 \tag{16.20}$$

where

$$c = \frac{y}{X} = \frac{a_1 g}{b_1} \exp\left[(Q_1 - Q_L)/RT\right]$$

where $a_1 g/b_1$ approximates to unity.

Substituting equation (16.20) in equation (16.14):

$$\frac{V}{V_m} = \frac{c \sum\limits_{i=1}^{i=\infty} i X^i}{1 + c \sum\limits_{i=1}^{i=\infty} X^i} \tag{16.21}$$

The summation in the denominator is merely the sum of an infinite geometric progression:

$$\sum_{i=1}^{i=\infty} X^i = \frac{X}{1-X} \tag{16.22}$$

while that in the numerator is:

$$\sum_{i=1}^{i=\infty} iX^i = X \frac{d}{dX} \sum_{i=0}^{i=\infty} X^i = \frac{X}{(1-X)^2} \tag{16.23}$$

Therefore

$$\frac{V}{V_m} = \frac{cX}{(1-X)(1-X+cX)} \tag{16.24}$$

On a free surface the amount adsorbed is infinite. Thus at $P = P_0$, the saturation vapour pressure of the adsorbate at the temperature of adsorption, X, must be 1, in order to make $V = \infty$. Therefore, substituting $X = 1$ and $P = P_0$ in equation (16.18) and dividing the resultant by equation (16.18) gives:

$$X = \frac{P}{P_0} \quad \text{(i.e. } X = x) \tag{16.25}$$

Substituting in equation (16.24):

$$V = \frac{V_m cP}{(P_0 - P)\,[1 + (c-1)P/P_0]} \tag{16.26}$$

which transforms to:

$$\frac{P}{V(P_0 - P)} = \frac{1}{V_m c} + \frac{c-1}{V_m c}\frac{P}{P_0} \tag{16.27}$$

which is the commonly known form of the BET equation. A plot of $P/V(P_0 - P)$ against P/P_0 should yield a straight line having a slope $(c-1)/V_m c$ and an intercept $1/V_m c$.

This equation is capable of describing type 1, type 2 and type 3 isotherms (see section 16.2.6), depending upon the constant c. In general it has been found that only type 2 isotherms (i.e. those with high values of c) have well-defined 'knee-bends', which are essential for accurate V_m values. The preference for using nitrogen at liquid nitrogen temperature is due to the fact that at liquid nitrogen temperatures, with all solids so far reported, this gas exhibits higher c values than the c values for alternative gases.

For type 2 isotherms, the BET equation has been found to be valid generally between 0.05 and 0.3 relative pressure, but examples have also been reported where this range has been extended or shortened [21].

The internal consistency of the BET method has been demonstrated by many authors [2, 13] by their measurements on several solids. The degree of correspondence between the specific surfaces obtained with several adsorbates allows confidence to be placed in the method.

The intercepts obtained are usually very small. Negative values of intercepts in the BET plot have also been reported [22]. MacIver and Emmett [23] find that this can be accounted for by the BET equation not fitting the results at $P/P_0 > 0.2$.

When $c \gg 1$ equation (16.25) takes the form:

$$(P_0 - P)v = P_0 v_m \tag{16.28}$$

Hence, it may be assumed that for high c values the BET plot passes through the origin and the slope is inversely proportional to the monolayer capacity. Thus only one experimental point is required. This simplification is frequently applied for routine analyses. The error in v_m depends on the relative pressure and value of c:

$$\text{Error} = 1 - \frac{cx}{cx + (1 - x)}$$

Hence for $c = 100$, at $x = 0.1, 0.2, 0.3$ respectively, the error is 8%, 4%, 2%.

16.2.3 The n-layer BET equation

If, owing to special considerations, the number of layers cannot exceed n, the equation becomes:

$$V = \frac{V_m cx}{(1 - x)} \frac{1 - (n + 1)x^n + nx^{n+1}}{1 + (c - 1)x - cx^{n+1}} \tag{16.29}$$

which is referred to as the n-layer BET equation; this applies to adsorption in limited space such as a capillary.

When $n = 1$, it reduces, at all values of c (unlike the simple BET equation which only reduces when $x \ll 1$ and $c \gg 1$) to the Langmuir equation. The n-layer equation also reduces to the same form as the Langmuir equation when $n = 2$ and $c = 4$ [11], and if $n \gg 3$ it is analytically capable of reproducing the shapes of all five isotherm types provided c lies within certain narrow ranges of values [12], or x lies between 0.05 and 0.35 [98].

Brunauer, Emmett and Teller successfully applied this isotherm equation to a variety of experimental isotherms obtained by themselves and others.

The BET equation for adsorption of gases includes isotherms of types 1, 2 and 3 but not 4 and 5. However, Brunauer et al. derived a new isotherm equation to cover all five types [1]. The BET equation has also been derived by statistical reasonings by several authors [13].

The n-layer BET equation may be written:

$$\frac{V}{V_m} = \frac{c\phi(n, x)}{1 + c\theta(n, x)}$$

where

$$\phi(n, x) = \frac{x[(1 - x^n) - nx^n(1 - x)]}{(1 - x)^2}$$

$$\theta(n, x) = \frac{x(1 - x^n)}{(1 - x)}$$

Joyner et al. [106] compiled tables of ϕ and θ for increasing x. Using these tables the best straight line is selected for ϕ against θ since the linear form of the equation is:

$$\frac{\phi}{V} = \frac{1}{c V_m} + \frac{\theta}{V_m} \qquad (16.30)$$

This procedure is best carried out with a computer.

The BET plot of $[x/(1 - x)](1/V)$ against x for microporous materials is convex to the x axis since, as the pressure increases, the increments adsorbed are smaller than those for a non-porous material. This is due to micropore filling. The n-layer equation will produce a straight line with an increased intercept on the θ axis (i.e. a lower c value) and a higher specific surface. Low c values are improbable with microporous materials, hence the validity of this technique is questionable. Gregg and Sing [100, p. 54] go so far as to say that n is no more than an empirical parameter adjusted to give best fit to the experimental data. It would seem doubtful if the procedure possesses any real advantage over the more conventional BET method.

16.2.4 Discussion of BET theory

The BET model has been criticized on the grounds that although c is assumed constant there is a great deal of evidence that this is not so. For constant c it is necessary that the surface be energetically homogeneous, i.e. there are no high energy level sites. Experimental measurements of variation of heats of adsorption with coverage show that the first molecules to be adsorbed generate more energy than subsequent molecules. Brunauer, Emmett and Teller suggest that this is why their theory breaks down at low surface coverage.

A further criticism is that horizontal interactions are neglected, i.e. only the forces between the adsorbing molecule and the surface are considered. Thus the first molecule to be adsorbed is considered to generate the same energy as the final molecule to fill a monolayer although in the former case the molecule has no near adsorbate neighbours whereas in the latter it has six. These two effects must, in part, cancel each other out.

It is also questionable as to whether adsorbing molecules in the second and subsequent layers should be considered as being equal. One would not expect a sudden transition in the energy generated between the first and subsequent layers.

Cassel [14, 15] showed, using Gibbs' adsorption isotherm, that the surface tension of the adsorbed film at $P = P_0$ is negative, arising from the total disregard of the interaction forces. Since the BET model assumes the existence of localized adsorption at all layers, the molecules being located on top of one another, and since the adsorption can take place in the nth layer before the $(n - 1)$th layer is filled, the adsorbed phase is built up not as a series of continuous layers, but as a random system of vertical molecular columns. Halsey [16] has pointed out that the combinational entropy term associated with these random molecular piles is responsible for the stability of the BET adsorbed

layers at pressures below the saturation vapour pressure. This large entropy term is the cause of too large adsorptions observed when $P/P_0 > 0.35$.

Gregg and Jacobs [17] doubted the validity of the assumption that the adsorbed phase is liquid-like, and found that the integration constants of the Clausius–Clapeyron equation as applied to adsorption and vapour pressure do not show the interrelationship demanded by BET theory. They conclude that any constant can be used in the place of P_0 and that the correspondence between the adsorbed and the liquid phase is a loose one, arising out of the fact that the same type of forces is involved.

Halsey [18] pointed out that the hypothesis that an isolated adsorbed molecule can adsorb a second molecule on top, yielding the full energy of liquefaction, and that in turn the second molecule can adsorb a third, and so on, is untenable. If the molecules are hexagonally packed, one would be much more likely to find a second-layer molecule adsorbed above the centre of a triangular array of first-layer molecules. Applying this modification, the BET model results in very little second-layer adsorption when the first layer is one-third full and virtually no adsorption in the third layer except at high relative pressures. Values of V_m given by this modified theory, however, do not differ appreciably from that given by the simple BET theory.

The monolayer capacity of an isotherm occurs at the so-called point B on the VP isotherm where the slope of the isotherm changes at the completion of a monolayer. This is normally situated between $0.05 < x < 0.15$. The BET theory however predicts a point of inflection at variable coverage dependent on the c value.

At relative pressures greater than 0.35 the BET equation predicts adsorption greater than observed. A recent review has been presented by Dollimore et al. [178].

16.2.5 Mathematical nature of the BET equation

The BET equation is equivalent to the difference between the upper branches of two rectangular hyperbolas [151] and may be written:

$$\frac{V}{V_m} = \frac{1}{1-x} - \frac{1}{1+(c-1)x} \tag{16.31}$$

For c less than 2 and positive a type 3 isotherm is generated. Values of c less than unity infer that the cohesive force between the adsorbing molecules is greater than the adhesive force between the adsorbate and the adsorbent.

The point of inflection of the isotherm may be determined by differentiating twice with respect to x and equating to zero. This gives

$$x_I = \frac{(c-1)^{1/3} - 1}{(c-1)^{2/3}} \tag{16.32}$$

$$\left(\frac{V}{V_m}\right)_I = \frac{(c-1)^{2/3} - 1}{(c-1)^{2/3} - (c-1)^{1/3} + 1} \tag{16.33}$$

By plotting $(V/V_m)_I$ against x_I for variable c it can be seen that at high c values, monolayer coverage occurs at or near the point of inflection. For low c values the ratio

of V to V_m at the point of inflection becomes progressively smaller until at $c < 2$ the point occurs at negative values of x_I (figure 16.1).

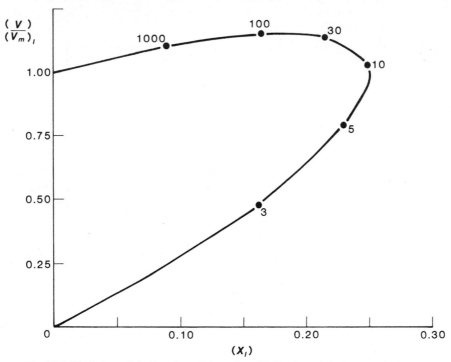

Fig. 16.1 Variation of the location of the point of inflection of the BET isotherm with c value (e.g. for a c value of 100 the point of inflection occurs at a relative pressure x_I equal to 0.169 when the volume coverage V equals 1.148 V_m).

Recently Genot [152] described a more accurate method of determining the monolayer capacity by considering the mathematical nature of the BET equation. His treatment was as follows:

At $V = V_m$, let $P = P_m$ and $x = x_m$ at point M on the isotherm. Then:

$$x_m = \frac{1}{1 + \sqrt{c}}$$

At $V = \frac{1}{2} V_m$ let $P = P_{0.5}$ and $x = x_{0.5}$; then

$$c = \frac{(1 + x_{0.5})^2}{x_{0.5}(1 + x_{0.5})}$$

For high c values,

$$x_{0.5} = \frac{1}{c + 3} \triangleq \frac{1}{c} = \frac{\tau_2}{\tau_1},$$

the inverse ratio of the lifetimes of adsorbed molecules in the first and subsequent layers.

The tangent at point M passes through the point G $(x = 1, V/V_m = 3)$ for all c. As

$$\left[\frac{d(V/V_m)}{dx}\right]_M = \left[\frac{d(\ln V)}{dx}\right]_M = \frac{2}{1-x_m} = -2\left[\frac{d(\ln(1-x))}{dx}\right]$$

or

$$\left[\frac{d(\ln V)}{d\ln(1-x)}\right]_M = \left[\frac{d(\log V)}{d\log(1-x)}\right]_M = -2$$

Hence, point M will result from the determination of the tangent of slope -2 on the graph $\log V$ against $[\log(1-x)]^{-1}$. Special graph paper will facilitate this operation and this treatment will produce more standardized results than using the conventional BET plot.

Hill [164] showed that when sufficient adsorption had occurred to cover the surface with a monolayer, some fraction of the surface $(\theta_0)_m$ remained bare. Hill established that

$$(\theta_0)_m = \frac{c^{1/2} - 1}{c - 1}$$

Lowell [165] extended the argument to show that the fraction of surface covered by molecules i layers deep is

$$(\theta_i)_m = c\left(\frac{c^{1/2} - 1}{c - 1}\right)^{i+1}$$

For example, for $c = 100$:

i	0	1	2	3	4	5
$(\theta_i)_m$	0.0909	0.8264	0.0751	0.0068	0.0006	0.0001

i.e. 82.64% of the surface is covered with a monolayer, 9.09% is bare and so on.

16.2.6 Shapes of isotherms

The majority of adsorption isotherms may be grouped in the six types shown in figure 16.2. The first five types were given by Brunauer, Deming and Teller [1]; type 6 was identified later [106].

Type 1 isotherm is characterized by a rapid initial rise in amount adsorbed at low pressures followed by a flat region. In some cases the curve is reversible and the amount adsorbed approaches a limiting value. In others the curve approaches the line $P/P_0 = 1$ asymptotically and the desorption curve may lie above the adsorption curve shown down to very low pressures. For many years it was thought that the shape of this iso-

therm was due to adsorption being limited to a monolayer and the isotherm was interpreted on the basis of the Langmuir theory [7, 8] (this type of isotherm is still known as a Langmuir isotherm). It is now generally accepted that the shape is characteristic of micropore filling and the limiting amount adsorbed is a measure of micropore volume rather than monolayer surface. Type 1 isotherms may also occur for adsorption on high energy level surfaces [107].

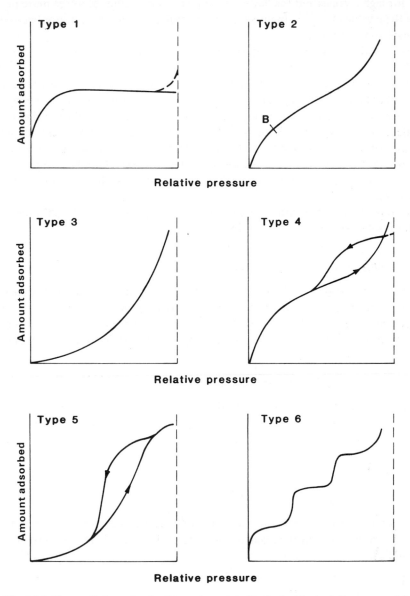

Fig. 16.2 Types of adsorption isotherms (amount adsorbed against relative pressure).

The reversible type 2 isotherm is obtained by adsorption on many non-porous or mesoporous powders and represents unrestricted multilayer adsorption on a hetero-geneous substrate. Although layers at different levels may exist simultaneously, mono-layer completion occurs at the point of inflection of the isotherm. This is known as point B and was first identified by Emmett and Brunauer [108]. They subsequently developed a theory, containing a constant c, to locate this point. Type 2 isotherms occur for high c values and the 'knee' at the point of inflection becomes more pronoun-ced as the c value increases. Increasing c values indicate increasing affinity between the adsorbate and adsorbent.

Type 3 isotherms arise when adsorbate–adsorbent interaction is weak and c is less than 2.

A characteristic feature of type 4 isotherms is the hysteresis loop. This has been explained as being due to capillary phenomena [2, 3] and this part of the isotherm is used for pore size distribution evaluation. As the pressure is reduced from the saturat-ion value, the gas molecules condensed in the capillary cracks of adsorbents do not evap-orate as readily as they would from the bulk liquid due to the lowering of the vapour pressure over the concave meniscus formed by the condensed liquid in the pores.

Type 5 isotherms are similar to type 4 except that the adsorbate–adsorbent inter-action is weak.

Type 6 isotherms arise with stepwise multilayer adsorption of noble gas molecules on a uniform substrate [109, 130, 154, 155].

16.2.7 Modifications of the BET equation

Pickett [99] modified the n-layer BET treatment to take into account the decrease in probability of escape from an elemental area covered with n layers as adjacent elemental areas also become covered with n layers. This leads to the result that there can be no evaporation from such an area. With this assumption Pickett derived the equation:

$$\frac{V}{V_m} = \frac{cx(1 - x^n)}{(1 - kx)(1 + (c - 1)kx)} \tag{16.34}$$

Anderson [90] made the assumption that the heat of adsorption in the second to about the ninth layer differs from the heat of liquefaction by a constant amount. The resultant equation is similar to the BET equation but contains an additional parameter k $(k < 1)$

$$\frac{V}{V_m} = \frac{kcx}{(1 - kx)(1 + (c - 1)kx)} \tag{16.35}$$

Another modification took into account the decrease in the area available in success-ive layers, a situation likely to prevail in porous solids [111, 112]. This leads to the equation:

$$\frac{V}{V_m} = \frac{kcx}{(1 - ikx)(1 + (c - i)kx)} \tag{16.36}$$

where $i = A_n/A_{n-1}$, where A_n is the number of molecules required to fill the nth layer and A_{n-1} is the number required to fill the $(n-1)$th layer.

Brunauer et al. [110] introduced a parameter K into the BET equation, where K is a measure of the attractive field force of the adsorbent. Although the derived equation is identical with Anderson's, the model on which it is based is different.

Barrer et al. [113] derived eighteen analogues of the BET equation by making various assumptions as to the evaporation–condensation properties of the molecules in each adsorbed layer.

16.2.8 The Hüttig equation

Several modifications have been derived to extend the scope of the BET equation. Hüttig [19, 20] derived a formulation of the localized multilayer adsorption on a uniform surface. The difference between the BET and the Hüttig theories is in the assumption by the latter that the evaporation of the ith layer molecule is entirely unimpeded by the presence of molecules in the $(i + 1)$th layer, whereas the BET contention is that $(i + 1)$th layer molecules are completely effective in preventing the evaporation of underlying molecules. Hüttig's final equation is:

$$\frac{V}{V_m} = \frac{(cP/P_0)}{(1 + cP/P_0)} \, (1 + P/P_0) \tag{16.37}$$

i.e.

$$\frac{P}{V} \, (1 + P/P_0) = \frac{P_0}{c V_m} + \frac{P}{V_m} \tag{16.38}$$

so that a plot of $(P/V)(1 + P/P_0)$ against P (or more conventionally P/P_0) should be linear with a slope of $1/V_m$ and intercept of $P_0/c V_m$. Theory and experiment agree up to $P/P_0 = 0.7$, but, at higher pressures, Hüttig's equation predicts too low an amount adsorbed whereas the BET prediction is larger than the amount observed. For the majority of gas–solid systems the V_m values calculated from Hüttig's equation exceed the BET values by 2 to 20% depending upon the value of c [13]. Compromise equations between Hüttig and BET have also been attempted [13].

16.2.9 The relative method of Harkins and Jura (HJr)

Harkins and Jura [25] derived an isotherm, by analogy with condensed liquid layers, independent of V_m, hence avoiding any explicit assumption of the value of the molecular area of the adsorbate in the calculation of the surface area.

Condensed monolayers on water are characterized by the fact that they exhibit a linear PV – area relationship.

$$\pi = b - a\sigma \tag{16.39}$$

where π and σ are the pressure and mean area per molecule, while a and b are constants.

This linear-pressure relationship persists up to high pressures where the film is several molecules thick. They transformed this equation into an equivalent equation

$$\ln P = B^1 - A^1/V^2 \tag{16.40}$$

$$\ln\left(\frac{P}{P_0}\right) = B - A/V^2 \tag{16.41}$$

where B is a constant of integration and

$$A = \frac{10^{20}\, aS^2 M^2}{2RTN} \tag{16.42}$$

where a is a constant, S the surface area of the solid, M the molar gas volume, R the gas constant, T the absolute temperature and N Avagadro's number.

Equation (16.41) involves only the quantities P and V which are measured directly in the experimental determination of adsorption. Harkins and Jura reported that this simple isotherm was valid over more than twice the pressure range of any other two-constant adsorption isotherms.

A plot of $\log P/P_0$ against $1/V^2$ should give a straight line of slope $-A$. From equation (16.42), it is evident that the surface area of the solid is related to the slope of the line by the relation:

$$S = k\sqrt{A} \tag{16.43}$$

where k is a constant for a given gas at a constant temperature. For convenience, the values of k are so determined that when V is in cm^3 g^{-1} at standard conditions, S is in m^2 g^{-1}.

The constant k has to be determined by calibration, using an independent surface area method. For this reason, the method is usually referred to as HJ relative or HJr method. The original determinations were carried out using anatase whose area had been evaluated from heat of wetting measurements.

Orr and Dallavalle [2] have listed the values of k for some gases (Table 16.1).

It was tacitly assumed that k was a function solely of the temperature and the nature of the adsorbate and independent of the nature of the solid. Fundamentally the HJr method is empirical.

Table 16.1 The value of HJr constants.

Gas	Temp ($^\circ$C)	k
Nitrogen	-195.8	4.06
Argon	-195.8	3.56
Water vapour	25	3.83
n-Butane	0	13.6
n-Heptane	25	16.9
Pentane	20	12.7
Pentane-1	20	12.2

If the relation between $\log P/P_0$ and $1/V^2$ is expressed by two or more segments with different slopes then, according to Harkins and Jura, the slope for the lower-pressure region is always to be taken, since this is the one in which the transition from a monolayer to a poly-layer always occurs. The result of more than one straight line is attributed to the existence of more than one condensed phase.

It may be noted that the BET method yields a value for V_m from which the surface area is to be calculated whereas the HJr method yields the surface area directly without giving the value of V_m.

16.2.10 Comparison between BET and HJr methods

Livingstone [26, 27] and Emmett [28] have found that in the linear BET region (P/P_0 = 0.05 to 0.3) a linear HJr plot is obtained only when $50 < c < 250$. For c = 10, 5 or 2, and $P/P_0 < 0.4$, there is no linear relationship, while for c = 100 the range of mutual validity of the two equations is limited to the region 0.01 to about 0.13 relative pressure. Smith and Bell [29] extended this inquiry to the n-layer BET equation.

Both the BET and HJr methods are open to criticism in that they involve the arbitrary selection of constants (k and σ) which undoubtedly depend on the nature of the solid surface.

Of the two, the HJr method is relatively inferior for the following reasons [13]:

(a) The quantity $1/V^2$ is sensitive to slight errors in V.
(b) The range of relative pressure over which a linear HJr plot obtains is variable, depending upon the value of c in the BET equation. For each new solid, a large number of experimental points may be needed in order to locate this linear region.
(c) Some systems yield HJr plots with more than one linear section.

However, the adsorption of nitrogen at $-195°C$ on the majority of solids is characterized by c values in the range 50–240, so that the surface areas obtained from the two equations are in agreement.

16.2.11 The Frenkel–Halsey–Hill equation (FHH)

Hill [129] took into account the decrease in interaction energy for molecules adsorbed in second and subsequent layers and derived the equation

$$\ln x = -b\left(\frac{V}{V_m}\right)^s \tag{16.44}$$

where b and s are constant. This equation has been discussed by Sing [109] and an analysis of the two constants has been carried out [153].

16.2.12 The Dubinin–Radushkevich equation (D-R)

Dubinin [121, 132–134] derived an equation, based on the Polanyi potential theory, for the determination of micropore volume. In this theory it is postulated that the force

of attraction at any given point in the adsorbed film is given by the adsorption potential (ϵ), defined as the work done by the adsorption forces in bringing a molecule from the gas phase to that point. Polanyi expressed the adsorption potential as the work required to isothermally compress one mole of gas from the equilibrium pressure P to the vapour pressure P_0.

$$\epsilon = RT \ln \left(\frac{P_0}{P} \right) \tag{16.45}$$

Dubinin considered that the adsorption potential varies according to the nature of the adsorbate and adsorbent. Thus the characteristic curves for different adsorbates on the same adsorbent are homologous and can be superimposed by use of an affinity coefficient (β) where:

$$\beta = \frac{\epsilon_1}{\epsilon_2} \tag{16.46}$$

where the suffixes refer to the respective vapours. If adsorbate 2 is taken as an arbitrary standard then (ϵ/ϵ_0) $= \beta$ where the suffix zero refers to the standard vapour and that without suffix to the other vapour. Assuming that the adsorption volume may be expressed as a Gaussian function of the corresponding adsorption potential, then for the standard vapour:

$$\frac{V}{\rho V_p} = \exp \left(- k\epsilon_0^2 \right) \tag{16.47}$$

where V is the volume of vapour adsorbed at standard temperature and pressure, ρ is the density of the liquid adsorbate, V_p is the micropore volume and k is a constant.

This equation can be written in the linear form:

$$\log \left(\frac{V}{\rho} \right) = \log V_p - D \log^2 \left(\frac{P_0}{P} \right) \tag{16.48}$$

where $\qquad\qquad D = 2.303 k R^2 \, T^2 / \beta^2.$ $\tag{16.49}$

Equation (16.48) is known as the **Dubinin–Radushkevich** equation. The micropore volume can be evaluated from the intercept on the adsorption axis where $P_0/P = 1$. Dubinin found the equation held for the adsorption of nitrogen, saturated hydrocarbons, benzene and cyclohexane over the relative pressure range 10^{-5} to 0.2 for truly microporous solids.

This equation has been widely used for determining the micropore volume of carbons [114–120] and the thermodynamic parameters of adsorption [121–123, 168]. It has also been shown that many isotherms of vapours on non-porous solids can be linearized in the sub-monolayer region by applying the D-R equation [135–136]. It was later extended by Kadlec [166].

Marsh and Rand [124–125] critically appraise the D-R equation and state that it predicts a Rayleigh distribution of adsorption-free energy and only when this distribution is present in microporous solids will a completely linear D-R plot result. Adsorption of carbon dioxide, nitrogen and argon at 77 K on various microporous carbons is

examined and in no case is the complete Rayleigh distribution found to apply. In order to obtain meaningful parameters they recommend that the experimental data should be extended to as near as possible to unit relative pressure.

In a later paper [143] Rand shows that the data can be linearized by using the more general Dubinin–Astakhov (D-A) equation:

$$\frac{V}{\rho V_p} = \exp\left[-k\left(\frac{\epsilon}{E}\right)^n\right] \tag{16.50}$$

The significance of this equation is discussed and it is shown that for carbon dioxide and nitrogen on microporous carbons the constants E and k are independent of the physical nature of the adsorbent and the temperature of adsorption. The energetic heterogeneity of the surface is therefore described by one parameter n.

The adsorption of carbon dioxide at $T = 273$ K is frequently used to determine V_p and since, at this temperature, pressurized apparatus is required to reach P_0 it has been suggested [144] that adsorption should be determined at pressures less than 1 atmosphere and extrapolated using the D-R equation. Since, in some cases, the change of slope occurs at $\epsilon = 10$ kJ mole^{-1} and at 1 atmosphere with $T = 273$ K, $\epsilon = 8$ kJ mole^{-1} these plots cannot be treated this way. They conclude that the D-A equation is empirical and should be used with caution when the $V-t$ method cannot be used.

Kaganer [126] modified Dubinin's treatment and developed a method for the calculation of specific surface by assuming a distribution of adsorption energy over the surface rather than the volume in the region of monolayer coverage. Hence, the D-R equation becomes:

$$\ln\left(\frac{V}{V_m}\right) = -D\left(RT\ln\frac{P_0}{P}\right)^2 \tag{16.51}$$

The D-K equation has the same form as the D-R equation except that the fractional coverage (V/V_m) replaces the fractional filling of the pore volume.

Kaganer examined a range of powders, using nitrogen as adsorbate, at relative pressures between 0.0001 and 0.01 and obtained good agreement with BET.

Jovanovic [127] derived two equations for monolayer and multilayer adsorption

$$V_m = V(1 - \exp(-ax)) \tag{16.52}$$

$$V_m \exp(bx) = V(1 - \exp(-ax)) \tag{16.53}$$

where a and b are constants which describe adsorption in the first and in subsequent layers respectively where

$$a = \sigma\tau P_0/\sqrt{(2\bar{u}mkT)} \tag{16.54}$$

$$b = \sigma\tau_i P_0/\sqrt{(2\bar{u}mkT)} \tag{16.55}$$

where τ is the molecular residence time in the first layer
τ_i is the molecular residence time in subsequent layers.

For isotherms having sharp 'knees', $\exp(-ax) \ll 1$, and the equation for multilayer adsorption simplifies to:

$$V = V_m \exp(bx) \tag{16.56}$$

The above equation has been used to interpret experimental data with some success [128].

Ramsay and Avery [145] investigated nitrogen adsorption in microporous silica compacts for $10^{-2} > x > 10^{-4}$ and found that $\epsilon < 7$ kJ mole^{-1} for loose powder and increased with compaction and decrease in pore size. They found close agreement between the volume adsorbed at monolayer coverage and the intercept volume using the D-K equation which indicated surface coverage as opposed to volume filling.

Rosai and Giorgi [161] applied the D-R equation to the adsorption of argon and krypton on to barium at 77.3° C and state that this enables one to distinguish between barium films of different surface characteristics and gives a qualitative description of the dynamics of sintering.

Stoeckli [163] used a generalized form of the D-R equation for the filling of a heterogeneous micropore system. He found that the constant 'k' was not constant but decreased with increasing $[(T/\beta) \log (P_0/P)]^2$.

16.2.13 The V_A-t method

Schüll et al. [85] showed that for a number of non-porous solids, the ratio between the adsorbed volume V and the volume of the unimolecular layer V_m, if plotted as a function of x, could be represented approximately by a single curve. With the aid of this function the thickness of the adsorbed layer as a function of x could be calculated.

The film thickness for nitrogen is given by the equation (section 18.4)

$$t = 0.354 \left(\frac{V}{V_m} \right) \text{ nm} \tag{16.57}$$

(the constant is the thickness of one layer of nitrogen molecules) [86].

This equation can be written

$$S_t = 15.47 \left(\frac{V}{t} \right) \tag{16.58}$$

with V cm^3 g^{-1} (vapour at STP), t is the film thickness in nm and S_t is the specific surface in m^2 g^{-1}.

Lippens et al. [88] have published tables of t against x, for the construction of the V–t curve, from $t = 0.996$ nm at $x = 0.76$ to $t = 0.351$ nm at $x = 0.08$. Brockhoff [91] extended the curve to $x = 0.92$.

These values cannot be used for all substances and other t-curves are available. One of the most widely used is de Boers' common t-curve [84, 87] (figure 16.3). The experimental points, for a variety of substances, deviate by 10% or more from the average curve. When the t-curve is being used for pore size analysis these errors are small enough to be neglected. For surface-area evaluation t-curves with small deviations

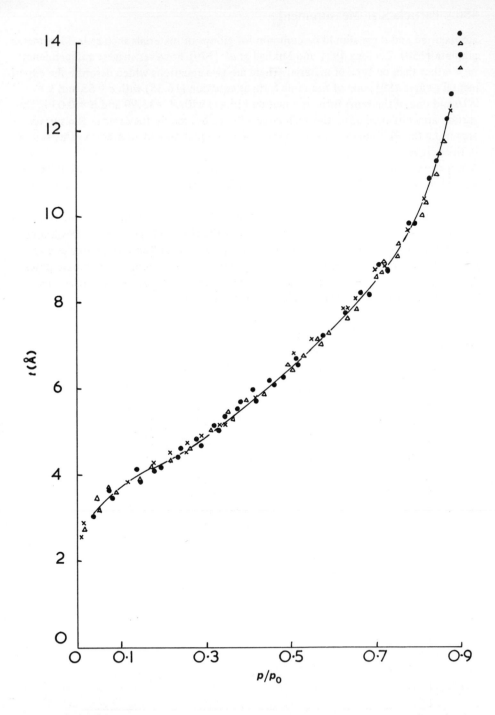

Fig. 16.3 The common *t*-curve of multi-molecular nitrogen adsorption on oxidic and hydroxidic adsorbents [84, 87].

are required and these should be common for groups of materials such as halides, metals, graphite [150]. Lecloux [92] and Mikhail *et al.* [179] however suggest a dependency on c rather than on type of material. There are two equations which describe the experimental t-curve [89], one of the same form as equation (16.35) with $c = 53$ and $k = 0.76$ and one of the same form as equation (16.41) with $A = 13.99$ and $B = 0.034$. The surface area obtained using this technique will not be exactly the same as the surface area using the BET equation, since c varies in this equation and an average value is used in the t-curve.

Experimentally, values of V measured as a function of relative pressure are transformed to functions of t. This gives a straight line passing through the origin and the specific surface may be obtained from the slope.

The t-method is based on the BET conception, but yields additional information. For non-porous solids a graph of V against t yields a straight line (figure 16.4). Deviations from the straight line are interpreted as (a) decrease in accessible surface area due to blocking of micropores, and (b) onset of capillary condensation in intermediate pores. A second linear portion gives the surface area of what is left, that is, the surface area of the wider capillaries or of the outside area of the granules.

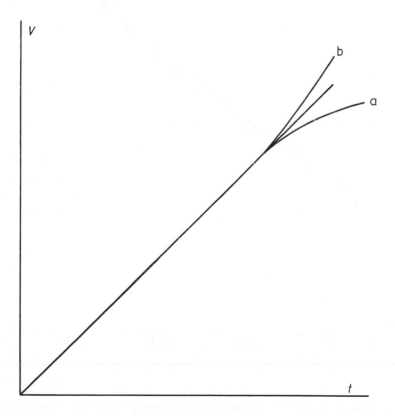

Fig. 16.4 The V–t curve.

Sing [137] suggests that it is necessary to have standard t-curves of non-porous reference materials of known structure in order to obtain an understanding of the true nature of the adsorption process.

The application of the t-curve method has been challenged by Pierce [138] and Lamond and Marsh [139]. Marsh and Rand [140] state that once a molecule is adsorbed into a micropore it fills spontaneously thus leading to the unrealistically high surface area values found with some activated carbons. Kadlec [166] states that the t-method yields incorrect values for the specific surface of mesopores at pressures below the closure pressure of the hysteresis loop since at these pressures microfilling has not been completed. These limitations were overcome with the t/F method where:

$$v_c/F = v_{micro} + (S_{me} + S_{ma})t/F \qquad (16.59)$$

where v_c is the condensed volume adsorbed, F is the degree of volume filling of the micropores, v_{micro} is the volume of micropores and S_{me} and S_{ma} are the surface areas of meso- and macropores. He also proposes a method of evaluating F. This technique was later applied by Dubinin et al. [180].

16.2.14 Kiselev's equation

A criterion as to the correctness of the BET equation is its agreement with other models, such as those derived from the Kelvin equation. This has been applied to a porous adsorbent as follows:

$$RT\ln x = \frac{-2\gamma V_m}{r_k} \qquad (16.60)$$

which is the Kelvin equation for pressure lowering over a concave meniscus. If a_H is the number of moles adsorbed at the beginning of the hysteresis loop when the relative pressure is x_H, a small increase of pressure will lead to Δa moles of adsorbate being adsorbed, where:

$$\Delta a = \frac{\Delta V}{V_m} \qquad (16.61)$$

where ΔV is the liquid volume of adsorbate adsorbed. This volume will fill the core of pores of surface area ΔS and core radius r_c. For cylindrical pores:

$$r_c \Delta S = 2\Delta V \qquad (16.62)$$

hence

$$\Delta S = -\frac{RT}{\gamma} \ln x \, \Delta a$$

This may be written:

$$S = \frac{RT}{\gamma} \int_{a_H}^{a_S} \ln x \, da \qquad (16.63)$$

where a_S is the number of moles adsorbed at saturation pressure.

This is known as Kiselev's equation [101].

Brunauer [98] uses the following form of equation (16.62), which he claims reduces the equation to a modelless equation as opposed to the adoption of cylindrical pores:

$$r_H \Delta S = \Delta V \tag{16.64}$$

where r_H is the hydraulic radius.

For the adsorption branch of the isotherm, a cylindrical pore model yields: $r_k = 2r_c = r_H$. For the desorption branch $r_c = r_K = 2r_H$.

There is, therefore, no difference between Brunauer's modelless method (see section 18.7.1) and the cylindrical pore method without correction for residual film (t correction). They both yield the surface and volume of the cores of the pores which will approach the BET surface area for materials having only large-diameter pores. Brunauer *et al.* [148] argue however that the graphical integration of equation (16.60) is more accurate than the previous tabular methods.

In a more rigorous derivation Kadlec [166] replaces γ in equation (16.63) by $\gamma_{LV} - \gamma_{SV} + \Delta\pi$ where LV and SV refer to the liquid–vapour and solid–vapour interfaces and $\Delta\pi$ is the difference between the spreading pressure at P_0 and at the given pressure. He further states that the beginning of the hysteresis is not connected with the lower size limit for mesopores [167].

16.3 Experimental techniques – factors affecting adsorption

16.3.1 Degassing

A most important preliminary to the accurate measurement of an adsorption isotherm is the preparation of the adsorbent surface. In their usual state all solid surfaces are covered with a physically adsorbed film which must be removed or 'degassed' before any quantitative measurements are made. As the binding energy in physical adsorption is due to weak van der Waals' forces, this film should be readily removed if the solid is maintained at a high temperature while under vacuum.

The degree of degassing attained is dependent therefore on three variables: pressure, temperature and time. In test and control work, the degassing conditions may be chosen empirically and maintained identical in all estimations since only reproducibility is required. For more accurate measurements, conditions have to be chosen more carefully.

16.3.2 Pressure

Although it is advisable to degas at as low a pressure as possible, due to considerations of time and equipment, the degassing pressure is kept as high as is consistent with accurate results. The pressures usually recommended are easily attainable with a diffusion pump. Emmett [30], for example, recommends 10^{-5} mm Hg, while Joy [31] recommends 10^{-4} mm Hg, since under this condition the rate of degassing is controlled largely by diffusion from the interior of the particle.

For routine analysis, Bugge and Kerlogue [32] found that a vacuum of 10^{-2} to 10^{-3} mm Hg is sufficient and the difference in surface areas so obtained was smaller by less than 3% than those obtained at pressures of 10^{-5} mm.

16.3.3 Temperature and time

Recommended temperatures and times for degassing vary considerably in the literature, and it is difficult to establish any single degassing condition acceptable for all solids. However, Orr and Dallavalle [2] give an empirical relationship which they suggest is acceptable as a safe limit for ordinary degassing at pressure of less than 5×10^{-6} mm Hg.

$$\theta = 14.4 \times 10^4 \, t^{-1.77} \tag{16.65}$$

where θ is in hours and t in $°C$ (applicable between 100 and 400°C). This can only be taken as a general safe limit as many others have found that the necessary time is much less [2].

McBain [33] has recommended an adsorption–desorption cycle to reduce the time of degassing. He flushes the surface with adsorbate at the temperature of the forthcoming measurements, followed by heating in vacuum. Holmes et al. [174] determined the surface area of zirconium oxide using argon, nitrogen and water vapour as adsorbates. They found that the surface properties depended upon the amount of irreversibly adsorbed water vapour which was far in excess of a monolayer. Degassing at 500°C resulted in a 20% decrease in area.

16.3.4 Adsorbate

Any gas may be used for surface area determination at a relative pressure determined by the equation to be used. The standard technique is the adsorption of nitrogen at liquid nitrogen temperature, evaluation being by the BET equation used in the approximate relative pressure range $0.05 < x < 0.35$. For powders of reasonably high surface areas (greater than $10 \, \text{m}^2 \, \text{g}^{-1}$) the proportion of the admitted gas that is adsorbed is high and the consequent changes in pressure, using the volumetric method, can be measured accurately with a mercury manometer. However, for powders of low surface areas, the proportion is low: most of the gas admitted to the sample bulb remains unadsorbed in the 'dead-space' and this leads to considerable error in the determined surface area. In such cases Beebe et al. [36] recommended the use of krypton at liquid nitrogen temperature which, due to its low saturation vapour pressure, reduces the amount of unadsorbed gas in the gas phase. It is arguable as to the correct value to use for the area occupied by a krypton molecule but Beebe's original value of $18.5 \, \text{Å}^2$ is preferred by most investigators [154–156]. There is some disagreement over the correct saturation vapour pressure of krypton, P_0. The use of the solid saturation vapour at 77.5 K, 1.76 torr usually results in the production of markedly curved plots [157]. Later investigators tend to use the extrapolated vapour pressure of krypton, 2.59 torr. Ethylene at liquid oxygen temperature has been used for low surface area determination [34, 35].

Young and Crowell [13] have listed the molecular areas of many adsorbates. In practice, to get consistency, the molecular areas are corrected on the basis of the area of

nitrogen molecules σ = 16.2 Å^2 for nitrogen at liquid nitrogen temperature. In some cases the area occupied by the molecule depends upon the nature of the surface and calibration for that particular solid may be necessary [24].

For example, with argon, the area covered by the molecule is governed by adsorbent–adsorbate interaction, hence the area varies from system to system. Nitrogen adsorption is governed by adsorbate–adsorbate interaction, particularly near the completion of a monolayer. This lateral interaction pulls the molecules together to form a close-packed liquid-like monolayer [149].

An exception [138] is with graphitized carbon on which the nitrogen molecule occupies 20 Å^2, i.e. one nitrogen molecule to three carbon hexagons. The lateral interaction, in this case, is not strong enough to pull the nitrogen molecules together. This can only happen on high energy surfaces.

For the adsorption of argon on boron carbide, Knowles and Moffat [162] found that the application of the BET theory gave more consistent results using liquid argon pressure rather than solid-vapour pressure. The specificity in the adsorption of nitrogen and water on hydroxylated and dehydroxylated silicas was investigated by Baker and Sing [175]. Non-, meso- and micro-porous silicas were analysed by BET, FHH and α_s methods.

16.3.5 Interlaboratory tests

The goal of analysts is high precision within a laboratory and high reproducibility between laboratories. Desbiens and Zwicker [158] carried out interlaboratory tests with alumina and found that the degassing temperature and time were critical. AFNOR [159] also carried out interlaboratory tests with alumina and found wide disparity between laboratories. It cannot be too strongly stressed that commercial equipment should be calibrated against standard equipment at regular intervals. An alternative is to calibrate with standard powders [160].

16.4 Experimental techniques – volumetric methods

16.4.1 Principle

A great variety of volumetric apparatus has been described in the literature, the earlier ones of which have been reviewed by Joy [31]. In all the volumetric methods, the principle underlying the determination is the same. The pressure, volume and the temperature of a quantity of adsorbate are measured and the amount of gas present is calculated. The material is then brought into contact with the adsorbate, and when the constant pressure, volume, temperature conditions show the system to have attained equilibrium, the amount of gas is again calculated. The difference between the amount of gas present initially and finally represents the adsorbate 'lost' from the gas phase to the adsorbed phase. The accurate determination of the amount of gas unadsorbed at equilibrium depends upon a precise knowledge of the 'dead-space' or the space surrounding the adsorbent particles. The dead-space volume is usually determined by expansion measurements using helium, whose adsorption can be assumed to be negligible. Estimation of

the quantity of unadsorbed gas is often complicated by the fact that part of the dead-space is at room temperature and part at the temperature of the adsorbent.

Since the amount adsorbed represents the difference between the amount admitted to the dead-space and the amount remaining in the dead-space at equilibrium, it can only be evaluated with confidence if these two quantities are of unlike magnitude. To achieve this, the apparatus is so designed as to minimize the dead-space volume. In practice, it is convenient to fix the volume and temperature and measure the changes in pressure.

Regardless of the particular design, the basic apparatus must provide means for removing gases and vapours which all materials pick up when exposed to the atmosphere. The apparatus must also provide means for permitting readsorption of known quantities of the gas on to the material. It should also have evacuating systems, gauges to measure the vacuum, a gas storage part, and the analytical part.

16.4.2 Volumetric apparatus for high surface areas

Conventional types of nitrogen adsorption apparatus invariably follow the assembly originally described by Emmett [37], shown in figure 16.5. Adsorbate gas is taken into the burette and its pressure measured on the manometer. The stopcock between the sample and the burette is then opened and the new pressure, after allowing time for the equilibrium to be established, is read on the manometer. The volume of the gas admitted to the sample bulb is proportional to the difference in the pressures before and after opening the stopcock. This later pressure is also the equilibrium adsorption pressure. The volume adsorbed is equal to the volume admitted less the volume of gas required to fill the dead-space in the sample bulb and the burette connections. To obtain more adsorption points, the mercury level is raised to the next volume mark and a new pressure established. Helium is used to calibrate the dead-space. In all adsorption calculations a correction for the non-ideal behaviour of nitrogen at liquid nitrogen temperature is

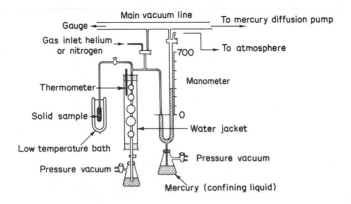

Fig. 16.5 Emmett's apparatus for surface-area determination by gas adsorption.

included. In a method like this, the total volume admitted is found by summing the separate doses.

The main disadvantage of the original design is that the sample tube is not connected directly to the vacuum line and hence any powder flying from the tube is likely to contaminate the whole apparatus. Joyner [38] has described a full account of the basic principles and a step by step description of the construction and assembly of a complete apparatus. Elaborations of the same apparatus have also been described [39–41]. Vance and Pattison [42] have also described a similar apparatus in detail.

A number of refinements have been suggested either to increase the accuracy or reduce the tedium of measurements. For example, Vance and Pattison [42] used a magic-eye electrical zero-point device for the manometer. Harkins and Jura [43] used a narrow-bore mercury cut-off to serve as a null-point instrument, the absolute pressure being measured on a wide-bore manometer. Several authors have shown [44, 45] how the functions of a manometer and burette can be combined in a single device. Cathetometers have also been used to improve the accuracy [46].

It has been suggested that the adsorption of mercury vapour can affect adsorption and to overcome this problem Dollimore et al. [142] devised a doser unit incorporating a pressure transducer to replace the mercury manometer.

Bugge and Kerlogue [32] simplified the apparatus by using only one bulb instead of several, but with a loss of versatility. They also gave a simplified method of calculation to eliminate the dead-space determination. It appears that this procedure is satisfactory only when the product $V_m c$ is large enough to cause the BET plot to pass through the origin [31]. Several authors have tried to use oxygen or nitrogen thermometers for the accurate measurements of the saturation vapour pressure of nitrogen [32, 42]. Loebenstein and Deitz [46] developed an apparatus not requiring a vacuum system by adsorbing nitrogen from a mixture of nitrogen and helium. They degassed the sample in a current of helium.

Lippens, Linsen and de Boer [131] state that none of the above apparatus is suitable for accurate determination of the complete adsorption–desorption cycle for pore size distribution evaluation. They describe an apparatus which fulfils the basic criteria.

(a) Rapid removal of heats of adsorption and supply of heats of desorption to give rapid equilibrium;
(b) clear establishment of equilibrium;
(c) in order to check for leakage, the total amount of nitrogen adsorbed must be recovered on desorption;
(d) changes in temperature of the liquid nitrogen bath must be continuously measured;
(e) it must be possible to arrest the measurements at certain points in order to eliminate the need for overnight supervision.

Recently [146] an automatic apparatus for surface area measurement and pore size analysis of fine powders from nitrogen adsorption isotherms has been described. The Isothermegraphe is a volumetric apparatus with a calibrated tube which draws complete adsorption–desorption isotherms using a piston of mercury which modifies the pressure slowly at a programmed speed.

16.4.3 Volumetric apparatus for low surface area

Surface areas down to 1 m^2 g^{-1} can be determined using nitrogen adsorption provided great care is taken. A semi-micro unit has been described for surface area evaluation down to 2 m^2 g^{-1} [141] and a capillary differential manometer to keep the dead-space low [88]. With decreasing surface area, the error in the dead-space factor makes nitrogen unsuitable. Since the amount of gas in the dead-space is proportional to the absolute pressure it is preferable to use gases with low saturation vapour pressures. Krypton gas with a saturated vapour pressure of 1.76 mmHg at − 195°C and ethylene gas with 0.1 mmHg saturated vapour pressure at liquid air temperatures lend themselves readily for low-pressure measurement.

Wooten and Brown [34] used this low-pressure method to measure the surface areas of oxide-coated cathodes, about 100 cm^2, by adsorption of ethylene and butane at −183° C and −116° C respectively. Because of the very low pressures involved in the technique, no leaks can be tolerated in the system. The apparatus was, therefore, made entirely of glass and used mercury cut-offs instead of stopcocks. The sample chamber was welded on to the system to eliminate any possibility of leaks due to a ground-glass joint. A dry-ice trap between the sample and the mercury cut-offs served to prevent mercury vapour from reaching the sample. Equilibrium pressure measurements were made with a highly sensitive McLeod gauge. Lister and McDonald [35] have described in detail the construction and calibration of low-temperature ethylene adsorption apparatus.

In measurements of such low pressures, two obvious risks must be considered, namely, the desorption of water and other vapours from the glass walls of the apparatus and thermal transpiration [13]. By heating the entire system for a short while, or by permanently keeping the system under vacuum, most of the adsorbed vapours from the glass walls should be removed. Otherwise, the slowly desorbing vapours will increase the pressure in the system during adsorption measurements leading to erroneous results.

When low-pressure measurements are made on a gauge held at a different temperature from that part of the apparatus where the adsorption takes place, correction for thermal molecular flow should be considered. To obtain accurate results, Lister and MacDonald [35] prepared and used correction data.

In most low-pressure measurements, the correction for unadsorbed gas is very small, even negligible, so that no effort need be made to minimize the volume of the dead-space.

Krypton at liquid air temperatures has a vapour pressure of about 2 mmHg, intermediate between those of ethane and nitrogen. Krypton is thus suited for the measurements of a much smaller surface area than is possible with nitrogen. In addition, the pressures encountered in krypton adsorption at the temperature of liquid nitrogen are low enough for the deviations from perfect gas relations to be neglected.

The adsorption equipments are similar to those already described, the only difference being the pressure range of the gauge. Several types of apparatus have been described in the literature [47–49, 171].

Krypton has the additional advantage that the pressure within the range related to krypton adsorption can be detected precisely with thermistors, hence avoiding the use of McLeod gauges. Rosenberg [50], Dollimore [51] and Leipziger and Altamari [52]

have used thermistor gauges successfully. Leipziger [52] has discussed the design, construction and precision of thermistors.

Aylmore and Jepson [80] used a novel method of krypton adsorption. They used labelled krypton (^{85}Kr) as adsorbate and from the measurement of the activity they calculated pressures.

Among the various gases used for the adsorption measurements on low surface areas, krypton at liquid nitrogen temperature appears to be the popular choice. This field has been reviewed by Choudhary [172].

16.5 Experimental techniques – gravimetric methods

16.5.1 Principle

The gravimetric techniques have the great advantage over the volumetric methods in that the volume of the adsorption system is quite immaterial and the amount of gas adsorbed is observed directly by measuring the increase in the weight of the solid sample upon exposure to a gas or vapour. The tedious volume calibration and dead-space determinations are thus eliminated.

The main disadvantages of the method are:

(a) The apparatus is much less robust and correspondingly more difficult to construct and maintain than volumetric apparatus.
(b) The apparatus has to be calibrated by placing known weights in the adsorbent-pan, and the method is hence subject to the errors always attached to determination which are dependent on the constancy of calibrations of easily fatigued and strained mechanical systems.
(c) Buoyancy corrections have to be made.

16.5.2 Single-spring balances

McBain and Bakr [33] introduced a sorption balance, the essential features of which are a quartz helical spring supporting a small gold or platinum bucket in which the sample is placed. The spring is calibrated by adding small known weights to the bucket and measuring the increase in length of the spring with a reading microscope. These calibrations must be done over the entire range of temperatures at which adsorption measurements are made. The liquid adsorbate, free from dissolved gases, is sealed in a small glass bulb and placed with a magnetic hammer in a glass or quartz envelope ('balance case'). The balance is heated in vacuum for outgassing. Finally the case is sealed off, the lower end cooled and the bulb broken. Adsorption measurements are carried out with the lower end in one thermostat bath to control the temperature of the liquid and the upper end in another to control that of the solid. The equilibrium pressure can be calculated from the temperature of the liquid, provided vapour pressure data are available. The amount adsorbed is proportional to the spring extension and the correction for buoyancy is significant at higher pressures.

This type of balance is restricted in use to condensable adsorbates and is especially useful at higher pressures. Morris and Maass [53], Dunn and Pomeroy [54] and McBain and Sessions [55, 56] have used a similar apparatus.

Several others have used similar apparatus with improvements and modifications to suit their applications. Boyd and Livingstone [57] used mercury cut-offs in the vapour-handling and compressing system. The pressure was controlled by compressing the gas in the dosing bulb, and it was read on a mercury manometer or a McLeod gauge depending upon the range. Seborg, Simmons and Baird [58] dried the sample in a current of dry air, and obtained the adsorption points subsequently by passing the air through saturators filled with solutions of known vapour pressure. Dubinin and Timofeev [59] used a magnetically operated greaseless doser for the precise admission of adsorbate increments. Automatic recording techniques have also been attempted [60].

16.5.3 Multiple-spring balances

Gravimetric methods have the additional advantage that several determinations can be carried out simultaneously by connecting several balance cases to the same gas or vapour manifold and observing the individual spring extensions.

Seborg and Stamm [61] connected five or six simple spring units in series. Pidgeon and Maass [62], Mulligan et al. [63] and Stamm and Woodruff [64] have all described similar multiple-spring balances. Mulligan et al. connected as many as fifteen springs to the same apparatus.

16.5.4 Beam balances

Beam-type vacuum balances have greater sensitivity than the helical-spring balances and also the troublesome buoyancy correction at higher pressures is eliminated, at least partially if not completely.

Beam balances can be of either high sensitivity at very low total loads or of medium sensitivity at large total loads, which is in contrast to the normal short-spring balances which have a medium sensitivity at low total loads.

The majority of the high-sensitivity low-load balances are based on those originally designed by Barrett, Birnie and Cohen [65] and by Gulbransen [66]. Barrett et al. used a glass beam 40 cm long supported on a tungsten torsion wire and enclosed the whole assembly in a tubular glass casing connected to the vapour and vacuum manifolds. Calibration was effected by moving a small soft-iron rider along the beam by means of a magnet outside the case. Gulbransen's balance was constructed from glass rod, quartz fibres and metal wires on the same principles as an ordinary chemical balance.

Rhodin's microbalance [67-69] is essentially a modification of these, in which some stability is sacrificed for increased sensitivity by the use of thinner and lighter wires. This balance has been adopted by Bowers and Long [70] for adsorption at liquid helium temperatures. Rhodin's balance was made as symmetrical as possible, in order to eliminate buoyancy corrections and to minimize thermal eddy currents. The adsorbent and counter weights were matched to within 10^{-5} g and immersed to the same depth in

identical thermostatic baths and the outgassing was done at 400°C in a vacuum of 10^{-7} mm Hg. With this balance, it was possible to observe a vertical displacement of 10 mm to better than 0.01 mm and with loads up to 1 g, it was possible to observe weight changes of 10^{-7} g $\pm$ 20% in a reproducible manner.

Beam balances have also been operated as null-point instruments. The beam is acted upon by a solenoid outside the balance housing, the current through the solenoid being adjusted to restore the beam to its horizontal position. One such balance by Gregg [71] uses two concentric solenoids, the inner one suspended from the beam and the outer one fixed to the envelope. The original balance had a sensitivity of 0.3 mg, the range of load being as high as 10 to 20 mg. In an automatic version of this instrument described by Gregg and Wintle [72], a photoelectrically operated relay adjusts a potentiometer slide-wire contact which is connected to the solenoid on the balance.

Although so many different types of gravimetric apparatus have been reported, they have not become popular due to their delicate nature and the difficulty of accurately compensating for buoyancy effects.

16.6 Continuous-flow gas chromatographic methods

In recent years, a continuous-flow method, based on the gas chromatographic technique, has been introduced for the measurement of surface area of fine powders by gas adsorption. A scheme was first proposed by Loebenstein and Deitz [46] for reducing the vacuum requirements by using a flowing mixture of the adsorbate and an inert gas, such as helium, in order to obtain the low adsorbate gas pressure required.

The method, a modification of gas-adsorption chromatography in which the column packing is the sample itself and the mobile gas phase is a mixture of a suitable adsorbate and an inert gas, was developed by Nelsen and Eggertsen [73]. They used nitrogen as the adsorbate and helium as the carrier gas in the following manner.

A known mixture of nitrogen and helium is passed through the sample and then through a thermal conductivity cell connected to a recording potentiometer. When the sample is cooled in liquid nitrogen, the sample adsorbs the nitrogen from the mobile phase; this is indicated by a peak on the recorder chart, and after equilibrium is established, the recorder-pen resumes its original position. Removing the coolant gives a desorption peak equal in area and in the opposite direction to the adsorption peak and either peak may be used to measure the nitrogen adsorbed.

Calibration for such a system may be either absolute (by injecting a known amount of nitrogen into the mobile phase at the point normally occupied by the sample and obtaining a factor for the amount of nitrogen per unit peak area on the resulting recorded curve), or by comparison with a sample of known surface area.

A schematic diagram of their apparatus is shown in figure 16.6. Nitrogen flow control was achieved by two capillary tubes in parallel, 0.25 mm inner diameter and 150 and 300 mm long. The capillaries were used independently or together to give three nitrogen flow rates in the range of 5–20 ml min^{-1} with a pressure head of 2 lb in^{-2} g. The helium flow was controlled by needle valves. The flow measurements were made by rotameter and soap-film meter.

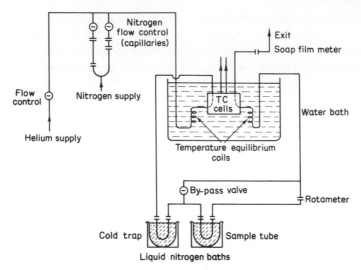

Fig. 16.6 Schematic diagram of Nelsen and Eggertsen's apparatus.

The mobile phase was first passed through the reference arm of the thermal conductivity cell, on to the sample and then, again, to the thermal conductivity cell, which was immersed in a temperature-controlled bath.

Nelsen and Eggertsen measured adsorption at three flow rates, i.e. at three partial pressures. The sample was outgassed at the desired temperature (up to 500° C) while being purged with He at 20 ml min^{-1} for 1 h. Nitrogen relative pressures in the range of 0.05–0.3 and a total nitrogen and helium flow of 50 ml min^{-1} were used. The desorption peaks were used for measurement because they were relatively free from tailing effects.

Calculation is essentially the same as for the pressure–volume method but is much simpler since no dead-space corrections are needed. The authors assumed complete linearity of the thermal conductivity cell over the concentration range employed. They analysed samples ranging in surface area from 3 to 450 m^2 g^{-1} and the results obtained on continuous-flow and pressure–volume methods were in good agreement.

The main advantages of the continuous-flow method over the conventional BET method are:

(a) Elimination of fragile and complicated glassware.
(b) Elimination of a high-vacuum system.
(c) Permanent records obtained automatically.
(d) Speed and simplicity.
(e) Elimination of dead-space correction.

Ellis, Forrest and Howe [74] made some modification and improvements to the original technique for their specific applications. The schematic diagram of their apparatus is shown in figure 16.7. All flow controls were done by needle valves and all flow measurements were made by rotameters. They used a helium flow rate of 50 ml min^{-1} and a nitrogen flow rate of 3, 5 or 10 ml min^{-1}.

By taking more care (when chilling the sample to avoid shock effects in the gas stream), they could extend the method to surface areas in the region of 0.01 m^2 g^{-1}. They obtained good linearity in the BET plot, even with surface areas as low as 0.02 m^2 g^{-1}. They analysed samples with surface areas of 14.2 to 0.005 m^2 g^{-1}.

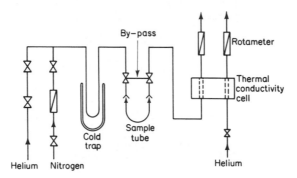

Fig. 16.7 Schematic diagram of Ellis, Forrest and Howe's apparatus.

Below 0.01 m^2 g^{-1} conventional adsorption methods with nitrogen are not practical and hence no results using the PVT methods were given by them. Above that range agreement between the continuous flow and PVT methods was good, but in their results the surface-area value by the new method was slightly higher than that given by the PVT method.

Ellis *et al.* [74] also developed a shortened method using only one flow rate, i.e. a single-point method. Since, usually, the BET intercept is very small, the intercept term can be ignored and $S_w \propto V(P_0 - P)/P$. For constant P values, i.e. N$_2$/He flow-rate ratio constant in the flow stream, $S_w \propto V$. Since $V \propto$ area of recorder chart, a plot of S_w against area for fixed sample weight and fixed N$_2$/He ratio should be linear. They analysed a number of samples (300 mg each) and determined the adsorption peak areas for 10 ml min^{-1} flow of nitrogen and 50 ml min^{-1} helium and obtained a linear graph from which the surface areas of subsequent samples were obtained. Results by this method were also comparable to those obtained by the normal BET method based on PVT measurements. This method was especially good compared with multipoint methods for routine measurements, as one sample can be analysed per hour.

Atkins [75] developed a precision instrument for use in the carbon black industry. For precision measurements he stated that it is necessary to consider the effect of changes in ambient temperature, barometric pressure, liquid nitrogen temperature and nitrogen concentration in the gas mixture. Correction for the non-linearity of the katharometer was also necessary, and was achieved either by adjusting the sample size so that the desorption and calibration peaks were of the same height, or by using a correction factor; this was determined as a function of relative peak areas, i.e. desorption peak area divided by the calibration peak area. Atkins used heat exchanger coils in the detector circuit in addition to the temperature control of the detector. Two different premixed gases were connected to the apparatus so that either of them could be used or both

together. The apparatus also had special calibration valves and calibration loops to select the volume of gas nearest to the desorption volume for calibration so that similarly shaped calibration peaks and heights of similar magnitude were obtained. He has also tabulated the variables like percentage nitrogen in the mixture, calibration loop temperature, barometric and saturated vapour pressure of the nitrogen and the corresponding percentage changes in calculated surface area due to their variation and has discussed the errors due to the non-linearity of the katharometer.

Haley [76] extended the continuous-flow measurement to include the size distribution in pores in the 10 to 300 Å radius range. He used 10% nitrogen in helium (as mobile phase) at various pressures up to 150 lb in^{-2}, causing the nitrogen partial pressure to reach its liquefaction point, consequently varying the nitrogen relative pressure in the sample tube in the range 0.16 to 1. Nitrogen adsorbed or desorbed, by increasing or decreasing the pressure in the tube, was measured continuously. He also measured surface areas and obtained a variation of approximately ± 2.5% in the range of 40 to 1250 m^2 g^{-1}.

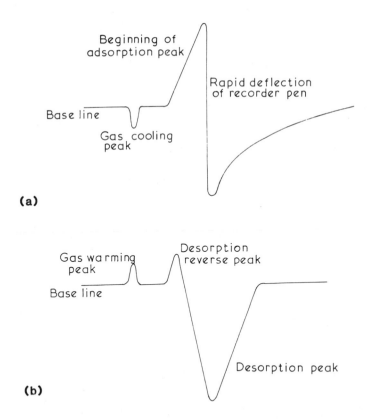

Fig. 16.8 Shape of peaks obtained with Nelsen and Eggertsen-type equipment with powder of low surface area [82]. (a) Adsorption reverse peak and (b) desorption reverse peak.

Since helium is extremely expensive, it may be replaced by other gases which are not adsorbed under the experimental conditions, e.g. hydrogen. Whitehead [77] used hydrogen as a carrier gas at a flow rate of 50 ml min^{-1} and found it was very satisfactory.

Several problems arise when the Nelsen and Eggertsen-type apparatus is used for measuring surface areas less than 500 cm^2 g^{-1}, the adsorption and desorption peaks having the shapes shown in figure 16.8.

During adsorption a peak is produced when the sample tube is immersed in liquid nitrogen. This immersion causes a contraction of the gas inside the sample tube and a reduction in the gas flow through the katharometer; the thermistor on the measurement side warms up causing a change in its resistance and a peak on the recorder chart. The adsorption peak results from a cooling of the thermistor due to removal of nitrogen from the stream which causes an increase in the thermal conductivity of the gas mixture.

A reverse peak occurs midway through the adsorption. During desorption a gas-warming peak occurs and, immediately prior to desorption, a desorption-reverse peak.

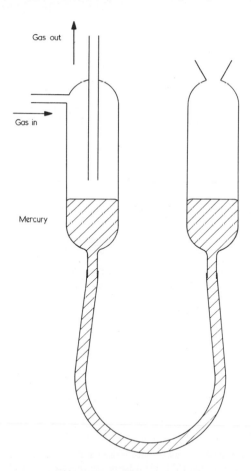

Fig. 16.9 The expansion chamber.

Since the desorption peak is used for measurement purposes, effort has been mainly directed at finding an explanation for the desorption reverse peak and correcting for it.

Lovell [81] considered that the reverse peak on desorption was due to transverse thermal diffusion. When the sample tube is removed from the coolant, pre-cooling of the gas stream in the inlet section quickly ceases and gas enters the sample-catching section at very nearly room temperature. Since the sample is still cold, partial separation of the gases takes place due to transverse thermal diffusion, nitrogen moving to the walls of the container and helium to the centre. Gas flow is more rapid at the centre and helium-rich gas is carried to the katharometer, giving rise to a thermistor cooling peak. Immediately afterwards, when the temperature of the sample tube has risen considerably, the separated nitrogen diffuses into the gas stream and produces a peak in the opposite direction and necessarily of the same area. It can be seen that, when a sample is present in the container, the area of the desorption peak will be increased by the area of the reverse peak. Thus, the true area of the desorption trace may be obtained by subtracting the area of the reverse peak from that of the observed peak.

Lovell's experimental refinement was to allow the desorbed gases to expand into a vessel whose volume could be adjusted by altering the amount of mercury in it (figure 16.9).

As the gas expands, the mercury level is forced down and the pressure is equalized by lowering the mercury level in a second container. When the desorption and expansion process is complete and enough time has elapsed for mixing to take place, the adsorbed gas is swept through the thermal-conductivity cell and the area of the desorption peak measured with no interference from a reverse peak. Tucker [147] removed the anomalous peaks by using an interrupted flow technique. Bhat and Krishnamoorthy [176] describe a simple continuous gas flow adsorption apparatus and Yen and Chang [177] introduced a modified system.

16.6.1 Commercially available continuous-flow apparatus

Perkin-Elmer Ltd manufacture a continuous-flow apparatus called the Perkin-Elmer shell sorptometer, model 212C. The manufacturers claim that surface areas can be determined from approximately 0.1 to 1000 m^2 g^{-1}. A typical three-point surface-area determination can be carried out in 20 to 30 min using pre-mixed gases. Degassing is carried out by heating samples with a gas purge [96].

Atkins [75], using the above apparatus, obtained a relative standard deviation, varying from 1.76% to 2.99% according to sample material, with ten single-point determinations, each with a new sample. With his own equipment, the comparable deviation varied from 0.25% to 1.35%.

Similar instruments for one-point or multipoint evaluations are available from Quantachrome. Calibration is accomplished with precision gas syringes to inject known amounts of adsorbate into the gas flow. These instruments are claimed to be particularly useful for low surface-area powders.

Although elaborate precautions and the use of more complicated apparatus improved the precision of the technique, the accuracy of the commercial equipment should suffice for most normal applications.

16.7 Standard volumetric gas-adsorption apparatus

Nitrogen gas-adsorption apparatus is fully described in BS 4359 (1969): Part 1. The standard apparatus is rather time-consuming in operation and several commercial versions are available. With these equipments, operator involvement and operating time are reduced, usually with a reduction in accuracy and versatility, and it is recommended that these should always be calibrated against the standard equipment.

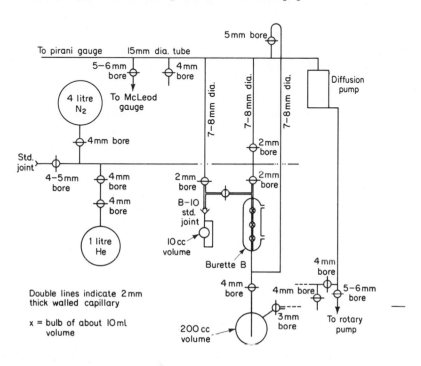

Fig. 16.10 Gas-adsorption apparatus, schematic diagram.

In the apparatus illustrated (figure 16.10), the main vacuum line consists of a 15-mm bore glass tube to which are attached the adsorption unit, a McLeod gauge and a Pirani gauge. A 4-l flask containing nitrogen and a 1-l flask containing helium are connected to a secondary line which is joined to a gas burette, a sample tube and a mercury manometer. The gas burette consists of three carefully calibrated bulbs enclosed in a water jacket. The volume in the burette can be adjusted by raising the level of the mercury to any one of three calibration marks and the pressure may be read on the manometer. The sample tube is connected to the gas burette through a ground-glass joint. The glass tube connection between the water-jacketed burette and the liquid-nitrogen thermostated sample tube is made of a short length of 2 mm capillary to keep the volume of dead-space not thermostated to a minimum. The sample tube, of about 10-ml volume, is specially designed to prevent loss of powder by 'spitting' during degassing (figure 16.11).

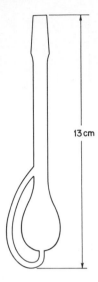

Fig. 16.11 The sample tube for the static BET method.

A third vacuum line controls the vacuum in the mercury reservoirs of the gas burettes and the McLeod gauge which may also be opened to the atmosphere to raise or lower the level of the mercury. The entire system is evacuated by a mercury diffusion pump backed by a rotary pump capable of an ultimate vacuum of 10^{-6} mmHg. A small electrical furnace is used to heat the sample tube while the sample is degassed.

16.7.1 Worked example

The predetermined cumulative volumes of the gas burettes by filling with mercury and weighing are:

$$V_1 = 9.19 \text{ ml}; \quad V_2 = 17.56 \text{ ml}; \quad V_3 = 26.85 \text{ ml}.$$

The sample is weighed in the sample tube and fitted to the apparatus, and the system is then evacuated. The heating furnace is then placed round the sample tube and degassing proceeds under vacuum. The tube is then immersed in liquid nitrogen and helium is drawn into the burette, and the pressure noted. The helium is then allowed to expand into the sample tube and the pressure again noted. The volume of gas in the burette at STP is calculated as follows, using the experimental data presented in Table 16.2:

$$V_0 = \frac{P}{P_0} \frac{T_0}{T} V = \frac{163.5}{760} \frac{273}{285} 9.19 = 1.892 \text{ ml}$$

After expansion:

$$V_0 = \frac{46.0}{760} \frac{273}{285} 9.19 = 0.533 \text{ ml}$$

Table 16.2 Observed and calculated data for a sample.

Sample: Cement, Hoogovern cement (cemy) Klasse – B; *Weight*: 10.2028 g; *Degassing*: 300° C, 3 h; *Temperature of the water jacket*: 12° C; *Burette used*: B.

Sample	Gas	Burette volume	Pressure (mm)	Volume of the gas in the burette (ml)	Volume of the gas in the sample tube (ml)	Volume of the gas in the free space (ml)	Volume of the total gas (ml)	V_{ad}	$V_{ad/s}$
Off	He	1	163.5	1.892		1.359	Free space factor		
On	He	1	46.0	0.533		1.359	$\dfrac{1.359}{46} = 0.02956 \dfrac{ml}{mm}$		
Off	N₂	3	142.5	4.830			4.830		
On	N₂	1	27.0	0.312	4.518				
Off	N₂	3	159.0	5.380			9.898		
On	N₂	3	86.5	2.930	6.968	2.575		4.393	0.431
On	N₂	2	103.5	2.290	7.608	3.090		4.518	0.443
On	N₂	1	125.0	1.448	8.450	3.722		4.728	0.464
Off	N₂	3	174.5	5.920			14.370		
On	N₂	3	147.0	4.980	9.390	4.390		5.000	0.490
On	N₂	1	211.5	2.445	11.925	6.350		5.575	0.546

Thus, the sample tube contains (1.892–0.533) ml of gas under a pressure of 46.0 mm Hg.

$$\text{The free-space factor} = \frac{1.892 - 0.533}{46.0} \text{ ml mm}^{-1} \text{ pressure}$$

$$= 0.029\ 56 \text{ ml mm}^{-1}$$

After the free-space factor determination the helium is pumped out of the system and nitrogen admitted into the burette. Time is allowed for the gas to reach equilibrium temperature and the pressure is then noted. The gas is then admitted to the sample tube and the pressure again noted. All volumes are reduced to STP.

(a) Volume of nitrogen admitted $= \dfrac{142.5}{760} \cdot \dfrac{273}{285} \cdot 26.85 = 4.830$ ml

Volume remaining in burette after expansion =

$$\frac{27.0}{760} \cdot \frac{273}{285} \cdot 9.19 = 0.312 \text{ ml}$$

Volume in free space $= 0.02956 \times 27.0 = 0.789$ ml

Volume adsorbed $= 4.830 - (0.312 + 0.798) = 3.720$ ml.

(b) More nitrogen admitted to the burette $= 5.380$ ml
Nitrogen already in the sample tube $= 4.518$ ml
Total nitrogen in the system $= 9.898$ ml.

The calculations are carried out cumulatively in steps in the same manner. The observed and calculated data are given in Table 16.2.

A BET plot of $P/V(P_0 - P)$ against P/P_0 gives a value of V_m from which the surface is determined from the equation:

$$S_w = 4.35\,V_m \text{ m}^2 \text{ g}^{-1}$$

A Harkins and Jura plot of P/P_0 against $1/V^2$ on semi-log paper has a negative slope A from which the surface area is determined using:

$$S_w = 4.06A \text{ m}^2 \text{ g}^{-1}$$

16.8 Commercially available volumetric- and gravimetric-type apparatus

There is available a wide range of commercial equipment. These generally offer rapid surface-area evaluation with some loss of accuracy. Glass equipment can be produced cheaply and is capable of providing results at a similar rate to commercial equipment (6 analyses per day), but skilled, careful operators are required and an analysis rate of 1 or 2 per day is more usual.

BET and HJr data are given in Table 16.3. BET and HJr plots are shown in figures 16.12 and 16.13 respectively.

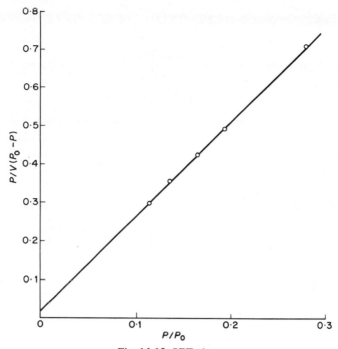

Fig. 16.12 BET plot.

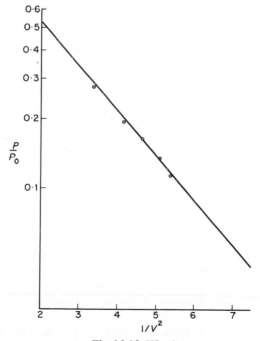

Fig. 16.13 HJr plot.

Table 16.3 BET and HJr data.

P	V ml g^{-1}	P/P_0	$P/V(P_0 - P)$	$1/V^2 P$ per g
86.5	0.431	0.114	0.298	5.38
103.5	0.443	0.136	0.356	5.09
125.0	0.464	0.165	0.424	4.63
137.0	0.490	0.194	0.489	4.15
211.5	0.546	0.278	0.706	3.36

BET plot

Slope   = 2.47
Intercept = 0.02
$1/V_m$ = 2.49
V_m = 0.405 ml
S_w = 1.76 m^2 g^{-1}

HJr plot

Slope = 0.196
S_w = 1.79 m^2 g^{-1}

Micromeritics, for example, manufacture several instruments, one of the simplest of which is the model 2200 (figure 16.14). This is designed for single-point BET, using fixed pressures and variable volume.

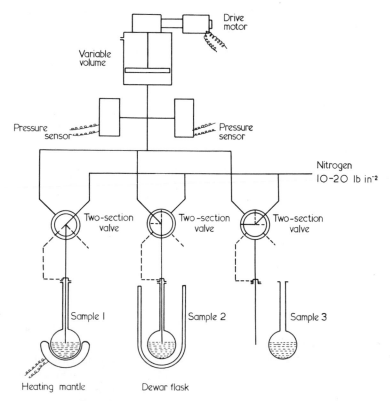

Fig. 16.14 Schematic diagram of model 2200.

A known volume of gas is introduced into the sample tube at a fixed higher pressure. The bulb is then immersed in liquid nitrogen and the pressure drops to a second fixed point. The volume reading is indicative of amount of gas adsorbed. An empirical correction is added to compensate for using the origin for the BET line instead of determining an intercept.

The technique used by the Carlo Erba Sorptomatic (figure 16.15) is as follows. A special metering pump allows the injection into a previously outgassed cuvette kept at constant temperature of known and reproducible volumes of gas $V1$, $V2$, $V3$ etc. The cuvette pressure will reach pressures P, $2P$, $3P$, etc. accordingly. The maximum pressure reachable by the system will be equal to the saturation pressure (gas in equilibrium with the liquid phase). By repeating the test with a cuvette containing a solid adsorbent, after each injection a part of the gas introduced will be adsorbed and the equilibrium pressure will result $P'P$, $2P'2P$, $3P'3P$. Knowing the volumes of the injected gas, it is possible to calculate the volumes of gas adsorbed at the various equilibrium pressures and then plot the adsorption isotherm. In a recent investigation of this equipment Nikishova *et al.* [173] found errors of 24% for a surface area of 10 m^2 g^{-1} falling to 6% at 100 m^2 g^{-1}.

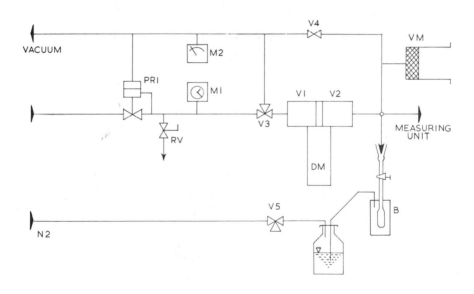

Fig. 16.15 Schematic diagram of Carlo Erba Sorptomatic.

Numinco market the Orr surface-area pore-volume analyser with a claimed precision of 1% surfaces from 0.1 to 1000 + m^2 g^{-1} using the conventional BET volumetric method. A less accurate, but simpler and more versatile, apparatus is available as the Numinco surface-area and density apparatus with which density may be determined to better than 1% and surface may be determined by multiple-point and single-point BET, and by a single-point flow method due to Innes [78].

The Ströhline areameter (figure 16.6) consists of an adsorption vessel with the sample and a reference vessel on either side of a mercury manometer [79, 83]. Degassing is carried out by purging with dry measuring gas with the sample tube in a heating block. Both vessels are then filled with nitrogen at ambient temperature and atmospheric pressure, after which they are immersed in a liquid-nitrogen bath. Adsorption of nitrogen on to the sample creates a pressure difference across the manometer from which the surface area of the sample may be determined, correction being applied for the sample volume and the effect of the change in height of the manometer level. The surface may be determined by a simple nomogram or by calculation. This is essentially a one-point method of determining specific surface and may be used for surfaces in the range 0.2 to 1000 m^2 g^{-1}. The reproducibility is of the order of 1% and the error introduced by assuming that the BET-plot passes through the origin produces specific surfaces about 10% lower than those obtained using the conventional multipoint method. Using one-point determinations, up to 16 analyses per day may be performed. Alternatively, the nitrogen may be fed in under pressure and a full isotherm determined [95].

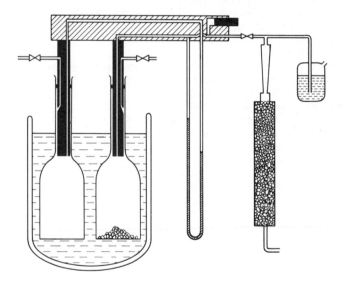

Fig. 16.6 The Ströhline areameter.

The Gravimat is a beam-balance for surface-area and pore size distribution measurement. A full description of this instrument is given in [93] and a description of the measuring technique for nitrogen adsorption in [94, 97]. It is possible to measure and control pressures from atmospheric down to 10^{-3} mm Hg, hence the instrument can be used with a range of adsorbates.

The balance has a sensitivity of 0.1 μg, but thermal effects and buoyancy reduce the measuring accuracy; a monolayer of 34.8 cm^2 of nitrogen weighs 1.0 μg giving a lower limit of surface-area measurement of 100 cm^2. Conditions are more favourable with krypton of which 18.7 cm^2 weighs 1.0 μg.

In 1970 AB Atom-energie, Sweden, constructed a low temperature static volumetric gas-adsorption apparatus, the Alpha-mec [169]. This instrument needs no dead-space factor calibration but only a static mechanical dead-space factor when used specifically to measure the surface area of active nuclear fuel powder. This equipment was tested by Johansson and Stanley-Wood [170] who found that calibration could only be dispensed with for powders of surface area greater than 30 m².

References

1   Brunauer, S. *et al.* (1940), *J. Am. Chem. Soc.,* **62**, 1723.
2   Orr, C. and Dallavalle, J.M. (1959), *Fine Particle Measurement,* Macmillan, NY.
3   Adamson, A.W. (1960), *Physical Chemistry of Surfaces,* Interscience, NY.
4   Langmuir, I. (1918), *J. Am. Chem. Soc.,* **40**, 1361.
5   Volmer, M. (1925), *Z. Phys. Chem.,* **115**, 253.
6   Fowler, R.H. (1935), *Proc. Camb. phil. Soc.,* **31**, 260.
7   Brunauer, S. and Emmett, P.H. (1937), *J. Am. Chem. Soc.,* **59**, 2682.
8   Emmett, P.H. and Dewitt, T.W. (1937), *ibid.,* 2682.
9   Brunauer, S., Emmett, P.H. and Teller, E. (1938), *ibid.,* **60**, 309.
10 Baly, E.G.G. (1937), *Proc. R. Soc.,* **A160**, 465.
11 Jones, D.C. and Birks, E.W. (1950), *J. Chem. Soc.,* 1127.
12 Jones, D.C. (1951), *ibid.,* 1461.
13 Young, D.M. and Crowell, A.D. (1962), *Physical Adsorption of Gases,* Butterworths.
14 Cassel, H.M. (1944), *J. Chem. Phys.,* **12**, 115.
15 Cassel, H.M. (1944), *J. Phys. Chem.,* **48**, 195.
16 Halsey, G.D. (1952), *Adv. Catalysis,* **4**, 259.
17 Gregg, S.J. and Jacobs, J. (1948), *Trans. Faraday Soc.,* **44**, 574.
18 Halsey, G.D. (1948), *J. Chem. Phys.,* **16**, 931.
19 Hüttig, G.F. (1948), *Mschr. Chem.,* **78**, 177.
20 Ross, S. (1953), *J. Phys. Chem.,* **53**, 383.
21 Corrin, M.L. (1953), *J. Am. Chem. Soc.,* **75**, 4623.
22 Loeser, E.H., Harkins, W.D. and Twiss, S.B. (1953), *J. Phys. Chem.,* **57**, 591.
23 MacIver, D.S. and Emmett, P.H. (1956), *ibid.,* **60**, 824.
24 Dzisko, V.A. and Krasnopolskaya, V.N. (1952), *Zh. Fiz. Khim.,* **26**, 1841.
25 Harkins, W.D. and Jura, G. (1943), *J. Chem. Phys.,* **11**, 430, 431.
26 Livingstone, H.K. (1947), *ibid.,* **15**, 617.
27 Livingstone, H.K. (1944), *ibid.,* **12**, 466.
28 Emmett, P.H. (1946), *J. Am. Chem. Soc.,* **68**, 1784.
29 Smith, T.D. and Bell, R. (1948), *Nature,* **162**, 109.
30 Emmett, P.H. (1941), *ASTM Symp. New Methods for Particle Size Determinations in the Sub-sieve Range,* pp. 95–105.
31 Joy, A.S. (1953), *Vacuum,* **3**, 254.
32 Bugge, P.E. and Kerlogue, R.H. (1947), *J. Soc. chem. Ind., Lond.,* **66**, 377.
33 McBain, J.W. and Bakr, A.M. (1926), *J. Am. Chem. Soc.,* **48**, 690.
34 Wooten, L.A. and Brown, C. (1943), *ibid.,* **65**, 113.
35 Lister, B.A.J. and MacDonald, L.A. (1952), UK AERE Report C/R 915.
36 Beebe, R.A., Beckwith, J.B. and Honig, J.M. (1945), *J. Am. Chem. Soc.,* **67**, 1554.
37 Emmett, P.H. (1940), 12th Report of the Committee on Catalysis, *Physical Adsorption in the Study of the Catalysis Surface,* Wiley, NY, Ch. 4.
38 Joyner, L.G. (1949), *Scientific and Industrial Glass Blowing and Laboratory Techniques,* Instruments Publ. Co., Pittsburgh, Pennsylvania, USA.
39 Emmett, P.H. (1944), Colloid Chem., V, Reinhold, NY.

40 Harvey, E.N. (1947), *ASTM Symp. Paint and Paint Material.*
41 Schubert, Y. and Kopelman, B. (1952), *Powder Metall. Bull.*, **6**, 105.
42 Vance, R.F. and Pattison, J.N. (1954), *Special Report on Apparatus for Surface Area Determination and Other Adsorption Studies on Solids*, Battelle Memorial Institute, Ohio, USA.
43 Harkins, W.D. and Jura, G. (1944), *J. Am. Chem. Soc.*, **66**, 1366.
44 Thompson, J.B., Washburn, E.R. and Guildner, L.A. (1952), *J. Phys. Chem.*, **56**, 979.
45 Bensen, S.W. and Ellis, D.A. (1948), *J. Am. Chem. Soc.*, **70**, 3563.
46 Loebenstein, W.V. and Deitz, V.R. (1951), *J. Res. Nat. Bur. Stand.*, **46**, 51.
47 Pickering, H.L. and Eckstrom, H.C. (1952), *J. Am. Chem. Soc.*, **74**, 4775.
48 Tomlinson, L., UKAEA Report 1, G.R.- TN/S - 1032.
49 Haul, R.A.W. (1956), *Angew. Chem.*, **68**, 238.
50 Rosenberg, A.J. (1956), *J. Am. Chem. Soc.*, **78**, 2929.
51 Dollimore, J. (1963), *Chemy. Ind.*, (18), 742.
52 Leipziger, F.D. and Altamari, L.A. (1960), *Nucl. Sci., Engng*, **8**, 312.
53 Morris, H.E. and Maass, P. (1933), *Can. J. Res.*, **9**, 240.
54 Dunn, R.C. and Pomeroy, H.H. (1947), *J. Phys. Colloid Chem.*, **51**, 981.
55 McBain, J.W. and Britton, H.T.S. (1930), *J. Am. Chem. Soc.*, **52**, 2198.
56 McBain, J.W. and Sessions, R.F. (1948), *J. Colloid Sci.*, **3**, 213.
57 Boyd, G.E. and Livingstone, H.K. (1942), *J. Am. Chem. Soc.*, **64**, 2838.
58 Seborg, C.O., Simmons, F.A. and Baird, P.K. (1936), *Ind. Engng Chem. ind. Edn*, **28**, 1245.
59 Dubinin, M.M. and Timofeev, D.P. (1947), *Zh. Fiz. Khim.*, **21**, 1213.
60 Lemcke, W. and Hofmann, U. (1934), *Angew. Chem.*, **47**, 37.
61 Seborg, C.O. and Stamm, A.J. (1931), *Ind. Engng Chem. ind. Edn*, **23**, 1271.
62 Pidgeon, L.M. and Maass, O. (1950), *J. Am. Chem. Soc.*, **52**, 1053.
63 Mulligan, W.O. *et al.* (1951), *Analyt. Chem.*, **23**, 739.
64 Stamm, A.J. and Woodruff, S.A. (1941), *Ind. Engng Chem. analyt. edn*, **13**, 386.
65 Barrett, H.M., Birnie, A.W. and Cohen, M. (1940), *J. Am. Chem. Soc.*, **62**, 2839.
66 Gulbransen, E.A. (1944), *Rev. scient. Instrum.*, **15**, 201.
67 Rhodin, T.N. (1950), *J. Am. Chem. Soc.*, **72**, 4343.
68 Rhodin, T.N. (1950), *ibid.*, 5691.
69 Rhodin, T.N. (1953), *Adv. Catalysis*, **5**, 39.
70 Bowers, R. and Long, E.A. (1955), *Rev. scient. Instrum.*, **26**, 337.
71 Gregg, S.J. (1946), *J. Chem. Soc.*, **561**, 564.
72 Gregg, S.J. and Wintle, M.F. (1946), *J. scient. Instrum.*, **23**, 259.
73 Nelsen, F.M. and Eggertsen, F.T. (1958), *Analyt. Chem.*, **30**, 1387.
74 Ellis, J.F., Forrest, C.W. and Howe, D.D. (1960), UK AERE Report D–E, G.R. 229 (CA).
75 Atkins, J.H. (1964), *Analyt. Chem.*, **36**, 579.
76 Haley, A.J. (1963), *J. appl. Chem.*, **13**, 392.
77 Whitehead, R.C. (1966), private communication.
78 Innes, W.B. (1951), *Analyt. Chem.*, **23**, 759.
79 Haul, R. and Dümbgen, G. (1960), *Chem. Ingr. Tech., Part 1*, **32**, 349; (1963), *Part 2*, **35**, 586.
80 Aylmore, D.W. and Jepson, W.B. (1961), *J. scient. Instrum.*, **38**, 4, 156.
81 Lovell, G.H.B. (1970), Surface area determination, *Proc. Conf. Soc. Chem. Ind.*, Bristol, 1969, Butterworths.
82 Hardman, J.S. (1971), Ph.D. thesis, Leeds Univ.
83 Gall, L. (1964), *Angew. Mess., Regeltech.*, **4**, 12, 107–11.
84 de Boer, J.H., Linsen, B.G. and Osinga, J. (1965), *J. Catalysis*, **4**, 643.
85 Schüll, C.G., Elkin, P.B. and Roess, L.C. (1948), *J. Am. Chem. Soc.*, **70**, 1405.
86 Barret, E.P., Joyner, L.G. and Halenda, P.P. (1951), *ibid.*, **73**, 373.
87 Lippens, B.C. and de Boer, J.H. (1965), *J. Catalysis*, **4**, 319.
88 Lippens, B.C., Linsen, B.G. and de Boer, J.H. (1964), *ibid.*, **3**, 32.
89 de Boer, J.H. (1970), Surface area determination, *Proc. Conf. Soc. Chem. Ind.* Bristol, 1969, Butterworths.

90 Anderson, R.B. (1946), *J. Am. Chem. Soc.*, **68**, 686.
91 Brockhoff, J.C.P. (1969), thesis, Delft Univ., (*cit.* [89]).
92 Lecloux, A. (1970), *J. Catalysis*, **81**, 22.
93 Robens, E. (1969), *Lab. Pract.*, **18**, 3, 292.
94 Robens, E. and Sandstede, G. (1967), *Z. Instrum.*, **75**, 167–78.
95 Roth, R. (1971), *Staub-Reinhalt Luft*, **31**, 8, 320–2.
96 Lapointe, C.M. (1970), *Can. Mines Br. Tech. Bull.*, TB119, 27.
97 Robens, E. and Sandstede, G. (1969), *J. Phys. E*, **2**, 4, 365–8.
98 Brunauer, S. (1970), Surface area determination, *Proc. Conf. Soc. Chem. Ind.* Bristol, 1969, Butterworths.
99 Pickett, G. (1945), *J. Am. Chem. Soc.*, **67**, 1958.
100 Gregg, S.J. and Sing, K.S.W. (1967), *Adsorption, Surface Area and Porosity,* Academic Press, NY.
101 Kiselev, A.V. (1945), *USP Khim*, **14**, 367.
102 Halsey, G. and Taylor, H.S. (1947), *J. Chem. Phys.*, **15**, 9, 624–30.
103 Burevski, D. (1975), Ph.D. thesis, Univ. Bradford.
104 Sips, R. (1948), *J. Chem. Phys.*, **16**, 490.
105 Sips, R. (1950), *ibid.*, **18**, 1024.
106 Joyner, L.G., Weinberger, E.B. and Montgomery, C.W. (1945), *J. Am. Chem. Soc.*, **67**, 2182–8.
107 Wade, W.H. and Blake, T.D. (1971), *J. Phys. Chem.*, **75**, 1887.
108 Emmett, P.H. and Brunauer, S. (1937), *J. Am. Chem. Soc.*, **59**, 1553.
109 Parfitt, G.D. and Sing, K.S.W. (eds) (1971), *Characterisation of Powder Surfaces,* Academic Press.
110 Brunauer, S., Skalney, J. and Bodor, E.E. (1969), *J. Colloid Interfac. Sci.*, **30**, 546.
111 Anderson, R.B. and Hall, W.K. (1948), *J. Am. Chem. Soc.*, **70**, 1727.
112 Keenan, A.G. (1948), *ibid.*, 3947.
113 Barrer, R.M., MacKenzie, N. and McLeod, D. (1952), *J. Chem. Soc.*, 1736.
114 Lamond, T.G. and Marsh, H. (1963), *Carbon*, **1**, 293.
115 Walker, P.L. and Shelef, M. (1967), *ibid.*, **5**, 7.
116 Freeman, E.M. and Marsh, H. (1970), *ibid.*, **8**, 19.
117 Toda, Y. *et al.* (1970), *ibid.*, **8**, 565.
118 Freeman, E.M. *et al.* (1970), *ibid.*, **8**, 7.
119 Dubinin, M.M. (1972), *Adsorption–Desorption Phenomena*, Academic Press, p. 3.
120 Dovaston, N.G., McEnaney, B. and Weedon, C.J. (1972), *Carbon*, **10**, 277.
121 Bering, B.P., Dubinin, M.M. and Serpinsky, V.V. (1966), *J. Colloid Interfac. Sci.*, **21**, 378.
122 Bering, B.P. *et al.* (1971), *Izv. Akad. Nauk SSSR, Ser. Khim.* (English Trans.), 17.
123 Shram, A., (1965), *Nuovo Cimento,* Suppl. 5, 309.
124 Marsh, H. and Rand, B. (1970), *J. Colloid Interfac. Sci.*, **33**, 101–16.
125 Marsh, H. and Rand, B. (1971), *3rd Conf. Industrial Carbon and Graphite,* Soc. Chemy. Ind., p. 212.
126 Kaganer, M.G. (1959), *Russ. J. Phys. Chem.*, **33**, 352.
127 Jovanovic, D.S. (1969), *Kolloid Z.Z. Poly.* **235**, 1203.
128 Jovanovic, D.S. (1969), *ibid.*, 1214.
129 Hill, T.L. (1952), *Adv. Catalysis*, **4**, 211.
130 de Boer, J.H., Kaspersma, J.H. and Van Dongen, R.H. (1972), *J. Colloid Interfac. Sci.*, **38**, 1, 97–100.
131 Lippens, B.C., Linsen, B.G. and de Boer, J.H. (1964), *J. Catalysis*, **3**, 32–37.
132 Dubinin, M.M. (1955), *Q. Rev. (London)*, 9, 101.
133 Dubinin, M.M. (1967), *J. Colloid Interfac. Sci.*, **23**, 487–99.
134 Dubinin, M.M. (1966), *Chemistry and Physics of Carbon,* (ed. P.L. Walker), **2**, Arnold, London, p. 51.
135 Hobson, J.P. and Armstrong, P.A. (1959), *J. Phys. Chem.*, **67**, 2000.
136 Hobson, J.P. (1967), *The Solid–Gas Interface* (ed. E.A. Flood), **1**, Arnold, London, p. 14.

137 Sing, K.S.W. (1970), *Proc. Int. Symp. Surface Area Determination* (eds D.H. Everett and R.M. Ottewill), Butterworths, London.
138 Pierce, C. (1968), *J. Phys. Chem.*, 72, 3673.
139 Lamond, T.G. and Marsh, H. (1964), *Carbon*, 1, 281, 293.
140 Marsh, H. and Rand, B. (1970), *J. Colloid Interfac. Sci.*, 33, 3, 478–9.
141 Harris, M.R. and Sing, K.S.W. (1955), *J. appl. Chem.*, 5, 223.
142 Dollimore, D., Rickett, G. and Robinson, R. (1973), *J. Phys. E.*, 6, 94.
143 Rand, B. (1976), *J. Colloid Interfac. Sci.*, 56, 2, 337–46.
144 Marsh, H. and Siemiensiewska, T. (1965), *Fuel*, 44, 335.
145 Ramsay, J.D.F. and Avery, R.G. (1975), *J. Colloid Interfac. Sci.*, 51, 1, 205–8.
146 Rasneur, B. and Charpin, J. (1975), *Fine Particle Int. Conf.* (ed. W.E. Kuhn), Electrochem. Soc. Inc., Princeton, NJ.
147 Tucker, B.G. (1975), *Analyt. Chem.*, 47, 4, 78–9.
148 Mikhail, R. Sh. and Brunauer, S. (1975), *J. Colloid Interfac. Sci.*, 52, 3, 626–7.
149 Mikhail, R. Sh. and Brunauer, S. (1975), *ibid.*, 572–7.
150 Parfitt, G.D., Sing, K.S.W. and Urwin, D. (1975), *ibid.*, 53, 2, 187–93.
151 Gregg, S.J. and Jacobs, J. (1948), *Trans. Faraday Soc.*, 44, 574.
152 Genot, B. (1975), *J. Colloid Interfac. Sci.*, 50, 3, 413–8.
153 Alzamora, L. and Cortes, J. (1976), *ibid.*, 56, 2, 347–9.
154 Singleton, J.H. and Halsey, G.D. (1954), *J. Phys. Chem.*, 58, 1011.
155 de Boer, J.H. *et al.* (1967), *J. Catalysis*, 7, 135–9.
156 Sing, K.S.W. and Swallow, D. (1960), *J. appl. Chem.*, 10, 171.
157 Jaycock, M.J. (1977), The Krypton BET Method, Chemistry Dept., Univ. Tech. Loughborough, Leics. LE11 3TU, UK.
158 Desbiens, G. and Zwicker, J.D. (1976), *Powder Technol.*, 13, 15–21.
159 Anon. (1976), Measure of Powder Fineness resulting from Inter-Laboratory Studies, Assoc. Franc. Normalisation, Tour Europe, Cedex 7–92080 Paris, La Defense.
160 Wilson, R. (1977), *Particle Size Analysis*, (ed. M.J. Groves) (1978), Chem. Soc. Analyt. Div., Heyden.
161 Rosai, L. and Giorgi, T.A. (1975), *J. Colloid Interfac. Sci.*, 51, 2, 217–24.
162 Knowles, A.J. and Moffatt, J.B. (1972), *ibid.*, 41, 1, 116–23.
163 Stoeckli, H.F. (1977), *ibid.*, 59, 1, 184–5.
164 Hill, T.L. (1946), *J. Chem. Phys.*, 14, 268.
165 Lowell, S. (1975), *Powder Technol.*, 12, 291–3.
166 Kadlec, O. (1976), *Dechema Monogr.*, 79 (1589–1615), 181–90.
167 Kadlec, O. and Dubinin, M.M. (1969), *J. Colloid Interfac. Sci.*, 31, 479.
168 Ricca, F., Medana, R. and Bellarda, A. (1967), *Z. Phys. Chem.*, 52, 276.
169 Larsson, E. and Grönroos, B. (1970), Workreport AE–MB–183, Aktiebolaget Atom-energie, Studvik, 611 01 Nyköping, Sweden.
170 Johansson, M.E. and Stanley-Wood, N.G. (1977), *Powder Technol.*, 16, 1, 145–8.
171 Carden, J.L. and Pierotti, R.A. (1974), *J. Colloid Interfac. Sci.*, 47, 2, 379–94.
172 Choudhary, V.R. (1974), *J. Sci. Ind. Res.*, 33, 12, 634–41.
173 Nikishova, N.I. and Landa, Ya.A. (1975), *Proizvod Ognev Porov.*, 4, 117–25.
174 Holmes, H.F., Fuller, E.L. Jr. and Beh, R.A. (1974), *J. Colloid Interfac. Sci.*, 47, 2, 365–71.
175 Baker, F.S. and Sing, K.S.W. (1976), *ibid.*, 55, 3, 605–13.
176 Bhat, R.K. and Krishnamoorthy, T.S. (1976), *Indian J. Technol.*, 14, 4, 170–1.
177 Yen Chi-Min, Chang, Chi-Yuan (1977), *Hua Hsueh Hsueh Pao*, 35, 3–4, 131–40.
178 Dollimore, D., Spooner, P. and Turner, A. (1976), *Surface Technol.*, 4, 2, 121–60.
179 Mikhail, R.Sh., Guindy, N.M. and Ali, I.T. (1976). *J. Colloid Interfac. Sci.*, 55, 2, 402–8.
180 Dubinin, M.M. *et al.* (1975), *Izv. Akad. Nauk SSSR, Ser. Khim.*, 6, 1232–9.

17 Other methods for determining surface area

17.1 Introduction

The most widely used method for surface-area determination is low-temperature gas adsorption, particularly nitrogen and krypton at liquid-nitrogen temperature. Most gases can and have been used, and these include water vapour at room temperature and carbon dioxide at room temperature and at $-78°C$. The problems that arise when one deviates from the standard conditions are: what is the applicable molecular area and what is the correct theoretical model to use? The first question is usually resolved by accepting published values or carrying out experiments to determine molecular area by comparison with nitrogen adsorption at liquid-nitrogen temperature. Since there is no unanimity in published data, the second procedure is probably preferable. The second question is usually resolved from an examination of the isotherm, the BET or the Langmuir equation being then used. When coverage is very low, as with carbon dioxide at room temperature, the Freundlich equation may be applicable.

Permeametry and, to a lesser extent, gas diffusion are used for comparison purposes due to their ease of operation and simplicity. Surface areas may also be calculated from size distribution data and this transformation is the subject of a British Standard (BS 4359 (1972): Part 3).

Other adsorption techniques include adsorption from solution, and here the problem is one of determining the amount adsorbed since this is usually very small. Adsorption studies have been described with fatty acids, polymers, ions, dyestuffs and electrolytes using a range of analytical techniques. The most usual experimental method of determining a single point on the adsorption isotherm of a binary solution is to bring a known amount of solution of known composition into contact with a known weight of adsorbent in a vessel at the required temperature and stir for several hours. After equilibrium an aliquot part of the bulk liquid is separated and the concentration change determined by some suitable method. The amount adsorbed will then be some function of the final concentration.

Surface areas may also be determined from heats of adsorption and this technique has been greatly simplified with the introduction of the flow microcalorimeter. This instrument can be used with gas or liquid mixtures to determine heats of adsorption and amount adsorbed; it thus provides information on molecular areas as well as energies of adsorption.

17.2 Calculation from size distribution data

If the fractional weight of powder of measured mean size d_r is x_r, then:

$$x_r W = \alpha_v \rho_s n_r d_r^3 \qquad (17.1)$$

where W is the total weight of powder, ρ_s the powder density, α_v the volume-shape factor and n_r the number of particles of size d_r.

The surface of this fraction is:

$$\Delta S = \alpha_s n_r d_r^2 \qquad (17.2)$$

where α_s is the surface shape factor.

The specific surface is equal to S/W which, for a weight distribution, is:

$$S_w = \sum \frac{\alpha_s}{\rho_s \alpha_v} \frac{x_r}{d_r}$$

$$= \sum \frac{\alpha_{sv}}{\rho_s} \frac{x_r}{d_r} \qquad (17.3)$$

α_{sv} and ρ_s are usually considered to be constant over a limited size range making:

$$S_w = \frac{\alpha_{sv}}{\rho_s} \sum \frac{x_r}{d_r} \qquad (17.4)$$

For a number distribution:

$$S_w = \frac{\alpha_{sv}}{\rho_s} \frac{\sum n_r d_r^2}{\sum n_r d_r^3} \qquad (17.5)$$

It is usual to use an additional suffix to denote the method of measurement. Thus, for a sieve analysis, equation (17.4) becomes;

$$S_w = \frac{\alpha_{sv,A}}{\rho_s} \sum \frac{x_r}{d_{r,A}} \qquad (17.6)$$

Alternatively:

$$S_{w,A} = \frac{6}{\rho_s} \sum \frac{x_r}{d_{r,A}} \qquad (17.7)$$

S_w is the weight specific surface determined using a previously obtained value for the surface volume shape coefficient. $S_{w,A}$ is the weight specific surface by sieving assuming spherical particles. The former is suitable for comparison with other techniques such as permeametry; the latter, for comparison between powders using the same technique. This type of conversion is dealt with in BS 4359 (1972): Part 3.

Surface area may be obtained more readily if the distributions are log-normal or Rosin–Rammler, since the equations developed in Chapter 4 may be used. Surface area may also be determined by turbidity using equation (9.19).

17.3 Adsorption from solution

The accumulation of one molecular species at the interface between a solid and a solution is governed by complex phenomena. The molecules may accumulate at the interface as a result of interfacial tension, may attach on to the solid surface through strong chemical valency forces or may attach on to the solid surface through relatively weak physical, van der Waals', attractive forces. In physical adsorption, the desorption isotherm is essentially the same as the adsorption isotherm, whereas for chemisorption the molecules are not easily removed by merely lowering the equilibrium concentration of the solution.

Molecules are adsorbed on solid surfaces by interaction of the unsatisfied force fields of the surface atoms of the solid with the force fields of the molecules striking the surface. In this way the free energy of the solid surface is diminished [1]. The type of interaction, if any, that occurs between the solute molecules and the solid will be dependent on the nature of the surface and of the solute molecules.

In adsorption from solution a complicating factor arises, that of the possibility of competition between the solvent and solute molecules for the sites on the surface. The competition between the components of a solution depends mostly on the difference in the strength of interaction between adsorbent and the adsorbates.

17.3.1 Orientation of molecules at the solid–liquid interface

The idea of molecular orientation at interfaces was conceived by Benjamin Franklin who in 1765 spread olive oil on a water surface and estimated the thickness of the resulting film at one ten-millionth of an inch. In subsequent work with films of oil on water, Lord Rayleigh [2] in England and Miss Pockels [3] in Germany established that the films were only one molecule thick. Langmuir [4] introduced new experimental methods of great importance which resulted in new conceptions concerning these films.

Instead of working with oils, Langmuir used pure substances of known constitution and observed the effect of varying this constitution. He measured the outward pressure of the films directly by use of a floating barrier with a device to measure the force on it. The clearest results were obtained with normal saturated fatty acids and alcohols. Langmuir found that as the area on which the film was spread was reduced, no appreciable surface pressure developed until the area per molecule had been reduced to approximately 0.22 nm^2, at which point the pressure increased very rapidly with further decrease in area. One of the most striking facts illustrated by Langmuir's work is that the area is independent of the number of carbon atoms in the molecules. This would indicate that the molecules are orientated vertically to the surface of liquid and are orientated in the same manner in all the films regardless of chain lengths. According to Adam [1] each molecule occupies an area of 0.205 nm^2 on the surface of the substrate.

Some investigators suggest that the effective area occupied by fatty acid molecules at solid–liquid interfaces is the same as that occupied by these molecules in films on water. The area for stearic acid for example ranges from 0.205 to 0.251 nm^2, the former

being the area for closest packing of ellipses and the latter the area for free rotation [5, 6].

However, a greater variation than this is expected for an immobile interface since the adsorbate is not constrained to take up any definite orientation. In adsorption on carbon blacks, Kipling and Wright [7] suggest that stearic acid is adsorbed with the hydrocarbon chain parallel to the surface, the effective areas of each stearic acid molecule being calculated as 1.14 nm^2. Kipling and Wright [8] also suggest that this is true of other acids in homologous series and adsorption of these acids by non-polar adsorbents indicates that the major axis of the hydrocarbon chain is parallel to the surface. This value was also adopted by Roe [113] who applied the multilayer theory of adsorption to stearic and other aliphatic acids dissolved in cyclohexane at a volume concentration of 0.01235.

McBain and Dunn's [9] results for adsorption of cetyl alcohol by magnesium oxide are also probably best interpreted in terms of orientation parallel to the surface. Smith and Hurley [37] determined the surface area of fatty acid molecules adsorbed on to carbon black from cyclohexane and arrived at a value of 0.205 nm^2, which suggested a perpendicular orientation.

Ward [11] suggested a coiling into a hemispherical shape, and Allen and Patel [12, 13] found that the surface increased from 0.192 to 0.702 nm^2 with chain length irrespective of the adsorbent, while for alcohols the increase for long-chain alcohols was 0.201 to 0.605 nm^2 [14]. These values were explained in terms of coiling of the chains.

In early work on oxides, it was suggested by Harkins and Gans [15, 16], that oleic acid and butyric acids adopted the perpendicular orientation on titania, as did stearic acid on aluminium hydroxide [17]. In these experiments it was not clear whether adsorption was physical or chemical in nature. This now seems an important distinction to draw, especially with basic solids. In chemisorption, the orientation of the solute generally presents no problem, as the functional group determines the point of attachment. Thus the long-chain fatty acids are attached to the surface by the carboxyl group, $-COOH$, with the hydrocarbon chain perpendicular to the surface.

Harkins and Jura [18] have shown that the mean molecular area of nitrogen when adsorbed is not a constant value, but varies with the nature of the substrate. The molecular area of the adsorbed molecule was reported to vary from 0.136 to 0.169 nm^2 per molecule. While this may introduce some uncertainty into the values for areas of some very polar solids, it does not seriously reduce the utility of adsorption for determining surface-area value.

17.3.2 Polarity of organic liquids and adsorbents

Generally the organic liquids and solid adsorbents are classified according to their polarity, i.e. as to whether they are essentially polar or non-polar in character.

Polar molecules are defined as uncharged molecules in which the centres of gravity of positive and negative charges do not coincide, and these therefore show dipole moments. The larger the dipole moment, the more polar the molecule. The term *polar*

group is applied to a portion of a molecule with polar characteristics, such as —OH, —COOH, —COONa, —COOR and similar groups.

Non-polar molecules have an equal number of positive and negative charges with coinciding centres of gravity. Dipole moment is zero for non-polar molecules. The term *non-polar* may be applied to a portion of a large molecule with non-polar characteristics such as, benzene, *n*-heptane, hexane and other hydrocarbons.

The general rule is that a polar adsorbate will tend to prefer that phase which is the more polar, i.e. it will be strongly adsorbed by a polar adsorbent from a non-polar solution. Similarly, non-polar adsorbate will be adsorbed strongly on non-polar adsorbent from a polar solution.

Freundlich [19] found that the order of adsorption of normal fatty acids from aqueous solution on to a blood charcoal to be formic, acetic, propionic, butyric in increasing order. The same order of adsorption isotherm for homologous series of fatty acids, formic through caproic acids, from water on to Noril charcoal was reported by Linner and Gortner [20]. These results agree well with Traube's rule [21]. Holmes and McKelvey [22] made a logical extension of Freundlich's statement by noting that the situation was really a relative one and that a reversal order should occur if a polar adsorbent and a non-polar solvent were used. They indeed observed the reverse sequence for fatty acids adsorbed on silica gel from toluene solution. This agrees with the general observation that silica gel and charcoal, water and toluene are opposite in their polarity. Langmuir [24] gave an instructive interpretation to this rule. The work W to transfer one mole of solute from solution to surface is (see [25], p. 95):

$$W = RT \ln \frac{C_s}{C} = RT \ln \frac{\Gamma}{\tau C} \tag{17.8}$$

where C_s is the surface concentration and is given by Γ/τ, where Γ denotes the moles of solute adsorbed per unit area and τ is the film thickness. For solutes of chain length n and $(n-1)$ the difference in work is then:

$$W_n - W_{n-1} = RT \ln \left(\frac{\Gamma_n}{\Gamma_{n-1}} \frac{C_{n-1}}{C_n} \right) \tag{17.9}$$

Traube found that for each additional CH_2 group the concentration required to give a certain surface tension was reduced by a factor of 3, i.e. if

$$C_{n-1} = 3C_n \quad \text{then} \quad \gamma_n = \gamma_{n-1}$$

and

$$W_n - W_{n-1} = RT \ln 3$$
$$= 2.68 \text{ kJ mol}^{-1} \text{ at } T = 20°C$$

The figure of 2.68 kJ mol^{-1} may be regarded as the work to bring one CH_2 group from the body of the solution to the surface region. Adamson [25, p. 95] assumed this to imply that the chains were lying flat on the surface, but suggested that this was undoubtedly an oversimplification.

Harkins and Dahlstrom [23] have shown that the oxides of titanium, tin and zinc act like water in attracting polar rather than non-polar groups. Thus in oils any —COOH, —OH, —COOR, —CN and other similar groups orient towards the particle of oxide powder and the hydrocarbon groups towards the oil.

17.3.3 Drying of organic liquids and adsorbents

In adsorption by solids from liquid phase, substances present in low concentration are often adsorbed preferentially. The presence of water and other impurities in the solution may therefore have an effect on the adsorption. The purification and drying procedure usually employed consists of a fractionation, after which the recovered solution is stored over metallic sodium or other drying agents such as silica gel, calcium sulphate, alumina, etc. A very useful method of purification of solvents is given by Weissberger and associates [26].

It has been reported by Harkins and Dahlstrom [23] that extremely small quantities of water in benzene increase the energy of immersion of the used oxides to about three times the value for pure benzene.

Most solid adsorbents are capable of adsorbing water vapour from the atmosphere and should therefore be dried. The drying of adsorbent is usually done by heating for 2 or 3 h at 120 to 130°C. Many workers claim [27, 28] that this temperature is sometimes not high enough to drive away vapours previously adsorbed by the solids. If a higher temperature is used, then care should be taken that the solid is not altered by being heated, e.g. that sintering or alteration of the nature of the surface does not take place. The temperature of drying must therefore be carefully chosen for each adsorbent.

Some authors consider that adsorbents should be outgassed before use and then be introduced to the solution in the absence of air. Others claim that such outgassing treatment does not affect the extent of adsorption. Thus, it was reported by Greenhill [29] and by Russell and Cochran [17] that adsorption was essentially the same on metals, metal-oxides, and non-porous alumina, whether the samples were degassed or not prior to exposure to the solutions; gases adsorbed on the solids being apparently displaced by the liquid phase [30]. No systematic effect was found in adsorption by charcoal from mixtures of carbon tetrachloride and methanol [31]. Hirst and Lancaster [32] examined the effect of very small quantities of water on the interaction of stearic acid with finely divided solids. For adsorbents such as TiO_2, SiO_2, TiC and SiC, the presence of water was found to reduce the amount of acid adsorbed to form a mono-layer, and with reactive materials such as Cu, Cu_2O, CuO, Zn and ZnO, water was found to initiate chemical reaction.

17.4 Methods of analysis of amount of solute adsorbed on to solid surfaces

In almost all studies of adsorption by solids from solution, it is necessary to measure the concentration of the solution before and after adsorption. A variety of analytical methods of analysis may be used to measure such changes in concentration, including the Langmuir trough, gravimetric, titrimetric, interferometry and precolumn methods.

17.4.1 Langmuir trough [4]

This technique can be useful where the adsorptive can easily be spread on an aqueous substrate to give a coherent film. The area occupied by the adsorptive film, after evaporation of the solvent, is proportional to the weight present. This method has been successfully applied to analysis of solutions of long-chain fatty acids by Hutchinson [33], Gregg [34], Greenhill [29] and of alcohols and phenols by Crisp [35] in organic solvents such as benzene. Equal volumes of solution before and after adsorption were spread on aqueous substrate.

17.4.2 Gravimetric method

If an involatile solute is dissolved in a volatile solvent, analysis can be effected by evaporating off the solvent from a sample of known weight and weighing the residual solute. This simple technique was adopted by Smith and Fuzek [36] and thereafter widely used in many laboratories. In their procedure, an estimated 0.5 to 1 g of adsorbent was placed in a glass sorption tube to which a vacuum source could be attached. Then 40 ml solution of fatty acid (0.15 g) in benzene was introduced into the sorption tube. The tube was then stoppered tightly and shaken for a definite period of time. The tube was then centrifuged in order to settle the adsorbent and 5 ml of clear liquid withdrawn. This liquid was delivered into a weighed container which was placed in an oven at a temperature just below the boiling point of the solvent. Evaporation of the solvent was speeded by means of a slow stream of filtered air, and the fatty acid which remained after the evaporation was determined by weighing. Blank runs established the dependability of this analytical procedure. The adsorption tube was restoppered, shaken again for a definite interval of time, and some of the liquid removed and analysed as just described. At the end of the experiment the adsorbent was filtered out, dried and weighed under CO_2. The weight of the adsorbent was corrected for the weight of adsorbed fatty acid.

17.4.3 Volumetric method

Many standard procedures are available for studying adsorption by volumetric or titrimetric method. Adsorption of fatty acids [37] has frequently been examined by titration with aqueous alkali, even if the fatty acid was originally dissolved in an organic solvent. The extraction of the acid from the solvent seems to cause no difficulty, especially if warm ethyl alcohol is added [38], but the validity of the method should be checked by titration of a known sample for each time.

Conductimetry [39] and potentiometry [40] titrations have been used as alternatives to those carried out with a coloured indicator.

17.4.4 The Rayleigh interferometer

This instrument is used to measure the differences in refractive index or in optical pathlength between two liquids being compared; this is obtained by a 'null' method. The

drum reading is converted into a difference in composition by means of a calibration curve. This curve is drawn by successive comparisons of a set of mixtures of accurately known composition, covering the relevant range of concentration values plotted; as successive points are cumulative, so considerable care is required in constructing the calibration curve. The use of the interferometer is restricted to systems with a small difference in refractive index, otherwise a large number of standard mixtures have to be made up for calibration.

Bartell and Sloan [41] and Ewing and Rhoda [42] have made successful use of the interferometer in measuring the change in concentration due to adsorption from non-aqueous solutions. Further details of the instrument are given by Candler [43].

17.4.5 The precolumn method

This was suggested by Groszek [44] for measuring the amount of solute adsorbed on to a solid surface using the flow microcalorimeter. This is described in detail later.

17.5 Theory for adsorption from a solution

Liquid phase adsorption methods depend on the establishment of an equilibrium between adsorbed and unadsorbed solute molecules. Adsorption of solute on to the surface of a solid will continue till it reaches a saturation point giving a clear plateau in the isotherm. As the isotherm usually tends towards a limiting value, the limit has often been taken to correspond to the coverage of the surface with a complete monolayer of the solute. The equation derived for monolayer coverage is:

$$\frac{x_1 x_2}{\Gamma_1^{(n)}} = \frac{1}{Kx_m} + \frac{K-1}{K} \cdot \frac{x_1}{x_m} \tag{17.10}$$

where $\Gamma^{(n)}$ is the Gibbs adsorption value, x_1 and x_2 are the mole fractions of the two components of a completely miscible solution and K is a constant.

For K much greater than unity and for low concentrations of component one, this reduces to Langmuir's equation. Alternatively, the Langmuir equation, replacing pressure P with concentration of solution C, has been used to determine the limiting value [12–14].

$$\frac{C}{x} = \frac{1}{Kx_m} + \frac{C}{x_m}; \quad S_w = \frac{N\sigma x_m}{M_v} \tag{17.11}$$

where x = amount of solute adsorbed per gram of adsorbent,
    x_m = solute monolayer capacity,
    K = constant,
    N = Avogradro's number,
    σ = area occupied per molecule,
    M_v = molar volume.
Thus a plot of C/x versus C should give a straight line of slope $1/x_m$ and intercept $1/Kx_m$.

For the determination of specific surface of the adsorbent, three things are required, namely:

(1) The area σ occupied by one molecule of the solute in a close-packed film on the surface of the adsorbent must be known.
(2) It must be possible to clearly locate a point on the isotherm which corresponds to a complete monolayer adsorbed.
(3) Any competitive adsorption on the adsorbent surface of solvent molecules must be compensated for.

17.6 Quantitative methods for adsorption from a solution

17.6.1 Adsorption of non-electrolytes

This is usually considered to be essentially monolayer adsorption with competition between solvent and solute. The non-electrolytes that have been studied are mainly fatty acids, aromatic acids, esters, and other single functionless group compounds plus a great variety of more complex species such as porphyrins, bile pigments, carotenoids, lipoids and dyestuffs.

17.6.2 Fatty acid adsorption

It has been known for a long time that when fatty acid molecules are closely packed on the surface of distilled water, each molecule occupies an area of 0.205 nm^2 irrespective of the length of the hydrocarbon chain [25].

This property was used by Harkins and Gans [15] for the determination of the surface area of titanium dioxide, using oleic acid; their results were in general agreement with microscopy. Since then the method has been used extensively and a detailed review is to be found in Orr and Dallavalle [10 (p. 21)].

Smith and Fusek employed the procedure described in section 17.4.2.

Another procedure of Smith and Fusek was to place 20 ml of solvent containing 0.2 to 0.4 g of fatty acid in a tube and mix for 24 h. After this time the tube was centrifuged, 10 ml of the liquid withdrawn and 10 ml of pure solvent added. The procedure was then repeated. The fatty acid content of the liquid samples withdrawn was determined as before [36].

Gregg [34] used the same solvent but determined the number of gram molecules adsorbed with a surface-tension balance. Smith and Hurley [37] however, recommended the use of cyclohexene as solvent, and stated that with some solvents multilayer adsorption takes place. Hirst and Lancaster [32], instead of adding more fatty acid to the solvent, increased C/C_s by decreasing the temperature of the solution.

The specific surface area determined by liquid-phase adsorptions will usually be low due to adsorption of the solvent. It is thus preferable that determinations should be carried out with a variety of solvents and comparisons made with a standard technique such as gas adsorption.

17.6.3 Adsorption of polymers

Adsorption isotherms of linear polymer molecules are found to be of the Langmuir type [47, 48]. Many workers assume the molecules are adsorbed in the shape of a random coil and [45, 46] have developed equations to give the area occupied by a molecule. Some workers assume a modified Langmuir equation to be necessary since polymers may occupy more than one site [49]. Others adopt a more empirical approach [50]. An estimate of the inner and outer surface areas of porous solids has also been obtained by using a set of polystyrene fractions having a narrow range of molecular weights [51].

17.6.4 Adsorption of dyes

Dyes have been used by many investigators for specific surface evaluation, but the results have not been widely accepted because of the inconsistency of the reported results both between different dyes and with other methods [52, 53]. Giles [54] attributes these inconsistencies mainly to injudicious choice of dyes and an incomplete understanding of the adsorption process.

Extensive studies of the adsorption of non-ionic [82] and cationic [83, 84] dyes on alumina and inorganic surfaces were carried out by Giles and his co-workers [85, 86] who consider that these dyes are adsorbed by an ion-exchange mechanism, largely as a monolayer of positively charged micelles or aggregates. The presence of aggregates on the substrate is inferred from the nature of the isotherm and its change with temperature.

Anionic and non-ionic compounds with free hydroxyl groups are adsorbed on silica mainly as a monolayer of single molecules or ions with the hydroxyl group bonded to the silica surface, but micelle adsorption may also occur. Leuco vat dyes [87] are adsorbed in aggregate form, but the evidence is not wholly unambiguous since it is based on the spectral characteristics of the adsorbed dye on the substrate and also its fluorescence characteristics [88]. Non-reactive anionic phthalocyanine blue is adsorbed in aggregate form to give an appreciable percentage of dimers [88].

Giles *et al.* [86] discussed the orientation of molecules at graphite surfaces. They conclude that non-ionic azo dye is adsorbed flat and a complete monolayer is formed by adsorption from aliphatic solvents but not from benzene due to competition from the benzene solvent molecules. With monosulphonates a condensed monolayer is formed; the molecules probably stand perpendicular to the surface, the sulphonated groups in the water and the unsulphonated ends at the graphite surface. Disulphonate molecules are oriented edge-on or end-on and trisulphonate molecules lie flat.

On anodic alumina films sulphonated dyes can be adsorbed both by covalent bonds between the sulphonate groups and alumina atoms and also by electrostatic attraction of anionic dye micelles by the positively charged surface [54].

The experimental technique is to tumble gently 0.05 to 0.50 g of sample with 10 ml aqueous solutions of dye at room temperature; 10 to 30 min is sufficient time for non-porous solids, but 12 to 48 h may be required for porous powders. The tubes are then centrifuged and the solutions analysed spectrophotometrically. With porous

powders, a rate curve develops; extrapolating this back to zero gives the surface concentration of dye; the saturated value represents total coverage. The isotherms usually have a long plateau, and this value is accepted as monolayer coverage; this feature makes the method attractive as a one-point technique.

If Y_m (in mmol g^{-1}) is the amount of dye adsorbed at monolayer coverage per g of adsorbent, σ the flat molecular area of the dye and N is Avogadro's number, the weight specific surface is given by:

$$S_w = \frac{Y_m N \sigma}{X} \tag{17.12}$$

where X is the coverage factor which is equal to the number of dye ions in a micelle.

For methylene blue Giles $et\ al.$ [72, 73] use $\sigma = 1.2$ nm^2 and $X = 2$. They found it preferable to buffer the dye solution at pH 9.2, by using buffer tablets, to reduce competition between the H$^+$ and the dye cations. In a later paper however [54] they advise against tests being made from solutions outside the range of pH $c.$ 5–8.

Giles $et\ al.$ [54] advise the avoidance of dyes which may be adsorbed in a dual manner and indeed any dyes which may form covalent bonds with the surface, since they may be liable to selective adsorption on particular sites. They recommend the use of methylene blue BP, brilliant basic red B, crystal violet BP, victoria pure lake blue BO, orange II or solway ultra-blue. The two BP dyes can be used as bought, the rest need some pretreatment.

The surface area of fibres was determined by Mesderfer $et\ al.$ [79] with Direct Red 23 cation dye in cold water. Parallel tests were carried out on carbon black of known specific surface and the resulting isotherms compared.

A review of dye adsorption has been presented by Padday [55] who concentrates particularly on the use of cyanine dyes for the surface area measurement of silver bromide and silver iodide.

17.6.5 Adsorption of electrolytes

There are several variations of this technique. The negative adsorption method is based on the exclusion of co-ions from the electrical double layer surrounding charged particles [56]. The ion-exchange method is based on the replacing of loosely held ions by others of the same sign [25, p. 593]. A large amount of work has been done on the adsorption of electrolytes by ionic crystals and the adsorption of ions from solution on to metals. Since the adsorption tends to be very small and the measurements rather tedious, these are not suitable as routine methods.

17.6.6 Deposition of silver

The surface area of paper pulp fibres has been determined by the deposition of a continuous film of metallic silver on the surface by use of the reducing properties of cellulose [81]. The deposited amount of metal was determined by its ability to decompose hydrogen peroxide catalytically. A standard surface, e.g. regenerated cellulose film, is required for calibration.

17.6.7 Adsorption of p-nitrophenol

Giles *et al.* [90, 91] have investigated the adsorption of PNP for surface area evaluation. It is normally used in aqueous solution but can be used in an organic solvent [70, 71]. The method is recommended as being suitable for a wide variety of solids, both porous and non-porous, provided they either form a hydrogen bond with PNP or have aromatic nuclei. With porous solids the results may or may not reveal a lower accessible surface than nitrogen adsorption depending on the pore size distribution. With porous solids a rapid estimate of the relative proportion of small and large pores can be made together with a measure of external surface only. PNP has also been used, by a modified technique, to measure the external surface, as opposed to the much greater inner surface, of fibres [92].

Normally PNP is adsorbed flatwise with an effective molecular area of 52.5 x 10^{-20} m^2. In some cases on polar inorganic solids it is adsorbed end-on with an effective area of 25 x 10^{-20} m^2. The mode of adsorption is indicated by the type of isotherm, an S-type curve indicating end-on adsorption.

The method is similar to that used for dye adsorption [93]. Since PNP does not adsorb on cellulose it can also be used for the measurement of specific surface of colourants in cellulose [94].

The use of PNP adsorption from aqueous solutions on to porous adsorbents such as silica gel or carbons has been queried by Sandle (see [54]) who states that competition of water molecules for the surface will yield erroneous results. Giles agrees that complete coverage is not always obtained with certain acidic solids and recommends the use of other solvents.

Padday [55] states that the accuracy of PNP is suspect since many isotherms show no clearly defined plateaux.

17.6.8 Other systems

Adnadevic [102] compared the sorption of polar molecules with the BET gas adsorption technique for the surface-area determination of microporous solids. He examined activated charcoal, zeolites and silica clay using MeOH, EtOH, CO_2 and C_6H_6. Larionov [106] reviewed specific area determination by adsorption from solution. Veselov and Galenko [111] used a 1 : 1 volume ratio of C_6H_6–heptane solution and found good agreement with BET. Koganovskii and Leuchenko [112] examined the adsorption of nitrobenzene, *p*-chloroaniline and *p*-nitroaniline on active carbons and found that the Langmuir and D–R monolayer isotherms were obeyed. They also determined pore size distribution using the *t*-method with C_6H_6 as the adsorbate.

Gata [103] used ethylene glycol to examine clay minerals. Madzen [104] determined total surface (internal and external) by removal of a previously adsorbed two-layer glycerol complex by assay of the weight loss on heating. Kulshreshtha *et al.* [107] estimated fibre porosity from glycerol retention value.

The temperature change and its dependence on time, during dissolution rate studies, have also provided information on particle size distribution and surface [105]. Ruzek

and Zbuzek [108] determined surface area from dissolution rates and compared their results with electron microscope data.

A rapid method of particle size evaluation of carbon black by iodine number has also been described [109]. Chetty and Naidu [110] determined the surface area of sulphides by the isotopic exchange method using tracer ^{203}Hg for CdS and HgS and ^{124}Sb for Sb_2S_6.

Corrin *et al*. [89] determined the adsorption isotherms of sodium dodecyl sulphate and potassium myristate on ash-free graphite. Calculations based on two extreme assumptions concerning the concentration of solvent in the surface region, yield values of specific adsorption which differ by less than the experimental errors. The isotherm of sodium dodecyl sulphate exhibits a discontinuity at the critical concentration for micelle formation. The adsorption of sodium dodecyl sulphate on polystyrene was also measured. The minimum area per molecule was 5.1 nm^2 for the sulphate and 3.66 nm^2 for the myristate. Experimental methods are discussed in some detail.

17.7 Theory for heat of adsorption from a liquid phase

17.7.1 Surface free energy of a fluid

A fluid has a surface energy only when it exists in a sufficiently condensed state. A gas has no surface energy. Because of the uniform energy distribution in a gas, no difference exists between an internal molecule in the centre of the gas volume and a molecule located near to a wall.

The theory for the forces acting between molecules was put forward by Lennard-Jones and Devonshire [57]. The liquid molecule is assumed located in a cage formed by the neighbouring molecules, and is constantly under the influences of their fields yet being sufficiently free to execute translatory and rotary movements.

Each molecule in a liquid volume is surrounded by other molecules on all sides, and hence is subjected to attractive forces acting in all directions. Generally speaking a uniform attraction in all directions is exerted by every molecule for a period of time which is relatively long compared with periods of vibration.

Very different conditions obtain at the surface. The molecules are attracted back towards the liquid and also from all sides by their neighbours, yet no attraction acts outward to compensate for the attraction towards the centre. Each surface molecule is subjected to a powerful attraction towards the centre acting, for symmetrical reasons, in a direction normal to the surface.

The work required to increase the area of the surface by an infinitesimal amount dA, at constant temperature, pressure and composition, is done against a tension γ, generally known as the surface tension. The surface tension can be defined from the point of view of energy involved as shown by Brillouin [58] and Michand [59]. The free-energy change dF is equal to the reversible work done:

$$dF = \gamma dA$$

$$\gamma = \frac{(\mathrm{d}F)}{(\mathrm{d}A)_{T,P,n}} = F_s \tag{17.13}$$

where F_s is surface free energy per unit area.

Surface tension and free energy are, in effect, two different aspects of the same matter. In the SI system, the number which, in newtons per metre indicates the surface tension will, in joules per metre^2, express the surface free energy of the liquid. The two equations above express a fundamental relationship in surface chemistry.

17.7.2 Surface entropy and energy

The entropy of a system at constant pressure, surface area and composition is:

$$-S = \frac{(\mathrm{d}F)}{(\mathrm{d}T)_{P,A,n}} \tag{17.14}$$

For a pure liquid, the surface entropy per square metre S_s is:

$$-S_s = \frac{\mathrm{d}\gamma}{\mathrm{d}T} \tag{17.15}$$

The total surface energy per square metre E_s for a pure liquid is:

$$E_s = F_s + TS_s \tag{17.16}$$

or as usually expressed:

$$E_s = \gamma - T\frac{\mathrm{d}\gamma}{\mathrm{d}T} \tag{17.17}$$

It is the work which must be done in order to remove from the bulk of the liquid and bring to the surface, a sufficient number of molecules to form a surface unit. Conversely, energy is liberated when a liquid surface disappears due to the return of the molecules from the surface to the centre of the liquid. The total surface energy of a pure liquid is generally larger than the surface free energy.

17.7.3 Heat of immersion

The present accepted theory of the heat of immersion is due to Bangham [60, 61], Razouk [62], Harkins [63] and their associates.

When a clean solid surface is immersed in or wetted by a liquid, it leads to the disappearance of the solid surface and the formation of a solid–liquid interface. As a result of the disappearance of the solid surface, the total energy of the solid surface is liberated. The formation of the solid–liquid interface leads to an absorption of energy equal to the energy of the interface. Thus from thermodynamic consideration the heat of immersion (E_{imm}) is equal to the surface energy of the solid–gas E_{sv}, minus the interfacial energy E_{sL} between the solid and the liquid, so:

$$E_{imm} = (E_{sv} - E_{sL})S$$

$$-q_0 = \frac{E_{imm}}{W} = S_w(E_{sv} - E_{sL}) \tag{17.18}$$

where q_0 is the heat evolved per gram of solid, S_w is the weight specific surface area of solid, and E the energy per unit area.

By analogy with equation (17.17), the surface energies when solid–liquid and solid–vapour are in contact are written as:

$$E_{sL} = \gamma_{sL} - T \frac{d\gamma_{sL}}{dT}$$

$$E_{sv} = \gamma_{sv} - T \frac{d\gamma_{sv}}{dT}$$

which on combination give:

$$(E_{sv} - E_{sL}) = \left(\gamma_{sv} - T \frac{d\gamma_{sv}}{dT} \right) - \left(\gamma_{sL} - T \frac{d\gamma_{sL}}{dT} \right)$$

$$= (\gamma_{sv} - \gamma_{sL}) - T \frac{d(\gamma_{sv} - \gamma_{sL})}{dT} \tag{17.19}$$

This may be simplified by the use of the adhesion tension relationship of Young and Dupre ([64], *cit.* [25]).

$$\gamma_{sv} = \gamma_{sL} + \gamma_{vL} \cos \theta$$

or

$$\gamma_{vL} \cos \theta = \gamma_{sv} - \gamma_{sL} \tag{17.20}$$

where θ is the contact angle between the solid and liquid.

If equation (17.20) is substituted in (17.19), it is found that upon simplification:

$$E_{sv} - E_{sL} = \gamma_L - T \frac{d\gamma_L}{dT} \cos \theta \tag{17.21}$$

Finally substituting equation (17.21) into (17.18) and considering that a liquid that wets a solid has zero contact angle, $\cos \theta = 1$.

$$-q_0 = S_w \gamma_L - T \frac{d\gamma_L}{dT} \tag{17.22}$$

or

$$S_w = \frac{-q_0}{\gamma_L - T \frac{d\gamma_L}{dT}} \tag{17.23}$$

17.8 Static calorimetry

The calorimetric method may be used in two ways: immersion of the bare outgassed solid in pure liquid [65], and immersion of the solid precoated with the vapour phase [66]. The former approach is not widely used because of problems due to the actual state of the surface (impurities, defects), although heats of immersion per unit area of a number of solid/liquid systems are known [65, p. 301; 96, p. 286]. In order to reduce their effects Chessick et al. [97] and Taylor [98] used liquid nitrogen; the method is limited since it is not applicable to microporous materials and a surface area in excess of 150 m² is required. If the solid is first equilibrated with saturated vapour, then immersed in pure liquid adsorbate, the solid–vapour interface is destroyed and the heat liberated per unit area should correspond simply to E_L, the surface energy of the pure liquid. The above assumption is made in what is termed the absolute method of Harkins and Jura [66] who obtained a heat of immersion of 1.705 kJ kg^{-1} for titanium dioxide which, when divided by the surface energy of the adsorbent, water (11.8 mJ kg^{-1}) gave a surface area of 14.4 m² g^{-1}. For a comprehensive bibliography and description of the calorimeters used, readers are referred to Adamson [25]. The validity of the H-J method may be questioned because in a saturating vapour capillary condensation occurs which reduces the available surface [99]. A correction must also be applied for the thickness of the adsorbed film (4% for H-J anatase).

The same technique was used by Clint et al. [67] for the determination of the surface of carbon blacks by the adsorption of n-alkanes. Equation (17.23) was used with the following correction for small particles where the thickness of the adsorbed layer t was not negligible in comparison with the particle radius r:

$$S'_w = \left(\frac{r}{r+t}\right)^2 S_w \tag{17.24}$$

For low surface areas this method gave reasonable agreement with other techniques, but the surfaces for smaller particles were too low. The method was essentially comparative since the entropy is obtained using a reference sample. The method was considered unsuitable for powders having a surface area smaller than 20 m² g^{-1}.

Partyka et al. [100] examined seven non-porous adsorbents with water and amyl alcohol as adsorbates and found agreement with BET for five. The heat of immersion values indicated the number of adsorbed layers were between 1 and 2 as opposed to the 1.5 to 2 nm (5 to 7 layers) found by Harkins and Jura. They later extended this work using mainly water but also a long polar–non-polar molecule, pentanol, butanol and the non-polar molecule decane [101]. They equated q_0 with enthalpy and showed that the curve of immersion enthalpy reached a plateau when the sample is equilibrated with 1.5 layers pre-adsorbed; from this coverage onwards the interface behaves like a bulk liquid surface. The surfaces using water were in good agreement with BET areas; there was long-range interaction with butanol so this proved unsatisfactory; decane was satisfactory; pentanol compared well with water and seemed suitable for hydrophobic surfaces.

17.9 Flow microcalorimetry

A variation of this technique is employed with the flow microcalorimeter (figure 17.1). The calorimeter consists of a metal block (1) surrounding a cylindrical cavity in which the calorimeter cell (2) is situated. The cell forms a continuation of the inlet tube for the carrier liquid and is joined to the outlet (4). The outlet tube is fitted with a 200-mesh stainless-steel gauze (3) on which the powdered solids (5) are placed.

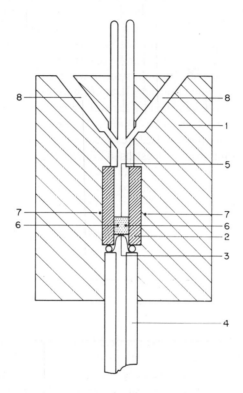

Fig. 17.1 Schematic diagram of the flow microcalorimeter: 1, metal block; 2, PTFE calorimeter cell; 3, fine stainless-steel gauze; 4, metal outlet tube; 5, powder bed; 6, two thermistors; 7, reference thermistors; 8, cavities to hold containers of carrier liquid or solution for analyses at elevated temperatures.

17.9.1 Experimental procedures – liquids

There are three main experimental methods of studying the heat of adsorption at a solid–liquid interface by using the flow microcalorimeter. These are generally referred to as:

(a) Pulse or injection adsorption.
(b) Equilibrium adsorption.
(c) Successive or incremental adsorption.

(a) Pulse adsorption
In the pulse method a micrometer syringe may be used to introduce small quantities of
surface-active substances (e.g. 1–100 μg as 0.1 to 1% solutions). These should be introduced into the stream of carrier liquid against the wall of the inlet tube below the point
at which the liquid leaves the flow-control capillary.

Any change in the calorimeter cell will be registered by the recorder due to adsorption of active agent from a solution on the solid surfaces. Normally adsorption is
accompanied by the evolution of heat in the bed with a corresponding recorder deflection. In the case of an irreversible (chemical) change, the recorder-pen will return to
the base-line, but if the change is reversible (physical) the pen will cross the base line
to describe a negative heat of desorption. A typical pulse adsorption is shown in figure
17.2(a).

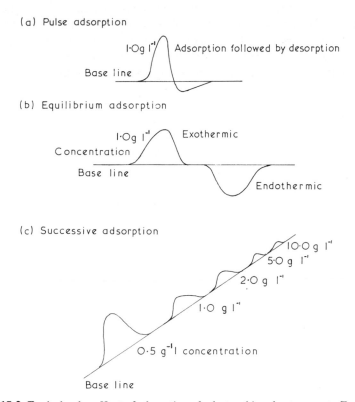

Fig. 17.2 Typical pulses. Heat of adsorption of *n*-butanol in *n*-heptane on to Fe$_2$O$_3$
adsorbent. (a) Pulse adsorption; (b) equilibrium adsorption; (c) successive adsorption.

(b) Equilibrium adsorption
In this method two reservoirs are prepared, one of which contains the pure carrier
liquid (solvent) and the second the solution of active agent. Initially a steady flow of
carrier liquid is allowed to flow through the adsorbent bed. When the calorimeter

comes to thermal equilibrium, giving a stable base-line on the recorder chart, the flow of carrier liquid may be interchanged for the flow of active solution. Care should be taken that the flow-rate of this solution should not differ from that of the carrier liquid by more than 0.01 ml min^{-1}.

Adsorption of solute, which is accompanied by a heat, hence temperature, change is measured by the thermistors which are connected via a Wheatstone bridge network to a potentiometric recorder. The result for an exothermic reaction (heat evolution typical of adsorption) is a positive pulse on the pen-recorder trace which then returns to the base line when the adsorption of solute is complete. Desorption may be carried out by returning to a flow of the carrier liquid. The result for an endothermic reaction (heat change typical of desorption) is a negative pulse on the recorder trace; see figure 17.2(b).

For physical desorption, the area under the pulse is the same as the area under the adsorption peak. The rate of desorption, which controls the shape of the desorption pulse, depends on the relative strengths of adsorption of the solute and solvent. In some instances a long time elapses before all the solute molecules are removed, and this results in the desorption pulse having a long trailing tail, making it difficult to determine when desorption is complete.

In chemical adsorption the heat generated will be greater than for physical adsorption (greater than, compared with less than, 42 kJ mol^{-1}). Further, the desorption peak is much smaller than the adsorption peak.

Adsorption–desorption of various solute concentrations may be studied on the same plug or different plugs of the same adsorbent. This way the form of adsorption isotherm can be determined.

In practice it is found that using the same plug of adsorbent and different concentrations of solution is not a good method to adopt for the adsorption process, due to the difficulty of determining when desorption is complete. It is therefore necessary to use a different plug for each point on the isotherm.

(c) Successive adsorption

In this method of using increasing concentrations of the solution for adsorption on to the adsorbent a series of heat effects occur (pulses), and these generally decrease for equal increments in solute concentration. When the surface of an adsorbent is completely saturated with the solute, further runs with increasing concentration do not give any heat effects. Adding the pulse areas gives the same results as using the technique described in the above section. This method is however simpler than the previous one, although errors in measuring pulse areas are compounded.

This method is therefore the preferred one. An example of the type of results obtained is shown in figure 17.2(c).

17.9.2 Calibration

Calibration is effected by replacing the standard outlet tube with one containing a heating coil. With a powder bed in position and the carrier liquid flowing, known

quantities of heat are injected into the system. A calibration line is then produced of area under the pulses against heat injected.

17.9.3 Determination of the amount of solute adsorbed: the precolumn method

In constructing adsorption isotherms, it is necessary to measure the amount of solute adsorbed at a range of concentrations. This is usually done by determining changes in the concentrations of solute before and after contact with a given mass of adsorbent. For the following however a variation of this technique known as the precolumn method is used.

The adsorbent is placed in a precolumn constructed from a glass tube with the same internal diameter as the cell in the calorimeter. The carrier liquid is then percolated through the adsorbent in the precolumn before it enters the adsorbent in the calorimeter. To determine the amount of solute adsorbed from a given solution, the flow of carrier liquid is stopped at time t_1 and at the same time the flow of solution is started. If the precolumn contained an inert solid, the solution would emerge at time t_0 and would then contact the adsorbent in the calorimeter, but when the precolumn contains an adsorbent, the solute is retained, so that only solvent emerges at time t_0, the solution emerging at a later time t_2. The difference between t_2 and t_0 is a measure of the amount of solute adsorbed.

In the above method the critical points are the accurate determination of the time at which the solution emerges from the precolumn and an accurate knowledge of the constant flow-rate of solution through the precolumn. It is also necessary to ensure, for all adsorbates, that the flow-rates used are sufficiently low to permit full saturation of the adsorbent in the precolumn. If the flow-rate of the solution is too high, solution may emerge from the precolumn before the adsorbent is fully saturated. In such a case the estimate for the amount of solute adsorbed would be lower than the equilibrium value.

However, if the retention time and constant flow rate are known, the amount of the solute adsorbed can be calculated from the equation:

$$x = \frac{tCf}{w}$$

(17.25)

where x = amount of solute adsorbed (mg/g),
$\quad t$ = retention time ($t_2 - t_0$),
$\quad C$ = concentration of solute (mg ml^{-1}),
$\quad f$ = flow rate of solution (ml min^{-1}),
and w = weight of adsorbent in the precolumn (g).

17.9.4 Gases

A similar procedure can be carried out with gases using the mark 2V flow microcalorimeter. This technique has been used with carbon dioxide as the adsorbate and nitrogen as the carrier gas [74]. Adsorbents used were γ-alumina, Gasil and Degussa silica;

experiments were performed at 22°C and compared with carbon dioxide adsorption at 22°C and −78°C. The carbon dioxide isotherm at 22°C on γ-alumina was found to obey Sips' equation [75] which is valid for adsorption, at low relative pressures, on a non-uniform surface.

$$\frac{V}{V_m} = \theta = \left(\frac{P}{P+B}\right)^c \tag{17.26}$$

The procedure to determine V_m is to select the value of B to give the best fit when $\log V$ is plotted against $\log P/(P+B)$.

The heat of adsorption of carbon dioxide on γ-alumina decreased with increasing coverage thus substantiating that the surface was non-uniform. The surface of Gasil silica was found to be slightly heterogeneous and that of Degussa silica was found to be homogeneous. In later papers Allen and Burevski examined the adsorption of gases on microporous carbons [76] and the adsorption of sulphur dioxide on powdered adsorbents [77].

17.9.5 Application to the determination of surface area

Using liquid-flow microcalorimetry, it is possible to determine the area occupied per molecule and the energy of adsorption per molecule. Allen and Patel [12, 13] investigated a range of long-chain fatty acids and long-chain alcohols and obtained information on molecular orientation at solid–solution interfaces. Surface areas may be evaluated using the following form of equation (17.23)

$$S_w = K_0 q_m \tag{17.27}$$

where q_m is the Langmuir monolayer value. For the adsorption of n-octoic acid from n-heptane, $K_0 = 16.7 \text{ m}^2 \text{ J}^{-1}$, and two-thirds of the determined surface areas agreed with BET nitrogen adsorption values to within 10%. Full details of experimental procedures and results are given in [68].

For the adsorption of gases, Burevski [69] found the energy of adsorption was not constant but varied with coverage. The manner in which it varied depended on the system under examination, rendering the method unsuitable for surace-area determination.

17.10 Density method

The use of the air comparison pyknometer for skeletal density measurements has been described in Chapter 8. When instruments, such as the Beckmann, are used with air, erroneous results are often obtained and sometimes even negative volumes are found. This indicates that air is being adsorbed and the amount adsorbed can be determined by comparison with helium density measurements. As nitrogen and oxygen at room temperatures obey Henry's law, the measured adsorption on increasing the pressure from 1 to 2 atm is practically equivalent to the air adsorption at 1 atm.

The formula is

$$wS_w = CV_a$$

$$= C(V_{\text{He}} - V_{\text{N}_2})$$

$$= C\left[\frac{1}{\rho_{\text{He}}} - \frac{1}{\rho_{\text{N}_2}}\right] \tag{17.28}$$

where S_w is the weight specific surface of the sample

V_a is the volume of air adsorbed as the pressure increases from 1 atm to 2 atm

w is the weight of powder

$V_{\text{He}}, V_{\text{N}_2}$ are the indicated powder volumes using helium and nitrogen respectively

$\rho_{\text{He}}, \rho_{\text{N}_2}$ are the indicated powder densities using helium and nitrogen respectively

C is a constant.

Tuul and Innes [95] reported a variation of C with substrate whereas Jäkel [78] stated that C was constant for a wide range of materials.

References

1 Adam, N.K. (1941), *The Physics and Chemistry of Surfaces*, Oxford University Press.
2 Rayleigh, Lord (1899), *Phil. Mag.*, **48**, 321.
3 Pockels, A. (1891), *Nature*, **43**, 437.
4 Langmuir, I. (1917), *J. Am. Chem. Soc.*, **39**, 1848.
5 Linnar, E.R. and Williams, A.P. (1950), *J. Phys. Colloid Chem.*, **54**, 605.
6 Vold, M.J. (1952), *J. Colloid Sci.*, **7**, 196.
7 Kipling, J.J. and Wright, E.H.M. (1963), *J. Chem. Soc.*, 3382.
8 Kipling, J.J. and Wright, E.H.M. (1964), *ibid.*, 3535.
9 McBain, J.D. and Dunn, R.C. (1948), *J. Colloid Sci.*, **3**, 308.
10 Orr, C. and Dallavalle, J.M. (1959), *Fine Particle Measurement*, Macmillan, NY.
11 Ward, A.F.H. (1946), *Trans. Faraday Soc.*, **42**, 399.
12 Allen, T. and Patel, R.M. (1971), *J. Colloid Interfac. Sci.*, **35**, 4, 647–55.
13 Allen, T. and Patel, R.M. (1971), *Particle Size Analysis*, Soc. Analyt. Chem., London.
14 Allen, T. and Patel, R.M. (1970), *J. Appl. Chem.*, **20**, 165–71.
15 Harkins, W.D. and Gans, D.M. (1931), *J. Am. Chem. Soc.*, **53**, 2804.
16 Harkins, W.D. and Gans, D.M. (1932), *J. Phys. Chem.*, **36**, 86.
17 Russell, A.S. and Cochran, C.N. (1950), *Ind. Engng Chem.*, **42**, 1332.
18 Harkins, W.D. and Jura, G. (1944), *Chem. Phys.*, **66**, 1366.
19 Freundlich, H. (1907), *Z. Phys. Chem.*, **57**, 385.
20 Linnar, E.R. and Gortner, R.A. (1935), *J. Phys. Chem.*, **39**, 35–67.
21 Traube, I. (1891), *Ann. Phys., Liepzig*, **265**, 27.
22 Holmes, H.N. and McKelvey, J.B. (1928), *J. Phys. Chem.*, **32**, 1522.
23 Harkins, W.D. and Dahlstrom, R. (1930), *Ind. Engng Chem.*, **22**, 897.
24 Langmuir, I. (1917), *Trans. Faraday Soc.*, **42**, 399.
25 Adamson, A. W. (1963), *Physical Chemistry of Surfaces*, Interscience, NY.
26 Weissberger, A. *et al.* (1955), *Organic Solvents*, Interscience, NY.
27 de Boer, J.H. (1953), *The Dynamical Characteristics of Adsorption*, Princeton University Press.
28 Berthier, P., Kerlan, L. and Courty, C. (1858), *C.r. Acad. Sci., Paris*, **246**, 1851.

29 Greenhill, E.B. (1949), *Trans. Faraday Soc.*, **45**, 625.
30 Krasnovskii, A.A. and Gurevich, T.N. (1949), *Chem. Abs.*, **43**, 728.
31 Innes, W.B. and Rowley, H.H. (1947), *J. Phys. Chem.*, **51**, 1172.
32 Hirst, W. and Lancaster, J.K. (1951), *Trans. Faraday Soc.*, **47**, 315.
33 Hutchinson, E. (1947), *ibid.*, **43**, 439.
34 Gregg, S.J. (1947), Symp. Particle Size Analysis, *Trans. Inst. Chem. Eng., London*, **25**, 40–6.
35 Crisp, D.J. (1956), *J. Colloid Sci.*, **11**, 356.
36 Smith, H.A. and Fusek, J.F. (1946), *J. Am. Chem. Soc.*, **68**, 229.
37 Smith, H.A. and Hurley, R.B. (1949), *J. Phys. Colloid Chem.*, **53**, 1409.
38 Kipling, J.J. and Wright, E.H.M. (1962), *J. Chem. Soc.*, **855**, 3382–9.
39 Maron, S.H., Ulevith, I.N. and Elder, M.E. (1949), *Analyt. Chem.*, **21**, 691.
40 Hanson, R.S. and Clampitt, B.H. (1954), *J. Phys. Chem.*, **58**, 908.
41 Bartell, F.E. and Sloan, C.K. (1929), *J. Am. Chem. Soc.*, **51**, 1637.
42 Ewing, W.W. and Rhoda, R.N. (1951), *Analyt. Chem.*, **22**, 1453.
43 Candler, C. (1951), *Modern Interferometers*, Hilger, London.
44 Groszek, A.J. (1968), *SCI Monograph, No. 28*, 174.
45 Flory, P.J. (1953), *Principles of Polymer Chemistry*, Cornhill University Press, Ithaca, p. 579.
46 Morawetz, H. (1965), *Macromolecules in Solution*, Interscience, NY.
47 Jenkel, E. and Rumbach, B. (1951), *Z. Electrochem.*, **55**, 612.
48 Habden, J.F. and Jellinek, H.H.G. (1953), *J. Polymer Sci.*, **11**, 365.
49 Frisch, H.C. and Simha, R. (1954), *J. Phys. Chem.*, **58**, 507.
50 Jellinek, H.H.G. and Northey, H.L. (1954), *J. Polymer Sci.*, **14**, 583.
51 Eltekov, Yu. A. (1970), *Surface Area Determination*, Butterworths, pp. 295–8.
52 Kolthoff, I.M. and MacNevin, W.N. (1937), *J. Am. Chem. Soc.*, **59**, 1639.
53 Japling, D. W. (1952), *J. Appl. Chem.*, **2**, 642.
54 Giles, G.H., Silva, A.P.D. and Trivedi, A.S. (1970), *Surface Area Determination* (D.H. Everett and R.H. Ottewill eds), Butterworths, p. 317.
55 Padday, J.F. (1970), *ibid.*, pp. 331–7.
56 Lyklema, J. and Van der Hul, H.J. (1970), *ibid.*, pp. 341–54.
57 Lennard-Jones, J.E. and Devonshire, A.F. (1937), *Proc. R. Soc.*, **163A**, 53.
58 Brillouin, L. (1938), *J. Phys.*, **9**, 7, 462.
59 Michand, F. (1939), *J. Chim. Phys.*, **36**, 23.
60 Bangham, D.H. and Razouk, R.I. (1937), *Trans. Faraday Soc.*, **33**, 1459.
61 Bangham, D.H. and Razouk, R.I. (1938), *Proc. R. Soc.*, **166**, 572.
62 Razouk, R.I. (1941), *J. Phys. Chem.*, **45**, 179.
63 Harkins, W.D. (1919), *Proc. natn. Acad. Sci.*, **5**, 562.
64 Dupre, A. (1869), *Mechanical Theory of Heat*, Paris, p. 368.
65 Gregg, S.J. and Sing, K.S.W. (1967), *Adsorption Surface Area and Porosity*, Academic Press, London.
66 Harkins, W.D. and Jura, G. (1944), *J. Am. Chem. Soc.*, **66**, 1362.
67 Clint, J.H. *et al.* (1970), *Proc. Int. Symp. Surface Area Determination*, Bristol, 1969, Butterworths.
68 Patel, R.M. (1971), Physical adsorption at solid–liquid interface, Ph.D. thesis, Univ. Bradford.
69 Burevski, D., private communications, Ph.D. project, Univ. Bradford.
70 Giles, C.H. and D'Silva, A.P. (1969), *Trans. Faraday Soc.*, **65**, 1943.
71 Giles, C.H., D'Silva, A.P. and Trivedi, A.S. (1970), *J. appl. Chem.*, **20**, 37.
72 Giles, C.H. *et al.* (1978), *Proc. Conf. Structure of Porous Solids*, Neuchatel, Switzerland, Swiss Chem. Soc.
73 Giles, C. H. and Trivedi, A.S. (1969), *Chem. Ind.*, 1426–7.
74 Allen, T. and Burevski, D. (1977), *Powder Technol.*, **17**, 3, 265–72.
75 Sips, R. (1950), *J. Chem. Phys.*, **18**, 1024.
76 Allen, T. and Burevski, D. (1977), *Powder Technol.*, **18**, 2, 139–48.
77 Allen, T. and Burevski, D. (1978), *ibid.*, **21**, 1, 91–96.

78 Jäkel, K. (1972), Beckmann Report 2, S.33-35, Beckmann Instruments GmbH, Frankfurter Ring 115, D-8000, Munchen 40, W. Germany.
79 Mesderfer, J.W. *et al*. (1952), *TAPPI*, 35, 374.
80 Giles, C.H. and Tolia, A.H. (1964), *J. appl. Chem*., 14, 186–95.
81 Clark, J. d'A. (1942), *Paper Trade J*., 115, 32.
82 Giles, C.H. *et al*. (1959), *J. Chem. Soc*., 535–44.
83 Giles, C.H., Greczek, J.J. and Nakhura, S.N. (1961), *ibid*., 93–95.
84 Giles, C.H., Easton, I.A. and McKay, R.B. (1964), *ibid*., 4495–503.
85 Allington, M.H. *et al*. (1958), *J. Chem. Soc*., 8, 108–16.
86 Giles, C.H. *et al*. (1958), *ibid*., 8, 416–24.
87 Wegmann, J. (1962), *Am. Dyest. Rep*., 51, 276.
88 Padhye, M.R. and Karnik, R.R. (1971), *Indian J. Technol*., 9, 320–2.
89 Corrin, M.L. *et al*. (1949), *J. Colloid. Sci*., 4, 485–95.
90 Giles, C.H. *et al*. (1960), *J. Chem. Soc*., 3793–973.
91 Giles, C.H. and Nakhura, S.N. (1962), *J. Appl. Chem*., 12, 266–73.
92 Giles, C.H. and Tolia, A.H. (1964), *ibid*., 14, 186–94.
93 Giles, C.H., D'Silva, A.P. and Trivedi, A.S. (1970), *ibid*., 20, 37–41.
94 Giles, C.H. *et al*. (1971), *J. Appl. Chem. Biotechnol*., 21, 5–9.
95 Tuul, J. and Innes, W.B. (1962), *Analyt. Chem*., 34, 7, 818–20.
96 Bickerman, J.J. (1970), *Physical Surfaces*, Academic Press, NY.
97 Chessick, J.J., Young, G.J. and Zettlemoyer, A.C. (1954), *Trans. Faraday Soc*., 50, 587.
98 Taylor, J.A.G. (1965), *Chemy. Ind*., 2003.
99 Wade, W.H. and Hackerman, N. (1960), *J. Phys. Chem*., 64, 1196.
100 Partyka, S. *et al*. (1975), *4th Int. Conf. Thermodynamic Chemistry* (CR), 7, 46–55.
101 Partyka, S., Rouquerol, F. and Rouquerol, J. (1979), *J. Colloid Interfac. Sci*.. 68, 1, 21–31.
102 Adnadevic, B.K. and Vucelic, D.R. (1978), *Glas. Hem. Drus., Belgrade*, 43, 7, 385–92.
103 Gata, G. (1975), *Tek. Econ. Inst. Geol., Rumania*, Ser. 1, 13, 13–19.
104 Madzen, F.T. (1977), *Thermochim. Acta*, 21, 1, 89–93.
105 Richter, V., Merz, A. and Morgenthal, J. (1975), *Konf. Metal. Proszkow Pol.* (mater. Konf.), Inst. Metal. Niezelaz, Glivice, Poland, pp. 123–34.
106 Larionov, O.G. (1976), *V. sb Adsorbtsiya i Poristot*., 122–6.
107 Kulshreshtha, A.K., Chudasama, V.P. and Dweltz, N.E. (1976), *J. appl. Polym. Sci*., 20, 9, 2329–38.
108 Ruzek, J. and Zbuzek, B. (1975), *Silikaty*, 19, 1, 49–66.
109 Kloshko, B.N. *et al*. (1974), *Nauch tekhu sb*, 6, 22–24.
110 Chetty, K.V. and Naidu, P.R. (1972), *Proc. Chem. Symp*., 1, 79–83, Dept. Atom. Energy, Bombay.
111 Vesolov, V.V. and Galenko, N.P. (1974), *Zh. Fiz. Khim*., 48, 9, 2276–9.
112 Koganovskii, A.M. and Leuchenko, T.M. (1976), *Dopou Akad. Ukr. SSR*, Ser. B, 4, 326–8.
113 Roe Ryong-Joon, (1975), *J. Colloid Interfac. Sci*., 50, 1, 64–69.

18 Determination of pore size distribution by gas adsorption

18.1 Miscellaneous techniques

Pore size and size distribution have significant effects over a wide range of phenomena from the adsorbency of fine powders in chemical catalysis to the frost resistance of bricks. Due to this, pore size measurements have been described using a wide range of techniques and apparatus. Pore surface area is generally accepted as being the difference between the area of the surface envelope of the particle and its total surface area. The pores may be made up of fissures and cavities in the particle: they may be V-shaped, i.e. wide-necked; or 'ink-bottle' pores, i.e. narrow-necked. In order that their volume distribution may be determined, it is necessary that they are not totally enclosed and that the molecules used for measurement purposes may enter through the neck. The presence and extent of small open pores may be determined by finding the volume of a powder by immersing it in mercury and finding the volume displacement, then finding the volume in a gas pyknometer, using helium as the gas [1]. Since mercury does not wet most solids it leaves the pores unfilled, and the difference between the two volumes is the pore volume. Closed pores may be evaluated by grinding the powder which opens out some of the pores, thus decreasing the apparent solid volume [2].

Total pore volume may be determined by boiling the powder in a liquid, decanting the liquid and determining the volume of liquid taken up by the solid after it has been superficially dried. Pore size distribution may be found by using a range of liquids of different molecular sizes [2]. Direct visual examination of particle sections under the optical and electron microscopes has also been used with a range of powders [3-6]. Porosity may also be measured from the absorption of gamma rays [67].

18.2 The Kelvin equation

The adsorption of a vapour on to a porous solid is governed by the Kelvin equation which may be derived as follows. Consider a liquid within a pore in equilibrium with its vapour. Let a small quantity, δa moles, be distilled from the bulk of liquid outside the pore, where its equilibrium pressure is P_0, into the pore where its equilibrium pressure is P. The total increase in free energy dG is the sum of three parts: evaporation of δa moles of liquid at pressure $P_0(\delta G_1)$; expansion of δa moles of vapour from

pressure P_0 to pressure $P(\delta G_2)$; condensation of δa moles of vapour to liquid at pressure $P(\delta G_3)$. Since the condensation and evaporation are equilibrium processes $\delta G_1 = \delta G_2 = 0$, whilst the increase in free energy during expansion $\delta G_3 = RT \ln (P_0/P) \cdot \mathrm{d}a$, the vapour being assumed to behave as a perfect gas.

The condensation of the vapour in the pores results in a decrease in solid–vapour interface and an increase in solid–liquid interface of area δS. The increase in free energy during this process is:

$$\delta G' = \delta S(\gamma_{SL} - \gamma_{SV})$$

where

$$\gamma_{SL} - \gamma_{SV} = \delta_{LV} \cos \theta$$

since

$$\delta G' = -\delta G$$

$$\delta a RT \ln \frac{P}{P_0} = -\gamma_{LV} \cos \theta \cdot \delta S$$

The volume condensed in the pores is:

$$\delta v_c = V_L \, \delta a$$

where V_L is the molar volume. Therefore,

$$\frac{\delta v_c}{V_L} \cdot RT \ln \frac{P}{P_0} = -\gamma_{LV} \cos \theta \cdot \delta S$$

the limiting case being:

$$\frac{\mathrm{d} v_c}{\mathrm{d} S} = \frac{-V_L \, \gamma_{LV} \cos \theta}{RT \ln (P/P_0)} \tag{18.1}$$

For cylindrical pores of radius r and length L:

$$v_c = \pi r^2 L$$

$$S = 2\pi r L$$

hence

$$\frac{v_c}{S} = \frac{r}{2}$$

Equation (18.1) may therefore be written:

$$RT \ln x = \frac{-2\gamma_{LV} \cdot V_L \cdot \cos \theta}{r} \tag{18.2}$$

where $x = P/P_0$.

For non-cylindrical pores having mutually perpendicular radii r_1 and r_2, equation (18.2) becomes [1, p. 141]:

$$RT \ln x = -\gamma_{LV} V_L \left[\frac{1}{r_1} + \frac{1}{r_2} \right] \cos \theta \qquad (18.3)$$

For nitrogen at liquid nitrogen temperature [7]

$$\gamma_{LV} = 8.72 \times 10^{-3} \text{ N m}^{-2}$$

$$V_L = 34.68 \times 10^{-6} \text{ m}^3 \text{ mole}^{-1}$$

$$R = 8.314 \text{ J mole}^{-1} \text{ K}^{-1}$$

$$T = 78 \text{ K}$$

$$\theta = 90°$$

$$r_K = -\frac{4.05}{\log x} \times 10^{-10} \text{ m}$$

where r_K is the Kelvin radius.

Gregg and Sing [1, p. 162] prefer to use $V_L = 36 \text{ cm}^3 \text{ mole}^{-1}$ and γ as $8.85 \times 10^{-3} \text{ N m}^{-2}$ making:

$$r_K = -\frac{4.10}{\log x} \times 10^{-10} \text{ m}$$

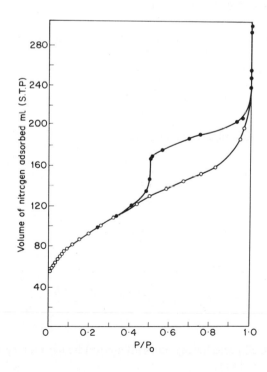

Fig. 18.1 Adsorption of nitrogen on activated clay catalyst. Open circles indicate adsorption; solid circles, desorption. Total (BET) area, $S_w = 339 \text{ m}^2 \text{ g}^{-1}$; monomolecular volume $v_m = 78.0 \text{ cm}^3$ gas at STP.

Pore volume and pore surface distributions may be determined from gas adsorption isotherms. If the amount of gas adsorbed on the external surface is small compared with the amount adsorbed in the pores, the total pore volume is the condensed volume adsorbed at the saturation pressure.

With many adsorbents a hysteresis loop occurs between the adsorption and desorption branches of the isotherm (figure 18.1). This has been explained as being due to capillary condensation augmenting multilayer adsorption at the pressures at which hysteresis is present, the radii of curvature (equation (18.3)) being different during adsorption from the radii during desorption.

18.3 The hysteresis loop

Consider a cylindrical pore open at both ends and of radius r_p (figure 18.2). During adsorption $r_1 = r_c, r_2 = \infty$ where r_c is the core radius:

$$r_c = r_p - t \qquad (18.4)$$

where t is the thickness of the condensed vapour in the pores.

During desorption the radii are $r_1 = r_2 = r_c$. Inserting in equation (18.3) gives:

$$RT \ln x_A = - \frac{\gamma_{LV} V_L \cos \theta}{r_c}$$

$$RT \ln x_D = - \frac{2\gamma_{LV} V_L \cos \theta}{r_c}$$

Hence:

$$x_A^2 = x_D \qquad (18.5)$$

where A = adsorption, D = desorption.

Hence, for a given volume v adsorbed $x_A > x_D$.

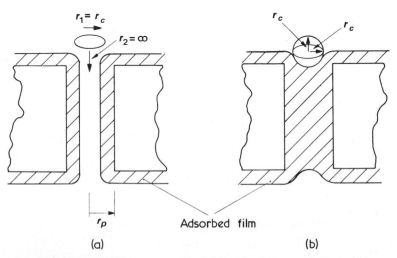

Fig. 18.2 (a) Adsorption in, and (b) desorption from a cylindrical pore.

Fifteen shape groups of capillaries were analysed by de Boer [2] from a consideration of five types of hysteresis loop (figure 18.3).

Type A   Both adsorption and desorption branches are steep at intermediate relative pressures.

Type B   The adsorption branch is steep at saturation pressure, the desorption branch at intermediate relative pressures.

Type C   The adsorption branch is steep, at intermediate relative pressures the desorption branch is sloping.

Type D   The adsorption branch is steep at saturation pressure, the desorption branch is sloping.

Type E   The adsorption branch has a sloping character, the desorption branch is steep at intermediate relative pressures.

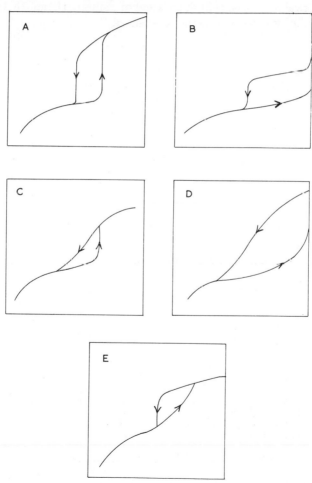

Fig. 18.3 de Boer's five types of hysteresis loop.

Type A   includes tubular capillaries open at both ends; tubular capillaries with slightly widened parts; tubular capillaries of two different main dimensions, wide-necked 'ink-bottle' pores provided $r_n < r_w < 2r_n$; tubular capillaries with one narrowed part, narrow-necked 'ink-bottle' pores; capillaries with wide parts and narrow short necks open at both ends, trough-shaped capillaries (r_n, r_w are the radii of the narrow and wide parts respectively).

Type B   includes open slit-shaped capillaries with parallel walls, capillaries with very wide bodies and narrow short necks.

Type C   typifies a heterogeneous distribution of pores of some of the following shape groups: tapered or doubly tapered capillaries and wedge-formed capillaries with closed sides and open ends.

Type D   loops occur for a heterogeneous assembly of capillaries with wide bodies of sufficiently wide dimension r_w and having a greatly varying range of narrow necks; wedge-shaped capillaries open at both ends.

Type E   loops can be formed by assemblies of capillaries of one of the shape groups for type A when the dimensions responsible for the adsorption branch are heterogeneously distributed and the dimensions responsible for the desorption branch are of equal size.

The theoretical treatment for open-ended cylindrical pores has already been covered, and results in equation (18.5). For tubular capillaries with narrow necks and wide bodies where $r_w \leqslant 2r_n$, the necks will fill when the relative pressure corresponds to r_n; this will produce a spherical meniscus in the wider parts and increase the pressure there to:

$$P = P_0 \exp \left[\frac{2\gamma V_L \cos \theta}{RTr_n} \right]$$

which is greater than the pressure required to fill the wider parts:

$$P = P_0 \exp \left[\frac{\gamma V_L \cos \theta}{RTr_w} \right]$$

The whole capillary will therefore fill at the adsorption pressure for the small capillary. On desorption it is emptied when the pressure is given by $RT \ln x_D = 2\gamma V_L / r_n \cos \theta$. Hence $x_D = x_A^2$, as before.

For parallel plates or open slit-shaped capillaries, a meniscus cannot be formed during adsorption, but during desorption a cylindrical meniscus is already present, hence desorption is delayed to produce hysteresis.

With 'ink-bottle' pores with wide closed bodies and open short necks, the necks are filled at a relative pressure corresponding to r_n, but it is only at a relative pressure corresponding to $\frac{1}{2}r_w$ that the whole capillary is filled. Emptying takes place at a relative pressure corresponding to $\frac{1}{2}r_n$. Hence $x_D = x_A^2$.

For cylindrical pores closed at one end there is no hysteresis.

For a full discussion readers are referred to the original paper [2].

The relative pressure at which the hysteresis loop closes depends on the nature of the adsorbate [7] being around 0.42 for nitrogen. Low pressure hysteresis is associated with inelastic distortion of the solid [28].

18.4 Relationship between the thickness of the adsorbed layer and the relative pressure
(see also section 16.2.13)

From the Kelvin equation one can calculate the value of r, say r_1, corresponding to any given point on the isotherm, i.e. for any given value of the relative pressure, x_1 and volume adsorbed v_1. If one were to neglect the amount adsorbed on the walls, then v_1 would be equal to the volume of all the pores which have radii up to and including r_1. One could then plot a graph of v against r and the pore size distribution would result.

If allowance is made for the thickness of the adsorbed film t, the relevant radius would be $r_1 - t$, and the volume adsorbed would be made up of two parts, the volume filling capillary cores and the volume which increases the thickness of the adsorbed layer on pores with radii greater than r_1. In order to determine the pore size distribution, it is therefore necessary to know t.

Oulton [3] assumed the thickness of the adsorbed layer remained constant over the whole pressure region. More accurately, t must be related to the amount adsorbed. If V_m is the monolayer capacity of a non-porous reference solid, the adsorption at any pressure can be converted into film thickness:

$$t = y \left(\frac{V}{V_m} \right) \qquad (18.6)$$

where y is the average thickness of one layer of molecules.

The value of y will depend upon the method of stacking of successive layers. For nitrogen: if a cubical packing is assumed, $y = \sqrt{16.2} = 4.02$ Å. Schull [5] and Wheeler [4] assumed a more open packing and arrived at a value $y = 4.3$ Å. Lippens et al. [30] assumed hexagonal close-packing for nitrogen to give:

$$y = \frac{MV_s}{N\sigma} \qquad (18.7)$$

where M is the molecular weight of the gas
V_s is the specific volume
N is the Avogadro number
σ is the area occupied by one molecule.

For nitrogen, $y = \dfrac{28 \times 1.235 \times 10^{-6}}{6.023 \times 10^{23} \times 16.2 \times 10^{-20}}$ m

$= 0.354$ nm

t may be obtained in terms of x by inserting equation (18.6) in the BET equation **to give:**

$$t = \frac{cyx}{(1 - x) [1 + (c - 1)x]} \tag{18.8}$$

Since the BET equation predicts too high a value for V in the high pressure region this equation will predict too high a value for t in this region.

Schüll [5] used the multilayer thickness from experimental data with non-porous solids. He then developed a simplified model for fitting experimental data to theoretical size distributions.

Wheeler [4] suggested that the adsorption on the walls of fine pores is probably greater than on an open surface at low relative pressures and proposed the use of Halsey's equation [14].

$$t^3 \ln \frac{P_0}{P} = 5y^3 \tag{18.9}$$

This equation was also used by Dollimore and Heal [11] and Giles *et al.* [64]. Barrett *et al.* [6] used a computation which is in effect a tabular integration of Wheeler's equation, but they introduce a constant C in a manner criticized by Pierce [8]. Their distributions do, however, tend to agree with those obtained by high-pressure mercury-intrusion methods [9].

Cranston and Inkley [10] derived a curve of thickness of adsorbed layer t against relative pressure x from published isotherms on fifteen non-porous materials by dividing the volume of nitrogen adsorbed by the BET surface area. They state that their method may be applied either to the adsorption or desorption branch of the iso-therm and that the indications were that the adsorption branch should be used, a proposal which was at variance with current practice. They assumed cylindrical pores with one end closed but stated that this assumption was unnecessary.

Pierce [8] begins with the sample saturated with vapour at P_0 and derives a pore size distribution by considering incremental desorption as the pressure drops. He uses a cylindrical pore model and applies the Kelvin equation, assuming the residual layers to be the same as on a non-porous surface at the same relative pressure. In a later paper [15] he uses the FHH equation with $s = 2.75$ and $b = 2.99$ for nitrogen adsorp-tion on carbons, oxides and ionic crystals.

Fifteen papers by de Boer and associates form a notable contribution to an under-standing of pore systems in catalysts [30–44]. Attempts to improve and simplify the earlier models have been carried out by Dollimore and Heal [11], who used a cylin-drical model; Innes [12], who used a parallel plate model; and Roberts [13] whose method is applicable to both the above models. It was found by de Boer *et al.* [35] that the amount of nitrogen adsorbed per unit of surface area of non-porous adsorb-ents is a unique function of the relative pressure for a large number of inorganic oxides and hydroxides as well as for graphitized carbon blacks. If it is assumed that the adsorbed nitrogen multilayer has the same molar volume as the bulk liquid at the same temperature, then the common isotherm may be represented in the form of a t-curve representing the thickness of the adsorbed layer (in Å) as a function of x (see figure 16.2).

Up to a relative pressure of 0.75 to 0.80 this t-curve may be represented by an empirical relation of the Harkins–Jura type [39] $(t < 10$ Å) [16]:

$$\log x = 0.034 - 13.99/t^2 \qquad (18.10)$$

For x greater than about 0.4 up to 0.96, the following empirical relationship holds $(t < 5.5$ Å)

$$\log x = -16.11/t^2 + 0.1682 \exp(-0.1137t) \qquad (18.11)$$

Alternatively, the isotherm may be represented by an equation of the Anderson type (section 16.2.7) with $c = 53$ and $k = 0.76$.

Brockhoff and de Boer [38, 39] state that the thickness of the adsorbed layer in a cylindrical pore is expected to be different from that on a flat surface at the same pressure and suggest the use of a modified form of the Kelvin equation.

$$RT \ln \left(\frac{P_0}{P}\right) = \frac{\gamma\sigma}{r_p - t} + 2.303RT \cdot F(t) \qquad (18.12)$$

where $F(t)$ is put equal to equation (18.10) or (18.11). This was later extended to 'ink-bottle'-type pores [40] and applied to cylindrical [41, 42] and slit-shaped pores [43].

Radjy and Sellevold [61] developed a phenomenological theory for the t-method of pore structure analysis for slit-shaped and cylindrical pores.

A comparison of adsorption and desorption methods for pore size distribution, with transmission electron microscopy using closely graded cylindrical pores in alumina closed at one end, confirms the superiority of the Brockhoff–de Boer equations over the Kelvin equation [23].

Lamond [24] found Lippen's t-silica values unsuitable for carbon blacks and proposed a new set based on adsorption data for fluffy blacks.

18.5 Classification of pores

Dubinin [18, 19] classified pores into three groups: (a) macropores having widths in excess of one or two thousand Angstroms; capillary condensation does not take place in these pores which are essentially an avenue of transport to smaller pores; (b) intermediate, also known as transition or mesopores having widths of between 15 and 16 Å up to 1000 to 2000 Å; these mark the limit of applicability of the Kelvin equation; (c) micropores, from 5 to 6 Å up to 13 to 14 Å. Since the concept of surface of a solid body is a macroscopic notion, surface area loses its significance when micropores are present. However, pore volume is still an applicable concept.

18.6 The α_s method

To avoid the problems associated with determining t-values, Sing [25–29] replaced t with $\alpha_s = V/V_s$ where V_s is the amount adsorbed at a selected relative pressure. In principle α_s can be made equal to unity at any chosen point on the isotherm; in prac-

tice Sing found it convenient to use a relative pressure of 0.4. Precision is increased by locating α_s in the middle range of the isotherm but higher relative pressures than 0.4 are unsuitable due to the onset of capillary condensation with its associated hysteresis loop which is located at $x > 0.4$.

Values of S_s are calculated from the slopes of the linear section of the α_s plot by using a normalizing factor obtained from the standard isotherm on a non-porous reference solid of known surface area. The micropore volume is obtained by the backward extrapolation of the linear branch of the α_s plot to $\alpha_s = 0$; the intercept on the x-axis gives the effective origin for the monolayer–multilayer adsorption on the external surface.

The α_s method has been used for the analysis of various gases on a range of solids [29, 45]. It has also been used for potential pore size reference materials and the results compared with mercury porosimetry [64]; α_s was calculated from the silica TK 800 isotherm [65].

18.7 Pore size distribution determination of mesopores

The technique that is usually applied is to consider the desorption isotherm at saturation vapour pressure with all the pores completely filled with liquid. A slight lowering of the pressure will result in the desorption of a measurable quantity of gas. Let this volume be ν_1, for a fall in pressure from P_1 to P_2, and assume that this empties all pores with Kelvin radii greater than r_{K_2}. Equation (18.1) may be written:

$$\ln\frac{P_1}{P_0} = -\frac{2\gamma V_L}{RTr_{K_1}} \tag{18.13}$$

$$\ln\frac{P_2}{P_0} = -\frac{2\gamma V_L}{RTr_{K_2}} \tag{18.14}$$

If the volume of gas desorbed is reduced to cm^3 per gram of adsorbate at STP ($\Delta\nu_c$), then the condensed volume is given by:

$$\Delta\nu_c = \frac{M\Delta\nu}{\rho V_L} \tag{18.15}$$

where M is the molecular weight of the adsorbate, ρ the density of the liquefied gas at its saturated vapour pressure and V_L the molar gas volume at STP. For nitrogen:

$$\Delta\nu_c = \frac{28}{22\ 400 \times 0.808}\Delta\nu$$

$$= 1.547 \times 10^{-3}\Delta\nu \tag{18.16}$$

18.7.1 Modelless method

The pore shape of very few adsorbents is known and it is unlikely that any one solid will have pores of only one given shape. In the modelless method no pore shape is assumed. The analysis is based on the hysteresis region of the isotherm [20, 21].

The pore volume and pore surface distributions are determined as functions of the hydraulic radius (r_H) which is defined as the ratio of the volume to the surface of the pores. Hence, from equation (18.1):

$$r_H = \frac{v_p}{S_p}$$

$$= -\frac{\gamma V_L \cos \theta}{RT \ln x} \quad [= \tfrac{1}{2} r_K] \qquad (18.17)$$

The cumulative pore size distribution by surface comprises a plot of condensed volume (v_c) against hydraulic radius (r_H). This may be differentiated graphically to produce a relative pore size distribution by surface or the calculation may be carried out in a tabular manner.

The cumulative pore size distribution by surface comprises a plot of S_p against r_H where S_p is obtained from equation (18.1).

$$S_p = -\frac{RT}{\gamma V_L \cos \theta} \int \ln x \cdot dv_c \qquad (18.18)$$

the limits of the integration being the relative pressures at which the hysteresis loop closes.

For nitrogen:

$$S_p = -\frac{8.314 \times 78 \times 2.303}{8.72 \times 10^{-3} \times 34.68 \times 10^{-6}} \int \log x \cdot dv_c$$

$$= -4.939 \times 10^9 \int \log x \cdot dv_c \qquad (18.19)$$

If the condensed volume (v_c) is in cm^3 g^{-1} (liquid), S_p is in m^2 g^{-1} with a constant 4939.

Applying equation (18.16) gives the equation in the alternative form:

$$S_p = 7.64 \int \log x \cdot dv \qquad (18.20)$$

Applying equation (18.17), for nitrogen, the hydraulic diameter is given by:

$$r_H = -\frac{2.05}{\log x} \times 10^{-10} \text{ m} \qquad (18.21)$$

Again, the relative frequency distribution may be determined graphically or in a tabular manner.

The desorption or adsorption branch of the isotherm may be used for this technique but the former is preferred due to thermodynamic considerations.

S_p may be evaluated by graphically integrating a plot of $\log x$ against condensed volume v_c or by a tabular method [22].

An application of this technique is shown in Table 18.1.

It is clear that S_p is not the surface of the walls of the pores but the surface of the cores (the Kelvin or core surface S_K) and v_c is the volume of liquid required to fill the cores (i.e. $v_c = v_K$ the Kelvin volume).

Table 18.1 Pore size distribution evaluation using the modelless method.

Relative pressure (x)	Volume desorbed v_D (cm³ g⁻¹) at STP	$-\log \bar{x}$	ΔS_p (m² g⁻¹)	S_p (m² g⁻¹)	Hydraulic radius r_H (nm)	Condensed volume v_c (cm³ g⁻¹)	Δr_H (nm)	$\dfrac{\Delta S_p}{\Delta r_H}$ (m² g⁻¹ nm⁻¹)	$\dfrac{\Delta v_c}{\Delta r_H}$ (cm³ g⁻¹ μm⁻¹)
0.99	220	0.0133	1.73		46.4	0.3403	37.3		1
0.95	203	0.034	2.60	1.7	9.09	0.3140	4.66	0.6	3
0.90	193	0.058	4.00	4.3	4.43	0.2986	1.56	2.6	9
0.85	184	0.084	6.40	8.3	2.87	0.2846	0.78	8.2	20
0.80	174	0.111	8.81	14.7	2.09	0.2692	0.47	18.7	34
0.75	163.6	0.140	13.47	23.5	1.62	0.2531	0.31	43.5	63
0.70	151	0.171	16.98	37.0	1.31	0.2336	0.23	73.8	87
0.65	138	0.204	21.87	54.0	1.08	0.2135	0.168	130	129
0.60	124	0.241	24.83	75.9	0.912	0.1918	0.133	187	157
0.55	110.5	0.280	28.91	100.7	0.779	0.1709	0.107	270	194
0.50	97	0.324	32.17	129.6	0.672	0.1501	0.088	366	230
0.45	84	0.357	13.62	161.8	0.584	0.1222	0.032	426	241
0.43	79			175.4	0.552	0.1162			

$\Delta S_p = 7.64 v_D \log \bar{x}$; $r_H = \dfrac{-2.025}{\log x}$; $v_c = 1.547 \times 10^{-3} v_D$.

Kiselev [50] employed this method successfully for the determination of the total surface areas of a number of adsorbents, having only wide pores, which were in good agreement with the BET surface areas. For narrow pores, core and pore surfaces vary considerably. In terms of volume distributions this technique is equivalent to plotting condensed volume desorbed (v_c) against half the Kelvin radius.

The condensed volume desorbed is related to the pore surface by the following equation

$$S_p = \frac{2\Delta v_c}{r_K} \tag{18.22}$$

This is identical with the cylindrical core model described below.

In the corrected modelless method [22] a correction is applied for the residual film thickness, i.e. the Kelvin radius is used as $r_K = r_p + t$ making $r_H = 2r_p + 2t$.

Brunauer [51] validates his method on the grounds that in industrial operations one could use the core structure analysis, which is completely modelless, in place of pore structure analysis. In the second place if one had only one experimental core parameter, the volume, the derived surface could be in considerable error [21]. The third reason is that in order to confirm a correct pore structure analysis, it is usual to compare the pore surface (S_{cum}) with the BET surface (S_{BET}) and if the two are comparable the analysis is considered good. To this Brunauer adds a second criterion, namely the cumulative pore volume had to agree with the volume adsorbed at saturation pressure. Earlier investigators could not use this criterion because the Kelvin radius is infinite at the saturation pressure, thus the largest pores were left out of consideration. In these pores, although the surface is small the volume is large. By using the hydraulic radius the whole isotherm can be covered.

Havard and Wilson [55] describe pore measurements on the SCI/IUPAC/NPL meso-porous silica surface area standard. Pore size distributions are presented using the modelless method and the Kelvin equation based on open-ended cylinders and spheres with coordinate numbers of 4, 6 and 8. The isotherm can be used to calibrate the BET apparatus over the whole range; samples are available from the National Physical Laboratory.

18.7.2 Cylindrical core model

For the adsorption isotherm, using a cylindrical pore model, open at both ends, the mutually perpendicular radii (equation (18.3)) are given by $r_1 = r_c$, $r_2 = \infty$. Hence, $r_K = 2r_c$; for the desorption isotherm $r_1 = r_2 = r_c$. Hence, $r_K = r_c$. Thus:

$$r_c \ln x_A = -\frac{\gamma V_L}{RT} \cos \theta \tag{18.23}$$

$$r_c \ln x_D = -\frac{2\gamma V_L}{RT} \cos \theta \tag{18.24}$$

To obtain the core size distribution either the adsorption or desorption branch of the isotherm may be used but, as before, it is preferable to use the desorption branch.

As the pressure falls from P_{r+1} to P_{r-1} a condensed volume v_{c_r} is desorbed where:

$$v_{c_r} = \pi r_{c_r}^2 \, L_r(r) \tag{18.25}$$

where $L_r(r)$ is the frequency (i.e. total length) of cores in the size range $r_{c_{r+1}}$ to $r_{c_{r-1}}$, mean size r_{c_r}.

The surface of the cores is given by:

$$S_{c_r} = 2\pi r_{c_r} \, L_r(r) \tag{18.26}$$

Hence,

$$S_{c_r} = \frac{2v_{c_r}}{r_{c_r}} \tag{18.27}$$

This is identical to equation (18.22).

18.7.3 Cylindrical pore model

If allowance is made for the thickness of the adsorbed film the true pore size distribution is obtained. This technique was pioneered by Wheeler who made partial correction [4]. The full correction is as follows:

As the pressure falls from P_0 (in practice a pressure close to P_0 is taken) to P_1, a condensed volume, v_c, is desorbed where v_{c_1} is the core volume of pores with radii greater than r_{P_1} and of average size $\bar{r}_{P_1}$ (see figure 18.4). Hence,

$$v_{c_1} = \pi \bar{r}_{c_1}^2 \, L_1(r)$$

$$v_{P_1} = \pi \bar{r}_{P_1}^2 \, L_1(r)$$

$$S_{P_1} = 2\pi \bar{r}_{P_1} \, L_1(r)$$

The volume of the first class of pores is given by:

$$v_{P_1} = \left(\frac{\bar{r}_{P_1}^2}{\bar{r}_{c_1}^2} \right) v_{c_1} \tag{18.28}$$

The surface of the first class of pores is given by:

$$S_{P_1} = \frac{2v_{P_1}}{\bar{r}_{P_1}} \tag{18.29}$$

$\bar{r}_{c_1}$ is given by the Kelvin equation:

$$\bar{r}_{c_1} = - \frac{\gamma V_L \cos \theta}{RT} \left[\frac{1}{\ln x_0} + \frac{1}{\ln x_1} \right] \tag{18.30}$$

where x_0 is approximately equal to unity, and:

$$\bar{r}_{P_1} = \bar{r}_{c_1} + \bar{t}_1 \tag{18.31}$$

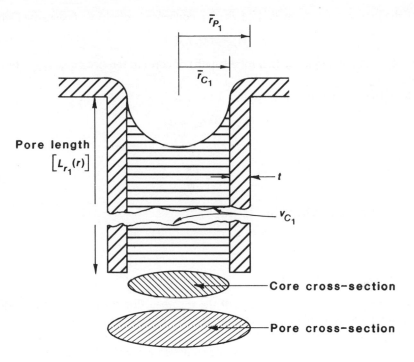

Fig. 18.4 The cylindrical pore model.

t is obtained from an appropriate t-curve. As the pressure falls from P_1 to P_2 a condensed volume v_{c_2} is desorbed. This consists of two parts:

(a) The core volume of the pores in the size range r_{P_1} to r_{P_2} having a mean size $\bar{r}_{P_2}$, i.e. the Kelvin volume v_K.

(b) An amount $S_{P_1} \Delta t_2$ desorbed from the exposed surface of the first group of pores ($\Delta t_2 = t_2 - t_1$).

Hence,

$$v_{c_2} = \pi \bar{r}_{c_2}^2 \, L_2(r) + S_{P_1} \, \Delta t_2$$

$$v_{P_2} = \pi \bar{r}_{P_2}^2 \, L_2(r)$$

$$S_{P_2} = 2\pi \bar{r}_{P_2} \, L_2(r)$$

Therefore,

$$v_{P_2} = \left(\frac{\bar{r}_{P_2}}{\bar{r}_{c_2}}\right)^2 \left[v_{c_2} - S_{P_1} \, \Delta t_2 \right] \qquad (18.32)$$

where

$$v_{K_2} = v_{c_2} - S_{P_1} \, \Delta t_2 \qquad (18.33)$$

In Wheeler's treatment the second term in the bracket is neglected. The core (or Kelvin) volume is given by:

$$v_{K_2} = [v_{c_2} - S_{P_1} \, \Delta t_2]$$
(18.34)

Also,

$$S_{P_2} = \frac{2}{r_{P_2}} v_{P_2}$$
(18.35)

For the general, or rth desorption step

$$v_{P_r} = \left(\frac{\bar{r}_{P_r}}{\bar{r}_{c_r}}\right) \left[v_{c_r} - \Delta t_r \sum_{x=0}^{r-1} S_{P_x}\right]$$
(18.36)

$$S_{P_r} = \frac{2}{r_{P_r}} v_{P_r}$$
(18.37)

An application of this procedure is given in Table 18.2 which is on a basis of 1 g of powder.

The core volume for the first group of pores having core (or Kelvin) radii between 93.9 and 18.4 nm with an average radius of 56.2 nm is equal to the volume desorbed as the relative pressure falls from 0.99 to 0.95; this equals 0.0263 cm³.

The pore volume is given by:

$$\left(\frac{57.9}{56.2}\right)^2 \times 0.0263 = 0.0279 \text{ cm}^3$$

The pore surface is given by:

$$S_{P_1} = \frac{2}{579} \times 279 = 0.96 \text{ m}^2$$

For the next group of pores of average pore radius $\bar{r}_{P_2} = 15.13$ nm, average core radius $\bar{r}_{c_2} = 13.7$ nm, the evaporation of 0.0154 cm³ of liquid nitrogen empties the cores in the radii range 18.4 to 8.96 nm leaving a residual film 1.16 nm thick. It also removes a layer 0.252 nm thick from the already exposed surface. Hence,

$$v_{P_2} = \left(\frac{15.13}{13.7}\right)^2 [154 - 2.52 \times 0.94] = 0.0185 \text{ cm}^3$$

$$S_{P_2} = \frac{2 \times 185}{151.3} = 2.45 \text{ m}^2$$

For the next group of pores:

$$v_{P_3} = \left(\frac{8.62}{7.38}\right)^2 [140 - 1.68 \times 3.41] = 0.0183 \text{ cm}^3$$

$$S_{P_3} = \frac{2 \times 183}{86.2} = 4.25 \text{ m}^2$$

Table 18.2 Evaluation of pore size distribution (cylindrical model).

Relative pressure x	Volume desorbed v_D (cm³ g⁻¹) at STP	Condensed volume v_c (cm³ g⁻¹)	Core radius r_c (×10⁻¹⁰ m)	Residual thickness t (×10⁻¹⁰ m)	Pore thickness r_p (×10⁻¹⁰ m)	$\bar{r}_c$	$\bar{r}_p$	Δt	Δv_c (cm³ g⁻¹)	Δv_p (cm³ g⁻¹)	ΔS_p (m² g⁻¹)	Δr_p (×10⁻¹⁰ m)	$\dfrac{\Delta v_p}{\Delta r_p}$ (cm³ m⁻¹ g⁻¹)	$\dfrac{\Delta S_p}{\Delta r_p}$ (m² m⁻¹ kg⁻¹)
0.99	220	0.3403	939	19.1	958	562	579	3.33	0.0263	0.0279	0.96	758	0.4	0.01
0.95	203	0.3140	184	15.8	200	137	151	2.52	0.0154	0.0185	2.45	97	1.9	0.25
0.90	193	0.2986	89.6	13.2	103	73.8	86.2	1.68	0.0140	0.0183	4.25	33.6	5.4	1.26
0.85	184	0.2846	58.0	11.6	69.6	50.1	61.0	1.23	0.0154	0.0214	7.03	17.1	12.5	4.11
0.80	174	0.2692	42.2	10.3	52.5	37.5	47.3	0.956	0.0161	0.0234	9.89	10.3	22.7	9.60
0.75	163.6	0.2531	32.8	9.4	42.2	29.7	38.6	0.776	0.0195	0.0297	15.4	7.1	41.8	21.70
0.70	151	0.2336	26.5	8.6	35.1	24.3	32.5	0.650	0.0201	0.0313	19.3	5.1	61.4	37.8
0.65	138	0.2135	22.0	8.0	30.0	20.2	27.9	0.560	0.0217	0.0351	25.1	4.1	85.6	61.2
0.60	124	0.1918	18.5	7.4	25.9	17.1	24.3	0.493	0.0209	0.0338	27.8	3.2	105.6	86.9
0.55	110.5	0.1709	15.8	6.9	22.7	14.7	21.4	0.440	0.0208	0.0336	31.4	2.6	129.0	121.0
0.50	97	0.1501	13.6	6.5	20.1	12.7	19.0	0.401	0.0202	0.0323	34.0	2.2	146.8	154.5
0.45	84	0.1299	11.8	6.1	17.9	11.5	17.5	0.151	0.0077	0.0116	13.3	0.8	145.0	166.3
0.43	79	0.1222	11.2	5.91	17.1	10.7	16.5	0.22	0.0062	0.0020	2.44	1.1	18.2	22.2
0.40	75	0.1162	10.3	5.69	16.0	9.6	15.1	0.35	0.0077	0.0051	6.80	1.7	30.0	40.0
0.35	70	0.1085	9.0	5.34	14.3	9.6	13.5	0.33	0.0066	0	0	1.5	0	0
0.30	66	0.1019	7.8	5.01	12.8	8.4								

$v_c = 1.547 \times 10^{-3} v_D$; $t = \sqrt{\left(\dfrac{13.99}{0.034 - \log x} \right)}$; $r_P = r_c + t$; → hysteresis loop closes. Total pore volume = 0.3240 cm³ g⁻¹

$r_c = r_K = -\dfrac{4.1}{\log x}$

$S_{cum} = 200 \text{ m}^2 \text{ g}^{-1}$

$(v_{c_r} = \Delta v_c, \ S_{p_r} = \Delta S_p, \ v_{P_r} = \Delta v_p)$

The calculation continues until the hysteresis loop closes (usually around a relative pressure of 0.4).

Continuing pore size distribution below the point at which the hysteresis loop closes, points to condensation in pores with shapes not leading to hysteresis phenomena as could be expected with wedge-shaped or conical pores. In the example, the hysteresis loop closes at $x = 0.43$ and the pore size distribution terminates at $x = 0.35$.

The V_A–t curve for this sample is linear up to $t = 0.55$ nm and is then convex to the t-axis indicating the onset of capillary condensation in mesopores, $S_t = 204$ m^2 g^{-1}. The BET surface $S_{BET} = 212$ m^2 g^{-1}. The surface by summing pore surfaces (S_{cum}) may be higher than S_{BET} due to the surfaces of intersecting pores being included. The volume of adsorbate condensed into ink-bottle pores is attributed to narrower pores thus increasing S_{cum}; the pore volume should still, however, be correct.

18.7.4 Parallel plate model

During adsorption a meniscus cannot be formed but during desorption a cylindrical meniscus is present. The Kelvin equation takes the form [32, 43]:

$$-RT \log x_D = \frac{2\gamma V_L}{r_K} = \gamma V_L \left[\frac{2}{d - 2t} + \frac{1}{\infty} \right] \qquad (18.38)$$

Thus, the plate separation is given by:

$$d = r_K + 2t$$

A similar argument applies to wedge-shaped pores (e.g. plate-shaped clay). Many workers suggest a minimum pore size to which pore size distributions may be determined on the basis of the Kelvin equation, 20 Å diameter for cylindrical pores and slightly smaller for slits [54]. In this case the procedure seems valid to relative pressures smaller than those pertaining when the hysteresis loop closes.

Assuming that at the highest measured pressure the pores are completely filled, v_c will be the core volume (v_{K_1}) of the first group of pores of pore volume v_{P_1} and surface area S_{P_1}.

$$v_{c_1} = v_{K_1}$$

where

$$v_{K_1} = \frac{S_{P_1}}{2} \bar{r}_{K_1}$$

and

$$v_{P_1} = \frac{S_{P_1}}{2} \bar{d}_{P_1}$$

From these equations the pore volume and pore surface may be calculated:

$$v_{P_1} = \frac{\bar{d}_{P_1}}{\bar{r}_{K_1}} v_{c_1} \tag{18.39}$$

$$S_{P_1} = \frac{2v_{c_1}}{\bar{r}_{K_1}} \tag{18.40}$$

When the pressure is lowered from P_2 to P_3 the desorbed volume will consist of two parts, the volume desorbed from the second group of pores plus the volume desorbed from the surface of the first group of pores:

$$v_{c_2} = v_{K_2} + \Delta t_2 S_{P_1}$$

where

$$v_{K_2} = \frac{S_{P_2}}{2} \bar{r}_{K_2}$$

$$v_{P_2} = \frac{S_{P_2}}{2} \bar{d}_{P_2}$$

Hence:

$$v_{P_2} = \frac{\bar{d}_{P_2}}{\bar{r}_{K_2}} [v_{c_2} - \Delta t_2 S_{P_1}] \tag{18.41}$$

$$S_{P_2} = \frac{2v_{P_2}}{\bar{d}_{P_2}} \tag{18.42}$$

In general, for a lowering of pressure from P_{r+1} to P_{r-1} emptying the cores of slits in the size range d_{r+1} to d_{r-1} with average separation d_r:

$$v_{P_r} = \frac{\bar{d}_{P_r}}{\bar{r}_{K_r}} \left[v_{c_r} - \Delta t_{r+1} \sum_{r=1}^{r} S_r \right] \tag{18.43}$$

$$S_{P_r} = \frac{2v_{P_r}}{\bar{d}_{P_r}}$$

The experimental data from Table 18.2 are recalculated using this model in Table 18.3.

Table 18.3   Evaluation of pore size distribution (slit-shaped pore model).

Relative pressure (x)	Volume desorbed v_D	Thickness of adsorbed layer t (Å)	Kelvin radius r_K (Å)	Slit width d_p (Å)	Condensed volume desorbed v_c (cm³ g⁻¹ ×10⁻⁴)	r_K (Å)	$\bar{d}_p$ (Å)	Δd_p (Å)	Δt (Å)	v_c (cm³ g⁻¹ ×10⁻⁴)	S_p (m² g⁻¹)	v_p (cm³ g⁻¹ ×10⁻⁴)	$\dfrac{\Delta v_p}{\Delta d_p}$ (cm³ g⁻¹ μm⁻¹)	$\dfrac{\Delta S_p}{\Delta d_p}$ (m² g⁻¹ μm⁻¹)	v_p (cm³ g⁻¹ ×10⁻⁴)	S_p (m² g⁻¹)
0.99	220	19.1	939	977	3403	562	597	761	2.53	263	0.9	279	0.4	0.002	279	0.9
0.95	203	15.77	184.1	215.6	3140	137	166	99.5	1.67	154	2.2	184	1.8	0.020	463	3.1
0.90	193	13.24	89.6	116.1	2986	73.9	98.7	34.9	1.23	140	3.7	182	5.2	0.11	645	6.8
0.85	184	11.57	58.1	81.2	2846	50.2	72.1	18.2	0.956	154	6.0	215	11.8	0.33	860	12.8
0.80	174	10.34	42.3	63.0	2692	37.6	57.3	11.4	0.776	161	8.3	237	20.8	0.73	1097	21.1
0.75	163.6	9.38	32.8	51.6	2531	29.7	47.7	7.9	0.650	195	12.7	303	38.4	1.61	1400	33.8
0.70	151	8.61	26.5	43.7	2336	24.3	40.8	5.9	0.560	201	15.9	325	55.1	2.69	1725	49.7
0.65	138	7.95	22.0	37.8	2135	20.3	35.6	4.5	0.493	217	20.6	368	81.8	4.58	2093	70.3
0.60	124	7.39	18.5	33.3	1918	17.2	31.5	3.7	0.440	209	23.3	366	99.0	6.30	2459	93.6
0.55	110.5	6.90	15.9	29.6	1709	14.8	28.1	3.1	0.401	208	26.9	378	122.0	8.68	2837	120.5
0.50	97	6.46	13.6	26.5	1501	12.7	25.2	2.6	0.370	202	30.1	379	146.0	11.58	3216	150.6
0.45	84	6.06	11.8	23.9	1299	11.1	22.8	2.2	0.347	77	9.4	111	123.0	10.50	3327	160.0
0.40	79	5.63	10.3	21.7	1232	9.65	20.7	2.0	0.332	77	14.2	147	73.5	7.10	3627	187.8
0.35	70	5.34	8.99	19.7	1083	8.42	18.8	1.8	0.322	62	13.6	127	70.6	7.56	3754	201.4
0.30	66	5.01	7.84	17.9	1021	7.33	17.0	1.7	0.321	62	15.7	134	78.8	9.24	3888	217.1
0.25	62	4.69	6.81	16.2	959	6.34	15.4	1.6	0.331	69	20.2	157	98.0	12.60	4045	237.3
0.20	57.5	4.37	5.87	14.6	890	5.43	13.8	1.5	0.360	109	37.7	260	173.0	25.10	4305	275.0
0.15	50.5	4.04	4.98	13.1	781	4.54	12.3	1.6	0.440	23	4.1	25	15.6	2.56	4330	279.1
0.10	49	3.68	4.10	11.5	758	3.63	10.6	1.87								
0.05		3.24	3.15	9.63												

$$t = \sqrt{\left(\frac{13.99}{0.034 - \log x} \right)}$$

$$r_K = -4.1/\log x$$

$$v_c = 15.47 \times 10^{-4}\, v_D$$

$$v_{P_r}\,\frac{d_{P_r}}{r_K}\left[v_c - \Delta t_{r+1} \sum_r S_{P_r} \right]$$

$$S_{P_r} = 2\,\frac{v_{P_r}}{d_{P_r}}$$

$$d_p = r_K + 2t$$

$$[\Delta S_p = S_{P_r},\ \Delta v_p = v_{P_r},\ \Delta v_c = v_{c_r}]$$

18.8 Analysis of micropores: the MP method

Brunauer [46, 52] realized that the V_A-t curve yielded more information than surface area. A plot of V_A against t for a non-porous solid produces a straight line, from
the slope of which the specific surface may be determined (section 16.2.13). With
microporous solids the curve becomes concave to the t-axis; a second linear portion
gives the surface area of the wider capillaries and the intercept of the second linear
portion on the V_A-axis gives the micropore volume [48].

   Brunauer [47] showed that the micropore isotherm looked very similar to a
Langmuir isotherm (figure 18.5). In the example the BET surface area, using the first
four points on the isotherm, is 793 m^2 g^{-1}. In figure 18.6 the V_A-t curve gives a
surface area, from the initial slope, of S_t = 792 m^2 g^{-1} and downward deviations begin
as one proceeds from t = 0.40 to 0.45 nm. The tangent between these two values
indicates a surface area of 520 m^2 g^{-1}. Thus a group of pores has become filled with
nitrogen and the surface of these pores is 272 m^2 g^{-1}. The volume of the first group
of pores is, therefore, given by:

$$V_P = 10^4(S_1 - S_2)\frac{t_1 + t_2}{2} \tag{18.44}$$

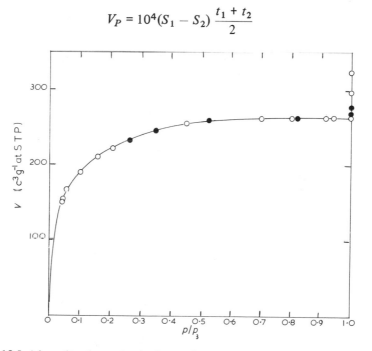

Fig. 18.5 Adsorption–desorption isotherm of nitrogen on silica gel Davidson 03 at
77.3 K. Empty circles adsorption; black circles desorption.

   One then proceeds in a similar manner to the second pore group with hydraulic
radii between 0.45 and 0.50 nm. The analysis continues until there is no further
decrease in the V_A-t slope which means no further blocking of pores by multilayer
adsorption. The pore volume distribution curve is shown in figure 18.7.

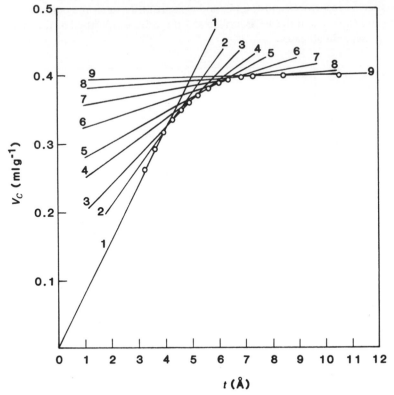

Fig. 18.6 Isotherm of figure 18.5 converted into a v_c-t plot. v_A has been converted to v_c using equation (18.6) and x has been converted to t using equation (18.10) so that $S_t = 10^4 (v_c/t)$. The surface areas of the pore walls of the different pore groups are obtained from differences between the slopes of straight lines 1 to 9, and the pore widths are obtained from the abscissa values. The figure illustrates the MP method of analysis of micropores.

The MP method is based on the use of the appropriate t-curve; the choice is far more important in the micropore region than in the mesopore since in this low pressure region the heats of adsorption affect the film thickness strongly. Far more important than this, the t-values constitute the total pore radius, whereas in the mesopore region they only appear as a correction term.

This approach has been criticized by Marsh and Rand [49] who state that in pores about two molecular diameters wide the influence of opposite walls is significant and once one molecule is adsorbed the pore is effectively reduced in size and spontaneously fills. Moreover pores three, four, or perhaps five diameters wide fill with adsorbate at relative pressures below that at which equivalent numbers of multilayers form on an open surface. This concept of volume filling was introduced in earlier papers by Dubinin *et al.* [18, 19].

In reply to the criticism, Brunauer [52, 53] analysed four silica gels, two containing no micropores and two containing micropores and mesopores. He used nitrogen, water

and oxygen as adsorbates and found good agreement between the cumulative pore volume and surface and the BET surface and the volume adsorbed at the saturated vapour pressure in all cases.

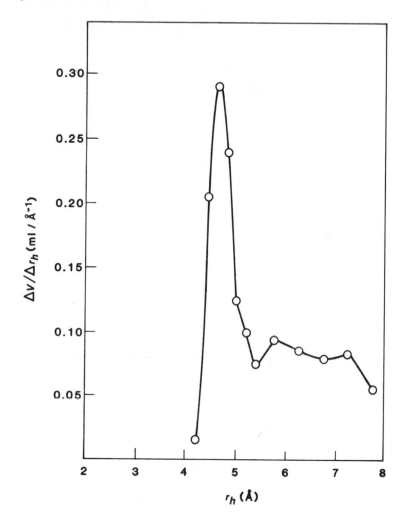

Fig. 18.7 Pore volume distribution curve of silica gel Davidson 03. (Figures 18.5 to 18.7 are taken from reference [46].)

18.9 Miscellaneous

A comparison of three methods of interpreting nitrogen adsorption isotherms on microporous spheron 6 was carried out by Rand [56]. He applied the D-R equation (section 16.2.12) and the Sing modification of de Boer's t-method [48]. The micro-

pore volume was determined by extrapolation of the D-R curve and by the Ross and Oliver procedure [57] and he found that the latter method best fitted the adsorption data. Tsunoda [66] discussed the validity of micropore volume, as determined from the D-R equation, in relation to the adsorption volume at the beginning of the hysteresis loop.

Mikhail and Shebl [58] examined microporous wide-pore only and mixed-pore samples of silica gel using nitrogen, water, methanol, cyclohexane and carbon tetrachloride as adsorbents. They conclude that nitrogen measures total surface whereas the other adsorbents measure a smaller surface either due to persorption or specific interaction.

Mieville [59] measured microporosity in the presence of mesopores and concluded that the intercept of the V_A-t curve gives the correct mesopore volume even in the presence of mesopores.

The pore structure of hydrated calcium silicate and Portland cement has been analysed by water adsorption using both the MP method and the corrected modelless method [60].

Hanna et al. [62] determined the t-curve for oxygen and applied this in conjunction with the MP method and the corrected modelless method. Their main aim was to find out if the same result obtained for adsorption at different temperatures. Using nitrogen adsorption as a standard they found that the area occupied by an oxygen molecule at 77 K was 14.3 Å (as compared to 13.6 Å for a liquid molecule) and 15.4 Å at 90 K.

Gregg et al. [63] studied the adsorption of water vapour on microporous carbon black at 25°C both before and after the micropores had been filled with n-nonane vapour. As expected the heat of adsorption in the micropores was greater than on an open surface but the carbon black remained hydrophobic despite the enhancement of the adsorption fields within the micropores.

References

1 Gregg, G.J. and Sing, K.S.W. (1967), *Adsorption, Surface Area and Porosity*, Academic Press, NY.

2 de Boer, J.H. (1958), *The Structure and Properties of Porous Materials*, Butterworths, p. 68.

3 Oulton, T.D. (1948), *J. Phys. Colloid Chem.*, 52, 1296.

4 Wheeler, A. (1955), *Catalysis*, 2, Reinhold, NY, p. 118.

5 Schüll, C.G. (1948), *J. Am. Chem. Soc.*, 70, 1405.

6 Barrett, E.P., Joyner, L.G. and Halenda, P.O. (1951), *ibid.*, 73, 373–80.

7 Burgess, C.G.V. and Everett, D.H. (1970), *J. Colloid Interfac. Sci.*, 33, 611–14.

8 Pierce, C. (1953), *J. Phys. Chem.*, 57, 149–52.

9 Joyner, L.G., Barrett, E.P. and Skold, R. (1951), *J. Am. Chem. Soc.*, 73, 3155.

10 Cranston, R.W. and Inkley, F.A. (1957), *Advances in Catalysis* (ed. A. Farkas), 9, Academic Press, NY, pp. 143–54.

11 Dollimore, D. and Heal, G.R. (1964), *J. appl. Chem.*, 14, 109.

12 Innes, W.B. (1957), *Analyt. Chem.*, 29, 7, 1069–73.

13 Roberts, B.F. (1967), *J. Colloid Interfac. Sci.*, 23, 266–73.

14 Halsey, G.D. (1948), *J. Chem. Phys.*, 16, 931.

15 Pierce, C. (1968), *J. Phys. Chem.*, 72, 3673.
16 de Boer, J.H. *et al.* (1966), *J. Colloid Interfac. Sci.*, 21, 405–14.
17 Mather, R.R. and Sing, K.S.W. (1977), *ibid.*, 60, 1, 60–66.
18 Dubinin, M.M. (1967), *ibid.*, 23, 487–99.
19 Bering, B.P., Dubinin, M.M. and Serpinsky, V.V. (1972), *ibid.*, 38, 1, 184–94.
20 Brunauer, S., Mikhail, R.S. and Bodor, E.E. (1967), *ibid.*, 24, 451.
21 Brunauer, S., Mikhail, R.S. and Bodor, E.E. (1967), *ibid.*, 25, 353.
22 Bodor, E.E., Odler, I. and Skalny, J. (1970), *ibid.*, 32, 2, 367–70.
23 Ihm, S.K. and Ruckenstein, E. (1977), *ibid.*, 61, 1, 146–59.
24 Lamond, T.G. (1976), *ibid.*, 56, 1, 116–22.
25 Sing, K.S.W. (1967), *Chemy. Ind.*, 829.
26 Sing, K.S.W. (1970), *Proc. Int. Symp. Surface Area Determination* (eds D.M. Everett and R.H. Ottewill), Butterworths, London, p. 25.
27 Sing, K.S.W. (1971), *J. Oil Colour Chem. Assoc.*, 54, 731.
28 Sing, K.S.W. (1973), Special Periodical Report, *Colloid Science*, 1, Chem. Soc., London, p. 1.
29 Sing, K.S.W. (1976), *Characterisation of Powder Surfaces* (eds G.D. Parfitt and K.S.W. Sing), Academic Press.
30 Lippens, B.C., Linsen, B.G. and de Boer, J.H. (1964), *J. Catalysis*, 3, 32–37.
31 de Boer, J.H. and Lippens, B.C. (1964), *ibid.*, 3, 38–43.
32 Lippens, B.C. and de Boer, J.H. (1964), *ibid.*, 44–49.
33 de Boer, J.H., Van Den Heuval, A. and Linsen, B.G. (1964), *ibid.*, 268–73.
34 Lippens, B.C. and de Boer, J.H. (1965), *ibid.*, 4, 319–23.
35 de Boer, J.H., Linsen, B.G. and Osinga, Th.J. (1965), *ibid.*, 643–8.
36 de Boer, J.H. *et al.* (1965), *ibid.*, 649–53.
37 de Boer, J.H. *et al.* (1967), *ibid.*, 7, 135–9.
38 Brockhoff, J.C.P. and de Boer, J.H. (1967), *ibid.*, 9, 8–14.
39 Brockhoff, J.C.P. and de Boer, J.H. (1967), *ibid.*, 15–27.
40 Brockhoff, J.C.P. and de Boer, J.H. (1968), *ibid.*, 10, 153–65.
41 Brockhoff, J.C.P. and de Boer, J.H. (1968), *ibid.*, 368–74.
42 Brockhoff, J.C.P. and de Boer, J.H. (1968), *ibid.*, 377–90.
43 Brockhoff, J.C.P. and de Boer, J.H. (1968), *ibid.*, 391–400.
44 de Boer, J.H. *et al.* (1968), *ibid.*, 11, 46–53.
45 Bhambhami, M.R. *et al.* (1972), *J. Colloid Interfac. Sci.*, 38, 1, 109–17.
46 Brunauer, S., Mikhail, R.S. and Bodor, E.E. (1968), *ibid.*, 26, 45.
47 Brunauer, S. (1970), *Chem. Engng Progr. Symp.*, 96, 65, 1–10.
48 Sing, K.S.W. (1968), *Chemy. Ind.*, 1520.
49 Marsh, H. and Rand, B. (1970), *J. Colloid Interfac. Sci.*, 33, 3, 478–9.
50 Kiselev, A.V. (1945), *USP Khim.*, 14, 367.
51 Brunauer, S. (1976), *Pure Appl. Chem.*, 48, 401–5.
52 Mikhail, R.S.L., Brunauer, S. and Bodor, E.E. (1968), *J. Colloid Interfac. Sci.*, 26, 45–53.
53 Mikhail, R.S.L., Brunauer, S. and Bodor, E.E. (1968), *ibid.*, 54–61.
54 Aylmore, L.A.G. (1974), *ibid.*, 46, 3, 410–6.
55 Havard, D.C. and Wilson, R. (1976), *ibid.*, 57, 2, 276–88.
56 Rand, B. (1974), *ibid.*, 48, 2, 183–6.
57 Ross, S. and Oliver, J.P. (1964), *On Physical Adsorption*, Wiley, NY.
58 Mikhail, R.S.L. and Shebl, F.A. (1972), *J. Colloid Interfac. Sci.*, 38, 1, 35–44.
59 Mieville, R.L. (1972), *ibid.*, 41, 2, 371–3.
60 Hagymassy, J. Jr. *et al.* (1972), *ibid.*, 38, 1, 20–34.
61 Radjy, F. and Sellevold, E.J. (1972), *ibid.*, 39, 2, 367–88.
62 Hanna, K.M. *et al.* (1973), *ibid.*, 45, 1, 27–54.
63 Gregg, S.J., Nashed, S. and Malik, M.T. (1973), *Powder Technol.*, 7, 15–19.
64 Giles, C.H. *et al.* (1978), *Proc. Conf. Structure of Porous Solids*, Neuchatel, Switzerland, Swiss Chem. Soc.

65 Everett, D.H. *et al.* (1974), *J. appl. Chem. Biotechnol.*, **24**, 199.
66 Tsunoda, R. (1978), *Bull. Chem. Soc. Japan*, **51**, 1, 341.
67 Kurz, H.P. (1972), *Powder Technol.*, **6**, 167–70.

19 Mercury porosimetry

19.1 Introduction

Mercury porosimetry is widely used for determining the pore size distribution of porous material and the void size distribution of tablets and compacts. The method is based on the capillary rise phenomenon whereby an excess pressure is required to cause a non-wetting liquid to climb up a narrow capillary. The pressure difference across the interface is given by the equation of Young and Laplace [1] and its sign is such that the pressure is less in the liquid than in the gas phase if θ is greater than 90° and more if θ is less than 90°.

$$\Delta P = \gamma \left[\frac{1}{r_1} + \frac{1}{r_2} \right] \cos \theta \qquad (19.1)$$

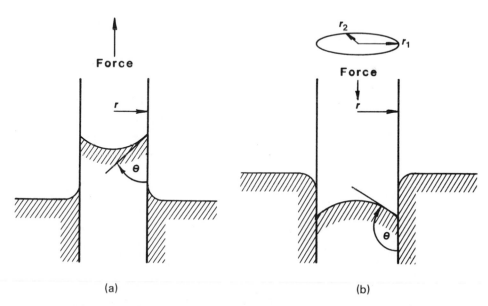

(a)   (b)

Fig. 19.1 (a) Capillary rise, when liquid wets the wall of the capillary ($\theta < 90°$). (b) Capillary depression, when liquid does not wet the wall of the capillary ($90° < \theta < 180°$).

where γ is the surface tension of the liquid, r_1 and r_2 are mutually perpendicular radii and θ, the angle of contact between the liquid and the capillary walls, is always measured within the liquid (figure 19.1).

If the capillary is circular in cross-section, and not too large in radius, the meniscus will be approximately hemispherical. The two radii of curvature are thus equal to each other and to the radius of the capillary. Equation (19.1) then reduces to the Washburn equation [2]:

$$\Delta P = -\frac{2\gamma}{r} \cos \theta \qquad (19.2)$$

where r is the radius of the capillary.

For angles of contact greater than $90°$, the pressure difference is negative and the level of the meniscus in the capillary will be lower than the level in a surrounding reservoir of liquid. Conversely, ΔP is the pressure required to bring the level of the liquid in the capillary up to the level of the surrounding liquid.

If one therefore considers a powder in an evacuated state, $\Delta P = P$, the absolute pressure, is the pressure required to force a non-wetting liquid into a pore of radius r.

One disadvantage of applying equation (19.2) is that the pores are not usually circular in cross-section and so the results can only be comparative.

Other disadvantages are:

(1) The presence of 'ink-bottle' pores or some other shape with constricted 'necks' opening into large void volumes. The pore radius calculated by the Washburn equation is not truly indicative of the true pore radius and capillaries are classified at too small a radius.
(2) The effect of compressibility of mercury with increasing pressure. This should be corrected for by carrying out a blank run.
(3) The compressibility of the material under test. This is a problem of particular importance for materials which have pores which do not connect with the surface, e.g. cork.
(4) The assumption of a constant value for the surface tension of mercury.
(5) The assumption of a constant value for the angle of contact of mercury.

The equipment must possess the facility to evacuate the sample, surround it with mercury and generate sufficiently high pressures to cause the mercury to enter the voids or pores whilst monitoring the pressure and the amount of intruded mercury.

In almost all porosimeters, the amount of mercury penetrating into the pores is determined by the fall in level of the interface between the mercury and the hydraulic fluid, correction being applied for compression of the mercury and distortion of the interface. This measurement may be carried out in several ways. For pressures less than 15 000 psi direct visual observation of the interface is possible through a window, called the 'sight-gauge', in the high pressure vessel. Remote read-out devices include: platinum–iridium resistance wire; a mechanical follower that maintains contact with the mercury manometer as it moves up the dilatometer stem under pressure and relates the linear distance travelled to the volume of mercury intruded; a capacitance bridge

to measure the change in capacitance between the column of mercury in the dilatometer stem and an external shield around it.

Commercially available instruments include the Aminco 60 000 psi porosimeter, the Carlo Erba Model 70 (3000 kg cm^{-2}) and the Micromeretics Model 905 (50 000 psi). These porosimeters work on an incremental basis and a full analysis takes several hours. The Quantachrome Scanning Porosimeter produces data on a continuous basis and a full curve can be produced in from under five minutes to an hour. The pressure range covered is from 0 to 60 000 psi and it is claimed to give highly reproducible results. It is however more expensive than conventional porosimeters. Unger *et al.* [49] have recently described an instrument for pressures up to 6000 bar.

19.2 Literature survey

The method was proposed by Washburn [2] in 1921 and the first experimental data were published in 1940 by Henderson, Ridgeway and Ross [3] who used compressed gas to obtain pressures in the range 30 to 900 psi. Ritter and Drake [4, 5] (1945) extended the range to 10 000 psi using a compressed oil pumping system which they considered safer than compressed gas. Drake [6] later extended the pressure to 60 000 psi corresponding to pore diameters greater than 35 Å. Further development was carried out by Burdine, Gournay and Reichertz [7] (1950) who used dry air at low pressures and cylinder nitrogen at high pressures. A simplified apparatus was described by Bucker, Felsenthal and Conley [8] (1956), an instrument for routine determination of pore size was designed by Winslow and Shapiro [9] (1959) and a commercial instrument for the automatic determination of pore size distribution by Guyer [10] (1959).

Many other modifications of the original equipment have been reported in the literature [11–16] but compressed nitrogen or air was always used to apply pressure to the mercury column. Winslow and Shapiro simplified the operation and improved the safety of mercury porosimeters by substituting isopropyl alcohol as the hydraulic fluid.

Mercury porosimetry is not applicable where the mercury will come into contact with metals with which it forms amalgams. An interesting alternative to mercury to use in these cases is glycerine [17]. These results were later compared with a method using a sedimentation balance and showed good agreement [18].

To gain maximum information it is necessary for the initial pressure to be as low as possible and a number of low pressure porosimeters have been developed for this purpose [19]. Leppard and Spencer [20] designed an apparatus for lump samples in the range $(200 > r > 3.5)$ μm; Reich [21] for $(500 > r > 0.05)$ μm.

The low pressure region is frequently the region where interparticle filling takes place (voids); pore distribution curves frequently have plateaux which form a demarcation between voids and pores. Problems can arise due to damage to the pores under pressure; fracture can occur thus opening up previously blind pores [22, 43].

Total pore volume may be determined by measuring the density of the material with helium and then with mercury. The difference between the respective specific volumes will give the pore volume.

The pore size distribution curves are frequently biased towards the small pore size because of the hysteresis effect caused by bottle-necked or ink-well type pores. Pores of these shapes have narrow-necked openings into large volumes, thus yielding erroneous information as to the true pore volume at the measured pore size. Meyer [23] attempted to correct raw data using probability theory and this considerably alters the distribution of the large pores.

Frevel and Kressley [50] proposed a model and derived expressions for the pressure necessary for mercury to intrude into a solid composed of a collection of non-uniform spheres. The treatment defines the pressure necessary to 'break through', in terms of the largest accessible opening, to the interior of the solid and then relates the size of the opening to the radius of the spheres. Their model was limited to a maximum porosity of 39.54% which was later extended to 47.64% by Meyer and Stowe [51]. This model was verified by Svata and Zabransky [52] who compared their results with microscopy and sedimentation. Rootare and Craig [48] suggested that pressurization and depressurization curves were required. Pressurization curves give size distributions according to the 'necks' and depressurization curves give size data according to the volume of the pores or voids behind the 'necks'. The Reverberi method [53] utilizes the difference between the ascending and descending branches of the curve for evaluating the broad and narrow parts of the pores independently of one another. The ascending branch is measured in the usual way. The descending branch is measured in steps; after each step the pressure is increased to the maximum pressure reached.

The procedure may best be seen in figure (19.2) [52]. After having reached the point X the pressure is decreased to point A value, corresponding to radius r_A. At this point, all pores of $r < r_A$ are emptied. If the pressure is increased again, the pores and necks in the interval $r_A, r_{A'}$ are filled from A to A'. From A' to A'' the pores of a radius $r_{A'}$ to $r_{A''}$ are filled together with the cavities of a radius between $r_{A'}$ and r_A having the neck size of the radius $r_{A''}$ to $r_{A'}$. After decreasing the pressure to point B, all pores of a radius $r < r_B$ are emptied. On increasing the pressure again the pores of a radius between $r_{B'}$ and r_B are filled first, then at point B'' the necks $r_{B''}$ to $r_{B'}$ are filled but at the same time the cavities of the size $r_{B'}$ to r_B having a neck in the interval $r_{B''}$ to $r_{B'}$. The volume increment between A and A' is the volume of 'cylindrical' pores of a radius $r_{A'}$ to r_A; the volume increment from B' to B'' stands for the sum of the cylindrical pore volumes of a radius from $r_{A'}$ to r_A and ink-bottle pores of a radius from $r_{B'}$ to $r_{B''}$. The difference between the two increments is thus the volume of the ink-bottle pores of a radius from $r_{B'}$ to $r_{B''}$. Continuing in this way through the whole interval measured, one obtains a series of radii and volumes whose sum must be identical to the total difference between X and F as it must be equal to the total volume of mercury released from the sample. This may be less than the total intruded into the sample. The difference, called the 'retention' varies a great deal according to the material.

Svata [54] tested this method using powder metallurgical compacts and found that

it was applicable if the powder was first coated with stearic acid to eliminate the interaction between the mercury and the metal. The compacts were coated by immersion in a 1% solution of stearic acid in chloroform.

An examination of the sintering process by mercury porosimetry reveals that, under certain circumstances, although the total pore volume decreases with sintering time the average pore size increases [39]. A review of microporosimetry in tablet micro-structure investigations has been presented by Gillard [55]. Many papers have been published reporting data on a wide range of materials. For further details readers are referred to various detailed reviews [22, 24–27].

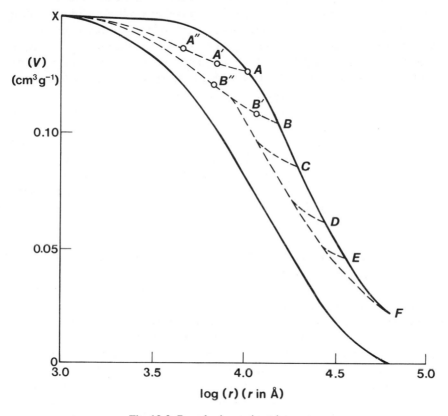

Fig. 19.2 Reverberi porosimetric curves.

19.3 Contact angle and surface tension for mercury

The correct value to be used for the contact angle is still a matter for debate. As an average value for many materials Drake and Ritter [5] used 140°. Juhola and Wiig [28] measured the pressure required to force mercury into a hole, 381 μm in diameter, and also into a fine calibrated capillary, in both cases 140° was found for the contact angle. Winslow and Shapiro [9] found 130° to be a valid value for the contact angle for

mercury intruding into a nickel block with 70 drilled holes of 560 μm diameter. A value of 125° was found for intrusion into Portland cement [41].

The use of an incorrect contact angle does not matter when porous materials of the same type are being compared. However, if an exact measure of the pore openings is required it is necessary to either measure the contact angle directly [1, p. 268] or look at the pores under a microscope to establish the relationship between the actual pore size openings to that measured by mercury intrusion.

In an extreme case, by choosing a value of 130° or 140° for the contact angle when the real value is either 160° or 110°, an error of 50% or 100% is introduced [25]. A commercial instrument, the Anglometer, is available for contact angle measurement.

Another variable factor arises from assuming a constant value for the surface tension of mercury (γ). According to Rootare [27, 29] the best of the published values is Roberts' [30] who found a value of 485 dynes cm^{-1} (0.485 N m^{-1}) at 25°C with a temperature coefficient dγ/dT of -0.21 dynes cm^{-1} °C^{-1}. He attributes the wide variation in published results to the use of contaminated mercury.

19.4 Principle

The principle of the technique may be understood by considering figure 19.3. The sample cell, which is constructed of glass, is shown in figure 19.4. It consists of two pieces, (a) the portion which contains the sample and (b) the cap with a precision bore tubing constituting the penetrometer (or dilatometer). The two parts are fitted together by means of ground surfaces. A weighed sample of powder is placed in the sample space, part (b) is then fitted on to part (a), and the sample holder is inserted in the pressure chamber. The pressure chamber is a heavy-walled steel vessel since it has to withstand very high pressures.

The sample is next subjected to vacuum to remove adsorbed gases. The time required for degassing depends upon the material to be tested. If the material is thoroughly dried beforehand it will usually suffice to reduce the pressure in the sample chamber to 20 μm of mercury. For high surface area materials which cannot withstand high temperatures it may be necessary to degas for two or three days in order to obtain reproducible results. This time can be greatly reduced if the sample can be heat treated under vacuum prior to analysis.

Mercury is next introduced into the sample chamber until it completely covers the sample so that the sample container is completely full of mercury. Excess mercury is then drained off.

The pressure in the sample chamber is then raised to 0.5 psi from which point the analysis has begun. The pressure is then raised in increments of 1 psi to atmospheric pressure. After each increment the reading on the digital counter corresponding to the void or pore volume penetrated is recorded along with the pressure.

Next, the chamber around the sample cell is filled with hydraulic fluid and the pressure increased in small increments to the upper pressure level. The applied pressure forces the mercury into sample voids and pores. Readings are taken at the various pressure levels to determine the volume of mercury penetrating the pores at each

pressure. A typical test will involve fifty or so separate points. One problem that arises with this particular equipment is mercury contamination. This is eliminated with the Carlo Erba and American Instrument Company's porosimeters in which the sample cell is filled with clean mercury under vacuum prior to inserting it in the pressure chamber.

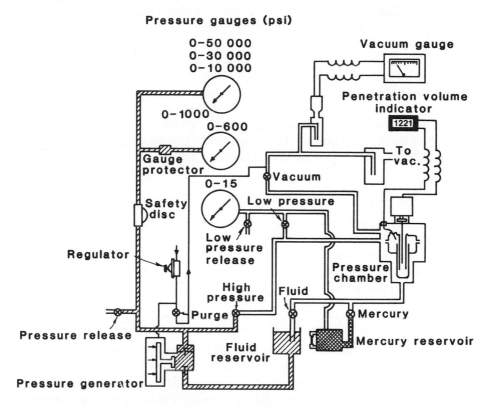

Fig. 19.3 Micromeretics' mercury porosimeter; schematic diagram of system.

The Micromeretics porosimeters have been specially designed to enable the sample to be evacuated to a pressure of approximately 2×10^{-2} torr inside the pressure chamber of the instrument. An alternative filling technique has been described [38] to allow evacuation, outside the pressure chamber, to a pressure of 10^{-5} torr. The advantages of this technique over the conventional filling method are:

(1) The process is under visual control and the fluidizing motion that can occur with very fine powders can be minimized.
(2) Under poorly controlled filling conditions air bubbles can be trapped in the cell leading to spurious results.
(3) Mercury is not contaminated by the hydraulic fluid during the filling operation and clean mercury is used for each sample.
(4) Samples can be evacuated for long periods at low pressures.

This filling process is at a disadvantage where pores are to be examined at less than atmospheric pressure. The errors introduced by adsorbed gases are less important for these large pores and such samples should be examined in the usual manner.

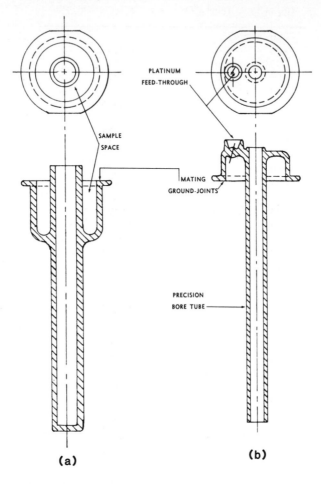

(a)   **(b)**

Fig. 19.4 Sample cell.

19.5 Theory for volume distribution determination

Experimental data are obtained in the form of P against volume intruded (Table 19.1). P is converted to pore radius using equation (19.2): assuming a surface tension for mercury of 0.480 N m^{-1} and a contact angle of 130°, with pressure P in psi and radius r in μm, equation (19.2) becomes:

$$r = \frac{89.5}{P} \qquad (19.3)$$

Table 19.1　Table of experimental pressurizing data.

Pressure P (lbf in^{-2})	Volume intruded V (cm^3 g^{-1})	Pore radius r (μm)
0		
400	0	0.224
500	0.0105	0.179
600	0.0181	0.149
750	0.0251	0.119
1000	0.0321	0.0895
1250	0.0495	0.0716
1600	0.0670	0.0559
1750	0.0809	0.0511
1900	0.120	0.0471
1950	0.148	0.0459
2000	0.191	0.0448
2050	0.218	0.0437
2100	0.240	0.0426
2150	0.254	0.0416
2200	0.275	0.0407
2250	0.289	0.0398
2350	0.318	0.0381
2500	0.352	0.0358
2650	0.374	0.0338
2750	0.388	0.0325
3000	0.423	0.0298
3250	0.445	0.0275
3500	0.458	0.0256
3750	0.472	0.0239
4000	0.479	0.0224
4500	0.493	0.0198
5000	0.500	0.0179
6000	0.508	0.0149
7000	0.512	0.0128
8000	0.515	0.0112
10000	0.515	0.00895

If the pore size range is narrow it is possible to plot the cumulative pore size distribution by volume on linear paper and to differentiate the curve to obtain the relative pore size distribution by volume (figure 19.6).

For a wide distribution (the radii may vary from 0.0015 to 180 μm) it is necessary to plot the data on log-linear paper with V on the linear axis (figure 19.5). This curve of V against P is called the pressurizing curve and from it the various distributions may be found.

If the total volume of pores having radii between r and $r + dr$ is dV, the relative pore frequency by volume is defined by:

$$D_3(r) = \frac{dV}{dr} \qquad (19.4)$$

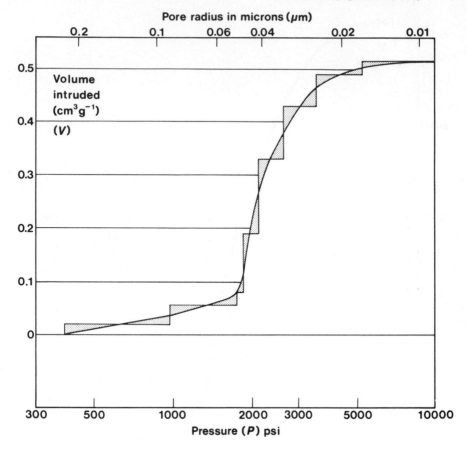

Fig. 19.5 Pressurizing curve.

From equation (19.2):

$$P\,dr + r\,dP = 0 \qquad (19.5)$$

Combining these two equations gives:

$$D_3(r) = -\frac{P}{r}\frac{dV}{dP}$$

$$= -\frac{1}{r}\frac{dV}{d(\ln P)} \qquad (19.6)$$

The application of this equation to the pressurizing curve is demonstrated in section 19.9. It is the convention to plot data as volume distributions and not percentage volume distributions.

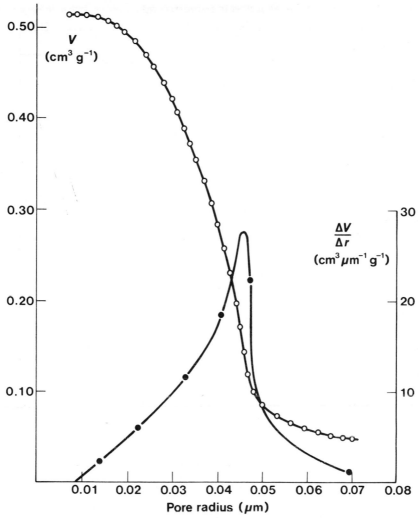

Fig. 19.6 Cumulative and relative pore size distribution by volume. The relative distribution is obtained by differentiating the cumulative curve in order to smooth out experimental errors. Tabular points are also included on the relative distribution curve (Table 19.2).

19.6 Theory for surface distribution determination

19.6.1 Cylindrical pore model

Using a cylindrical pore model:

$$\Delta V = \pi r^2 \, \Delta L \qquad (19.7)$$

$$\Delta S = 2\pi r \, \Delta L \qquad (19.8)$$

Defining the surface distribution as $D_2(r)$:

$$D_2(r) = \frac{dS}{dr}$$

$$= \frac{\Delta S}{\Delta V} \cdot \frac{dV}{dr}$$

$$= \frac{2}{r} \frac{dV}{dr} \qquad (19.9)$$

This equation may also be written, by combining with equation (19.5), as.

$$D_2(r) = -\frac{2P}{r^2} \frac{dV}{dP}$$

$$= -\frac{2}{r^2} \frac{dV}{d(\ln P)} \qquad (19.10)$$

19.6.2 Modelless method

Rootare and Prenslow [29] obtained surface areas from mercury intrusion data using no assumption of any specific pore geometry. The problem was approached from the point of view that work is required to force mercury into the pores, the work required to immerse an area dS of powder being:

$$dW = \gamma_{LV} \cos\theta \, dS = -P \, dV \qquad (19.11)$$

Therefore,

$$dS = \frac{-P \, dV}{\gamma_{LV} \cos\theta}$$

Integrating over the whole range of pressures:

$$S = -\int_{V_0}^{V} \frac{P \, dV}{\gamma_{LV} \cos\theta} \qquad (19.12)$$

For $\gamma_{LV} = 0.480$ N m^{-1}; $\theta = 130°$; m, the sample mass in grams and P in psi.

$$S_w = -\frac{0.0223}{m} \int_{V_0}^{V} P \, dV \quad \text{m}^2 \text{ g}^{-1} \qquad (19.13)$$

This is identical with the integral of equation (19.9), with substitution for r from equation (19.3). Alternatively, if S_w is known from BET gas adsorption, this equation may be used to determine $\cos\theta$.

$$\cos\theta = -\frac{0.01436}{m S_w} \int_{V_0}^{V} P \, d V \qquad (19.14)$$

Using this equation Rootare [27] found $121° < \theta < 160°$ for a range of powders.

19.7 Theory for length distribution determination

Defining the length distribution as $D_1(r)$:

$$D_1(r) = \frac{dL}{dr} \tag{19.15}$$

Combining with equation (19.7) gives:

$$D_1(r) = \frac{1}{\pi r^2} \frac{dV}{dr} \tag{19.16}$$

but Pr = constant, hence $P\,dr = -r\,dP$ and

$$D_1(r) = -\frac{P}{\pi r^3} \cdot \frac{dV}{dP}$$

$$= -\frac{1}{\pi r^3} \frac{dV}{d(\ln P)} \tag{19.17}$$

19.8 Worked example

The data in Table 19.1 were presented by Rootare and Spencer [35] in a paper in which they describe a computer programme for pore size and surface area determination.

For narrow pore size distributions V may be plotted against r to give the cumulative volume distribution and the curve differentiated to give the relative volume distribution (figure 19.5). Alternatively a tabular method may be used. Total surface area may be found from a graphical integration of the PV curve with the application of equation (19.13). This equation may also be used to determine the cumulative distribution in steps (Table 19.2); these data may be plotted to give the cumulative surface distribution (figure 19.7) and the curve can be differentiated to give the relative surface distribution.

Since a wide range of pore sizes can be measured it is often convenient to present pore radius on a logarithmic scale and evaluate $P\,dV$ from this. For the example presented in Table 19.1, figure 19.6, as the pressure rises from 400 to 1000 psi (average $\bar{P} = 700$ psi) the intruded volume of mercury rises from zero to 0.0321 cm^3 g^{-1} making $\bar{P}\,dV$ equal to 22. Taking the next increment as 1000 to 1800 psi, a further 0.0494 cm^3 g^{-1} of mercury is intruded at an average pressure of 1400 psi making $\bar{P}\,dV$ equal to 69. The increments should be small enough that, despite the logarithmic nature of the plot, the areas on either side of the curve should be substantially equal. This integration is carried out over the whole curve and the cumulative pore size distribution by surface determined (see Table 19.2).

Some investigators prefer to plot the relative volume distribution with $dV/d(\log r)$ as ordinate on an arithmetic scale versus r on a logarithmic scale as the abscissa.

Table 19.2 Calculation of cumulative and relative frequency distributions.

Pressure P (lbf in^{-2})	Volume V (cm^3 g^{-1})	Pore radius r (μm)	Mean pressure $\bar{P}$ (lbf in^{-2})	Volume increment ΔV (cm^3 g^{-1})	$\bar{P}\,\Delta V$	$\Sigma \bar{P}\,\Delta V$	Surface S_w (m^2 g^{-1})	Δr (μm)	$\Delta V/\Delta r$ (cm^3 μm^{-1} g^{-1})	$\Delta S/\Delta r$ (m^3 μm^{-1} g^{-1})
400	0	0.224				0	0			
1000	0.0321	0.0895	700	0.0321	22	22	0.5	0.1345	0.24	4
1800	0.0815	0.0497	1400	0.0494	69	91	2.0	0.0398	1.24	38
2000	0.191	0.0448	1900	0.1095	208	299	6.7	0.0049	22.3	959
2400	0.336	0.0373	2200	0.1390	306	605	13.5	0.0075	18.5	907
3120	0.430	0.0287	2710	0.1000	271	876	19.5	0.0086	11.1	698
5400	0.503	0.0166	4260	0.0730	311	1187	26.5	0.0121	6.0	573
8000	0.515	0.0122	6700	0.0120	80	1267	28.3	0.0044	2.2	409

Equation (19.13) $S_w = 0.0223 \times 1267$

$\qquad\qquad\quad = 28.3 \text{ m}^2 \text{ g}^{-1}$

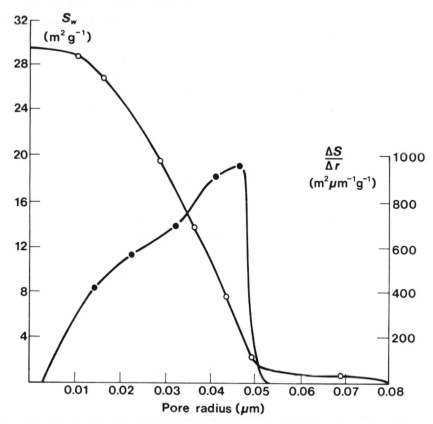

Fig. 19.7 Cumulative and relative pore size distribution by surface. The relative distribution is obtained by differentiating the cumulative curve in order to smooth out experimental errors. Tabular points are also indicated on the relative distribution curve (Table 19.2).

19.9 Comparison with other techniques

Cochran and Cosgrove [31] compared the analysis of aluminium using a porosimeter with a gas-adsorption analysis using n-butane (Table 19.3). The Kelvin equation (19.1) is used for the n-butane data, the lowest radius being calculated at the point where the adsorption branch of the hysteresis loop joins the desorption branch.

Table 19.3   Pore volume (ml g^{-1}) between pore radius limits (Å).

Alcoa H41 Pore										Re-main-der	Total pore volume
Radius	45 000	2000	1000	500	200	100	50	30	15		
Porosimeter	0.006	0.002	0.001	0.001	0.001	0.004	0.357				0.373
n-Butane				0.008	0.004	0.021	0.161			0.258	0.458

The n-butane yields a larger volume due to the presence of pores smaller than 3 nm in radius. The remainder is the volume of gas sorbed on to the sample and in the pores to a thickness of approximately two monolayers before capillary condensation starts.

Brown and Lard [44] found large discrepancies between nitrogen desorption and mercury porosimetry with high pore volume silicas. This they attribute to compression of the highly porous silica to reduce the pore volume and form smaller pores. Giles et al. [56] found marked hysteresis effects using a series of silica gels. These were explained in terms of the effect of pressure on the globular structure of the materials. Good agreement was found with nitrogen gas and methylene blue adsorption data.

Good correlation, in the range where nitrogen adsorption and mercury porosimetry overlap, has been found by others [32, 45–47]. Dullien and Dhawan [16] found that mercury intrusion gave the pore neck distribution for ink-bottle pores whereas Wood's metal gave the full distribution. Other relevant comparisons have been made by Dubinin and co-workers [33] who found good agreement particularly for benzene isotherms. Sneck [34] compared water adsorption and mercury porosimetry and found significant difference.

19.10 Correction factors

In order to calculate the true volume intrusion of mercury into the pores of a sample a correction must be made to account for the compression of the sample cell, mercury and sample. The usual procedure is to carry out a blank experiment in the absence of sample or on a non-porous sample. A theoretical analysis of the problem indicates that this is inadequate and equations have been derived which involve the mercury fill volume in the sample cell, the compressibilities of the mercury, glass and sample and the volume of the sample [36].

Working with the Micromeretics 903-1 porosimeter Lee and Maskell showed that the cell factor (F), which relates the digital counter reading to the penetration volume (see section 19.4), is not constant but decreases with increasing pressure since it is proportional to the cross-section of the constant-bore inlet tube which suffers a contraction.

$$f = F(1 - \tfrac{2}{3}\psi_g) \qquad\qquad (19.18)$$

where the compressibility of glass, $\psi_g = 2.6 \times 10^{-12}$ cm^2 dyn^{-1}.

The true blank correction at pressure P is given by:

$$j = \frac{P}{F}\left[- \Delta V_1(\psi_{Hg} - \tfrac{2}{3}\psi_g) + V_1(\psi_{Hg} - \psi_g)\right] \qquad\qquad (19.19)$$

$$= h - H$$

where ΔV_1 is the small unfilled volume at zero pressure as given by the initial digital counter reading (H)

$$\Delta V_1 = HF \qquad\qquad (19.20)$$

V_1 is the volume of the sample cell, taken as 30 cm^3.

The compressibility of mercury ψ_{Hg} = 3.55 x 10^{-12} cm^2 dyn^{-1}.

v_c the unfilled volume at pressure P is given by the counter reading h:

$$hf = v_c$$

For a non-porous material of volume V_N at zero pressure and ψ_N at pressure P.

$$J = \frac{P}{F} \left[- \Delta V_2 (\psi_{Hg} - \tfrac{2}{3} \psi_g) + V_1 (\psi_{Hg} - \psi_g) + V_N (\psi_N - \psi_{Hg}) \right] \quad (19.21)$$

where J is the intrusion count minus the count at zero pressure, $(h - H)$, ΔV_2 the initial unfilled volume and ψ_N the compressibility of the solid.

The corrected reading is given by:

$$J_{corr} = J - j$$

$$= \frac{P}{F} \left[(\Delta V_1 - \Delta V_2)(\psi_{Hg} - \tfrac{2}{3} \psi_g) + V_N (\psi_N - \psi_{Hg}) \right] \quad (19.22)$$

The first term represents the difference in fill between the blank and sample runs. If these are comparable the corrected reading is only a function of the net compressibility, pressure and sample volume.

The analysis can be extended to incorporate porous samples. For the case of a porous non-compressible solid, J_{corr} becomes:

$$J'_{corr} = \frac{P}{F} \left[(\Delta V_1 - \Delta V_2)(\psi_{Hg} - \tfrac{2}{3} \psi_g) + V_{pore} (1 - \alpha) \left(\frac{1}{P} + \tfrac{2}{3} \psi_g \right) \right.$$

$$\left. - \psi_{Hg}(V_N - V_{pore}) \right] \quad (19.23)$$

where V_{pore} is the pore volume and α the fraction of unfilled pores.

These equations have been found necessary with nylon materials in order to explain anomalous results [36, 40].

The errors which can arise from compression of the sample have also been considered by Ingles [42] who proposes the use of the following equation:

$$\Delta V_P = V_0 \left(1 - \exp(- \int \beta \, dP) \right) \quad (19.24)$$

where ΔV_P is the change in volume at pressure P, V_0 is the original volume and β the compressibility of the sample. The compressibility is the reciprocal of the bulk modulus, B, which is related to Young's modulus, E, by the equation:

$$E = 3B(1 - 2\sigma) \quad (19.25)$$

where σ is Poisson's ratio.

This correction is negligible for materials with a high Young's modulus, e g. glass $>$ 10 GN m^{-2}, but with materials such as polyvinyl chloride compression becomes important at pressures in excess of 7 mN m^{-2} [43].

Liabastre and Orr [57] examined graded series of controlled pore glasses and

Nuclepore membranes by electron microscopy and mercury porosimetry. Parameters were evaluated and correction for compressibility examined and the cause of hysteresis explored.

Stanley-Wood [58] examined non-porous spheres, microporous irregular shaped particles and meso- and macroporous powders. The sizes determined for the spheres (mean size 112 μm) were in fair agreement with the sieve size; the microporous particles were in agreement with the Stokes diameter but the particles having larger pores showed little agreement with Stokes.

The Carlo Erba (1500 kg cm^{-2}) porosimeter has been used to determine the porosity of filter media and distribution curves calculated [59].

References

1 Adamson, A.W. (1967), *Physical Chemistry of Surfaces*, 2nd edn, Interscience, NY.
2 Washburn, E.W. (1921), *Proc. natn. Acad. Sci.*, 7, 115–16.
3 Henderson, L.M., Ridgeway, C.M. and Ross, W.B. (1940), *Refin. Natur. Gas Mfr*, 19, 6, 69–74.
4 Ritter, H.L. and Drake, L.C. (1945), *Ind. Engng Chem. analyt. edn*, 17, 782–6.
5 Drake, L.C. and Ritter, H.L. (1945), *ibid.*, 17, 787–91.
6 Drake, L.C. (1949), *Ind. Engng Chem.*, 41, 780–5.
7 Burdine, N.T., Gournay, L.S. and Reichertz, P.P. (1950), *Trans. Am. Inst. Min. Metal. Engrs*, 189, 195–204.
8 Bucker, H.P., Felsenthal, M. and Conley, F.R. (1956), *J. Petrol. Technol.*, AIME, 65–66.
9 Winslow, N.M. and Shapiro, J.J. (1959), *ASTM Bull.*, TP49, 39–44.
10 Guyer, A., Boehlen, B. and Guyer, A., Jr. (1959), *Helv. Chim. Acta*, 42, 2103–10.
11 McKnight, T.S., Marchassault, R.H. and Mason, S.G. (1958), *Pulp Pap. Mag. Can.* 59, 2, 81–88.
12 Purcell, W.R. (1949), *J. Petrol. Technol.*, 1, 39–48.
13 Stromberg, R.R. (1955), *J. Res. NBS*, 54, 73–81.
14 Watson, A., May, J.O. and Butterworth, B. (1957), *Trans. Br. Ceram. Soc.*, 56, 37–50.
15 Plachenov, T.G. (1955), *J. appl. Chem., USSR*, 28, 223.
16 Dullien, F.A.L. and Dhawan, G.K. (1976), *J. Colloid Interfac. Sci.*, 52, 1, 129–36.
17 Svata, M. (1968), *Abh. Sächs Akad. Wiss.*, 49, 5, 191–6.
18 Svata, M. and Zabransky, Z. (1968/69), *Powder Technol.*, 2, 159–61.
19 Baker, D.J. (1971), *J. Phys. E.*, 4, 5, 388–9.
20 Leppard, C.J. and Spencer, D.M.T. (1968), *J. Phys. E*, 1, 573–5.
21 Reich, B. (1967), *Chem. Ing. Tech.*, 39, 22, 1275–9.
22 Spencer, D.J.M. (1969), *British Coal Utilisation Research Assoc. Bull.*, 33, 10, 228–39.
23 Meyer, H.I. (1953), *J. appl. Phys.*, 24, 510–12.
24 Rappeneau, J. (1965), *Les Carbenes*, 2, Masson, Paris, Chapter 14, pp. 134–40.
25 Scholten, J.J.F. (1967), *Porous Carbon Solids* (ed. R.L. Bond), Academic Press, pp. 225–49.
26 Diamond, S. (1970), *Clay Min.*, 18, 7.
27 Rootare, H.M. (1970), *Advanced Experimental Techniques in Powder Metallurgy*, VS: *Perspectives in Powder Metallurgy*, Plenum Press, NY, pp. 225–54.
28 Juhola, A.J. and Wiig, E.O. (1949), *J. Am. Chem. Soc.*, 71, 2078–80.
29 Rootare, H.M. and Prenzlow, C.F. (1967), *J. Phys. Chem.*, 71, 2734–6.
30 Roberts, N.K. (1964), *J. Chem. Soc.*, 1907–15.
31 Cochran, C.N. and Cosgrove, L.A. (1957), *J. Phys. Chem.*, 61, 1417.
32 Szeitering, P. (1958), *The Structure and Properties of Porous Solids*, Butterworths, p. 287.
33 Dubinin, M.M. *et al.* (1951), *Russ. J. Phys. Chem.*, 34, 959.
34 Sneck, T. and Cinonen, M. (1970), *Valtion Tek Tutkimaslaitos Julk*, 155, 60.
35 Rootare, H.M. and Spencer, J. (1972), *Powder Technol.*, 6, 1, 17–24.
36 Lee, J.A. and Maskell, W.C. (1973), *ibid.*, 7, 259–62.

37 Orr, C., Jr. (1970), *ibid.*, **3**, 117–23.
38 Palmer, H.K. and Rowe, R.C. (1975), *ibid.*, **11**, 195–6.
39 Whittemore, O.J., Jr. and Sipe, J.J. (1974), *ibid.*, **9**, 159–64.
40 Lee, J.A. and Maskell, W.C. (1974), *ibid.*, **9**, 165–71.
41 Winslow, D.N. and Diamond, S. (1970), *J. Mater.*, **5**, 564.
42 Ingles, O.G. (1959), *J. Aust. Appl. Sci.*, **10**, 484.
43 Palmer, H.K. and Rowe, R.C. (1974), *Powder Technol.*, **9**, 181–6.
44 Brown, S.M. and Lard, E.W. (1974), *ibid.*, **9**, 187–90.
45 Joyner, L.G., Barrett, E.P. and Skold, R. (1951), *J. Am. Chem. Soc.*, **73**, 3155.
46 Kamakin, N.M. (1953), *Akad. Nauk SSSR, Tr. Soveshch*, 47.
47 Brunauer, S. (1969), *Chem. Engng Progr. Symp. Sci.*, **65**, 1.
48 Rootare, H.M. and Craig, R.G. (1974), *Powder Technol.*, **9**, 199–211.
49 Unger, K., Schadow, E. and Fischer, H. (1976), *Z. Phys. Chem.*, **99**, 4–6, 245–56.
50 Frevel, L.K. and Kressley, L.J. (1963), *Analyt. Chem.*, **35**, 1492.
51 Meyer, R.P. and Stowe, R.A. (1965), *J. Colloid Sci.*, **20**, 893.
52 Svata, M. and Zabransky, Z. (1969/70), *Powder Technol.*, **3**, 296–8.
53 Reverberi, A., Ferraiolo, G. and Peloso, A. (1966), *Ann. Chim.*, **56**, 1552.
54 Svata, M. (1971/72), *Powder Technol.*, **5**, 345–9.
55 Gillard, J. (1975), *Lab-Pharma-Probl. Tech.*, **23**, 246, 789–99.
56 Giles, C.H. *et al.* (1978), *Proc. Conf. Structure of Porous Solids*, Neuchatel, Switzerland, Swiss Chem. Soc.
57 Liabastre, A.A. and Orr, C. (1978), *J. Colloid Interfac. Sci.*, **64**, 1, 1–18.
58 Stanley-Wood, N.G. (1979), *Analyst*, **104**, 1235, 97–105.
59 Mavrov, A. and Zvezdov, A. (1975), *Proc. 5th Int. Congr. Chemical Engineering Chemical Equipment and Design Automation*, Chisa, Prague, Czechoslovakia.

20 On-line particle size analysis

20.1 Introduction

Continuous monitoring of material in a disperse form is widely desired. This need may be divided into four broad categories but equipment may be applicable to more than one. Two categories, sampling and sizing dusty gases in gas streams and sampling and sizing from the atmosphere are dealt with elsewhere.

Equipment has also been developed for such purposes as monitoring contamination levels in parenteral liquids and hydraulic fluids. Where the same principle is embodied in both an airborne and a liquid-borne monitor, the principle is discussed in Chapter 3 and its particular application to liquid-borne monitoring is discussed below.

The demand for continuous control of milling circuits has led to the development of several on-stream sizing devices. Milling circuits are notoriously unstable and unwanted fluctuations in particle size, pulp density and volume flow rates can lead to the inefficient use of grinding capacity and to poor extraction of the valuable mineral.

In one gold mine studied [1–3] each 1% drop in sub-75 μm average daily grind could cause a drop of $4800 in monthly revenue. Since the rate of change may be as high as ±10% minus 75 μm per minute, over a period of a minute or two, average daily grind data are of little value, hence the demand for rapid response instrumentation.

Two solutions are available; one is to abstract a sample of the material and analyse it off-line; the problems that arise here are the obtaining of a representative sample from the bulk [4] and finding a rapid method of particle size assessment. The alternative is to analyse the bulk material, thus eliminating sampling problems; it must be realized however that if only a part of the bulk material is analysed on-line, sampling problems still arise.

Equipment may be divided into two types [5]: 'stream-scanning' for dilute systems in which the particles are sent in single file past a detecting device, and 'field-scanning' for concentrated systems in which the behaviour of the bulk material is monitored.

20.2 Stream-scanning

Stream-scanning is usually applied to dilute systems, hence it is most widely used for determining contamination levels.

A contamination limit test was introduced into the 1973 edition of *British Pharmacopoeia* for large volumes (500 cm^3 and greater) of most intravenous solutions. No

instrument was specified but the Coulter counter is the most widely used. Results of the test were to be reported in terms of the number concentration of particles greater than 2 and 5 μm. In the *US Pharmacopoeia* of 1975 the limit test was based on filtration and microscope counting. In Australia the standard is based on the HIAC [6]. The dominance of the Coulter is thus being challenged by the light-blocking and light-scattering instruments, in particular the HIAC particle counter.

Since these instruments operate on different principles one would expect a lack of agreement between them. The publication of two sets of comparisons between the three types of instrument: Coulter, HIAC and Royco highlights these differences [6, 7]. Standards would therefore be improved if tests had to be carried out on specified instruments, or correlated according to the method discussed by Groves.

The preference for the optical counters is due to their greater versatility. They can handle high volume throughputs more readily than the Coulter; they do not suffer aperture blockage; they are more readily adapted for on-line analysis; they can accurately analyse low concentration systems and an electrolytic carrier is unnecessary.

Lieberman [8, 9] stresses that optical instruments will produce accurate analyses provided the coincidence levels are not exceeded and that the calibration is valid. Optical counters are not primary measuring instruments and are always calibrated for a specific material. They are normally calibrated by the manufacturer using transparent spheres and before being used for other materials a new calibration must be performed. A calibration method for coal dust, using impactors, was described by Marple and Rubow [10] and a method for asbestos by Addingley [11].

The lower size limit with light attenuation instruments of about half a micrometre has been extended to a tenth of that value with the use of lasers [24, 32, 33].

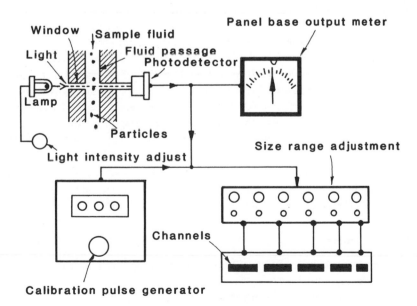

Fig. 20.1 The HIAC particle counter.

20.2.1 The HIAC particle counter

The HIAC particle counter (figure 20.1) was developed by Carver [12] as an alternative method to microscopy for counting and sizing contamination in hydraulic fluids. Since then the instrument has been used to examine parenteral liquids [13–15], drinking water [16, 21], viscous liquids [22], distilled alcoholic products [17] as well as hydraulic and transmission liquids [18–20].

The particles to be counted are suspended in a fluid having a refractive index different from the particles. The suspension is then forced through a sensor (figure 20.2) containing a small rectangular cell with windows on opposite sides. A collimated beam of light from a high intensity quartz halogen lamp is directed through the stream of liquid on to a sensor. The geometry of the sensor is such that the flow within the sensing zone is turbulent. This causes the particles to tumble as they traverse the zone thereby exposing all surfaces to the photodetector. The amplitude of the pulses produced is therefore proportional to the greatest projected area of the particles.

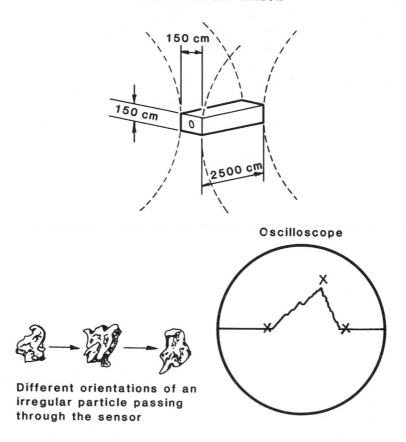

Fig. 20.2 Sensor window of the HIAC counter.

Earlier publications describe six sensors, each one capable of sizing particles in a 30 : 1 size range, ranging in particle size from 2 to 9000 μm. In a later paper [23] West describes five sensors, each one capable of sizing particles in the size range 45 : 1, ranging in particle size from 1 to 9000 μm. Twelve data thresholds may be dialled for each sensor giving twelve points on the size distribution curve. The instrument is also available with a microprocessor with 24 thresholds to give the cumulative and relative distribution by weight and by number [126].

The stated theoretical relationship between the size of the particle and the amplitude of the voltage pulse it produces is given as:

$$E_0 = \frac{a}{A} E_b$$

where E_0 = pulse amplitude from photodetector

a = maximum projected area of the particle

A = area of the window

E_b = 10 V, the base voltage from the photodetector.

The validity of this equation is highly questionable since it should include refractive index and light absorption factors. The instrument may be calibrated using this equation but, in practice, it is normally provided calibrated with spherical beads. ACFTD calibration [24] is also available.

West [25] discusses three calibration methods:

(a) AC Fine Test Dust [24]. This dust has been adopted as standard by ANSI, ISO and NFPA and resembles the type of contaminants usually found in hydraulic fluids.

(b) Monosize spherical latex using a half-count technique. The latices were 7.3, 10.3, 15.20 and 80 μm mean size by microscopy and Coulter count.

(c) NBS glass beads: Standard Reference Material 1003; size range 5 to 30 μm based on Stokes settling.

To determine the effects of coincidence and establish the maximum number concentration of particles that can be analysed he employed procedures outlined by the ISO [26]. These involved the analysis of continually increasing concentrations of ACFTD until the relationship between count and concentration was no longer linear. It is necessary to operate the instrument at the flow rate recommended by the manufacturers. Operating at too high a flow rate leads to anomalous results [27, 28].

The manufacturers state that a necessary requirement for the HIAC to operate is that there must be a difference in refractive index between the particles and the suspending liquid. Golden [28] found that the instrument had to be recalibrated for polymer beads in water due to the small difference in refractive index. Resin spheres are normally sized in a water-swollen state and Golden found that immersing them in sodium chloride, such as is used in the Coulter counter, caused them to shrink by 16%.

Behringer et al. [29] describe a novel application of the technique as an airborne particle counter. The particles were fed from a circular vibrator bowl and partially entrained in an airstream which transported them down through the sensor zone. The

particles were nickel shot of nominal size 100 μm and the results were in good agreement with optical microscopy. Calibration was based on theory, assuming the laws of geometric optics held.

20.2.2 The Climet particle counting systems

There are several systems in the Climet range, all operating on the same principle (see Chapter 3). The Liquid Particle Analyser Cl-220 incorporates an optical sensor which detects and measures individual particles. Sample preparation consists of extraction with a syringe which is then mounted on the sensor. Four particle size ranges are available via panel switches: 2 to 20, 5 to 50, 8 to 80 and 20 to 200 μm. Mean particle size can be read directly in all ranges at number concentrations in the range 1 to 10^8 cm^{-3}. The sampling rate may be varied in the wide range of 5×10^{-8} to 500 cm^3 min^{-1} in order to obviate the need to dilute for high concentration suspensions. Primary calibration is effected with monodisperse styrene–divinyl-benzene copolymer spheres. Particle data are presented on a two-channel five-digit solid state display of counts with adjustable limits.

The Cl-221 is an on-line particle analyser for high purity liquids. All particles are counted above the selected threshold in the range 2 to 200 μm for flow rates in the range 120 to 750 cm^3 min^{-1}. Particle concentrations up to 100 cm^{-3} are handled directly. Higher concentrations are sampled at proportionally lower flow rates. The control panel includes a five-digit display and a visual and audible alarm.

20.2.3 The Royco liquid-borne particle monitors

These employ one of three basic optical sensing techniques according to the model. The one recommended for liquid-borne applications uses light-adsorption/total scatter. Model 320 sizes at 10 to 25, 25 to 50, 50 to 100 and 100 to 500 μm; model 345 is a multiple size range instrument. The sensor may be plumbed directly into a liquid line or may be adapted for batch sampling.

Royco Instruments has recently reported on an optical particle counter developed for liquid-borne submicron particle counting and sizing [30]. A conceptual diagram of the system is shown in figure 20.3. A low powered He–Ne laser beam is passed through

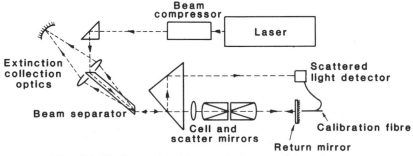

Fig. 20.3 The Royco liquid-borne submicron particle analyser.

a beam compressor to reduce the beam size. A portion of the direct beam is passed through the return mirror to a calibration fibre optic system, which passes chopped light pulses to the scattered light detector. These pulses are used to define the light level from the laser through the folding optics and cell, to the scattered light detector. As particles pass through the cell, a scattered light pulse is produced and collected by the scattered light detector. By virtue of the scatter mirror design, both forward and back scattered light is collected, and sent through the folding prism to the scattered light detector. An absorption signal is also produced by taking the return beam to the beam separator, collecting the separated beam in the lens at the periphery of the beam separator prism, and directing that beam to the extinction collection optics. Using a pulse height analyser, preliminary data taken from the scattered light portion of the system have shown it to be relatively noise free and capable of sensing particles of 0.5 μm diameter.

20.2.4 The Nuclepore Spectrex Prototron particle counter

This instrument [31] may be used to detect and count the number of particles, above a manually set threshold between 1 and 100 μm, in a bottled liquid and to monitor continuous flow through glass piping (figure 20.4). Although it has been found to be easy to operate, analysis time is protracted and reproducibility and sensitivity in the lower particle size ranges and calibration procedures have been found inadequate [16].

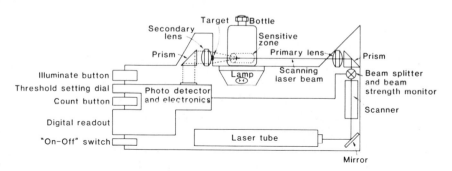

Fig. 20.4 The Prototron particle counter.

20.2.5 The Procedyne particle size analyser

In the Procedyne particle size analyser [35, 36] a He–Ne laser beam is passed through a flowing suspension via an oscillating mirror thus scanning the suspension horizontally and vertically (figure 20.5). Particles in the suspension aperiodically interrupt the light beam which is analysed by a fast response photodiode. The signal generated contains information related to particle size and size distribution. The instrument is designed for the on-line analysis of process slurries at a sampling rate of 250 cm^3 s^{-1} with a maximum particle concentration of 2% by weight and a particle size range of 5 to

2000 μm. There are five memory registers, four calibrated for particle size and the fifth is used for total particle count.

Hinde [87] reported on the use of this instrument for controlling mill circuits. Poor results were obtained, but this was explained by dispersion and agglomeration factors and the presence of a large proportion of fines in the slurry (cited in [123]).

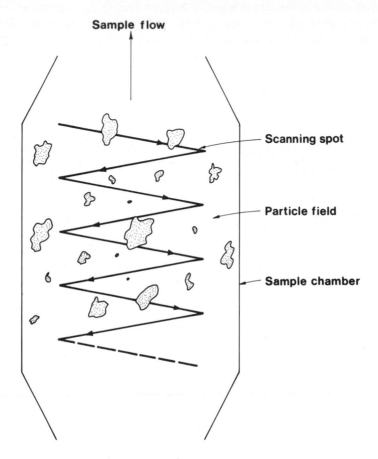

Fig. 20.5 Sample chamber of the Procedyne analyser.

20.2.6 Optical–electronic method

Talbot [37] designed an optical–electronic method to determine the settling speeds of particles suspended in a fluid. This instrument was designed for the quality control of mill pulps produced in the gold-mining industry where small variations in gold recovery can have large effects on profitability. The mill product is sampled automatically and diluted from about 100 g of solids in 1000 cm^3 to a solids–liquid ratio of about 10^{-4} by weight. The stream then flows into a transparent cell of rectangular cross-section at

a uniform velocity of 2 cm s⁻¹ (figure 20.6). A collimated beam of light forms an image of the suspended particles on a binary line grating consisting of alternating transparent and opaque bands of equal width and subsequently on a photovoltaic cell behind the grating. As a particle traverses the cell, its image falling on the grating produces a square wave component in the photo-cell current. The frequency of this wave is proportional to the transit speed of the particles and the amplitudes are proportional to the projected area of the particles. The signals from the photo-voltaic cell are sorted out into four channels equivalent to four size ranges (figure 20.7). The results are printed out every 30 s and the range of sizes measured is from 5 μm to 420 μm.

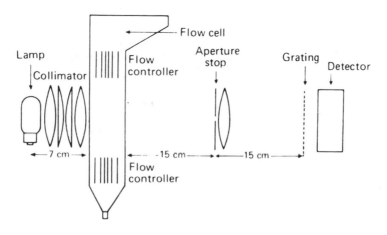

Fig. 20.6 The Talbot optical–electronic method: optical system and flow cell.

20.2.7 Miscellaneous optical methods

An optical–electronic instrument was designed by Rhodes et al. [38, 39] to measure the contamination levels in high pressure hydraulic systems in the NASA space shuttle. The system used a He–Ne laser source and a linear display of 50 photodiodes. Shadows of the particles are magnified and projected on to the photodiode array and particle size is measured by the number of photodiodes shadowed. 10 000 particles per second can be counted within the size range 5 to 100 μm.

In an apparatus described by Wu [40] two parallel laser beams shifted along the direction of particle motion and laterally from this direction are detected by two photo-transistors. After electronic examination of the dark signal given by moving particles the size and velocity can be deduced approximately. This method does not give high accuracy of particle size measurement and is more convenient for velocimetry.

A method has been described for evaluating the size of moving particles with a fibre optic probe. The average diameter is related to a characteristic length which is a product of pulse width and particle velocity. These are obtained experimentally using a correlation method [125].

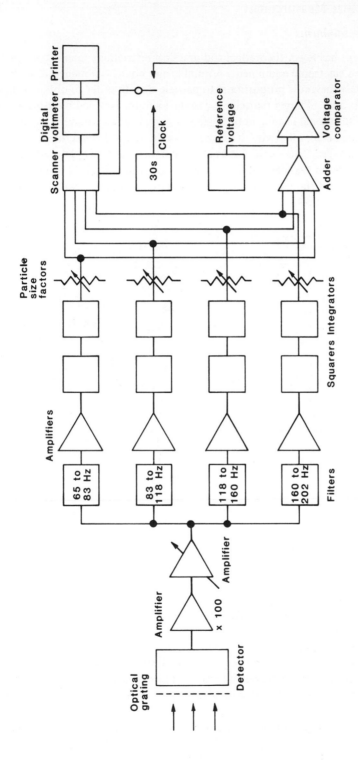

Fig. 20.7 The Talbot optical–electronic method: sorting and sizing system.

20.2.8 Echo measurements

Echo measurement has fewer theoretical and practical restrictions than ultrasonics but requires advanced electronic equipment. A number-size count is obtained directly since the echo pulse height is proportional to particle size. In order to obtain a statistically significant count of larger particles the sample volume is varied by using a divergent pulsed ultrasonic beam and time gates to sample at successively greater distances along the beam, otherwise the receiver would be overloaded due to the excess of small particles. This method is available commercially [41]. A de-aerated slurry is passed between two pairs of sensors; each pair, consisting of a transmitter and receiver, acts at a different frequency, one determining the size and the other the concentration (figure 20.8).

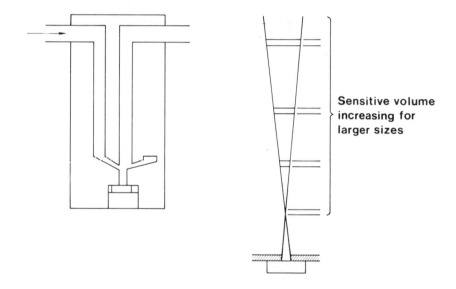

Sensitive volume increasing for larger sizes

Fig. 20.8 Ultrasonic on-stream size analyser.

20.2.9 The Langer acoustical counter (Erdco)

Audible sounds may be produced by particles exiting from a high velocity laminar flow tube into a lower velocity tube. This phenomenon was first reported by Langer [42] and has since been investigated by Langer and others [43–51] and now forms the basis for a commercial counter.

The acoustic sensor described by Langer (figure 20.9) consisted of a 15-cm long, 2.5-cm diameter glass tube which gradually tapered into a capillary 1.5 to 5.0 mm wide and 6 cm long. The capillary expanded rapidly back to the original 2.5 cm in the exit system. It was believed at first that the signal was generated as the particle entered the capillary [50, 51] although Hofmann suggested that it was formed as the particle

emerged into the jet [46]. Karuhn [49] found that as the particles enter the capillary section they interact with the boundary layer resulting in a toroidal vortex. The toroid moves into a shock front, which forms about 85% of the distance along the capillary and this produces the sound which is reflected back along the capillary.

In the commercial version of this instrument the microphone is replaced with a transducer at the outlet of the capillary. Since the toroid acts as a constriction as well as a sound generator the measurement of the exit jet pressure provides a more effective way of sensing the presence of particles. The original design was insensitive to particles smaller than about 30 μm; the modified design is claimed to operate down to 4 μm.

The patented transducer is of a novel design in that the displacement of the pressure sensing plate is measured optically. Light is fed to the rear of the pressure plate through a bundle of optical fibres and the reflected light picked up with another bundle, the amount being picked up reducing as the distance between the plate and the fibres decreases.

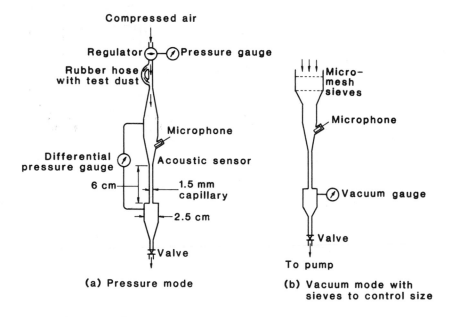

Fig. 20.9 Langer's acoustic particle counter for the detection of single particles.

20.2.10 The Coulter on-line monitor

The Coulter on-line monitor operates in exactly the same way as a normal Coulter counter in that particles suspended in electrolyte are passed through a small aperture through which an electric current is also passed. Each particle causes a displacement of electrolyte from within the aperature, modulating the current path, and this modulation is converted into a pulse which is subsequently displayed and counted on the appropriate Coulter counter.

The method of presenting the sample to the aperture and the operation of the sampling system require some detailed explanation and this is aided by the study of the schematic shown in figure 20.10.

The sample is taken continuously from a suitable outlet from a production plant or filtration system and can be sampled directly at pressures ranging from 0.3 to 3.3 N cm^{-2}. It is fed via a peristaltic pump through a coil at the bottom of the unit and then into the sample cell. The action of passing the sample through the coil effectively degasses it so that air bubbles are excluded and not counted.

Inset in a perspex block are three aperture tubes each with its own internal electrode and connected to a vacuum supply and two solenoid valves which operate independently of each other. The external electrode is common and surplus sample is exhausted through the top of the aperture block, to waste.

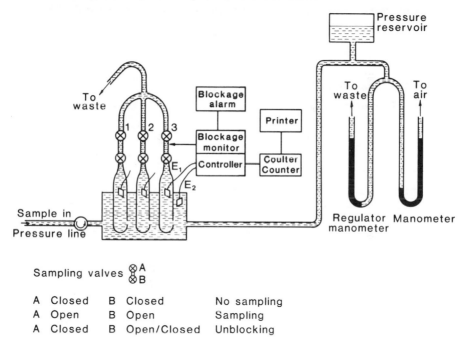

Fig. 20.10 The Coulter on-line monitor.

The rate of sample flow through the cell can be adjusted by a peristaltic pump and a dual manometer system on the right-hand side of the unit regulates this. Problems arising from the continuous flow of sample through the apparatus are prevented by an air-break system.

Sampling of the suspension is done by the first aperture, which can be of any diameter, e.g. typically 100 μm for particle monitoring between 2 and 40 μm. The sample volume being measured is controlled by a timer built into the control unit and this sample volume is variable. If an oversize particle passes into the aperture and a blockage results, this is detected electronically, the solenoid valves will close, no count

will be recorded, and the second set of valves will open and the analysis will recommence with the second aperture. In the event of the first and second apertures being simultaneously blocked, the third tube will automatically be selected.

The action of closing the solenoid valves on the first aperture has the effect of creating a pressure pulse which in most instances will clear the blockage so that this aperture is ready for use as and when the other apertures are, in their turn, blocked. In this way the on-line monitor never requires operating attention.

Sample that is drawn through the aperture is fed via heavily screened tubing into the waste air-break system and then allowed to run to drain or fed back into the process being monitored if so desired.

The control unit incorporates (i) a count time control to enable the operator to set the volume of sample he wishes to monitor, for example 1 ml, 5 ml etc. This is based on the time taken for a known volume of sample to pass through the selected aperture. (ii) An interval time (in minutes) control allowing the operator to set for himself the interval between each check on the system; this is variable in intervals from 2 to 64 min. (iii) A count alarm allowing the operator to pre-set the maximum count for the system. Once this is reached or exceeded an alarm is triggered, e.g., at 999 particle per ml above 2.0 μm for parenteral fluid applications.

The printer unit will print out each particle count and the index sample number will also advance one step each time so that a continuous record of the change in particle count can be obtained. A gap in the index sequence shows that a blockage has been cleared, itself a useful monitor of oversized particles.

The unit is supplied connected to a Coulter Counter Model ZB. With the ZB the monitor will provide a check on the particle count above a given size level when using the single threshold control or between two sizes when operating in the dual threshold mode.

With the Coulter Counter Model TA, the unit will provide information over a much wider size range, in fact the 2–40% size range covered by the chosen aperture tube. A choice of information can be obtained, either on a cumulative or frequency weight per cent versus size.

An application of this technique has been described by Barnett [52].

20.2.11 On-line automatic microscopy

The particles, dispersed in water, run in a laminar stream in front of a microscope connected to a TV camera [59]. A short light pulse (0.6 μs), from a stroboscope, 'captures' them and forms an image on a TV imaging device (a two-dimensional photodiode array). Since all information from the whole image area is stored in the TV camera simultaneously each light pulse provides one sample. Immediately after one image is processed another can be taken.

A silicon imaging device is used to produce electronic signals from a visual input. The image area of the device has dimensions 7.31 by 9.25 mm and is broken down into 241 by 320 parts; thus the size of a single image cell is approximately 0.03 by 0.03 mm.

The electronic signals are digitized and adapted for a computer by means of a camera to computer adapter unit. By changing the cut-off level during calibration an optimum value can be estimated to eliminate low energy signals representing particles too small and unwanted in the measurement to eliminate noise. A PDP-11 16-bit mini-computer is used to compute results.

This instrument has been used to monitor a flotation process in the size range 26 to 208 μm and the results are comparable with normal sieving practice.

20.2.12 Comparison between stream-scanning techniques

The Coulter counter is capable of high speed particle size distribution analysis and is therefore suitable for on-line analysis. Though it is widely used, a number of investigators have voiced concern over the limitations of the instrument [7, 14, 53, 54]. In particular its use is limited due to the requirement that the particles must be suspended in an electrolyte. There is also doubt as to whether the sensing units can function satisfactorily in remote on-stream mountings.

A comparison between the HIAC, the Coulter counter and microscope counting yielded highly comparable results considering the different parameters measured [56]; the HIAC counter was preferred for estimating the amount of particulate matter in parenteral solutions. A similar comparison has been carried out for hydraulic fluids [7].

The HIAC, the Coulter counter, the Royco counter and the Prototron and microscopy have been compared for water purity measurements and the HIAC was preferred [57]. The Coulter was discarded because of lack of sensitivity when dealing with low particle counts such as are found in filtered water; the need to use an electrolyte solution was a further factor in its rejection. The Prototron was rejected due to an apparent lack of sensitivity, problems associated with refractive index and inadequate calibration procedures. The conventional method of nephelometry was also rejected since it gave only a single value and this could be misleading, Di Grado compared the HIAC, Royco and Coulter, and found wide variation and little correlation [58].

20.3 Field-scanning

20.3.1 Some properties of the size distributions of milled products

It is commonly found that milled products have log-normal distributions of particle sizes [3]. Further grinding in the same mill produces a family of curves with the same geometric standard deviation (slope); thus a single point on the distribution curve characterizes the fineness of the mill product provided the nature of the feed material is unchanged. It also follows that if the percentage passing, say 100 mesh, is plotted against the percentage passing 200 mesh then the results will lie on a smooth curve (figure 20.11).

An alternative method for plotting such size distributions is the Gaudin-Schumann plot where the cumulative weight percentage finer than a given size is plotted against

that size, with each scale to a logarithmic base (figure 20.12). For the majority of ground material the relationship between the two is linear except at the coarse end of the distribution. The relationship may be characterized by two parameters: a distribution modulus, n (slope), and a size modulus, k. Again, n remains constant for consecutive grindings of the same material.

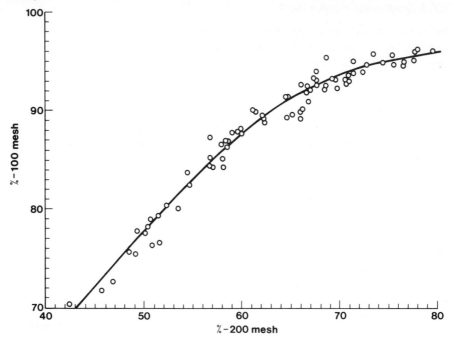

Fig. 20.11 Percentage passing 100 mesh versus the percentage passing 200 mesh.

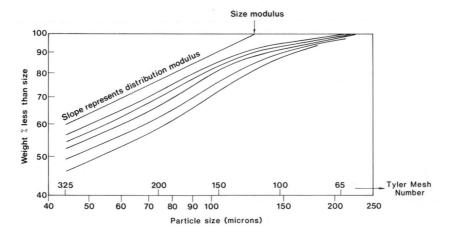

Fig. 20.12 Gaudin–Schumann plot.

Field-scanning techniques have been developed on the assumption that the properties defined above are applicable and thus a single point on the size distribution curve defines the distribution.

20.3.2 Static noise measurement

This technique has been applied to the measurement of the average size of milled silica powder (size range 2 to 5 μm) suspended in air [60]. A continuous sample from the product stream is drawn into a sampling probe and diluted by an air injector which also provides the driving force. The sample stream is then passed through a small 'uni-flow' cyclone which splits the sample into two streams, a low concentration 'fines' stream and a high concentration 'coarse' stream. As the relative mass flow rates of the two streams depend strongly on the size distribution of the feed (at a given flowrate) the average size may be found from a measure of the two concentrations.

Most particles suspended in air carry an electric charge, particularly if they have passed through a highly turbulent process. A probe inserted into the stream will detect this charge as an a.c. voltage which is strongly dependent on concentration. The system was calibrated by feeding in samples of known mean sizes and comparing these with the output of an analogue divider which was used to take the ratio of the a.c. signals from the two streams.

20.3.3 Ultrasonic attenuation measurements: the Autometrics PSM systems 100 and 200

The theory for attenuation methods shows that there are two main regions where direct calculation of size from the observed attenuation is possible.

If a pulsed beam of plane ultrasonic waves is transmitted through a sample slurry stream the intensity of the wave in the suspension is given approximately by:

$$I = I_0 \exp(-2\alpha x) \tag{20.1}$$

where 2α = absorption coefficient
I_0 = initial intensity of wave
x = distance travelled by wave.

The absorption coefficient is determined by two primary mechanisms, viscous and scattering losses. Viscous losses are associated with the relative movement of liquid and solid. The particles vibrate in response to the ultrasonic wave but with a phase lag and different amplitude. Extremely small particles tend to move in phase with the fluid and losses are very small. As size increases the particles tend to lag more and more behind the movement of the fluid and the loss per particle increases, but at the same time the solid–liquid interface area and therefore the loss per unit mass decrease. These opposing factors result in an absorption maximum [61].

The second loss mechanism is the scattering of energy due to the adsorption of a small amount of energy from the directed beam by each particle, and its subsequent

radiation away from the point of absorption. For particles of a discrete size and for $\lambda \gg 2\pi r$, it is possible to express the relationship between α and the properties of the solids and liquid by the equation:

$$2\alpha = c\left(\frac{k^4 r^3}{6} + k(\gamma - 1)^2 \frac{s}{s^2 + (\gamma + \tau)^2}\right) \tag{20.2}$$

where r = particle radius

λ = wavelength of ultrasonic wave

c = volume concentration of solids

k = angular wavenumber of ultrasonic wave in water

γ = ratio of densities of particles and water

ω = angular frequency of ultrasonic wave

$$s = \frac{9}{4\beta r}\left(1 + \frac{1}{\beta r}\right)$$

$$\beta = \left(\frac{\omega}{2\nu}\right)^{1/2}$$

ν = kinematic viscosity of water

$$\tau = \frac{1}{2} + \frac{9}{4\beta r}.$$

This equation is valid up to about 20% solids, above which the dependence on c becomes more complex.

The first term in equation (20.2) represents the attenuation due to scattering loss and the second due to viscous loss. For a given frequency, at very small particle sizes the viscous loss is predominant but as the size increases it becomes insignificant and the scattering loss becomes important.

For the case where $\lambda \ll 2\pi r$ equation (20.2) is no longer valid and must be replaced by the expression

$$2\alpha = \frac{3c}{2r} \tag{20.3}$$

In this range (known as the diffraction range) each particle can be considered as casting a simple shadow so that α is proportional to πr^2 per particle. In his thesis, Flammer [62] indicated that the scattering and diffraction terms in equations (20.2) and (20.3) were equivalent to similar expressions for the scattering of light by small particles. By analogy, he proposed that in the transition regime where $\lambda \simeq 2\pi r$ the value of α is independent of particle size. Figure 20.13 represents Flammer's interpretation of equations (20.2) and (20.3).

When the particles correspond to a wide distribution of sizes the integrated forms of equations (20.2) and (20.3) are very complex although the qualitative features displayed in figure 20.13 probably still apply. By choosing an appropriate frequency it is possible to find a regime where quite large changes in size distribution will not affect the attenuation of an ultrasonic wave. In such an instance the value of α is determined by the solids concentration only. By choosing a different frequency it is

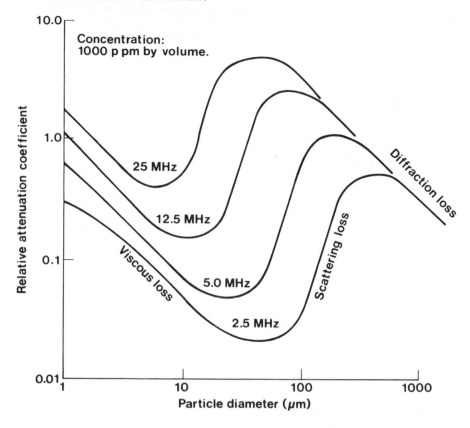

Fig. 20.13 The relationship given by Flammer [62] for the ultrasonic attenuation of slurries containing particles of discrete sizes.

possible to find a regime where even a small change in size distribution will result in a major change of the absorption coefficient. Since all particles contribute to the magnitude of the absorption coefficient it is imperative that the measured size distribution should be uniquely defined by a single point on the size distribution curve. If this were not the case, the results of attenuation measurements at only two frequencies would lead to ambiguous interpretations of the size distribution. However, commercial experience confirms this feature of many size distributions. In the Autometrics PSM system 100, figure 20.14, only two pairs of receiving and detecting transducers are necessary to produce adequate accuracy. Figure 20.15 shows some results for a gold mine [61]. The read-out of the size analyser is shown plotted as a function of the results of size analyses on small samples extracted from the sensor zone of the analyser. About 94% of the readings agree with the sieve analyses to within ± 2% at 200 mesh (78 data points).

A major problem area in the early development of the Autometrics system was that traces of air could lead to substantial attenuation losses. The air must be stabilized

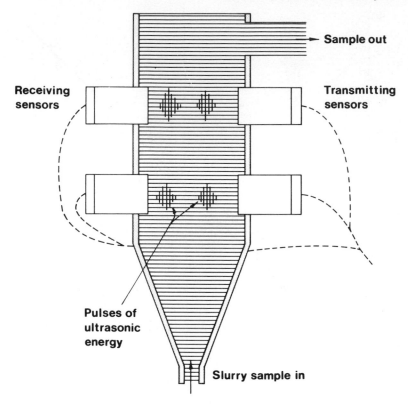

Fig. 20.14 The Autometrics PSM 100 analyser.

$$I = I_0 \exp(-2\alpha x) \quad \text{for } \lambda \triangleq 2\pi r$$

where $\alpha = f(c, r^3, \lambda \ldots)$, $\alpha \neq f(r)$.

or removed to allow accurate measurement of particle size. The problem was solved by removing the air with a device which utilizes a combination of centrifugal force and reduced pressure.

This need to remove air has increased the cost of the overall system significantly, and made it an expensive instrument when compared with other instrumentation often installed in grinding circuits. Nevertheless at present it appears to be perfectly competitive with other approaches when its inherent reliability and long-term stability as an accurate size analyser are taken into account.

Several articles have been written describing applications of the PSM system 100 [63-66]. The limitations of this system are (a) the particle size distribution should lie within the range 20 to 80% less than 270 mesh; (b) the percentage solids should be less than 60% by weight and (c) the slurry particles should not be magnetized. The PSM system 200 is a second generation instrument designed to overcome these limitations.

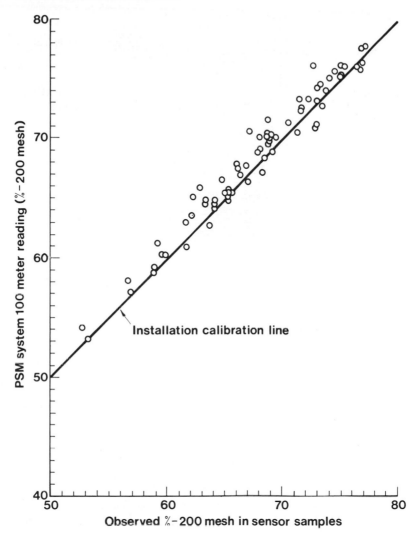

Fig. 20.15 Results obtained with the Autometrics size analyser.

20.3.4 β-ray attenuation: the Mintex/Royal School of Mines slurry sizer

The Mintex/RSM slurry sizer is based on the work of Holland-Batt [67–71] and has been subjected to rigorous investigation [72, 73]. A slurry, having a solids content in the range 10 to 30% by weight, flows through a rectangular tube, at flow rates in the range 100 to 200 cm³ s⁻¹, under constant pressure head conditions (figure 20.16). In the centre of the tube is a measuring cell fitted with windows and collimation plates top and bottom so that a beta-ray transmission measurement can be made to provide an index of its density.

After passing through this section the slurry flows through a single turn helix of the same dimensions. Centrifugal force causes the solid particles to separate according to size, the magnitude of the displacement being a function of particle size distribution. Two sets of windows and collimation slits are provided downstream of the helix to enable the solids concentration to be sensed; these are termed measure 1 and 2. The source and detector are designed to traverse the cell so that they can be stationed above and below any of four measuring positions: measure 1, measure 2, standard, density.

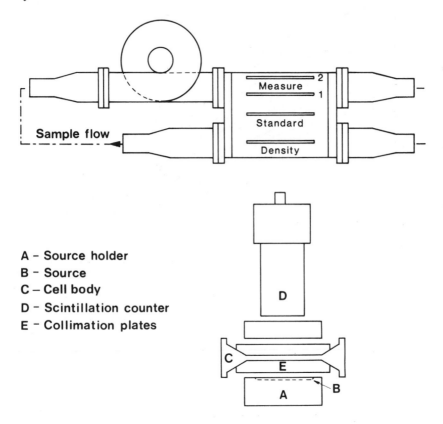

A - Source holder
B - Source
C - Cell body
D - Scintillation counter
E - Collimation plates

Fig. 20.16 The Mintex/RSM slurry sizer.

The accuracy of the device is limited by its sensitivity to changes in feed density [2]. Hence, in order to calibrate, it is necessary to recirculate slurry samples in a closed loop at a number of dilutions for each sample, sieve analyses being performed on a representative sub-sample.

A correlation exists between the solids content of the sample, the particle size of the sample represented by the percentage passing a 150 mesh sieve, and the difference reading (Δs) obtained from the two β-attenuation measurements. This is shown in

figure 20.17. By repeated cross-plotting at a number of constant densities a linear relationship between Δs and percentage passing through 150 mesh is obtained. The signals from the scintillation counter can be used to control mill feed rate so that changes in feed ore, grindability and particle size can be compensated for. Accuracies of 2 to 3% have been reported on Cornish granite, copper and iron pulps and nepheline systems in the size range 20 to 105 μm [74, 87].

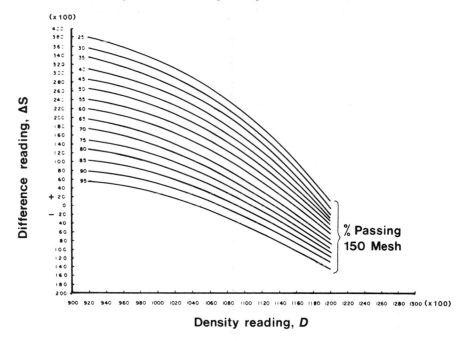

Fig. 20.17 Calibration nomogram for the Mintex/RSM slurry sizer.

20.3.5 X-ray attenuation and fluorescence

The use of X-ray absorption for on-stream has been investigated by Carr-Brion [75–77] (figure 20.18). The sensor is based on the comparison of the absorption of two X-ray beams, one of which is sensitive and the other insensitive to variations in particle size.

Each sensing head is specific to a particular system since the relationship between the two beams is dependent upon the composition of the solids in the stream. This technique is limited to materials that are opaque to X-rays.

The Courier 300 measures both X-ray scattering and X-ray fluorescence and is intended primarily as a composition monitor [74, 79]. The measured data can be analysed to give chemical composition, solids content and maximum size.

Von Alfthan [94] describes an on-stream X-ray fluorescence system to measure particle size. The analyser (figure 20.19) consists of two flow cells through which the slurry passes. In the classifying flow cell the slurry flows in a straight path behind a window; it then strikes an obstacle which causes slurry mixing as it enters the turbulent

flow cell. The slurry in both cells is excited by X-rays and the resulting fluorescent radiation is measured. The ratio of the radiation intensities was found to be a measure of the particle size.

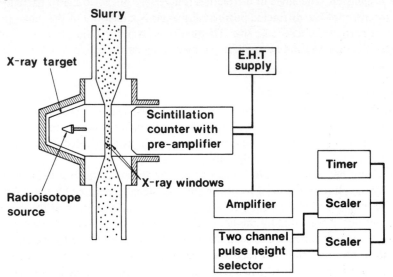

Fig. 20.18 The Carr-Brion X-ray absorption particle size analyser [75–77]. Schematic diagram of measuring head and associated electronics.

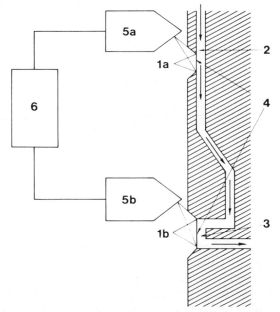

Fig. 20.19 X-ray on-line size analyser [94]. 1, X-ray tubes; 2, classifying flow cell; 3, turbulent flow cell; 4, thin plastic windows; 5, X-ray spectrometer, 6, data processing unit.

20.3.6 Laser diffraction

These methods are based upon measurements of the forward diffracted light from a disperse suspension. The angle of diffraction is inversely proportional to particle size and the intensity of the diffracted beam at any angle is a measure of the mean projec‐ ted areas of particles of a specific size. The theory is given by Talbot in a series of papers (see Chapter 14) and he applies it to measuring particles deposited on a micro‐ scope slide.

(a) The Cilas Granulometer 226

In this instrument [80–84, 127] the beam from a He–Ne gas laser is expanded and passed across a glass sample cell. The suspension of particles is stirred and ultrasonically dispersed in a separate chamber and circulated continuously by peristaltic pump from the chamber and through the sample cell.

The incident beam is cut off by a mask since its intensity is so strong that if it were received by the photodetector, it would swamp the signal generated by the diffracted image. The remaining light intensity is related to the size distribution of the suspended particles and this is evaluated by the use of eight apertures in a rotatable disc; the operation of the instrument consists in measuring the light intensity by means of a digital output photodetector for each aperture. The results are then processed by computer or desk calculator to give a seven-point size distribution at 2, 4, 8, 16, 32, 64 and 128 μm.

Particles greater than 128 μm cannot be detected due to the small angle of diffrac‐ tion by such large particles. The lower limit is imposed by the wavelength of the laser beam (0.63 μm).

In order that the photodetector is not saturated the recommended sample weight is 0.1 to 1.0 g, the lower limit applying to finer powders which diffract light more strongly than coarser materials.

Reproducibility is claimed to be of the order of 2% with an analysis time of about 10 min. The instrument is critically discussed by Stark [85] and the theory behind it and a discussion of its use in the ball milling of cement are presented by Meric et al. [86].

(b) The Leeds and Northrup Microtrac

In this instrument (figure 20.20) particles flow through a sample cell and are illumina‐ ted by a laser beam. Collected light passes through a rotating optical filter and is focused on to a solid-state detector. The resulting output is proportional to a selected mathematical function of particle size [88, 89].

Specifically designed Fraunhofer discs [124] pass diffracted light proportional to the second, third and fourth power of the radii of spherical particles. For a distribution the total volume can be monitored using the third power mask to produce a signal proportional to the sum of the cubes of particle radii. Ratios of signals passing through the discs provide statistical parameters of distribution such as volume, mean radius, surface mean radius and area standard deviation. The instrument can be used on-line or with batch samples.

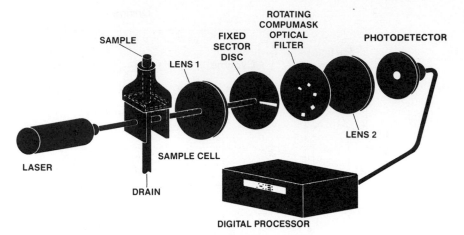

EXPLODED VIEW OF COMPLETE SYSTEM

Fig. 20.20 Leeds and Northrup Microtrac; exploded view of complete system.

(c) The Malvern Particle and Droplet Sizer

This apparatus consists of a small He–Ne laser fitted with a spatial filter and a collimating lens which provides a parallel, monochromatic, coherent incident beam. The particles are placed in this beam and the diffracted light is collected by a lens and brought to a focus on a special detector placed in the focal plane of the lens. This detector consists of 30 concentric, semicircular, photosensitive rings. The signal is transmitted to a computer and the whole system is controlled by a teletype. The acquisition of 200 data points for each detector ring takes two to three seconds. These data are then matched to a Rosin–Rammler distribution to give a best fit. Each detector/lens system gives approximately a 100 : 1 particle size ratio in the size ranges 2 to 197 μm, 5.7 to 560 μm and 12 to 1182 μm [90, 91].

20.3.7 Classification devices

(a) Counter-flow classifiers

Nakajima, Gotoh and Tanaka [92, 93] developed two classifiers for on-line measurement of the average size of a flowing powder coarser than 100 μm in size.

A side stream of solid particles from a process line is passed into an air elutriator. The powder is separated continuously into two streams at a certain cut-off size. The cut-off size is determined, according to the drag laws of Stokes, Allen and Newton, by the Reynolds number of the air (figure 20.21).

The cumulative oversize fraction which is equivalent to the ratio of the particle flow rate out (F_{out}) to the particle flow rate in (F_{in}) is controlled to a set value. If the signal from the outgoing flowmeter is reduced by 50% and the signal compared with the unaltered signal from the input flowmeter, when the difference is zero, the cut-off size of the elutriator is the average size of the powder.

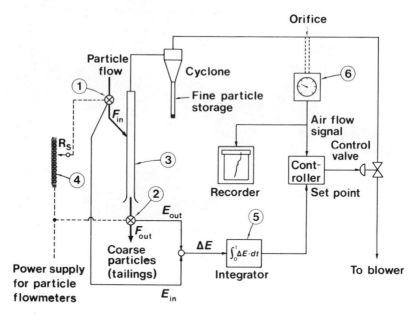

Fig. 20.21 Schematic diagram of the Tanaka and Nakajima elutriator for detecting average particle size. (1) particle flowmeter 1, (2) particle flowmeter 2, (3) elutriation tube, (4) variable transformer, (5) integrator (master controller), (6) orifice air flowmeter.

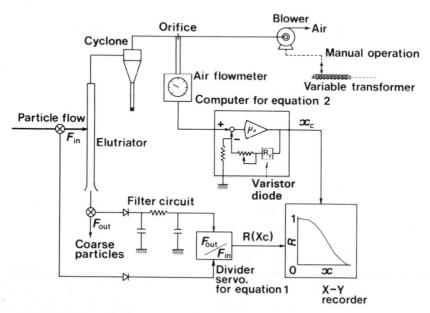

Fig. 20.22 Schematic diagram of the Tanaka and Nakajima elutriator for obtaining cumulative oversize distribution.

In their second instrument, figure 20.22, the cumulative oversize distribution was monitored by taking the signal ratio of the two flowmeters and inputting this as the Y-axis of an $X-Y$ recorder. The X-axis is obtained from the cut-off size analogue computed from the air velocity in the elutriator. A sweep time of 40–60 s at flow rates of 2.4–3.0 g s^{-1} gives a satisfactory cumulative size distribution curve for size in the range 700–100 μm.

(b) Cross-flow air classifier: the Humboldt PSA-type TDS

The principle, theory and experimental results for this technique have been severally described [95–98].

Particles are thrown into a planar flow of constant velocity and travel through the classification zone on size-dependent trajectories (figure 20.23). The size-dependent fan of particles remains stationary if the entry conditions are kept constant throughout an analysis, hence the size of the particles to be found at any point in the classification zone is fixed.

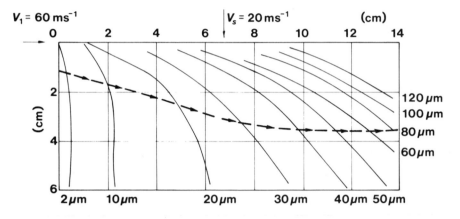

Fig. 20.23 Particle trajectories in a cross-flow air classifier. $- \rightarrow -$: path of photometer.

The solids to be analysed are fed from a hopper via a vibratory feeder to a rotating circular groove. The groove is completely filled with material, surplus material being removed with a skimmer. An injector, driven by pressurized air, sucks the material out of the groove, disperses and accelerates the particles and line-feeds them into the classification zone. The particle fan is scanned by the photometer and the position of the photometer and the attenuated light signal are fed to a computer. Analyses are performed at the rate of 10 to 20 per minute.

The injector works on the water-jet principle whereby the incoming pressurized air causes a partial vacuum at the inlet so that particles are sucked into the system, transported to the dispersion zone, accelerated in a rectangular duct and fed to the classification zone.

The instrument is precalibrated with similar material of known size distribution. A variation has also been described in which a blade system is incorporated to collect eight narrow size fractions [122].

20.3.8 Hydrocyclones

An ideal on-stream sizing device would sample the whole of the suspension and not include any special instrumentation. The nearest approach to this would be to use a classifying hydrocyclone as these are easily installed and often form part of industrial plants. One possible reason for attempting to use the operating parameters of a classifier is that sampling problems may be avoided since sampling often injects a large and unknown error [2].

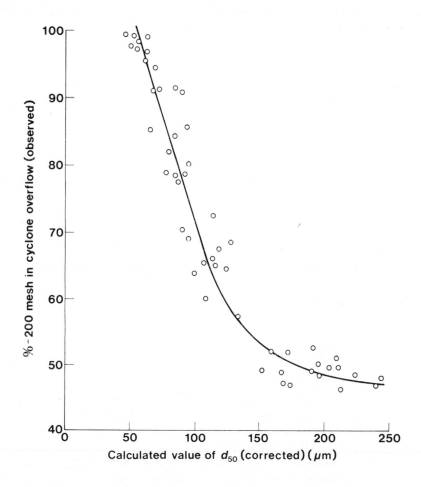

Fig. 20.24 Calculated values of d_{50} (corrected) size for a cyclone from the cyclone overflow characteristics.

The method of use depends on the theoretical relationship between the size distribution of the feed, the weight of solids in the overflow and the volume of water in the overflow. Thus, by measuring flow rates and pulp densities and assuming a size-distribution law for the feed a computer programme can be written to give the modulus and index of the feed. Under normal operating conditions the present state of the theory of cyclone operation renders this an impracticable proposition [78] although it can be used under favourable conditions [99].

Lynch [100] proposed that the percentage less than some chosen mesh size in the cyclone overflow can be related directly to the d_{50C} parameter of the cyclone provided that the size distribution of the feed to the cyclone does not change appreciably. The d_{50C} parameter is that size of classified particle which reports 50% to the overflow and 50% to the underflow and can be calculated from the operating parameters of the cyclone. Typical results of the approach are shown in figure 20.24.

In closed production circuits there may be marked changes in the size distribution of the cyclone feed and in order to compensate for the effect of such changes, an equation of the following form was proposed [101]:

$$\log_{10} (\%\text{--}200 \text{ mesh}) = a_1 - a_2 (d_{50C}) - a_3 T \tag{20.4}$$

where T is the tonnage of fresh feed to the circuit. Other work has also been carried out by Lees [102, 103] and Fewings [104] (cited in [2]).

The application of this technique requires very thorough analysis of the circuit and repeated checking of the resulting empirical equations. An alternative approach has been to accept the inherent difficulties of sampling and install smaller, more precise classifiers alongside the production classifiers [105].

20.3.9 Screening: the Cyclosensor

This device [106] has the disadvantage of being a batch size analyser. The basic principles behind its operation can be understood by referring to figure 20.25. An extremely dilute sample of the milled ore is introduced, at constant volume flow rate, to a coarse separator in the form of a tangentially fed cylindrical screen. The coarse fraction is allowed to settle and the finer fraction is further separated with an efficient hydrocyclone into a fine size fraction and a very fine fraction. The very fine fraction is discarded and the fine fraction is allowed to settle. The ratio of the times taken to fill the coarse and the fine fraction collection vessels to the indicated levels can be related directly to the particle size distribution. The Cyclosensor has a sensitivity whereby a change of ±1.8% passing 100 mesh can yield a 7% change in the ratio of the settling times. Its reproducibility is such that for the same feed rate of the same solids the ratio of the times remains constant to better than 1% and an increase in solids feed rate of 30% has no effect on the ratio. The use of hydrocyclones for on-line analysis has also been investigated by Tanaka [108].

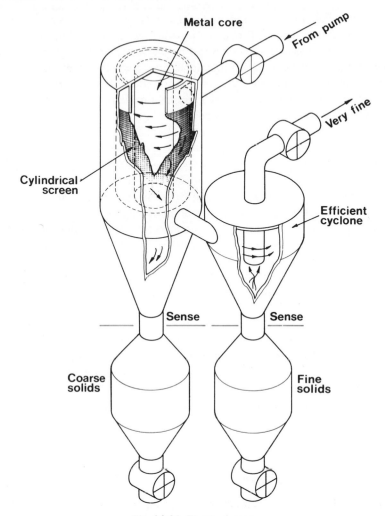

Fig. 20.25 The Cyclosensor.

20.3.10 Automatic sieving machines

Sieve analysis techniques have been accepted generally as the absolute standard for assessing the size distributions of products from milling circuits. Until about two years ago very little attempt had been made to automate and speed up sieving analysis.

Recently a prototype automatic wet sieving device was described [2, 107] which would determine a single point on the size distribution curve within a few minutes without any need to dry samples. The principle of operation of a modified design of the prototype can be understood by referring to figure 20.26. The sieving vessel (about one litre capacity) can be topped up with water or slurry to a specified level with the aid of solenoid valves activated by conductivity probes. The vessel is first

filled with slurry containing the particles to be sized and topped up with water to a precise level to allow an accurate determination of the mass of solids added. This weight (w_1) can be calculated easily by applying Archimedes' principle. The fine fraction is next removed from the vessel through the discharge valve. Screening is hastened by propeller agitation and with ultrasonics applied to maintain the sieve mesh completely free of pegged material. The weight of the residue (w_2) is determined by further application of Archimedes' principle, and the fraction coarser than the screen is given directly by w_2/w_1.

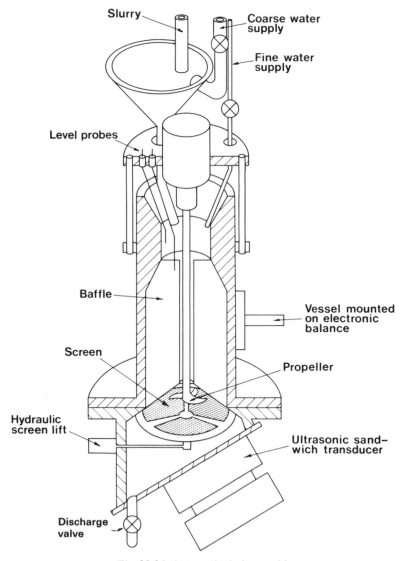

Fig. 20.26 Automatic sieving machine.

The ability to remove pegged material from the screen by the use of ultrasonics is of considerable importance. By this means, the free area of the sieve can be held constant throughout the sieving process, and the ideal rate of sieving can be approached. It is of some interest to note that the rate of sieving under these conditions is found experimentally to be independent of the mass of solids to be sieved.

A simple theory of the sieving process has been developed to account for these results. If $w_i(t)$ is the weight of particles retained by the screen at any one time, which lie in the size range x_i to x_{i+1}, and if P_i is the rate of passage of unit mass of particles in this same size range through a sieve of aperture size a, and if first-order kinetics on a mass basis are assumed, then

$$\frac{dw_i}{dt} = -w_i P_i \tag{20.5}$$

Integrating and summing over all sizes gives an equation for $W(t)$, the total mass on the screen at time t:

$$W(t) = \sum_{i=1}^{n} w_i(0)e^{-P_i t} \tag{20.6}$$

or in integral form

$$W(t) = W(0) \int_0^{x_{max}} e^{-P(x)t} \frac{dF(x)}{dx} dx \tag{20.7}$$

where x_{max} is the maximum size of particle in the feed and $F(x)$ is the cumulative weight size distribution of the feed.

A suitable form for $P(x)$ has been given previously by Rendell [109]:

$$P(x) = b \left(\frac{a}{x} - k \right)^m \tag{20.8}$$

with k and m being constants close to 1 and 2 in value, respectively; a being the screen aperture, in the same units as x, and b being a constant determined by the aperture shape, etc. Use of this function in equation (20.7) leads to a decrease in $W(t)$ with time very much as shown experimentally and the ratio $W(t)/W(0)$ is independent of $W(0)$.

It is interesting to note that the capacity of the rapid wet sieving device expressed as screen charge mass per unit screen area is more than an order of magnitude higher than that normally recommended for conventional dry test sieving. Comparisons with conventional sieving analysis using a 'Ro-Tap' sieve shaker indicated that the accuracy of a laboratory manually operated device was ± 2% 200 mesh with 95% confidence level (31 data points). The data correspond to samples containing 0 to 100% 200 mesh and for distribution moduli varying from 0.58 to 0.69. Pulp densities were between 10 and 30% solids by weight [2].

In a recent paper [110] Schönert, Schwenk and Steier describe a fully automatic sieving machine which can determine seven points on the size distribution curve. In

this machine, samples are wet sieved, using a technique involving a pulsating water column and the application of ultrasonics. Screen charges are dried and weighed automatically.

Bartlett and Chin [111] describe a technique involving a two-cell compartment divided by a screen. The slurry density in the two compartments is determined using nuclear gauges to provide a single point on the distribution curve.

20.3.11 Gas-flow permeametry

Tanaka and Nakajima [92, 93] found that, for particles finer than 100 μm, classification methods were unworkable due to the cohesiveness of powders and developed a frequency response method. Air, whose pressure is varying sinusoidally with a specific amplitude and frequency, is forced together with a steady air flow through a moving bed of powder. At a known height in the bed the attenuated and retarded air pressure is tapped by a pressure transducer so that the amplitude and pressure drop are measured after being separated into the pulsating and steady flow components.

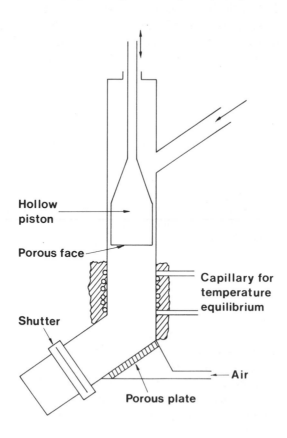

Fig. 20.27 Cell for automatic determination of surface area.

The amplitude attenuation of the pulsating pressure is related to bed porosity and specific resistance. Using the Carman–Kozeny relationship together with an analog computing circuit the average size of particle can be evaluated after the system has been calibrated with materials of known porosity and permeability.

The bed is packed from the top of the test cylinder and discharged by a vibratory feeder at the bottom after measurements are taken. This enables a new bed to be packed and measured within minutes.

The permeability method was also used by Papadakis [112] to evaluate the surface area of cements. A sample of cement was compressed by a porous piston into a cell. Air was then passed through the bottom porous plate, through the sample and porous piston to the atmosphere. The inlet pressure was automatically adjusted and recorded to give a known air flow rate. The surface area could then be evaluated from the inlet pressure. The cell was emptied automatically becoming ready for the next test (figure 20.27).

Wieland [113] used a similar idea but based on the Blaine permeability method. An automatic weigher produced a packed bed of known voidage in a standard cell. Air was drawn through the bed by the passage of water from one reservoir to another. After a certain volume of air had passed through the bed, measured by a certain weight of water flowing, the time required was converted to an electrical proportionality signal. These measurements were repeated every four minutes and the signals could be used to control the feeder to a grinding mill.

20.3.12 Pressure drop in nozzles

The flow rate of a gas stream can be obtained from the pressure drop across a pipe constriction. If particles are present the pressure drop across an orifice meter is unaffected but the pressure drop across a nozzle, however, is dependent on the size of the particles present [114]. No satisfactory theoretical analysis has been found to explain this fact but Leschonski is using this principle to develop an on-line size analyser [115]. A venturi [116] meter has also been used to measure gas flow rates with particles in the size range 10–200 μm but this method only meets with limited success due to unpredictable variations.

20.3.13 Non-Newtonian rheological properties

In the study of non-Newtonian slurries it has been found that the relative viscosity of the slurry increases with particle size [117]. This effect of particle size on viscosity is not predicted by Mooney, although the volume fraction of small particles in a mixture has some theoretical justification [118]. An on-line viscometer [119] using a differential pressure transducer has been designed for use with concentrations of 28–55% by volume of nickel, alumina, copper and glass solid of particle sizes between 10–70 μm. The system is also capable of handling liquid sodium suspensions at high temperature as well as suspensions containing high viscosity organic material.

20.3.14 Correlation techniques

Stanley-Wood, Beck and Lee [120, 121] showed that analysis of the autocorrelation function of the signal obtained from an alternating current conductivity transducer, initially designed for measuring mass flow rate, gave a measure of the particle size of a sand and water mixture. A measure of the mean size could thus be achieved by allowing the normalized signal from a correlator to be divided into two and passed through either high or low pass filters. This results in an inequality, due to the variation in frequencies from large and small particles; the ratio of this inequality can then be used to measure the mean particle size of a flowing slurry after calibration. The particle size range of sand used was between 70–2000 μm with a concentration of 10–30% by weight.

Correlation techniques can be used with signals received from attenuation of radiations such as light and lasers but are mainly used in low particle concentration systems.

References

1　Brittan, M.I. and van Vuuren, E.J.J. (1973), *J. S. Afr. Inst. Min. Metall.*, 73, 211–22.
2　Hinde, A.L. and Lloyd, P.J.D. (1975), *Powder Technol.*, 12, 1, 37–50.
3　Hinde, A.L. (1973), *IFAC Symp. Automatic Control in Mining, Mineral and Metal Processing, Sydney*, August 13th–17th, Inst. J. Engrs., Australia, 45–47.
4　Clarke, J.R.P. (1970), *Measurement and Control*, 3, 241–4.
5　Treasure, C.R.G. and Warren, M. (1968), *Proc. Soc. Analyt. Chem.*, 5, 5, 82–83.
6　Di Grado, J.C. (1970), *Bull. Parent. Drug Assoc.*, 24, 2, 62–67.
7　Groves, M.J. and Wana, D. (1977), *Powder Technol.*, 18, 215–23.
8　Lieberman, A. (1975), *J. Test. Eval.*, 3, 6, 3–98.
9　Lieberman, A. (1977), *Powder Technol.*, 8, 37–44.
10　Marple, V.A. and Rubow, K.L. (1976), *J. Aerosol Sci.*, 7, 425–33,
11　Addingley, C.G. (1966), *Ann. occup. Hygiene*, 9, 73–82.
12　Carver, L.D. (1961), *Hydraulics and Pneumatics*.
13　Lantz, R.J. Jr., Shami, E.G. and Lachman, L. (1976), *Bull. Parent. Drug Assoc.*, September/October, 30, 234–36.
14　Krueger, E.O. (1972), *ibid.*, 26, 2, 96–100.
15　Hopkins, G.H. and Young, R.W. (1974), *ibid.*, 28, 15–25.
16　Tate, C.H., Lang, J.S. and Hutchinson, H.L. (1977), *J. Am. Wat. Wks Assoc.*, 69, 7, 4245–50.
17　Shrunk, D.H., Timmel, B.M. and Andreasen, A.A. (1976), *J. Assoc. Official Analyt. Chem.*, 59, 31, 671–4.
18　Romine, J.O. and Gayle, J.B. (1964), *J. Am. Assoc. Contam. Control*, 10–18.
19　Karhnak, J.M. Jr. (1972), *Fluid Power Systems Contamination*, Soc. Automotive Engrs No. 720780, NY.
20　Karhnak, J.M. Jr. (1974), *Fluid Power Testing Symp.*, Milivankel, WI.
21　Beard, J.D. and Tanaka, T.S. (1976), *Proc. Am. Wat. Wks Assoc., Water Quality Techn. Conf.*
22　Dyer, J. and Smith, F.R. (1977), *Am. Chem. Soc., Symp.*, series No. 49, Textile and Paper Chemistry and Technology.
23　West, G.S. (1977), *Particle Size Analysis Conf.*, Salford (ed. M.J. Groves), Chem. Soc. Analyt. Div., London, Heyden.
24　AC Fine Test Dust, General Motors Corp., Flint, Michigan, USA.

25 West, G.C. (1975), *Hydraulics and Pneumatics*, 8, 1, July.
26 International Standards Organisation Report ISO/DIS 4402, *Hydraulic Fluid Power, Calibration of Liquid Automatic Particle Count Instruments; methods using air cleaner fine test dust contaminant*.
27 Moore, L.C. (1974), *Proc. 8th Annual Fluid Power Research Conf.*, Oklahoma State University, Paper number P74-48.
28 Golden, L.S. (1977), *Particle Size Analysis Conf.*, Salford (ed. M.J. Groves), Chem. Soc. Analyt. Div., London, Heyden.
29 Behringer, A. *et al.* (1977), *ibid.*
30 Lieberman, A. (1975), *Proc. 7th Ann. Conf. Fine Particle Soc.*, August, Chicago.
31 Porter, M.C. (1974), *Brewers Dig.*, February.
32 Schleusener, S.A. (1968/69), *Powder Technol.*, 2, 364.
33 Harris, F.S. and Morse, F.L. (1969), *J. opt. Soc.*, 1518; (1971), *ibid.*, 945.
34 Cormilliant, J. (1972), *J. appl. Opt.*, 11, 2.
35 Faxvog, F.R. (1972), *Int. Conf. Powder Technology*, Chicago (available from Illinois Institute of Technology Research Institute, IITRI, Chicago).
36 Ricci, R.J. and Cooper, H.R. (1970), *ISA Trans.*, 9, 28–36.
37 Talbot, J.H. and Jacobs, D.J. (1970), *Particle Size Analysis Conf.*, Bradford (eds M.J. Groves and J.L. Wyatt-Sargent), Soc. Analyt. Chem. London.
38 Rhodes, C.A. *et al.* (1976), *Powder Technol.*, 14, 2, 203–8.
39 Pettus, R.O., Rainsford, A.E. and Bonnell, R.D. (1975), *Proc. 7th Ann. Southeastern Symp. on System Theory*, Auburn University.
40 Wu Jin (1977), *Appl. opt.*, 16, 3, 596.
41 Kelsall, D.F. and Restarick, C.J. (1971), *Proc. Automatic Control Systems in Mineral Process Plants*, Brisbane, 361–6.
42 Langer, G. (1965), *J. Colloid Sci.*, 20, 6, 602–9.
43 Langer, G. (1968/69), *Powder Technol.*, 2, 307.
44 Langer, G. (1968), *Staub*, 28, 9, 13–14.
45 Hofmann, K.P. and Mohnen, V. (1968), *ibid.*, 15–19.
46 Hofmann, P. (1968), *ibid.*, 360.
47 Langer, G. (1968), *Proc. Conf. Powder Technology*, IITRI, Chicago, Illinois 60616, USA, pp. 198–9.
48 Sinclair, I., Lloyd, P.J. and Scarlett, B. (1970), *Proc. Particle Size Analysis Conf.*, Soc. Analyt. Chem., London, p. 61.
49 Karuhn, R.F. (1973), *Proc. Conf. Powder Technology*, IITRI, Chicago, Illinois 60616, USA, pp. 203–7.
50 Scarlett, B. and Sinclair, I. (1975), *Proc. 1st European Symp. Particle Size Measurement Techniques*, Nuremburg, *Dechema Monogram* 79, 1589–615, Part B, 393–403.
51 Langer, G. (1972), *Powder Technol.*, 6, 5–8.
52 Barnett, M.I., Sims, E. and Lines, R.W. (1976), *ibid.*, 14, 1, 125–30.
53 Davis, N.M., Turco, S. and Sivelly, E. (1970), *Am. J. Hosp. Pharm.*, 27, 823.
54 Kendall, C.E. (1968), *Ann. NY Acad. Sci.*, 158, 640.
55 Krueger, E.O. (1972), *Bull. Parent. Drug Assoc.*, 26, 2.
56 Hopkins, G.H. and Young, R.W. (1974), *ibid.*, 28, 1, 15–25.
57 Tate, C.H. and Trussel, B.R. (1976), *Water Quality Technology Conf.*, American Water Works Association, San Diego, CA.
58 Di Grado, C.J. (1970), *Bull. Parent. Drug Assoc.*, 24, 2.
59 Sadowski, J.W. and Byckling, E. (1978), *Powder Technol.* 20, 2, 273–84.
60 Svarovsky, L. and Hadi, R.S. (1977), *Proc. Particle Size Analysis Conf.*, Chem. Soc. Analyt. Div.
61 Urick, R.J. (1948), *J. acoust. Soc. Am.*, 20, 283–9.
62 Flammer, G.H. (1963), *The Use of Ultrasonics in the Measurement of Suspended Sediment Size Distribution and Concentration*, thesis, Univ. Microfilms Inc., Minnesota.

63 Diaz, L.S. and Webber, C.B. (1973), *Can. Mineral Processors Ann. Mtg*, January 23–25, O'Hawa.

64 Bassarear, J.H. and McQuie, G.R. (1971), *Mining Show*, Am. Min. Congr., October 11–14, Las Vegas, Nevada.

65 Dart, P.W. and Sand, J.A. (1973), *Conf. Metallurgists*, CIM, August 29, Quebec City.

66 Hathaway, R.E. and Guthals, D.L. (1976), *Can. Inst. Min. Metall. Bull.*, February 69, 766.

67 Holland-Batt, A.B. and Fleming, M.G. (1968), *8th Int. Mineral Process Congr.*, Leningrad, Inst. Mikhanobr, Preprint F4.

68 Holland-Batt, A.B. (1968), *Trans. Inst. Min. Metall.*, Sect. C77, 185–90.

69 Holland-Batt, A.B. (1969), *ibid.*, Sect. C78, 163–5.

70 Osborne, B.F. (1970), *ISA Conf.*, October, Philadelphia, Paper 844-70.

71 Osborne, B.F. (1972), *Can. Inst. Metall. Bull.*, 65, 97–107.

72 Holland-Batt, A.B. and Birch, M.W. (1974), *Powder Technol.*, 10, 4/5, 189–200.

73 Holland-Batt, A.B. (1975), *ibid.*, 11, 1, 11–26.

74 Stanley-Wood, N.G. (1974), *Control Instrum.*, 42, 43, 45, 47.

75 Carr-Brion, K.G. (1966), *Analyst*, 91, 289–90.

76 Carr-Brion, K.G. and Mitchell, P.G. (1967), *J. scient. Instrum.*, 44, 611.

77 Carr-Brion, K.G. (1968), BP Appl. 13598.

78 Joy, A.S. and Jenkinson, A. (1968), *Proc. Soc. analyt. Chem.*, 5, 5, 80–82.

79 Anon. (1971), *World Mining*, May.

80 Meric, J.P. (1972), *Bull. Soc. fr. Ceram.*, 95, 67–76.

81 Meric, J.P. (1971), *3rd European Symp. Comminution*, 5–8 Oct., Cannes, 1.12, pp. 345–59.

82 Meric, J.P. and Caron, J.F. (1973), *Rev. Matér. Construct. Trav.*, January, No. 676, CERILH Publ. No. 218, 30–34.

83 Zelwer, A. (1976), Paper pres. Commission *Breakage and Size Analysis*, March, ANRT CERILH.

84 Trang, Ngoc Lan and Dupas, M. (1973), *Bull. Liais. Labor. P. Ch.*, 68, 17–22.

85 Stark, H. (1972), *Appl. opt.*, 11, 7, 1655–6.

86 Meric, J.P., le Jean, Y. and Caron, J.F. (1974), *Rev. Matér. Construct.* No. 688, CERILH Publ. No. 238, pp. 179–83.

87 Hinde, A.L. (1973), *J. S. Afr. Inst. Min. Metall.*, 73, 8, 258–68.

88 Weiss, E.L. and Frock, H.N. (1976), *Powder Technol.*, 14, 2, 287–93.

89 Anon. (1977), *Pit Quarry*, 84–86.

90 Felton, P.G. (1978), In-stream measurement of particle size distribution, in: *Int. Symp. In-Stream Measurement of Particle Solid Prop.*, August 22–23, Bergen, Norway.

91 Swithenbank, J. *et al.* (1977), In: *Experimental Diagnostics in Gas Phase Combustion Systems, Prog. Astronautics and Aeronautics*, 53 (ed. B.T. Zinn), AIAA Aerospace Sci. Mtg., Jan. 1976, Paper 76-69.

92 Nakajima, Y., Gotoh, K. and Tanaka, T. (1970), *IEC Fundamentals*, 9, 3, 489–95, (1971), *ibid.*, 10, 2, 318–20; (1967), *ibid.*, 6, 587.

93 Tanaka, T. and Nakajima, Y. (1973), *Proc. Conf. Particle Technology*, IITRI, Chicago.

94 von Alfthan, C. (1973), *Dechema Monogram* 1589-1615, 79, Part B, 61–76.

95 Rumpf, H. and Leschonski, K. (1967), *Chem. Ingr. Tech.*, 39, 1231-41.

96 Leschonski, K. (1971), *ibid.*, 43, 317–23.

97 Bernotat, S. (1974), Dissertation, Univ. Karlsruhe.

98 Leschonski, K., Metzger, K.L. and Schindler, U. (1977), *Proc. Particle Size Analysis Conf.*, (ed. M.J. Groves), 1978, Chem. Soc. Analyt. Div., Heyden, pp. 227–36.

99 Lynch, A.J. (1966), Australian Mineral Industries Research Association, Progress Report Number 7.

100 Lynch, A.J., Rao, T.C. and Whiten, N.J. (1967), *Proc. Aust. Inst. Min. Metall.*, 223, 71–73.

101 Draper, N. and Lynch, A.J. (1968), *Trans. Mech. Chem. Engng*, Nov., 207–17.

102 Lees, M.J. (1968), B.Sc. Hons. thesis, October, Faculty of Science, Univ. Queensland.

103 Lees, M.J. (1973), Ph.D. thesis, January, Dept. Min. Metall., Univ. Queensland.

104 Fewings, J.H. (1971), *Proc. Symp. Automatic Control Systems in Mineral Processing Plants*, May, Brisbane, pp. 333–57.

105 Putman, R.E.J. (1973), *Min. Congr. J.* SA, 68–74.

106 Kelsall, D.F. and Restarick, C.J. (1971), *Proc. Symp. Automatic Control Systems in Mineral Processing Plants*, May, Brisbane, pp. 361–6.

107 Hinde, A. (1973), *IFAC Symp. Automatic Control in Mining, Mineral and Metal Processing*, August 13–17, Sydney, Australia, Inst. Engrs, pp. 45–47.

108 Tanaka, T. (1976), *Technocrat*, 9, 9, 40.

109 Rendell, M. (1964), *Separation of Particles by Sieving and Screening*, Ph.D. thesis, Univ. College, London.

110 Schönert, K., Schwenk, W. and Steier, K. (1974), *Aufbereit Tech.*, 7, 368–72.

111 Bartlett, R.W. and Chin, T.H. (1974), *Trans. Soc. Min. Engrs*, AIME 256, 323.

112 Papadakis, M. (1963), *Rev. Matér. Construct. Trav.*, 570, 79–81.

113 Wieland, W. (1967), *Min. Proc.*, 8, Feb., 30.

114 Boothroyd, R.G. and Goldberg, A.S. (1970), *Br. Chem. Eng.*, 15, 3, 357.

115 Davies, R. (1974), *Am. Lab. Instrum.*, 47, Feb., 6, 1, 73–76, 78–80, 82–86.

116 Farbar, L. (1952), ASME, Paper No. 52-A-31.

117 Shaheen, E.I. (1971/72), *Powder Technol.*, 5, 245.

118 Moreland, C. (1963), *Can. J. Chem. Eng.*, 41, 24.

119 Jay, E.C., Nelson, P.A. and Armstrong, W.P. (1969), *Am. Inst. Chem. Engrs J.*, 15, 815.

120 Stanley-Wood, N.G., Lee, T. and Beck, M.S. (1973), *Proc. Soc. Analyt. Chem.*, 10, 282.

121 Beck, M.S., Lee, T. and Stanley-Wood, N.G. (1973), *Powder Technol.*, 8, 85.

122 Metzger, K.L. and Leschonski, K. (1976), *Dechema Monogram*, Nuremburg, 79, Part B, 1589–1615, 77–94.

123 Davies, R. and Turner, H.E. (1977), *Proc. Particle Size Analysis Conf.* (ed. M.J. Groves), 1978, Chem. Soc. Analyt. Div., Heyden, pp. 238–49.

124 Wertheimer, A.L. and Wilcock, W.L. (1977), *Appl. opt.*, 1616–20.

125 Oki, K., Akehata, T. and Shirai, T. (1975), *Powder Technol.*, 11, 1, 51–57.

126 Küpfer, H.A. (1978), *Powder Metall. Int.*, 10, 2, 96–97.

127 Andres, H. (1978), *ibid.*, 98–99.

Problems

1. Calculate the average weight of sand particles having the sieve analysis tabulated below if the density of the sand is 2500 kg m^{-3} and its volume shape factor, by sieving, is 0.50. Calculate the expected variance, var (P_i), if this sand is mixed, in a 4 to 1 ratio, with equisized salt particles having a particle weight of 2400 μg. Assume the bulk powder is very large compared with the abstracted samples of weight 28 g. What is the maximum percentage of salt one would expect in any sample?

Sieve size d_a (μm)	600	422	294	212	128
Fraction retained on each sieve (q_r)	0	0.23	0.47	0.21	0.09

Use the formulae:

$$\text{var}(P_i) = \frac{P_A(1-P_A)}{w}[P_A w_B + (1-P_A)w_A]$$

$$w_B = \alpha_v \rho_s \, \Sigma(d^3_{A,r}/q_r) \qquad P_{salt}\,(\text{maximum}) = P_s \pm 3\sigma_i$$

2. Discuss the basic requirements for good sampling and the difficulties in meeting them in the case of (a) powders, (b) dusty gases, (c) aerosols.

Calculate the collection efficiency, dust concentration at the outlet and the size grading of the emitted dust from a combined cyclone electrostatic precipitator plant under the conditions tabulated below. If the dust burden is 0.2 kg m^{-3} and the gas flowrate is 65 m^3 s^{-1}, calculate the weight of dust emitted in kg s^{-1}.

Size of grade (μm)	75	40	20	10	0
% in grade at inlet	35	20	15	30	
Grade efficiency { Cyclone	98.7	95.1	85.5	49	
Grade efficiency { ESP	98.9	96.6	95.7	83.8	

3. Discuss the advantages of the spinning riffler over the cone and quartering method of sampling. Sugar is mixed with sand, in the ratio of 2 to 3 by weight, and split into sixteen samples using a spinning riffler. If the average weight of the particles is 2.22 mg

in each case and the bulk weight is 800 g determine the efficiency of sampling and calculate the maximum error. The percentage composition of sugar in the sixteen samples is as follows: 41.3, 38.6, 39.2, 40.7, 40.9, 38.9, 39.8, 40.2, 41.1, 39.8, 38.7, 40.0, 41.2, 39.8, 39.9, 39.9.

4. The following table gives the sieve analysis of a powder of density 2500 kg m^{-3}.

Mesh number	52	72	100	150	200	300
Size (μm), d	295	211	152	104	75	53
Wt% retained (q_r)	0	5	25	40	25	5

Develop an expression for the weight specific surface of the powder (S_w), and by assuming the particles are spherical evaluate this function for the above powder. If a microscope count gives $S_{w,a}$ = 295 cm^2 g^{-1} determine the ratio of the mean projected area diameter of the powder and the mean sieve diameter.

5. Discuss particle shape as applied to powders. Show that Wadell's sphericity factor (ψ) may be obtained by comparing an Andreasen with a Coulter analysis. Determine the values of this factor for a cylinder of diameter D and length KD as its shape changes from needle to flake (i.e. K = 1/8, 1/4, 1/2, 1, 2, 4, 8).
    Find also how the Stokes diameter varies for a unit face diameter (D = 1) cylinder as K varies from 1/8 to 8. Hence find how the surface-volume shape coefficient with respect to settling, i.e. $\alpha_{SV,St}/6$, varies with K.

6. The tabulated data below are the result of a sieving operation using a 120 mesh sieve and a sample weight of 25 g. Plot the results as log percentage passing against number of taps and discuss the resultant curve. Plot also the curve log ($R_t - R_\infty$) against number of taps in order to find the ultimate residue. Discuss your results.

Number of taps	2	4	6	10	20	30
Weight passing (g)	0.7	1.6	2.5	7.2	10.0	13.0

Number of taps	40	60	80	100	120	140
Weight passing (g)	15.0	17.5	19.5	21.3	22.02	22.32

7. A solid cuboid has sides of length 2, 3 and 4 μm. What are its volume diameter, surface diameter, projected area diameter (random and stable orientation) and Stokes diameter?
    A sample of powder has the size analysis given below. Determine its mean diameters.

Mean size (μm)	2	4	6	8	10	12
Frequency (number)	5	8	12	8	6	4

8. A solid cuboid has sides of length 20, 20 and 40 μm. What are its volume, surface, projected area and Stokes diameter?
    How does the surface-volume shape factor vary as the length of the third side changes from 40 to 20 to 10 μm using microscopy as the measurement technique?

A sample of powder has the size analysis given below. Using Normal Probability paper determine the mean, median, mode and standard deviation.

Particle size (μm)	Cumulative percentage undersize
3	11.6
5	30.2
7	55.5
9	78.0
11	92.0

9. What methods would you suggest for measuring the particle sizes of the following materials and why?

(a) A sub-micron metallic oxide.
(b) A free flowing coarse powder.
(c) A sub 75 μm powder.
(d) A powder in the 75 to 20 μm size range which must be analysed dry.
(e) A sub 75 μm powder where the relevant property is the surface area.

10. In a certain manufacturing process zinc oxide is required having a mean size, by gas adsorption, of 0.100 μm, a narrow size distribution and a particle shape which is conducive to good flow characteristics. State the method of size analysis you would use and the reasons for your choice.

Two zinc oxides were examined by centrifugal sedimentation and the following data derived. Analyse the data and discuss the merits and demerits of the two powders for the above process.

Zincoid A (%)	97.4	90	72	46	19	7
Zincoid B (%)	98.4	88	62	28	5	0
Particle size (μm)	0.62	0.44	0.31	0.22	0.15	0.11

Surface area by gas adsorption (m^2 kg^{-1}): A, $S_w = 9150$. B, $S_w = 9850$.
Powder density, 5570 kg m^{-3}:
Relationship between the surface–volume mean diameter (x_{sv}) and the geometric mean diameter (x_g):

$$\log x_{sv} = \log x_g - 1.151 (\log \sigma_g)^2$$

where σ_g is the geometric standard deviation.

11. The table below contains the data from a microscope count. Plot the distribution on a number and weight basis on log-probability paper.

d_{min}	2.8	4.0	5.6	8.0	11.3	16.0	22.6	32.0	45
d_{mean}	3.3	4.7	6.6	9.4	13.2	18.8	26.4	37.6	53
d_{max}	4.0	5.6	8.0	11.3	16.0	22.6	32.0	45	64
Number frequency	36	57	94	108	98	60	32	11	4

Using a sedimentation technique the weight distribution is found to be log-normal with a mean size of 10 μm and a standard deviation of 1.84. Determine the particle shape factor and explain how it arises.

By considering the number of particles counted in each size range in the above analysis, suggest some errors that might have arisen and how they could have been avoided.

12. A sample of free flowing powder is to be analysed by sieving and by Andreasen pipette. Explain briefly how you would carry out the analysis and correlate the results. If the results were as shown below, determine and tabulate the cumulative percentage by weight undersize distribution. Comment on the shape of the distribution curve and by comparing medians determine the shape factor relating Stokes diameter and sieve diameter.

Sample weight = 187.8 g

Weight through 200 mesh (75 μm) = 82.7 g

Sieve analysis of coarse fraction:

Sieve aperture size (μm)	75	105	150	212	300	424
Weight passing through (g)	0	29.7	59.3	88.3	100.5	105.1

Andreasen analysis of sub-sieve fraction:

Size (μm)	13.3	18.8	26.5	37.5	53	75	105
Percentage undersize	5.0	6.4	10.9	20.9	38.2	68.2	100

13. It is recounted that a party of Martians visited Earth in order to survey human dwellings. This was considered to be a simple exercise since they lived in spherical homes with a diameter related to their position in society. They found igloos which were hemispherical, wigwams which were conical, and skyscrapers which were cuboidal.

Define quantitative methods of carrying out the above exercise evaluating the shape factors and representative diameters. Assume the base areas of the three dwellings to be identical and of unit projected area diameter, the height of the wigwam (cone) to be equal to its base diameter and the height of the cuboid to be one-third of this. Include the base area of the shapes in your calculations.

What are the significant diameters in optical microscopy? Discuss the difference in procedure in carrying out a number and weight distribution count.

14. The following table is a microscope analysis of silica. Determine its size distribution by weight and make an estimate of the accuracy.

Size limits (μm)	Area of sample fields (mm^2)	Number of sample fields	Number of particles counted
106–75	0.600 × 20 × 20	–	28
53	0.600 × 20 × 15	–	117
37	0.600 × 20 × 7	–	175
27	0.127	200	137
19	0.127	50	71
13	$\frac{1}{2}$ × 0.127	50	79
9.4	7.94 × 10^{-3}	50	18
6.6	7.94 × 10^{-3}	25	16
4.7	$\frac{1}{2}$ × 7.94 × 10^{-3}	25	14
3.3	$\frac{1}{2}$ × 7.94 × 10^{-3}	25	22
3.3–2.3	$\frac{1}{2}$ × 7.94 × 10^{-3}	25	34

15. Show that the change in resistance due to a particle entering the aperture in a Coulter counter is the sum of the elements $(\rho_f a \delta l / A^2)/[1 - (a/A)]$ and solve this for a cylindrical particle with its axis parallel to the axis of the aperture.

The geometric solution for a spherical particle is, approximately,

$$\Delta R = -\tfrac{4}{3}\pi b^3 \frac{\rho_f}{A^2}\left[1 + \frac{4}{5}\frac{\pi b^2}{A} + \frac{24}{25}\frac{\pi^2 b^4}{A^2} + \ldots\right]$$

What is the error involved in assuming that ΔR is proportional to particle volume?

Determine the numerical value for particles having cross-sectional areas of 40% and 50% of the aperture cross-sectional area.

Enumerate any other uncertainties that exist in the operation of this instrument.

ρ_f is the resistivity of electrolyte; a the cross-sectional area of element of the particle; A the cross-sectional area of aperture; and $2b$ the diameter of spherical particle.

16. There are two methods of calibrating a Coulter counter:

(a) against a standard narrowly classified material of known size; and
(b) using the following equations.

$$v_p = \frac{v}{V}\frac{w}{\phi}$$

$$v_p = \frac{\pi}{6} k^3 \Sigma(\Delta n)\bar{l} \times 10^{-12}$$

Explain these equations and explain why the calibration constant, k, using the two techniques may be different and the use to which you would put this difference.

If 2.64×10^{-3} g of polymist of density 1.1 g cm^{-3} was dispersed in 250 cm^3 of saline giving a summated volume $\Sigma \Delta n \bar{l} = 56\,300$ units when 0.5 cm^3 was withdrawn, determine the Coulter calibration constant. Briefly annotate other methods of finding the 'end point' of a Coulter analysis.

17. Show that a cylindrical particle of cross-sectional area a and length l changes the resistance across a Coulter orifice of cross-section A by an amount:

$$\Delta R = \frac{P_f v}{A^2} \left(1 + \frac{a}{A} \right) \quad \text{where } v = al$$

when simplifying assumptions are made. Show that the size attributed to a particle is + 4% in error for a particle of size 0.4 of orifice diameter.

Determine the calibration constant (k) for an instrument if a particle of size 12.5 μm gives a threshold reading $t = 30$.

If $\Sigma(\Delta n)\bar{t} = 1 \times 10^5$ when sampling 0.5 ml from a suspension of concentration 0.2 g in 100 ml, determine the probable density of the powder. Give reasons why your answer may be in error.

18. The Walther air classifier was tested under certain operating conditions with silica powder of density 2600 kg m^{-3} and a total (coarse grade) efficiency of 81.3% was obtained. The feed material and the coarse product were analysed for particle size distribution using the Andreasen pipette method with the following results:

Particle size (equivalent Stokes diameter) (μm)	Cumulative size distribution	
	The feed (%)	The coarse product (%)
42.0	99.0	85.0
29.6	81.5	63.5
21.6	67.5	46.5
14.5	49.6	25.0
10.4	35.2	8.9
7.1	25.4	2.9
5.2	17.5	0.7

Using graphical differentiation, determine the grade efficiency for each of the given sizes. Estimate the equiprobable size (also called cut size) from your results and calculate its equivalent for materials of density 100 kg m^{-3}.

Discuss the importance of the grade efficiency for equipment selection and list all basic requirements for an air classifier.

19. Discuss critically the elutriation method of size analysis.

A sample of silica, density 2500 kg m^{-3}, is to be classified into two grades using a Gonell elutriator. If the cut is to be at 40 μm determine the air velocity required. If the histogram of the size distribution of the feed is rectangular from 10 to 70 μm estimate the distribution of the 'fines' at 10 μm intervals.

Viscosity of air = 1.86 $\times$ 10^{-5} N s m^{-2}.

20. Derive the coarse grade efficiency function for gravitational elutriation in a vertical tube of radius R, assuming Stokes' law and a velocity profile of the power-law form, i.e.

$$\frac{v(r)}{v_{max}} = \left(1 - \frac{r}{R}\right)^n$$

Explain why the practical performance of such an elutriator would be very different from the theoretical results which you have obtained. What assumptions could lead to more reliable results?

21. Give an explanation for the breakdown in Stokes' equation as the Reynolds number approaches one.

Determine the terminal settling velocity for a spherical particle settling in water under laminar flow conditions using the data given below.

Derive an expression for the time a particle will take to attain 99% of its terminal velocity under these conditions, starting from rest and apply this to the case under consideration.

Particle diameter $D = 60 \ \mu m$; particle density $\rho_s = 5450 \ kg \ m^{-3}$; liquid density $\rho_f = 1000 \ kg \ m^{-3}$; liquid viscosity $\eta = 0.001 \ N \ s \ m^{-2}$; $g = 9.81 \ m \ s^{-2}$.

22. Describe the vertical motion of a body placed in a fluid of density lower than the particle density.

Assuming that laminar flow conditions prevail, find the time taken and distance covered by a bronze particle of size $100 \ \mu m$ in attaining 99% of its terminal velocity starting from rest. Density of solid, $8800 \ kg \ m^{-3}$; density of fluid (water), $1000 \ kg \ m^{-3}$; viscosity of fluid, $0.001 \ N \ s \ m^{-2}$.

23. Enumerate the errors and limitations of sedimentation analysis and suggest how they may be reduced to a minimum.

The experimental data from a cumulative sedimentation analysis are given below. Determine the size frequency distribution by both a tabular and non-tabular method. Height of fall, 10 cm; density of solid, $2650 \ kg \ m^{-3}$; density of fluid, $1000 \ kg \ m^{-3}$; viscosity of fluid, $0.001 \ N \ s \ m^{-2}$.

Time (min)	1	2	4	8	16	32	64
% settled	20	38	51	64	78	90	100

24. The particle size distribution of a powder can be evaluated by monitoring the changing concentration of a suspension of this powder as it settles out under a gravitational or centrifugal field. Briefly describe some of the available techniques for carrying out such determinations.

Determine the size distribution of a powder having the following concentration changes with time and height of fall stating its geometric mean and standard deviation.

Time (min)	0	1	2	4	8	12	16	20
Height (cm)	20	20	20	20	20	16	12	8
Concentration (%)	100	100	99.5	93.5	65.0	26.2	6.3	0.6

Powder density, $3200 \ kg \ m^{-3}$; liquid density, $1000 \ kg \ m^{-3}$; liquid viscosity, $0.001 \ N \ s \ m^{-2}$; $g = 9.81 \ m \ s^{-2}$.

25. Show that the density of a sedimenting suspension at depth h and time t from the commencement of sedimentation $[\phi(h,t)]$ is related to the size distribution of the solids in suspension as follows:

$$F = \frac{\phi(h,t) - \rho_f}{\phi(h,0) - \rho_f}$$

where F is the cumulative frequency undersize Stokes diameter for t and depth h in the fluid and ρ_f the density of the fluid. Discuss and critically compare density methods of size analysis paying particular attention to their resolution.

Determine the resolution for silica particles of specific gravity 2.5 falling in water of viscosity 0.001 N s m^{-2} at time 10 min if the settling height is 10 cm and sampling height 5 cm. If the initial concentration was 0.5% by volume of solids in suspension and the measured density at 10 min is 1002.5 kg m^{-3} determine the percentage under-size the evaluated Stokes diameter.

26. Show that the changing density of a settling suspension with height and time may be related to the size distribution of the particulate material in the suspension. What are the methods which use this property for size analysis and what errors are they subject to?

Determine the size distribution of a powder having the following density changes with time and height. ρ_s, 2600 kg m^{-3}; ρ_f, 1000 kg m^{-3}; η, 0.001 N s m^{-2}.

h (cm)		6	9	10	6	4.5
t (min)	0	2	4	8	16	32
Density, ϕ (kg m^{-3})	1000	1062	1045	1029	1010	1000

27. Describe the Andreasen pipette method of size analysis and its limitations. What do you understand by the 'resolution' of an incremental sedimentation technique?

Determine the size distribution of a powder having the following concentration changes with time and height of fall. Find also the mean, mode, median and standard deviation.

Time (min)	Height (cm)	Concentration (g in 10 cm^3)
0	20	0.1300
1	20	0.1220
2	19.6	0.1060
4	19.2	0.015
8	18.8	0.0486
16	18.4	0.0234
32	18	0.0088

Powder density, 2600 kg m^{-3}; liquid density, 1000 kg m^{-3}; liquid viscosity, 0.001 N s m^{-2}; g, 9.81 m s^{-2}.

28. A pipette and photosedimentation analysis are carried out on a powder. The weight percentage undersize by the pipette technique and the optical density by the photosedimentation technique are tabulated below. Determine how the extinction coefficient varies with particle size assuming that it equals 2 at a particle size of 10 μm.

Particle size (μm)	4.5	5.5	6.5	7.5	8.5	9.5	10.5
Weight percentage undersize	0	10	25	50	75	90	100
Optical density	0	0.12	0.26	0.446	0.596	0.669	0.709

29. Determine the size distribution of a powder having the following density changes with time and height. Density of solid, 2600 kg m^{-3}; density of liquid, 1000 kg m^{-3}; viscosity of liquid, 0.001 N s m^{-2}.

Height of fall (cm)		6	9	10	6	4.5
Time (min)	0	2	4	8	16	32
Density (kg m^{-3})	1080	1062	1045	1029	1010	1000

30. The following table gives a wide angle scanning photosedimentation analysis of molybdenum, density 10 200 kg m^{-3}, settling in water of viscosity 0.914 g m^{-1} s^{-1} and density 1000 kg m^{-3}. Determine the surface and weight distribution together with the specific surface given that the initial powder concentration was 1.23 kg m^{-3}.

How does this specific surface compare with that obtained by summing the specific surface incrementally from the distribution? The length of the light path in the container is 5 cm.

Time (min)	Height of fall (cm)	Optical density
0	16.3	0.56
1		0.55
2		0.52
4		0.466
8		0.373
16		0.244
18	15.4	0.226
20	13.4	0.190
22	11.4	0.157
24	9.4	0.123
26	7.4	0.093
28	5.4	0.05

31. Determine the weight specific surface of the powder analysed below using equation (1) and compare this with the specific surface using equation (2). How do you account for the difference between these answers?

$$S_w = \frac{6}{\rho_s} \frac{\Sigma \Delta D/K}{\Sigma \Delta Dd/K} \tag{1}$$

$$S_w = 460 \left(\frac{D}{C}\right) \frac{1}{K_m} \quad \text{SI units} \tag{2}$$

Optical density	0.45	0.40	0.35	0.30	0.25	0.20	0.15	0.10	0.05	0.00
	0.50	0.45	0.40	0.35	0.30	0.25	0.20	0.15	0.10	0.05
Mean particle size, d (μm)	22.6	16	11.3	8	5.66	4	2.83	2	1.41	1.00
Mean extinction coefficient, K	1.2	1.4	1.6	1.8	2.0	2.2	2.4	2.6	2.4	2.2

$C = 0.25$ kg m^{-3}, $\rho_s = 1500$ kg m^{-3} and $K_m = 1.98$.

32. Determine the size distribution by weight and the specific surface of a powder having the analysis tabulated below. Knowing that the powder concentration to give an optical density of 0.67 is 0.50 kg m^{-3} and the length of the light beam in the suspension is 5 cm another estimate of surface area can be made. Evaluate this and suggest why it may differ from the value obtained from the tabulated results.

Define mean diameter and determine the surface–volume and volume–moment diameters for this sample (density of solid, 2610 kg m^{-3}).

What advantage over conventional instruments is claimed for the wide angle photosedimentometer and what advantages does this technique have over other size and surface determination methods?

Particle size, d (μm)	2	4	8	16	32
Optical density, D (%)	0	25	50	75	100

33. Compare and contrast the wide and narrow angle photosedimentometers. The analyses given below have been obtained using these techniques. By comparing the analyses determine the values of the extinction coefficient. If the maximum optical densities were 0.50 in each case determine also the specific surface of the powder and the mean extinction coefficient.

Optical densities, D (%)	Wide angle	0	10	20	30	40	50	60	70	80	90	100
	Narrow angle	0	11	23	46	48	59	69	78	86	93.5	100
Average diameter, d (μm)		1	1.4	2	2.8	4	5.6	8	11.3	16	22.6	

Powder concentrations, for the wide and narrow angle photosedimentometers respectively, 0.250 and 0.125 kg m^{-3}. Length of light path in the suspension, 5 cm.

34. Describe the photosedimentation method of particle size analysis. Discuss the limitations of this technique and how some of them are partially overcome with the wide angle scanning photosedimentometer.

Evaluate the specific surface of the powder for which the data below were obtained. Find also the particle size distribution by weight and the variation of extinction coefficient with particle size. Comment on your results.

Optical density (D)		Mean particle size, d (μm)
WASP	EEL	
0.45–0.50	0.470–0.50	22.6
0.40–0.45	0.435–0.470	16
0.35–0.40	0.39 –0.435	11.3
0.30–0.35	0.345–0.39	8
0.25–0.30	0.295–0.345	5.66
0.20–0.25	0.24 –0.295	4
0.15–0.20	0.18 –0.24	2.83
0.10–0.15	0.115–0.18	2
0.05–0.10	0.055–0.115	1.41
0.00–0.05	0.00 –0.055	1.00

Initial concentration: EEL C_e = 0.25 kg m^{-3}, L_e = 1.00 cm; WASP C_w = 0.10 kg m^{-3}, L_w = 5.00 cm.

35. Discuss the problems involved in the determination of particle size distribution by the attenuation of a light beam passing through a settling suspension and suggest means of reducing the inherent errors in the technique.

What are the advantages of this system? Suggest other means of determining particle size using a light source and detectors.

The optical density of a settling suspension is given below against the size of the largest particle present in the light beam. Assuming the laws of geometric optics to hold, determine the cumulative percentage undersize by weight by both a tabular and graphical method and the weight specific surface of the powder. Concentration, 0.016 g in 50 cm^3; length of light beam, 2.0 cm; particle sizes (μm), 126, 90, 63, 45, 32, 22.6, 16, 11.3, 8, 5.6, 4; optical densities D, 0.62, 0.61, 0.56, 0.50, 0.42, 0.33, 0.24, 0.16, 0.10, 0.05, 0.

36. 3.78 g of powder were dispersed in an ICI sedimentometer and the incremental amounts collected are given below.

X_{89}	γ_{63}	$\gamma_{44.5}$	$\gamma_{31.5}$	$\gamma_{22.3}$	$\gamma_{15.7}$	$\gamma_{11.1}$	$\gamma_{7.8}$	$\gamma_{5.6}$	$\gamma_{3.9}$	
0.017	0.017	0.028	0.036	0.045	0.049	0.050	0.042	0.034	0.026	(g)

Assuming that t_{89} is the time required for an 89 μm particle to fall the length of the sedimentation column, derive the size distribution by a tabular method and also by plotting weight deposited (P) against the natural log of relative time (ln t).

(*Note.* In the tabular method, all particles in the size range 75 to 105 μm are assumed to be 89 μm in size, hence W is plotted against the lower limit of 75 μm. In the graphical method W is plotted against the mean value.)

37. A homogeneous suspension is settling in a disc centrifuge. Calculate the size of the largest particle in the measurement zone at time 32 min and show that the correspon-ding cumulative percentage undersize is 10.7 given the following data.

The particles of density 2500 kg m^{-3} are settling in a liquid of density 1000 kg m^{-3} and viscosity 10^{-3} N s m^{-2} in a centrifuge rotating at 750 rev min^{-1}. The liquid surface and measurement zone are at radii 2.84 and 5.00 cm respectively from the centre of the disc.

Applying Kamack's equation gives

$$F_1 = \tfrac{1}{2}(y + 1)Q_1$$

and

$$F_2 = \tfrac{1}{2}(y + y_{12})Q_2 + \left(\frac{y + y_{12}}{y_{22} + y_{12}} - \frac{y + y_{12}}{y_{12} + y_{02}} \right) F_1$$

where $Q_1 = 2.5\%$ and $Q_2 = 6.0\%$ are the recorded concentrations at 64 min and 32 min and y is the square of the ratio of the measurement radius to the surface radius and

$$y_{12} = y^{1/2}; \quad y_{22} = y; \quad y_{02} = 1$$

38. A homogeneous suspension is settling in a disc centrifuge. Calculate the size of the largest particle in the measurement zone at time 2048 s and show that the correspond-ing cumulative frequency undersize is 21.4% given the following data.

The particles, of specific gravity 5.14, are settling in a liquid of specific gravity 1.00 and viscosity 0.001 N s m^{-2} in a centrifuge rotating at 1500 rev min^{-1}. The liquid surface and the measurement zone are 2.84 and 5.00 cm respectively from the centre of the disc.

Note. Kamack's equation has to be applied twice to solve this problem.

$$F_i = \tfrac{1}{2}(y + y_{i-1,i})Q_i + \sum_{j=1}^{i-1} \left[\frac{y + y_{i-1,i}}{y_{j+1,i} + y_{ji}} - \frac{y + y_{i-1,i}}{y_{ji} + y_{j-1,i}} \right] F_j$$

and $Q_1 = 5\%$, $Q_2 = 12\%$ are the recorded concentrations at times 4096 s and 2048 s. y is the square of the ratio of measurement diameter to surface diameter and D the particle diameter.

$$y_{ij} = y^{(D_i/D_j)^2}; \quad y_{0i} = 1; \quad y_{ii} = y$$

39. Develop an expression for the settling times of solid, spherical particles in a centri-fuge. How would you apply the above expression to a centrifugal particle size analysis?

Determine the time it would take for a (a) 1 μm, (b) 5 μm particle of quartz to settle from the surface of a liquid 5 cm from the axis of a centrifuge to a measurement zone at

10 cm radius. Particle density, 2650 kg m^{-3}; liquid density, 1000 kg m^{-3}; liquid viscosity, 0.001 kg m^{-1} s^{-1}; and centrifuge speed, 500 rev min^{-1}.

40. 'The most important single property of a powder is its specific surface.' Discuss this statement and indicate how specific surface may be determined. Determine the specific surface of cement if 1 g of powder adsorbs 15 cm^3 of nitrogen at STP, at a relative pressure of 0.20 and liquid nitrogen temperature. Discuss the origins of any equations used. Avogadro's number, 6.023 x 10^{23} molecules (mole)$^{-1}$; molar volume, 0.02241 m^3 (mole)$^{-1}$; and area occupied by one nitrogen molecule, 16.2 x 10^{-20} m^2.

41. Discuss the assumptions made in deriving the following equation and show how it may be used for surface area determination. (*Do not derive the equation.*)

$$\frac{1}{A}\frac{dV}{dt} = \frac{\Delta P}{k\eta L\rho_s^2 S_w^2}\frac{\epsilon^3}{(1-\epsilon)^2}$$

Symbols having their usual meaning.

Using Lea and Nurse's equipment the pressure drop across a packed bed is found to be 8.32 water gauge for a flow rate of 13.2 cm^3 min^{-1}. If the bed dimensions are 4.00 cm by 2.52 cm^2 and it contains 15.1 g of powder of density 2650 kg m^{-3} determine the weight specific surface of the powder. ($\eta = 18 \times 10^{-6}$ N s m^{-2}.)

42. The velocity of a fluid through a packed bed of powder is given by the equation

$$u = \frac{\Delta P}{k\eta L S_V^2}\frac{\epsilon^3}{(1-\epsilon)^2}$$

ΔP is the pressure drop across the bed; k, the aspect factor, is a function of particle size distribution, particle shape and the tortuosity of the voids through the packed bed; η is the viscosity of the fluid; L is the length of the bed; S_V is the volume specific surface of the powder in the bed; ϵ is the porosity of the bed.

Determine the volume specific surface of a granular powder if a pressure drop of 2 cm Hg is required to force air having a viscosity of 1.7 x 10^{-5} N s m^{-2} through a bed of cross-sectional area 1 cm, porosity 0.4 and length 5 cm at a flow rate of 2.5 cm^3 min^{-1}.

Discuss some other method of determining specific surface and explain the industrial importance of this parameter.

43. A mass W_p = 16.73 g of silver powder of density 10 520 kg m^{-3} is loosely packed to a depth of 6.03 cm in a bed of cross-sectional area 1.267 cm^2. Helium, at 23°C, diffuses through the bed from a container of total volume V = 1107 cm^3 and the time taken for the inlet pressure to fall from 43 to 10 torr is 283 s.

Determine the mass specific surface of the powder.

$[R = 8.314 \text{ J mol}^{-1} \text{ K}^{-1}, M = 4.]$

$$S_w = \frac{8\epsilon^2}{3} \frac{A^2}{W_p V} \sqrt{\frac{2RT}{\pi M}} \left(\frac{1}{P} \frac{dP}{dt} \right)^{-1}$$

44. Taking the average velocity of a fluid in a capillary to be that given by Poiseuille's equation:

$$v = \frac{\Delta p}{8\eta L} r^2$$

show that for a packed bed of powder

$$S_v = \frac{\epsilon}{1 - \epsilon} \frac{2}{r}$$

and

$$v = \frac{\Delta p}{k\eta S_v^2 L} \frac{\epsilon}{(1 - \epsilon)^2}$$

Assuming that slippage at the capillary walls occurs when the mean free path (λ) of the fluid (gas) molecules is of the same order of magnitude as the pore size and that this increases the effective radius of the pores to $(r + \lambda)$ show that

$$v = \frac{\Delta p}{k\eta S_v^2 L} \frac{\epsilon^3}{(1 - \epsilon)^2} + \frac{\Delta p}{k\eta L} \frac{\epsilon^2}{(1 - \epsilon)} \frac{2\lambda}{S_v}$$

Show also that:

$$S = \frac{S_m}{2} \pm \sqrt{\left(\frac{S_m^2}{4} + S_k^2 \right)}$$

All symbols have their usual meaning.

45. Air is passed through a porous plug of diameter 1.25 cm and length 10.0 cm at the rate of 1.00 cm^3 min^{-1}. If the density of the powder is 3000 kg m^{-3} and the weight of the sample is 18.4 g determine the porosity of the plug. If the pressure drop across the plug is 2.55 cm of water, determine the weight specific surface applying the permeametry equation.

If the inlet pressure is 30 cm of mercury determine the true weight specific surface of the powder.

$$u = \frac{1}{k\eta \rho_s^2 S_w^2} \frac{\Delta p}{L} \frac{\epsilon^3}{(1 - \epsilon)^2} + \frac{0.96}{\rho_s S_w} \frac{\epsilon^2}{(1 - \epsilon)} \frac{\Delta p}{\bar{p} L} \sqrt{\left(\frac{2RT}{\pi M} \right)}$$

$R = 8.314 \text{ kJ kmol}^{-1} \text{ K}^{-1}, T = 20°C, M = 29.37, k = 5, \eta = 1.8 \times 10^{-5} \text{ N s m}^{-2}.$

Compare the second term on the RHS with the value for S obtained from the expression below, using the transformations in equation (15.40), and evaluate Z.

$$S_x = \frac{1}{k\eta\rho_s u} \frac{\epsilon^2}{(1-\epsilon)} \frac{\Delta p}{L} Z\lambda$$

Enumerate some of the errors involved in this method of analysis. All symbols have their usual meaning.

46. The equation for the flow of a gas through a packed bed of powder is given by the Carman-Kozeny equation together with a slip correction term:

$$u = \frac{\Delta p}{k\eta LS_v^2} \frac{\epsilon^3}{(1-\epsilon)^2} + \frac{\Delta p}{k\eta LS_v} \frac{\epsilon^2}{(1-\epsilon)} 3.4\lambda$$

Symbols have their usual meaning.

Show that for air at 20°C, 760 mm Hg, the second term equals five times the first term for $d_{sv} = 26.6\,\mu m$ ($\lambda = 6.53 \times 10^{-6}$ m, $\epsilon = 0.50$). Show also that neglecting this second term will, in this case, lead to a $2\frac{1}{2}\%$ error in the calculated value of S_v.

A cylindrical bed of powder, 4.00 cm by 2.52 cm², is formed using 15.1 g of powder of density 2650 kg m⁻³. The flow rate of air through the bed is 2.2 × 10^{-7} m³ s⁻¹ when the pressure drop is 816 N m⁻². Neglecting the slip term, determine the weight specific surface of the powder ($\eta = 18 \times 10^{-6}$ N s m⁻², $k = 5$).

47. Under conditions of low pressure the rate of transfer of a gas through a packed bed of powder may be calculated by an equation containing a viscous and a molecular term.

$$\frac{1}{A}\frac{dV}{dt} = \frac{\Delta p}{5\eta LS_v^2} \frac{\epsilon^3}{(1-\epsilon)^2} + \frac{\Delta p}{5\eta LS_v} \frac{\epsilon^2}{(1-\epsilon)} z\lambda$$

From the given data evaluate the volume specific surface of the powder in the packed bed.

What meaning do you attach to the specific surface determined and what are the limitations of the technique?

Data: $\epsilon = 0.4$, $z = 3.8$, $\eta = 1.8 \times 10^{-5}$ N s m⁻², $\lambda = 9 \times 10^{-6}$ m, $\Delta p = 5$ cm Hg, $L = 5$ cm, $A = 2.5$ cm², $dV/dt = 10.8$ cm³ s⁻¹, density of mercury = 13 600 kg m⁻³, $g = 9.81$ m s⁻².

48. Under conditions of low pressure, the rate of transfer of a gas through a packed bed of powder is governed by an equation containing a viscous and a molecular term.

$$\frac{1}{A}\frac{dV}{dt} = \frac{p}{5\eta LS_v^2} \frac{\epsilon^3}{(1-\epsilon)^2} + \frac{p}{5\eta LS_v} \frac{\epsilon^2}{(1-\epsilon)} 3.4\lambda$$

From the data given below evaluate the volume-specific surface using each term separately and then show how these two values may be combined to give the true volume-specific surface of the powder.

$\epsilon = 0.4,$ $L = 5$ cm,

$\eta = 19.4 \times 10^{-6}$ N s m^{-2}, $A = 1.25 \times 10^{-4}$ m^2,

$\lambda = 2.75 \times 10^{-7}$ m, $dV/dt = 10^{-6}$ m^3 s^{-1},

$\Delta p = 0.5$ cm Hg, $g = 9.81$ m s^{-2},

density of mercury $= 13\,600$ kg m^{-3}.

49. Show that, for a variable pressure permeameter, the Carman–Kozeny equation may be written:

$$S_k = \left[\frac{Kte^3}{L(1 - \epsilon)^2} \right]^{1/2}$$

and that $K = 4.575 \times 10^8$ m^{-1} s^{-1} if the aspect factor $k = 5$, gas (air) viscosity $\eta = 18.6 \times 10^{-6}$ N s m^{-2}, $g = 9.81$ m s^{-2} and other variables are as given below.

A powder bed contains 25 g of powder of density 2600 kg m^{-3} packed to a length of 3.16 cm in a cell of cross-sectional area 5.067 cm^2. A manometer of cross-sectional area 4.75 cm^2, containing oil of density 880 kg m^{-3}, is used to force air through the bed. Determine the volume specific surface area of the powder if it takes 300 s for the oil level to fall from 11.1 cm to 7.2 cm above its equilibrium position (i.e. Δp varies from 22.2 to 14.4 cm oil), using the above equation.

What contribution does slip flow make to the surface area if the mean free path of air $\lambda = 9 \times 10^{-8}$ m^2?

The slip term may be written:

$$S_M = \frac{1.9Kt\lambda}{L} \frac{\epsilon^2}{(1 - \epsilon)}$$

Compare the surface–volume mean diameters derived above with the value obtained using the composite equation containing both terms.

50. Using British Standards gas adsorption apparatus and a sample weight of 1 g it is found that 12.945 cm^3 of nitrogen at STP is adsorbed at a relative pressure of 0.15.

Evaluate the specific surface of the sample explaining carefully the origin of any equations used. Area of nitrogen molecule, 16.2×10^{-20} m^2; Avogadro's number, 6.023×10^{23} molecules (mole)$^{-1}$; ideal gas volume, 0.02241 m^3 (mole)$^{-1}$.

51. Using a single point method for the determination of surface area it was found that at a pressure of 258 mm mercury the volume of nitrogen gas adsorbed on 0.25 g of solids at 77 K was 0.896 cm^2 at STP. Evaluate the specific surface explaining carefully the origin of any equations used.

What factors must be considered to ensure that adequate degassing has taken place?

Comment on the validity and accuracy of the evaluated specific surface by the single point method.

Saturated vapour pressure of nitrogen, 760 mm mercury; area occupied by a nitrogen molecule, 16.2×10^{-20} m^2; ideal gas volume, 0.0224 m^3 (mole)$^{-1}$.

52. Helium is admitted to a burette previously under vacuum until the pressure is 75 mmHg. It is then allowed to expand into a sample tube immersed in liquid nitrogen containing 0.732 g of cement and the pressure falls to 37.6 mmHg.

The system is then re-evacuated and nitrogen is admitted to the burette at a pressure of 450.0 mmHg which falls to 152.0 mmHg on expansion into the sample tube.

Find the volume of nitrogen at STP adsorbed by the cement and estimate the surface area of the cement.

Discuss this phenomenon. (Burette volume 18.5 cm^3 and temperature 292 K.)

53. The volume of nitrogen adsorbed at different pressures by a sample of cement using conventional BET equipment is tabulated below. Determine the surface area using the BET and HJr equations and comment on your results.

Pressure P (mm Hg)	33.8	60.7	71.9	84.9	108.1	129.6	153.0
Volume adsorbed V (cm^3 g^{-1} at STP)	0.6846	0.8116	0.8278	0.8753	0.9162	0.9742	1.0200
$x = P/P_0$	0.0445	0.0798	0.0946	0.1117	0.1422	0.1705	0.2013

54. The total amount of nitrogen in a two-volume burette static BET nitrogen adsorption apparatus was 4.08 cm^3 at STP. This was expanded into a sample tube containing 1.5 g of degassed zinc oxide cooled to −196°C. The pressure recorded at volumes 2 and 1 were 57.2 and 68.0 mm mercury, respectively. A further quantity of nitrogen was introduced to bring the total volume in the system to 7.175 cm^3 at STP. Upon expansion into the sample tube the pressures were 104.0 and 124.5 mm mercury for volumes 2 and 1, respectively.

Volume factors for the burettes: volume 1, 0.011 55 cm^3 at STP per mm Hg; volume 2, 0.0221 cm^3 at STP per mm Hg; helium dead space factor, 0.0391 cm^3 at STP per mm Hg.

Find the surface area of the zinc oxide (m^2 g^{-1}).

55. 1.62 g of titanium dioxide gave a dead space factor of 0.043 cm^3 (mm Hg)$^{-1}$ with a three-volume static BET apparatus. The volume factors were $A_1 = 0.011\,25$, $A_2 = 0.021\,50$, $A_3 = 0.033\,60$ cm^3 (mm Hg)$^{-1}$, the suffixes denoting the burette number.

9.502 cm^3 of nitrogen at STP were admitted to the system and gave a pressure, when expanded to the sample tube, of 136.0 and 161.5 mm Hg for volumes 2 and 1 respectively.

The total volume of nitrogen in the system was increased to 15.99 cm^3 at STP, when expanded to the sample tube and resulting pressures were 198.5, 235.0, 229.5 mm Hg for volumes 3, 2, 1 respectively.

Find the surface area of the titanium dioxide.

56. The nitrogen gas adsorption data for a sample of carbon are given below. It is found that the adsorption isotherm follows the n-layer BET equation with n equal to three.

Find the weight specific surface of the powder and its cumulative pore size distribution by surface using the MP method.

Discuss the equations used.

Relative pressure x	Volume adsorbed, V (cm^3 g^{-1} at STP)	Thickness of adsorbed layer, t (nm)	θ	$\dfrac{\phi}{V}$
0.05	225.0	32.4	0.053	2.44×10^{-4}
0.10	258.4	36.8	0.101	4.83×10^{-4}
0.15	261.4	40.4	0.176	7.46×10^{-4}
0.20	271.1	43.7	0.248	10.33×10^{-4}
0.25	277.8	46.9	0.328	13.50×10^{-4}
0.30	281.0	50.1	0.417	17.08×10^{-4}
0.35	284.9	53.4	0.515	20.88×10^{-4}

$$\left[n\text{-layer BET equation } \frac{\phi}{V} = \frac{1}{CV_m} + \frac{\theta}{V_m} \right]$$

$S_w = 4.35V_m$ m^2 g^{-1}; $S_t = 154.7(V/t)$ m^2 g^{-1};

S_w = weight specific surface, BET equation;

S_t = weight specific surface, V–t curve;

V_m = volume adsorbed at monolayer coverage, cm^3 g^{-1} at STP;

C = constant for a given system.

57. Helium admitted to a burette, previously under vacuum, produced a pressure of 77.5 torr. On expansion into a sample tube containing 0.75 g of solid immersed in liquid nitrogen, the pressure observed was 57.3 torr.

After re-evacuation nitrogen was admitted to the same burette at a pressure of 459.4 torr. This fell to a pressure of 121.2 torr when expanded into the sample tube.

Calculate the volume of nitrogen adsorbed at STP on the solid and evaluate the surface area of the solid.

Burette volume 31.5 cm^3 and temperature 292 K.

Comment on the evaluation of surface area of the solid by the above method.

Appendix 1
Equipment and suppliers

The names given in brackets are those of agents. The equipment is listed in chapter order as it occurs in the text.

Sampling devices (Chapter 1)

Spinning Riffler (1 litre and 20 litre)	Ladal
Sample Divider	Pascal
Rotary Riffler	Freeman Laboratories
Rotary Sample Divider	Glen Creston
Sample Splitter	Fritsch (Christison)
Limpet Autosampler	Simon
Rotary Microriffler	Quantachrome
Sieving Riffler	Quantachrome

Flue samplers (Chapter 2)

Smoke Dust Monitor (SEROP)	Airflow Development
BCURA Gas Flow Monitor	
Smoke Density Meter (chimneys/ducts)	Bailey Meters and Controls
In-Stack Samplers	Anderson
Dust Samplers	Anderson
Smoke Density Measuring Instruments	Erwin Sick, Pearson Panke
CERL Flue Dust Monitor	Foster Instruments
High Sensitivity Air Monitor	Photoelectronics
Nebetco Continuous Smoke Monitor	Nebetco
Leeds and Northrup Smoke Sampler	Leeds and Northrup
Bailey Smoke/Dust Monitor	Bailey
Edison Visibility Monitor	Edison
Stack Monitor	Gelman
Radioactive Air Particle Monitors	Nuclear Measurement Corporation
MAP 1B Continuous Monitor for Airborne Particulate Radioactivity	Trapelo
Velocity Sampling Nozzles	Babcock & Wilcox
Null Type Nozzles	Joy
CEGRIT	Airflow
Dust and Fume Determination Assembly Models D-1000 and D-1027	Joy

Dust Difficulty Determinator	Environeering
Environeering Stack Sampling Equipment	Environeering
Isokinetic Dust Sampler	Day
Staksampler	Research Appliance Co.
Automatic Stack Monitor	Research Appliance Co.

Aerosol samplers (Chapter 3)

Thermal Precipitators (Standard and Long Term)	Casella
Gravimetric Dust Sampler	(US Agents MSA)
Cascade Impactor, Hexhlet	(US Agents MSA)
Wrights TP (Ch. 3 ref. 94)	Adams
Hamiltons TP (Ch. 3 ref. 98)	Adams
Konisampler Thermal Precipitators	Ficklen
British Standard Deposit Gauge	Glass Developments
Continuous Oscillating and Gravimetric TP	American Instruments
Thermopositor	American Instruments
Dräger Dust Sampler	Drägerwerk
Thermal Precipitators, Konimeter, Dust	Sartorius (Howe)
Collectors, Gravicon, Porticon Dust Sampler (filters)	MSA (USA)
Electrostatic Air Sampler kit	Bendix
Olin Particle Mass Monitor (Ch. 3 refs. 131, 132)	Thermo-Systems, Proner
Electrostatic Air Sampler	Thermo-Systems, Proner
Smoke Pollution Sampler	Charles Austin
Cascade Centripeter	Bird and Tole
Membrane Filters	Nuclepore
Membrane Filters	Gelman, Millipore
Settlement Dust Counter	Casella, MSA
Hexhlet Gravimetric Personal Sampler	Casella, MSA
Settlement Dust Sampler	Research Appliances
Cascade Impactor	Research Appliances
Portable Dust Sampler	Rotheroe and Mitchell
Periodic Air Sampler	Rotheroe and Mitchell
Personal Sampler	Anderson
Cascade Sampler	Anderson
Aerosol Gravimetric Spectrometer	Fleming
Air Pollution Monitors	Fleming
Particle Sampling Unit	Fleming
Dust Sampling Unit	Fleming
Millipore Sampling Set	Thermal Control
Konimeter	Carl Zeiss Jena also Sartorius (Brinkman)
AERA Portable Air Sampler	Addy Products
Light Scattering Counters	High Accuracy Products (Air Supply)
	Royco (Hawksley)
	Bausch and Lomb (Applied Research Laboratories)
Light Reflectance Monitor	Research Appliance
Aerosol Spectrometers	Sartorius (Brinkman)
Sigrist Dust Measuring Equipment	(Howe)
Electricon Smoke Monitor	Ronald Trist

Coulter Contamination Counter	Coulter
LIDAR, Smoke Plume Tracking	Laser Associates
Laser Light Scattering Particle Counter	Procedyne
Cascade Impactor	Shimadzu
Andersen Samplers	Andersen
Rotorod	Metronics
Frieseke and Hoepfner Dust Monitor	TEM Sales
Sinclair-Phoenix	Sinclair (Techmation)
Spray Droplet Analyser	Ratteyon
Saab Photometer	Saab
S-3000 Specscan	Optronics
S-3400 Photomation Mx 1v	Optronics
H-900 Holoscan	Optronics
Nuclepore Filters	General Electric Co.
Thermal Precipitator M1C 501	Numinco
Electrostatic Precipitator	MSA (USA)
Electrostatic Precipitator	Del Electronics
Leap Electrostatic Precipitator	(ERC)
Litton Electrostatic Precipitator	Littons Systems Inc.
OWL	Process and Instrument Corp.
Nephelometer	Meteorology Research Inc.
Cascade Impactor	Unico
Thermal Precipitator	Sartorius (Brinkman)
Bacteria Sampler	Casella CF (Willson)
Hirst Spore Trap	Casella CF (Willson)
MSI Sampler	Environmental Research Corp.
Bacteria Collector	Gelman
Moving Slide Impactor	Meteorology Research Inc.
Airborne Bacteria Monitor	2000 Inc.
Brink Model B Cascade Impactor	Monsanto
Greenburg-Smith Impinger ($1 \text{ ft}^3 \text{ min}^{-1}$)	Willson
Midget Impinger ($0.1 \text{ ft}^3 \text{ min}^{-1}$)	Gelman Hawksley
Midget Impinger ($1 \text{ ft}^3 \text{ min}^{-1}$)	Mines Safety Appliance USA
Midget Impinger ($1 \text{ ft}^3 \text{ min}^{-1}$)	Unico
Micro Impinger ($0.1 \text{ ft}^3 \text{ min}^{-1}$)	Unico
Multi-Stage All Glass Liquid Impinger	Dixon
BCURA Size Selecting Gravimetric Personal Sampler	Willson (C.F. Casella)
'Respirable' Dust Sampler	Unico
Bausch & Lomb Dust Counter	Environmental Research Corp. (V. A. Howe)
Climet Particle Analysis Systems	Climet
GE Condensation Nuclei Counter	General Electric Co.
Small Particle Detector	Gardner
Condensation Nuclei Counter	Research Appliances Co.
Whitby Aerosol Analyser	Thermo-Systems
Holographic Particle Analyser	Laser Holography Inc.
Sartorius Scintillation Particle Counter	Sartorius (Brinkman)
Eberline Model A1M-3 Air Monitor	Eberline
Radioactive Air Monitor, Type RAM 1	Nuclear Enterprises
Classical Scattering Aerosol Spectrometer Probe	Particle Measurement Systems
Active Scattering Aerosol Spectrometer Probe	Particle Measurement Systems
Forward Scattering Aerosol Spectrometer Probe	Particle Measurement Systems
Axially Scattering Aerosol Spectrometer Probe	Particle Measurement Systems

Sieving equipment (Chapter 5)

Woven Wire Sieves	Endecottes, Pascall, Greenings
Inclyno Sieve Shaker, Turbine Sifter	Pascall
Electroformed Sieves	Buckbee Mears (Production Sales and Services)
Electroformed Sieves	Veco
Electroformed Sieves	Endecottes
Fisher Wheeler Sieve Shaker	Fisher
Sieve Shakers	American Instruments, Pascall
Sieve Shakers	Endecottes
Allen-Bradley Sonic Sifter	ATM Corporation (Kek)
Small Portable Sieve Shaker	La Pine
Tyler Ro-Tap	La Pine
Alpine Air Jet	Alpine, Lavino
Wet and Dry Sieve Shakers	Fritsch (Howe)
Gallie-Porritt Apparatus	Gallenkamp
Microsieve Shaker	Givliani

Microscopes (Chapter 6)

Watson Image Splitting Eyepiece	Watson
Push Button Counter	Casella
Vicker Image Shearing Eyepiece	Vickers
Zeiss Endter PSA	Carl Zeiss (Degenhardt)
Metals Research PSA	Metals Research
Quantimet	Image Analysing Computers
MC	Millipore
Spri Analyser	Sondes Place Research
Superscope Electron Microscope (also SEMs)	JEOL (Delviljem)
Magnifiers	Polaron
e.m. grids	Mason and Morton
Aids for e.m.	Alan Agar
The QMS System Omnicon	Bausch and Lomb
Cargill Series H Compound	R. G. Cargill
Classimat	Leitz
Coulter/Timbrell/Shearicon	Coulter
Fleming Shearing Microscope	Fleming
Optomax	Micromeasurements
Microvideomac	Zeiss
Digiplan	Kontron

Miscellaneous dispersing equipment (Chapter 8)

Ultrasonic Dispersers	Mullard, Ultrasonics
Pyknometers	Numinco
Ultrasonic Cleaner	Fritsch (Christison)
Helium Air Pyknometer	Micromeretics (Coulter UK)
Anti-Static Agent M441	ICI
Spray Gun for Electron Microscopy	Aerograph
Autopyknometer	Micromeretics (Coulter UK)
Gas Pyknometer	Beckman
Null Pyknometer	Quantachrome

Fluorochemical Surfactants	3M (Page)
Anti-Foaming Agent (SWS-211)	Stauffer
Standard Powders	General Motors
Standard Powders	Particle Information Service
Standard Powders	Dow Chemicals
Standard Powders	Duke Standards
Standard Powders	General Technical Institute
Standard Powders	National Physical Laboratory
Standard Powders	National Bureau of Standards
Standard Powders	American Society of Mechanical Engineers

Sedimentation equipment (Chapters 9 and 10)

Pipettes and Hydrometers	Gallenkamp
Granumeter	Brezina
WASP Photosedimentometer	Ladal
Wagner Turbidimeter	La Pine
EEL Photosedimentometer	Evans
Bound Brook Photosedimentometer	Goring Kerr
Micromerograph	Val-Dell
Shimadzu Sedimentograph	Shimadzu
	Northgate Traders
Sartorius Sedimentation Balance 4600	Sartorius
Sartorius Sedimentation Balance 4135	Sartorius
Micron Particle Distributometer	Bush GF
Pola-Travis Particle Size Apparatus	Pola
Fisher Dotts Apparatus	Fisher
Travis Method of Two-Layer Analysis	Schaar
Ladal X-Ray Sedimentometer	Ladal
Sedigraph X-Ray Sedimentometer	Micromeretics (Coulter UK)
Recording Sedimentometer	Shimadzu
Photomicronizer	Seishin (ECI)
Cahn Microbalance	Cahn USA
Hydrometer	Becker, H., Rechmann, H.
Balance	Mettler (Switzerland)
Martin Recording Balance	Prolabo (France)

Classifiers (Chapter 11)

Walther	Walther
Cascade Elutriator	American Instruments
Microsplit Separator	British Rema
Major Classifier	Donaldson
Centrifugal Classifier	Micromeretics
Andrews Kinetic Water	Griffin and George
Gonell	Chemisches Laboratorium
Haultain Infrasizer	Infrasizers
Roller	American Instruments
Bahco	Neu (Dietert)
Hexhlet Collector (Walton's Horizontal)	Casella
Microplex Classifier	Lavino
Nauta Hosokawa Classifier	Nautamix

Analysette 8	Fritsch (Christison)
Donaldson Classifier	Donaldson
Warmain Cyclosizer	Warmain (Simanacco)
Alpine Multiplex Zig-Zag Classifier	Alpine (M and M)
British Rema Microsplit	British Rema

Centrifuges (Chapter 12)

Simcar	Simon Carves
Joyce Loebl Disc	Joyce Loebl
Kaye Disc	Martin Sweeny
Sharples Centrifuge	Pennwalt Appliances (USA)
Whitby Apparatus	Mines Safety
Ladal X-Ray	Ladal
Modified Pipette	Ladal
Centrifugal PSA	Shimadzu
Technord Laser Photocentrifuge	Northey
Muschelnantz Centrifuge SW	

Streaming principle (Chapter 13)

Coulter Counter	Coulter
Celloscope	Lars
PD Analyser	Berg
Particle Volume Detector	Telefunken
TOA Microcell Counter	Toa
Granulometer	VEB
Microscale	Medicor
Picoscale	Kutes
Biotronics	Tricon
Digicell	Siemens

Light scattering (see also Aerosol samplers) (Chapter 14)

Shimadzu Light Scattering Photometer	Shimadzu
Sinclair Phoenix Forward Scattering Photometer	Phoenix
Scattering Light Photometer	Phoenix
Brice-Phoenix LSP	Phoenix
Absolute LSP	American Instruments
Photo-Nephelometers	Coleman
Recording Turbidimeter	General Electric Company
Scattermaster	Manufacturing Engineering Company
Photometers	Shimadzu, Société Française, Nethreler, Polymer Consultants
Differential Light Scattering Photometers	Science Spectrum
Atlas of Light Scattering Curves	Science Spectrum
Sartorius Aerosol Photometer	Sartorius (Brinkman)
Laser Light Scatterers, Biological Field	Biophysics
Nano-sizer	Coulter

Permeametry (Chapter 15)

Fisher Sub-Sieve Sizer	Kek, Paris Labo
Rigden Apparatus	Gallenkamp, Contest Instrum.
Knudsen Flow Permeameter	Micromeretics
Griffin Surface Area of Powders Apparatus	Griffin and George
Alstan Diffusion Areameter	Allen and Stanley-Wood
Blaine Permeameter	Normandie Labo, Precision Scientific Co., Ateliers Cloup, Prolabo, Perrier et Cie, Technotest
Permaran	Outokumpu (Harrison Cooper Assoc.)
Lea and Nurse Direct Reading Permeameter	Matelam

Gas adsorption (surface area and porosimetry) (Chapters 16, 17 and 18)

Perkin-Elmer Shell Sorptometer	Perkin-Elmer
Ströhline Areameter	Ströhlin
Quantasorb and Monosorb	Ameresco
	Quantachrome Corporation
Gravimat	Sartorius
A range of instruments	Micromeretics (Coulter UK)
Sorptomatic	Carlo Erba (Systems and Components) (Erba Science) (Allied Sci.)
Areatron	Leybold
Air Displacement Porometer	Numinco
Surface-Area and Gas-Adsorption Equipment	Numinco
Alpha-mec-4	AB Atomenergie
Isorpta Analyser	Engelhard

Mercury porosimetry (Chapter 19)

0–10 000 psa	Micromeretics (Coulter UK)
0–30 000	
0–50 000	
Anglometer	
0–3000 kg cm^{-1} 50 000 psa	Carlo Erba (Erba Science)
0–15 000	American Instruments Co.
0–60 000	(Aminco), Silver Springs Maryland 20910, USA
Scanning Porosimeter	Quantachrome (Northey)

On-line particle sizers (Chapter 20)

Nuclepore Spectrex Prototron Particle Counter	Spectrex/Nuclepore (International Trading)
HIAC Particle Counter. High Accuracy	High Accuracy Products, Pacific Scientific Co. (Particle Data Ltd.)
Climet Particle Counter	Climet
Royco Liquid-Borne Particle Monitor	Royco

Procedyne Particle Size Analyser	Procedyne
Autometric PSM 100 Analyser	Autometrics
Mintex/RSM Slurry Sizer	Mintex
Courier 300	Outokumpu Oy
Cilas Granulometer	Compagnie Industrielle des Lasers (Specfield Ltd.)
Microtrac	Leeds and Northrup
Malvern Particle and Droplet Sizer	Malvern
Prototron	Nuclepore
Langer Acoustical Counter	Erdco
Coulter On-Line Monitor	Coulter
Humboldt PSA	KHD Industrie

Appendix 2
Manufacturers' and suppliers' addresses

AB Atomenergie, Studvik, Sweden
Adams, L. Ltd, Minerva Road, London NW10
Addy Products Ltd, Solent Industrial Estate, Botley, Hampshire SO3 2FQ
Aerograph Co., Lower Sydenham, London SW26
Agar, Alan W., 127 Rye Street, Bishop's Stortford, Hertfordshire
Air Techniques Inc., 1717 Whitehead Rd., Baltimore, Maryland 21207, USA
Air Technology Inc., 2108 Carterdale Rd., Baltimore, Maryland 21209 USA
Airflow Development, 31 Lancaster Rd., High Wycombe, Buckinghamshire
Air Supply International, Gateway House, 302–8 High Street, Slough, Berkshire
Allen and Stanley-Wood, Powder Characterisation Systems, 7 Hall Close, Leeds,
    Yorkshire LS16 9JG
Allied Scientific Company Ltd, 2220 Midland Avenue, Scarborough, Canada
Alpine, 89 Augsburg 2, Postfach 629, W. Germany
Ameresco Inc., 101 Park Street, Montclair, New Jersey 07042, USA
American Instruments Co., 8030 Georgia Avenue, Silver Springs, Maryland 20910,
    USA
American Society of Mechanical Engineers, Powder Test Code Committee Number 28,
    345 East 47th Street, New York, NY 10017, USA
Anderson 2000 Inc., P.O. Box 20769, Atlanta, Georgia 30320, USA
Applied Research Laboratories, Wingate Rd., Luton, Bedfordshire
Ateliers Cloup, 46, boulevard Polangis, 94500 Champigny-sur-Marne, France
ATM Corporation, Sonic Sifter Division, P.O. Box 2405, Milwaukee, Wisconsin 53214,
    USA
Autometrics, 4946N, 63rd Street, Boulder, Colorado 80301, USA

Babcock and Wilcox Ltd, Cleveland House, St James's Square, London SW14 4LN
Bailey Meters and Controls, 218 Purley Way, Croydon, Surrey
Bailey Meter Co., Wickliffe, Ohio 44092, USA
Bausch and Lomb Inc., 820 Linden Avenue, 30320 Rochester, NY 14625, USA
Beckmann Instruments Inc., Fullerton, California 92634, USA
Beckmann Instruments Ltd, Glenrothes, Fife, Scotland
Bendix Corporation, Environmental Science Division, Dept. 81, Taylor Ave., Baltimore,
    Maryland 21204, USA

Bendix Corporation, Scientific Instruments and Equipment Division, 1775 Mt. Read Blvd., Rochester, NY 14603, USA

Bendix Vacuum Ltd, Scientific Instruments and Equipment Division, Easthead Ave., Wokingham, Berkshire RG11 2PW

Berg, R., Particle Data Inc., P.O. Box 265, Elmhurst, Illinois 60126, USA

Berkley Instruments Inc., 2700 Dupont Drive, Irwine, California 92715, USA

Biophysics, Baldwin Place Rd., Mahopac, NY 10541, USA

Brezina, J., Hauptstrasse 68, D-6901 Waldhilsbach, W. Germany

Brinkman Instruments, Canitague Rd., Westbury, NY 11590, USA

Bristol Industrial and Research Associates Ltd (BIRAL), P.O. Box 2, Portishead, Bristol BS20 9JB

British Rema, P.O. Box 31, Imperial Steel Works, Sheffield S9 1RA

Buckbee Mears Co., 245 East 6th Street, St Paul 1, Minnesota, USA

Bush G. F. and Associates, Princeton, New Jersey, USA

(BCR), Mr. A. Pozzo, Community Bureau of Reference (BCR), Directorate-General XII, Commission of the European Communities, 200 rue de La Loi, B-1049 Brussels

Cahn, 27 Essex Rd., Dartford, Kent

Cahn Division of Ventron Instruments Corporation, 7500 Jefferson Street, Paramount, California 90723, USA

Cargill, R. P., Laboratories Incorporated, Cedar Grove, New Jersey, USA

Carl Zeiss, 444 Fifth Ave., New York, NY 10018, USA

Carl Zeiss, 7082 Oberkochen, W. Germany

Carl Zeiss Jena Ltd, VEB Carl Zeiss Jena W. Germany: also England House, 93-7 New Cavendish Street, London W1

Carlo Erba, via Carlo Imbonati 24, 20159, Milan, Italy

Casella, C. F. and Co., Regent House, Britannia Walk, London N1 7ND

Celsco Industries Ltd, Environmental and Industrial Products, Costa Mesa, California

Central Technical Institute, CTI-TNO, P.O. Box 541, Apeldoorn, The Netherlands

Charles Austin Pumps, Petersham Works, 100 Royston Rd., Byfleet, Surrey

Chemische Laboratorium für Tonindustrie, Goslar, Harz, W. Germany

Christison A. (Scientific Equipment) Ltd, Albany Rd., East Gateshead Industrial Estate, Gateshead 8

Clay Adams, Division of Becton Dickinson, Parsippany, New Jersey, USA

Climet Instruments, 1320 West Colton Avenue, Redlands, California, 92373, USA

Climet Instruments Inc., 1240 Birchwood Drive, Sunnyvale, California, USA

Coleman Instruments Inc., 42 Madison Street, Maywood, Illinois, USA

Compagnie Industrielle des Lasers, Route de Nozay, 91 Marcoussis, France

Contest Instruments Ltd., Downmill Rd., Bracknell, Berkshire RG12 1QE

Datametrics Division, CGS Scientific Corporation, 127 Coolidge Hill Rd., Watertown, Massachusetts 02172, USA

Day Sales Company, 810 Third Ave. N.E., Minneapolis 13, Minnesota, USA

Degenhardt and Co. Ltd, 6 Cavendish Square, London W1

Del Electronics Corporation, 616-T Adams Street, Steubenville, Ohio 43952, USA
Delaran Manufacturing Co., West Des Moines, USA
Delviljem (London) Ltd, Delviljim House, Shakespeare Rd., Finchley, Middlesex
Dietert H. and Co., 9330 Roselawn Ave., Detroit, Michigan, USA
Dixon A.W. and Co., 30 Anerly Station Rd., London SE20
Donaldson Co. Inc., 1400 West 94th Street, Minneapolis, Minnesota 55431, USA
Donaldson Europe SV, Interleuvenlaan 1, B-3044, Leuven, Belgium
Dow Chemicals, Midland, Michigan, USA
Draeger Normalair Ltd, Kitty Brewster, Blythe, Northumberland
Drägerwerk Lubeck, D-24 Lubeck 1, P.O. Box 1339, Moislinger Allee 53-55,
    W. Germany
Dynac Corporation, Thompsons Point, Portland, Maine, USA
Dynac, Division of Dieldstone Corporation, P.O. Box 44209, Cincinnati, Ohio, USA

Ealing Beck Ltd, Greycaine Rd., Watford WD2 4PW
Eberline Instruments Corporation, P.O. Box 2108, Sante Fe, New Mexico 87501, USA
Edison, Thomas A., Industries, Instruments Division, West Grange, New Jersey, USA
Electronic Design A/S, Chr. Holms Parkvej 26, 2930 Klampenborg, Denmark
Endecottes Ltd, Lombard Rd., London SW19
Engelhard Industries Ltd, Newark, New Jersey, USA
Environeering Inc., 9933 North Lawler, Skokie, Illinois 60076, USA
Environmental Control International (ECI) Inc., 409 Washington Ave., P.O. Box
    10126, Baltimore, Maryland, USA
Environmental Research Corp. (ERC), 3725 N. Dunlap Street, St Paul, Minnesota
    55112, USA
Erba Science (UK) Ltd, 14 Bath Rd., Swindon SN1 4BA
Erdco Engineering Corp., 136 Official Rd., Addison, Illinois 60101, USA
Erwin Sick Optik-Elektronik, D-7808 Waldkirch, An der Allee 7-9 Postfach 310,

ERRATUM   , California 91104, USA
The following manufacturer's name and address
should be added to the list given in this Appendix.   lshire
Coulter Electronics Ltd, Northwell Drive, Luton   )
Beds LU3 3RH

                                                      Illinois 60018, USA
                                                      ny

Gallenkamp Ltd, Portrack Lane, Stockton-on-Tees, Co. Durham
Gardner Association Inc., 3643 Carman Rd., Schenectady Rd., NY 12303, USA
Gardner Laboratory, Bethesda, Maryland, USA
Gelman Hawksley, 12 Peter Rd., Lancing, Sussex
Gelman Instruments Co., 600 South Wagner Rd., Ann Arbor, Michigan 48106, USA

General Electric Co., Schenectady, NY, USA
General Electric Ordnance Systems, 100 Plastics Ave., Pittsfield, Massachusetts 01201,
      USA
General Motors Corporation, Flint, Michigan, USA
General Sciences Corp., Bridgeport, Connecticut 06604, USA
Giuliani, via Borgomanero 49, Turin, Italy
Glass Developments Ltd, Sudbourne Rd., Brixton Hill, London SW1
Glen Creston, The Red House, Broadway, Stanmore, Middlesex
Goring Kerr Ltd, Hanover Way, Windsor, Berkshire
Greenings, Britannia Works, Printing House Lane, Hayes, Middlesex
Griffin and George Ltd, Wembley, Middlesex

Hamamatsu Corporation, 120 Wood Ave., Middlesex, New Jersey 08846, USA
Hamamatsu TV Co. Ltd, 1126 Ichino-cho, Hermamatsu, Japan
Hamamatsu TV Europe GmbH, D8031 Hechendorf/Pilsensee, Haupstrasse 2, Postfach
      7, W. Germany
Harrison Cooper Associates, Salt Lake City, Utah, USA
Hawksley and Sons Ltd, 12 Peter Rd., Lancing, Sussex
High Accuracy Products Corporation, 141 Spring Street, Claremont, California 91711,
      USA
Hird-Brown Ltd, Lever Street, Bolton, Lancashire BL3 6BJ
Howe, V. A. and Co. Ltd, 88 Peterborough Rd., London SW6

Image Analysing Computers Ltd, Melbourne Rd., Royston, Hertfordshire SG6 6ET
Imanco, 40 Robert Pitt Drive, Monsey, NY 10952, USA
Imperial Chemical Industries Ltd, Nobel Division, Stevenston, Ayrshire, Scotland
Infrasizers Ltd, Toronto, Ontario, Canada
International Trading Co. Inc., 406 Washington Ave., P.O. Box 5519, Baltimore,
      Maryland 21204, USA; also at Orchard House, Victoria Square, Droitwich,
      Worcestershire WR9 8QT

Japan Electron Optics Ltd, Jealco House, Grove Park, Edgeware Rd., Colindale,
      London NW9; also at 477 Riverside Ave., Medford, Massachusetts 02155, USA
Joy Manufacturers Co., Western Precipitation Division, 100 West 9th Street, Los
      Angeles, California 90015, USA
Joyce Loebl Ltd, Princes Way, Team Valley, Gateshead 11, Co. Durham

KHD Industrie-anlagen GmbH, D-5000, Köln, W. Germany
Kek Ltd, Hully Rd., Hurdsfield Industrial Estate, Macclesfield, Cheshire SK10 2ND
Kontron D-8 München 50, Lerchenstrasse 8-10, W. Germany
Kontron Messgerate GmbH, D-8057, Eching, W. Germany

La Pine Scientific Co., Chicago 29, Illinois, USA
Ladal (Scientific Equipment) Ltd, 'Warlings', Warley Edge, Warley, Halifax, Yorkshire

Lars A.B. Ljungberg and Co., Stockholm, Sweden
Laser Associates Ltd, Paynes Lane, Warwickshire
Laser Holography Inc., 1130 Channel Drive, Santa Barbara, California 93130, USA
Lavino, Garrard House, 31–45 Gresham Street, London EC2
Leeds and Northrup Co., 4907 Stenton Ave., Philadelphia, Pennsylvania 19144, USA
Leitz, Ernst, D-633 Wetzlar GmbH, 1 Postfach 210, W. Germany; also at Rockleigh,
    New Jersey 07647, USA; also at 30 Mortimer Street, London W1N 8BB
Leitz, Ernst Inc., 468 Park Ave., South New York, NY 10016, USA
Litton Systems Inc., Applied Science Division, 2003 East Hennegin Ave., Minneapolis,
    Minnesota 55413, USA

M and M Process Equipment Ltd, Fir Tree House, Headstone Drive, Wealdstone,
    Harrow, Middlesex HA3 5QS
Malvern Instruments Ltd, Spring Lane Trading Estate, Malvern, Worcestershire WR14
    1AL
Manufacturing Engineering and Equipment Corporation, Warrington, Pennsylvania,
    USA
Mason and Morton Ltd, 32–40 Headstone Drive, Wealdstone, Harrow, Middlesex
Matelem., Les Cloviers, rue d'Argenteuil, 95110 Sannois, France
Meteorology Research Inc., 474 Woodbury Rd., Altaydena, California 91001, USA
Metals Research Ltd, 91 King Street, Cambridge
Metronics Associates Inc., 3201 Porter Drive, Stamford Industrial Park, P.O. Box 637,
    Palo Alto, California, USA
Mettler (Switzerland) Instruments A.G. CH-8606, Greifensee-Zurich, Switzerland
Micromeretics Instrument Corporation, 800 Goshen Springs Rd., Norcroft, Georgia
    30071, USA
Microscal Ltd, 20 Mattock Lane, Ealing, London
Millipore Corp., Ashby Rd., Bedford, Massachusetts 01730, USA
Mines Safety Appliances Co. Ltd, Greenford, Middlesex
Mines Safety Appliances Co. Ltd, 201 Braddock Ave., Pittsburg 8, Pennsylvania, USA
Mintex Division, Cartner Group Ltd, Stirling Rd. Trading Estate, Slough, Buckingham-
    shire
Monsanto Company, Engineering Sales Dept., 800N Lindberg Boulevard, St Louis,
    Missouri 63166, USA
Mullard Equipment Ltd, Manor Royal, Crawley, Sussex

National Bureau of Standards, Washington, USA
National Physical Laboratory, Division of Chemical Standards, Teddington, Middlesex
    TW11 0LW (Tel: 01-977 3222 ext. 3351)
Nautamix, N.V., P.O. Box 773, Haarlem, Holland
Nebetco Engineering, 1107 Chandler Ave., Raselle, New Jersey 07203, USA
Nethreler and Hinz., GmbH, Hamburg, W. Germany
NEU Engineering Ltd, 32–4 Baker Street, Weybridge, Surrey
NEU Etablissement, P.O. Box 28, Lille, France

Ni On Kagaka Kogyo Co. Ltd, 4168 Yamadashimo, Suita, Osaka, Japan
Normandie-Labo, 76210 Lintot, France
Northey International Systems Ltd, 5 Charles Lane, St John's Wood, High Street,
    London NW8 7SB
Northgate Traders Ltd, London EC2
Nuclear Enterprises Ltd, Sighthill, Edinburgh 11, Scotland
Nuclear Measurements Corporation, 2460 North Arlington Ave., Indianapolis, Indiana,
    USA
Nuclepore Corporation, 7035 Commerce Circle, Pleasanton, California 64566, USA
Numek Instruments and Controls Corporation, Appolo, Pennsylvania, USA
Numinco, 300 Seco Rd., Monroeville, Pennsylvania 15146, USA

Optronics International Inc., Chelmsford, Massachusetts, USA
Outokumpu Oy Forskningslaboratoriet, Töölönkatu 4, Helsinki, Finland

Pacific Scientific Co., P.O. Box 3007, 4719 West Brooks Street, Montclair, California
    91763, USA
Page (Charles) and Co. Ltd, Acorn House, Victoria Rd., London W3 6XU
Paris-Labo, 49 rue De France, 94300, Vincennes, France
Particle Data Inc., P.O. Box 265, Elmhurst, Illinois 60126, USA
Particle Data Ltd, 39 Tirlebank Way, New Town, Tewkesbury, Gloucestershire GL20
    5RX
Particle Information Service Inc., P.O. Box 702, Grant Pass, Oregon 97526, USA
Particle Measurement Systems, Boulder, Colorado, USA
Pascall Ltd, Gatwick Road, Crawley, Sussex RH10 2RS
Pearson Panke Ltd, 1–3 Halegrove Gardens, London NW7
Penwalt Ltd, Doman Rd., Camberley, Surrey
Perkin Elmer Ltd, Beaconsfield, Buckinghamshire
Perrier et Cie, 20 rue Marie-Debos, 92120 Montrouge, France
Phoenix Precision Instruments, Gardiner, New York, USA
Photoelectronics Ltd, Arcail House, Restmor Way, Hockbridge, Wallington, Surrey
Pola Laboratories Supplies Inc., New York 7, USA
Polaron, 4 Shakespeare Rd., Finchley, London N3
Procedyne Corporation, 221 Somerset Street, New Brunswick, New Jersey, USA
Process and Instruments Corporation, Brooklyn, New York, USA
Production Sales and Services Ltd, New Malden, Surrey
Prolabo (France), 12 rue Pelee, 75011 Paris XI, France
Prosser Scientific Instruments Ltd, Lady Lane Industrial Estate, Hadleigh, Ipswich,
    Suffolk IP7 6DQ

Quantachrome Corporation, 337 Glen Cove Rd., NY 11548, USA

Rao Instruments Co. Ltd, Brooklyn, NY, USA
Rattreyon Learning Systems, Michigan City, Indiana, USA
Research Appliance Co., Route 8, Gibsonia, Pennsylvania 15044, USA
Reynolds and Branson, Scientific Equipt, Dockfield Rd., Shipley, Yorkshire

Ronald Trist Controls Ltd, 6–8 Bath Rd., Slough, Berkshire
Rotheroe and Mitchell Ltd, Aintree Rd., Greenford, Middlesex UB6 7LJ
Royco Instruments, 141 Jefferson Drive, Menlo Park, California 94025

Saab Scania AB, Gelbgjutargarten 2, Fack 581-01-oe, Linkoping, Sweden
Sartorius Instruments Ltd, 18 Avenue Rd., Belmont, Surrey
Sartorius Werke, GmbH, D-34 Gottingen, W. Germany
Schaar and Co., Chicago, Illinois, USA
Schaeffer, K., Sprendlingen, W. Germany
Science Spectrum, 1216 State Street, P.O. Box 3003, Santa Barbara, California, USA
Seishin Enterprise Co. Ltd, Sotobori Sky Bldg., 13 Honmura-cho, Ichigaya, Shinjuku-
     ku, Tokyo, Japan
Sharples Centrifuges Ltd, Camberley, Surrey
Shimadzu Seisakusho Ltd, Kanda, Mitoshirocho, Chiyodra-Ku, Tokyo, Japan
Siemens Ltd, Siemens House, Windmill Road, Sunbury-on-Thames, Middlesex TW16
     7HS
Simon Carves Ltd, Stockport, Lancashire
Simon Henry Ltd, Special Products Division, P.O. Box 31, Stockport, Cheshire
Simonacco Ltd, Durranhill Trading Estate, Carlisle, Cumbria
Société Française D'Instruments de Contrôle et d'Analyses, Le Mesnil, Saint Denise,
     France
Sondes Place Research Institute, Dorking, Surrey
Spatial Data Systems Inc., Galeta, California
Specfield Ltd, 1a Jennings Bldg., Thames Ave., Windsor, Berkshire
Spectrex, 3594 Haven Ave., Redwood City, California 94063, USA
Stauffer Chemical Corporation, Westport, Connecticut 06880, USA
Ströhlein, Dusseldorf, D-4000, W. Germany
Systems and Components Ltd, Broadway, Market Lavington, Devizes, Wiltshire

TEM Sales Ltd, Gatwick Rd., Crawley, Sussex
Techecology Inc., Sunnyvale, California, USA
Techmation Ltd, 58 Edgware Way, Edgware, Middlesex
Technotest, 65 rue Marius Auffan, 94300, Levallois, France
Telefunken A.E.G., 79 Ulm, Elisabethstrasse 3, W. Germany; also at Laraterstrasse 67,
     8027, Zurich, Switzerland
Thermal Control Co. Ltd, 138 Old Shoreham Rd., Hove, Sussex
Thermo-Systems Inc. (TSI), 2500 Cleveland Ave., North St Paul, Minnesota 55113,
     USA
3M Company, Commercial Chemicals Division, 3M Centre, St Paul, Minnesota, USA
Touzart et Matignon, 3 rue Arnyot, 75005, Paris, France
Trapelo Division, LFE Corporation, 1601 Trapelo Rd., Waltham, Massachusetts 02154,
     USA
TOA Electric Co., Kobe, Japan
2000 Inc., Box 20769, Atlanta, Georgia 33010, USA

Ultrasonics Ltd, Otley Road, Bradford, Yorkshire
Unico Environmental Instruments Inc., P.O. Box 590, Fall River, Massachusetts, USA

Val-Dell Company, 1339E Township Line Rd., Norristown, Pennsylvania 19403, USA
VEB Transformratoren und Rontegemwerk, 48 Overbeckstrasse 8030, Dresden,
      W. Germany
Veco N.V. Zeefplatenfabrick, Eerbeck (Veluive), The Netherlands
Vickers Instruments Ltd, Haxby Rd., York

Walther and Co. Aktiengesellschaft, 5 Köln-Dellbruck, W. Germany
Warmain International Pty Ltd, Artarman, NSW, Australia
Warmain International Pty Ltd, Melbourne, Victoria, Australia
Warmain, Simon, Ltd, Halifax Road, Todmorden, West Yorkshire
Watson, W. and Sons Ltd, Barnet, Hertfordshire
Weathes Measure Corp., P.O. Box 41257, Sacramento, California 95841, USA
Wessex Electronics Ltd, Stoves Trading Estate, Yate, Bristol BS17 5QP
Wild-Heerbrug Ltd, CH-9435, Heerbrugg, Switzerland
Willson Products Division, ESB Inc., P.O. Box 622, Reading, Pennsylvania 19603, USA

Zimney Corporation, Monrovia, California, USA

Author index

The numbers in brackets give the chapter numbers followed by the number of the reference in the chapter.

Subject index

The numbers shown in bold indicate that a section about the subject follows that page.